Slow Light
Science and Applications

OPTICAL SCIENCE AND ENGINEERING

Founding Editor
Brian J. Thompson
University of Rochester
Rochester, New York

1. Electron and Ion Microscopy and Microanalysis: Principles and Applications, *Lawrence E. Murr*
2. Acousto-Optic Signal Processing: Theory and Implementation, *edited by Norman J. Berg and John N. Lee*
3. Electro-Optic and Acousto-Optic Scanning and Deflection, *Milton Gottlieb, Clive L. M. Ireland, and John Martin Ley*
4. Single-Mode Fiber Optics: Principles and Applications, *Luc B. Jeunhomme*
5. Pulse Code Formats for Fiber Optical Data Communication: Basic Principles and Applications, *David J. Morris*
6. Optical Materials: An Introduction to Selection and Application, *Solomon Musikant*
7. Infrared Methods for Gaseous Measurements: Theory and Practice, *edited by Joda Wormhoudt*
8. Laser Beam Scanning: Opto-Mechanical Devices, Systems, and Data Storage Optics, *edited by Gerald F. Marshall*
9. Opto-Mechanical Systems Design, *Paul R. Yoder, Jr.*
10. Optical Fiber Splices and Connectors: Theory and Methods, *Calvin M. Miller with Stephen C. Mettler and Ian A. White*
11. Laser Spectroscopy and Its Applications, *edited by Leon J. Radziemski, Richard W. Solarz, and Jeffrey A. Paisner*
12. Infrared Optoelectronics: Devices and Applications, *William Nunley and J. Scott Bechtel*
13. Integrated Optical Circuits and Components: Design and Applications, *edited by Lynn D. Hutcheson*
14. Handbook of Molecular Lasers, *edited by Peter K. Cheo*
15. Handbook of Optical Fibers and Cables, *Hiroshi Murata*
16. Acousto-Optics, *Adrian Korpel*
17. Procedures in Applied Optics, *John Strong*
18. Handbook of Solid-State Lasers, *edited by Peter K. Cheo*
19. Optical Computing: Digital and Symbolic, *edited by Raymond Arrathoon*
20. Laser Applications in Physical Chemistry, *edited by D. K. Evans*
21. Laser-Induced Plasmas and Applications, *edited by Leon J. Radziemski and David A. Cremers*
22. Infrared Technology Fundamentals, *Irving J. Spiro and Monroe Schlessinger*
23. Single-Mode Fiber Optics: Principles and Applications, Second Edition, Revised and Expanded, *Luc B. Jeunhomme*
24. Image Analysis Applications, *edited by Rangachar Kasturi and Mohan M. Trivedi*
25. Photoconductivity: Art, Science, and Technology, *N. V. Joshi*
26. Principles of Optical Circuit Engineering, *Mark A. Mentzer*
27. Lens Design, *Milton Laikin*
28. Optical Components, Systems, and Measurement Techniques, *Rajpal S. Sirohi and M. P. Kothiyal*

29. Electron and Ion Microscopy and Microanalysis: Principles and Applications, Second Edition, Revised and Expanded, *Lawrence E. Murr*
30. Handbook of Infrared Optical Materials, *edited by Paul Klocek*
31. Optical Scanning, *edited by Gerald F. Marshall*
32. Polymers for Lightwave and Integrated Optics: Technology and Applications, *edited by Lawrence A. Hornak*
33. Electro-Optical Displays, *edited by Mohammad A. Karim*
34. Mathematical Morphology in Image Processing, *edited by Edward R. Dougherty*
35. Opto-Mechanical Systems Design: Second Edition, Revised and Expanded, *Paul R. Yoder, Jr.*
36. Polarized Light: Fundamentals and Applications, *Edward Collett*
37. Rare Earth Doped Fiber Lasers and Amplifiers, *edited by Michel J. F. Digonnet*
38. Speckle Metrology, *edited by Rajpal S. Sirohi*
39. Organic Photoreceptors for Imaging Systems, *Paul M. Borsenberger and David S. Weiss*
40. Photonic Switching and Interconnects, *edited by Abdellatif Marrakchi*
41. Design and Fabrication of Acousto-Optic Devices, *edited by Akis P. Goutzoulis and Dennis R. Pape*
42. Digital Image Processing Methods, *edited by Edward R. Dougherty*
43. Visual Science and Engineering: Models and Applications, *edited by D. H. Kelly*
44. Handbook of Lens Design, *Daniel Malacara and Zacarias Malacara*
45. Photonic Devices and Systems, *edited by Robert G. Hunsberger*
46. Infrared Technology Fundamentals: Second Edition, Revised and Expanded, *edited by Monroe Schlessinger*
47. Spatial Light Modulator Technology: Materials, Devices, and Applications, *edited by Uzi Efron*
48. Lens Design: Second Edition, Revised and Expanded, *Milton Laikin*
49. Thin Films for Optical Systems, *edited by Francoise R. Flory*
50. Tunable Laser Applications, *edited by F. J. Duarte*
51. Acousto-Optic Signal Processing: Theory and Implementation, Second Edition, *edited by Norman J. Berg and John M. Pellegrino*
52. Handbook of Nonlinear Optics, *Richard L. Sutherland*
53. Handbook of Optical Fibers and Cables: Second Edition, *Hiroshi Murata*
54. Optical Storage and Retrieval: Memory, Neural Networks, and Fractals, *edited by Francis T. S. Yu and Suganda Jutamulia*
55. Devices for Optoelectronics, *Wallace B. Leigh*
56. Practical Design and Production of Optical Thin Films, *Ronald R. Willey*
57. Acousto-Optics: Second Edition, *Adrian Korpel*
58. Diffraction Gratings and Applications, *Erwin G. Loewen and Evgeny Popov*
59. Organic Photoreceptors for Xerography, *Paul M. Borsenberger and David S. Weiss*
60. Characterization Techniques and Tabulations for Organic Nonlinear Optical Materials, *edited by Mark G. Kuzyk and Carl W. Dirk*
61. Interferogram Analysis for Optical Testing, *Daniel Malacara, Manuel Servin, and Zacarias Malacara*
62. Computational Modeling of Vision: The Role of Combination, *William R. Uttal, Ramakrishna Kakarala, Spiram Dayanand, Thomas Shepherd, Jagadeesh Kalki, Charles F. Lunskis, Jr., and Ning Liu*
63. Microoptics Technology: Fabrication and Applications of Lens Arrays and Devices, *Nicholas Borrelli*
64. Visual Information Representation, Communication, and Image Processing, *edited by Chang Wen Chen and Ya-Qin Zhang*
65. Optical Methods of Measurement, *Rajpal S. Sirohi and F. S. Chau*
66. Integrated Optical Circuits and Components: Design and Applications, *edited by Edmond J. Murphy*
67. Adaptive Optics Engineering Handbook, *edited by Robert K. Tyson*

68. Entropy and Information Optics, *Francis T. S. Yu*
69. Computational Methods for Electromagnetic and Optical Systems, *John M. Jarem and Partha P. Banerjee*
70. Laser Beam Shaping, *Fred M. Dickey and Scott C. Holswade*
71. Rare-Earth-Doped Fiber Lasers and Amplifiers: Second Edition, Revised and Expanded, *edited by Michel J. F. Digonnet*
72. Lens Design: Third Edition, Revised and Expanded, *Milton Laikin*
73. Handbook of Optical Engineering, *edited by Daniel Malacara and Brian J. Thompson*
74. Handbook of Imaging Materials: Second Edition, Revised and Expanded, *edited by Arthur S. Diamond and David S. Weiss*
75. Handbook of Image Quality: Characterization and Prediction, *Brian W. Keelan*
76. Fiber Optic Sensors, *edited by Francis T. S. Yu and Shizhuo Yin*
77. Optical Switching/Networking and Computing for Multimedia Systems, *edited by Mohsen Guizani and Abdella Battou*
78. Image Recognition and Classification: Algorithms, Systems, and Applications, *edited by Bahram Javidi*
79. Practical Design and Production of Optical Thin Films: Second Edition, Revised and Expanded, *Ronald R. Willey*
80. Ultrafast Lasers: Technology and Applications, *edited by Martin E. Fermann, Almantas Galvanauskas, and Gregg Sucha*
81. Light Propagation in Periodic Media: Differential Theory and Design, *Michel Nevière and Evgeny Popov*
82. Handbook of Nonlinear Optics, Second Edition, Revised and Expanded, *Richard L. Sutherland*
83. Polarized Light: Second Edition, Revised and Expanded, *Dennis Goldstein*
84. Optical Remote Sensing: Science and Technology, *Walter Egan*
85. Handbook of Optical Design: Second Edition, *Daniel Malacara and Zacarias Malacara*
86. Nonlinear Optics: Theory, Numerical Modeling, and Applications, *Partha P. Banerjee*
87. Semiconductor and Metal Nanocrystals: Synthesis and Electronic and Optical Properties, *edited by Victor I. Klimov*
88. High-Performance Backbone Network Technology, *edited by Naoaki Yamanaka*
89. Semiconductor Laser Fundamentals, *Toshiaki Suhara*
90. Handbook of Optical and Laser Scanning, *edited by Gerald F. Marshall*
91. Organic Light-Emitting Diodes: Principles, Characteristics, and Processes, *Jan Kalinowski*
92. Micro-Optomechatronics, *Hiroshi Hosaka, Yoshitada Katagiri, Terunao Hirota, and Kiyoshi Itao*
93. Microoptics Technology: Second Edition, *Nicholas F. Borrelli*
94. Organic Electroluminescence, *edited by Zakya Kafafi*
95. Engineering Thin Films and Nanostructures with Ion Beams, *Emile Knystautas*
96. Interferogram Analysis for Optical Testing, Second Edition, *Daniel Malacara, Manuel Sercin, and Zacarias Malacara*
97. Laser Remote Sensing, *edited by Takashi Fujii and Tetsuo Fukuchi*
98. Passive Micro-Optical Alignment Methods, *edited by Robert A. Boudreau and Sharon M. Boudreau*
99. Organic Photovoltaics: Mechanism, Materials, and Devices, *edited by Sam-Shajing Sun and Niyazi Serdar Saracftci*
100. Handbook of Optical Interconnects, *edited by Shigeru Kawai*
101. GMPLS Technologies: Broadband Backbone Networks and Systems, *Naoaki Yamanaka, Kohei Shiomoto, and Eiji Oki*
102. Laser Beam Shaping Applications, *edited by Fred M. Dickey, Scott C. Holswade and David L. Shealy*
103. Electromagnetic Theory and Applications for Photonic Crystals, *Kiyotoshi Yasumoto*

104. Physics of Optoelectronics, *Michael A. Parker*
105. Opto-Mechanical Systems Design: Third Edition, *Paul R. Yoder, Jr.*
106. Color Desktop Printer Technology, *edited by Mitchell Rosen and Noboru Ohta*
107. Laser Safety Management, *Ken Barat*
108. Optics in Magnetic Multilayers and Nanostructures, *Štefan Višňovský*
109. Optical Inspection of Microsystems, *edited by Wolfgang Osten*
110. Applied Microphotonics, *edited by Wes R. Jamroz, Roman Kruzelecky, and Emile I. Haddad*
111. Organic Light-Emitting Materials and Devices, *edited by Zhigang Li and Hong Meng*
112. Silicon Nanoelectronics, *edited by Shunri Oda and David Ferry*
113. Image Sensors and Signal Processor for Digital Still Cameras, *Junichi Nakamura*
114. Encyclopedic Handbook of Integrated Circuits, *edited by Kenichi Iga and Yasuo Kokubun*
115. Quantum Communications and Cryptography, *edited by Alexander V. Sergienko*
116. Optical Code Division Multiple Access: Fundamentals and Applications, *edited by Paul R. Prucnal*
117. Polymer Fiber Optics: Materials, Physics, and Applications, *Mark G. Kuzyk*
118. Smart Biosensor Technology, *edited by George K. Knopf and Amarjeet S. Bassi*
119. Solid-State Lasers and Applications, *edited by Alphan Sennaroglu*
120. Optical Waveguides: From Theory to Applied Technologies, *edited by Maria L. Calvo and Vasudevan Lakshiminarayanan*
121. Gas Lasers, *edited by Masamori Endo and Robert F. Walker*
122. Lens Design, Fourth Edition, *Milton Laikin*
123. Photonics: Principles and Practices, *Abdul Al-Azzawi*
124. Microwave Photonics, *edited by Chi H. Lee*
125. Physical Properties and Data of Optical Materials, *Moriaki Wakaki, Keiei Kudo, and Takehisa Shibuya*
126. Microlithography: Science and Technology, Second Edition, *edited by Kazuaki Suzuki and Bruce W. Smith*
127. Coarse Wavelength Division Multiplexing: Technologies and Applications, *edited by Hans Joerg Thiele and Marcus Nebeling*
128. Organic Field-Effect Transistors, *Zhenan Bao and Jason Locklin*
129. Smart CMOS Image Sensors and Applications, *Jun Ohta*
130. Photonic Signal Processing: Techniques and Applications, *Le Nguyen Binh*
131. Terahertz Spectroscopy: Principles and Applications, *edited by Susan L. Dexheimer*
132. Fiber Optic Sensors, Second Edition, *edited by Shizhuo Yin, Paul B. Ruffin, and Francis T. S. Yu*
133. Introduction to Organic Electronic and Optoelectronic Materials and Devices, *edited by Sam-Shajing Sun and Larry R. Dalton*
134. Introduction to Nonimaging Optics, *Julio Chaves*
135. The Nature of Light: What Is a Photon?, *edited by Chandrasekhar Roychoudhuri, A. F. Kracklauer, and Katherine Creath*
136. Optical and Photonic MEMS Devices: Design, Fabrication and Control, *edited by Ai-Qun Liu*
137. Tunable Laser Applications, Second Edition, *edited by F. J. Duarte*
138. Biochemical Applications of Nonlinear Optical Spectroscopy, *edited by Vladislav Yakovlev*
139. Dynamic Laser Speckle and Applications, *edited by Hector J. Rabal and Roberto A. Braga Jr.*
140. Slow Light: Science and Applications, *edited by Jacob B. Khurgin and Rodney S. Tucker*

Slow Light
Science and Applications

Edited by
Jacob B. Khurgin
Rodney S. Tucker

CRC Press
Taylor & Francis Group
Boca Raton   London   New York

CRC Press is an imprint of the
Taylor & Francis Group, an **informa** business

CRC Press
Taylor & Francis Group
6000 Broken Sound Parkway NW, Suite 300
Boca Raton, FL 33487-2742

© 2009 by Taylor & Francis Group, LLC
CRC Press is an imprint of Taylor & Francis Group, an Informa business

No claim to original U.S. Government works
Printed in the United States of America on acid-free paper
10 9 8 7 6 5 4 3 2 1

International Standard Book Number-13: 978-1-4200-6151-2 (Hardcover)

This book contains information obtained from authentic and highly regarded sources. Reasonable efforts have been made to publish reliable data and information, but the author and publisher cannot assume responsibility for the validity of all materials or the consequences of their use. The authors and publishers have attempted to trace the copyright holders of all material reproduced in this publication and apologize to copyright holders if permission to publish in this form has not been obtained. If any copyright material has not been acknowledged please write and let us know so we may rectify in any future reprint.

Except as permitted under U.S. Copyright Law, no part of this book may be reprinted, reproduced, transmitted, or utilized in any form by any electronic, mechanical, or other means, now known or hereafter invented, including photocopying, microfilming, and recording, or in any information storage or retrieval system, without written permission from the publishers.

For permission to photocopy or use material electronically from this work, please access www.copyright.com (http://www.copyright.com/) or contact the Copyright Clearance Center, Inc. (CCC), 222 Rosewood Drive, Danvers, MA 01923, 978-750-8400. CCC is a not-for-profit organization that provides licenses and registration for a variety of users. For organizations that have been granted a photocopy license by the CCC, a separate system of payment has been arranged.

Trademark Notice: Product or corporate names may be trademarks or registered trademarks, and are used only for identification and explanation without intent to infringe.

Library of Congress Cataloging-in-Publication Data

Slow light : science and applications / editors, Jacob B. Khurgin and Rod Tucker.
       p. cm. -- (Optical science and engineering ; 140)
    Includes bibliographical references and index.
    ISBN 978-1-4200-6151-2 (alk. paper)
    1. Light--Transmission. 2. Optical communications. 3. Physical optics. I. Khurgin, Jacob B. II. Tucker, Rod. III. Title. IV. Series.

QC389.S64 2008
621.36--dc22   2008035802

Visit the Taylor & Francis Web site at
http://www.taylorandfrancis.com

and the CRC Press Web site at
http://www.crcpress.com

Contents

Introduction xiii
Editors xv
Contributors.. xvii

PART I Fundamental Physics of Slow Light in Different Media

Chapter 1 Slow Light in Atomic Vapors 3

Ryan M. Camacho and John C. Howell

Chapter 2 Slow and Fast Light in Semiconductors 13

Shu-Wei Chang and Shun Lien Chuang

Chapter 3 Slow Light in Optical Waveguides... 37

Zhaoming Zhu, Daniel J. Gauthier, Alexander L. Gaeta,
and Robert W. Boyd

Chapter 4 Slow Light in Photonic Crystal Waveguides 59

Thomas F. Krauss

PART II Slow Light in Periodic Photonic Structures

Chapter 5 Periodic Coupled Resonator Structures.. 79

Joyce K.S. Poon, Philip Chak, John E. Sipe, and Amnon Yariv

Chapter 6 Resonator-Mediated Slow Light: Novel Structures, Applications, and Tradeoffs .. 101

Andrey B. Matsko and Lute Maleki

Chapter 7 Disordered Optical Slow-Wave Structures: What Is the Velocity
of Slow Light? 119

Shayan Mookherjea

PART III   Slow Light in Fibers

Chapter 8   Slow and Fast Light Propagation in Narrow Band Raman-Assisted Fiber Parametric Amplifiers 149

Gadi Eisenstein, Evgeny Shumakher, and Amnon Willinger

Chapter 9   Slow and Fast Light Using Stimulated Brillouin Scattering: A Highly Flexible Approach 173

Luc Thévenaz

PART IV   Slow Light and Nonlinear Phenomena

Chapter 10   Nonlinear Slow-Wave Structures 195

Andrea Melloni and Francesco Morichetti

Chapter 11   Slow Light Gap Solitons 223

Joe T. Mok, Ian C.M. Littler, Morten Ibsen, C. Martijn de Sterke, and Benjamin J. Eggleton

Chapter 12   Coherent Control and Nonlinear Wave Mixing in Slow Light Media 235

Yuri Rostovtsev

PART V   Dynamic Structures for Storing Light

Chapter 13   Stopping and Storing Light in Semiconductor Quantum Wells and Optical Microresonators 257

Nai H. Kwong, John E. Sipe, Rolf Binder, Zhenshan Yang, and Arthur L. Smirl

Chapter 14   Stopping Light via Dynamic Tuning of Coupled Resonators 277

Shanhui Fan and Michelle L. Povinelli

PART VI   Applications

Chapter 15   Bandwidth Limitation in Slow Light Schemes 293

Jacob B. Khurgin

Chapter 16	Reconfigurable Signal Processing Using Slow-Light-Based Tunable Optical Delay Lines	321
	Alan E. Willner, Bo Zhang, and Lin Zhang	
Chapter 17	Slow Light Buffers for Packet Switching	347
	Rodney S. Tucker	
Chapter 18	Application of Slow Light to Phased Array Radar Beam Steering	367
	Zachary Dutton, Mark Bashkansky, and Michael Steiner	
Index		381

Introduction

SLOW LIGHT—FASCINATING SCIENCE, REMARKABLE APPLICATIONS

The topic of this book—*Slow Light: Science and Applications*—has a long history and a bright future. The physics of the phenomenon of light propagation in media and structures with reduced group velocity, for which the term "slow light" had been coined, can be traced to the nineteenth century when the classical theory of dispersion of the electromagnetic waves had been first formulated in the works of Lorentz [1] and others. Slow wave propagation has also been observed and widely used in the microwave range since as early as 1940s [2]. Building on this history [3,4], slow light is expected to become instrumental in enabling applications on the cutting edge of twenty-first century technology—high-capacity communication networks, quantum computing, ultrafast all-optical information processing, agile microwave systems, and so on.

This transformation of slow light from a scientific curiosity to a rapidly growing field with many potential applications has been made possible by rapid developments in the technologies required for practical implementation of slow light. These technologies include high power and narrow linewidth light sources, low-loss optical waveguides, photonic crystal devices, microfabrication techniques, and many others. What makes slow light a particularly intriguing topic is its truly interdisciplinary nature. In fact, slow light propagation had been observed in a wide variety of media and structures, ranging from Bose–Einstein condensates and low-pressure metal vapors on one hand to optical fibers and photonic band-gap structures on the other. With the seeming diversity of slow light schemes, they can all be characterized by a single common feature—the existence of a sharp single resonance or multiple resonances. The resonance can be defined by a simple atomic transition, by a Bragg grating or other resonant photonic structures, or by an external laser as in the schemes involving various nonlinear processes—resonant scattering, spectral hole burning, or four wave mixing. The field of slow light is unique, for it brings together experts in nonlinear optics with integrated optics professionals, experts in laser cooling with condensed matter physicists, semiconductor laser engineers with researchers in quantum optics, and engineers working on optical fiber communications with spectroscopists. This multidisciplinary atmosphere is conducive to the emergence of novel ideas, combining the expertise in all the aforementioned areas with the whole being more than the sum of its parts. It is with the thought of stimulating the cross-fertilization of minds that this book has been conceived.

To achieve this goal, we have assembled contributions from 18 groups that have been actively involved in the slow light field and have all made significant contributions in recent years. While it is impossible to account for all the captivating developments that occurred in the course of the last few years, we believe that this book provides a comprehensive introductory survey of the current state of slow light field and to the main directions along which it is moving.

Part I, "Fundamental Physics of Slow Light in Different Media," consists of four chapters and provides an introduction to the physics of slow light in diverse media. Chapter 1 deals with slow light in atomic vapors, including electromagnetically induced transparency. Atomic vapors were the first media in which slow light was observed, and also the one in which the most spectacular results were achieved, and thus it is only appropriate to start with this medium. Chapter 2 deals with slow light in semiconductors. While the reduction in group velocity achieved in semiconductors is currently less than in atomic vapors, it is the ability of semiconductor schemes to operate at convenient telecom wavelengths, to provide broad signal bandwidth, and their ability to be integrated with other electronic and optical components that make them particularly attractive. In Chapter 3, the science of various slow light schemes in optical waveguides and fibers is explored. Once again, the scale reduction in group velocity in optical fibers is not nearly as high as in atomic vapors, but these devices offer low loss, can be compact, and can be integrated with existing technologies. In Chapter 4, slow light schemes relying upon photonic crystals and other periodic photonic structures are introduced. These schemes are most attractive for wide bandwidth application and also carry a promise of miniaturization.

Part II, "Slow Light in Periodic Photonic Structures," contains three chapters giving a more detailed picture of how coupled optical resonators can be used to achieve significant reduction in the group velocity of light. Chapter 5 provides mathematical apparatus for describing various types of coupled resonators and their comparative analysis. Chapter 6 introduces advanced coupled resonator structures—vertically coupled resonators and whispering gallery modes. In Chapter 7, the fascinating phenomena of light localization in disordered coupled resonators are described.

Part III, "Slow Light in Fibers," comprises two chapters that give an insight into two mechanisms that can provide tunable group delay in optical fibers. Chapter 8 describes Raman-assisted optical parametric amplification and Chapter 9 deals with the amplification via stimulated Brillouin scattering.

Many slow light schemes (EIT, Raman, coherent population oscillations, etc.) rely upon nonlinear processes. However, Part IV, "Slow Light and Nonlinear Phenomena," deals specifically with the enhancement of nonlinear effects associated with slow light. Chapter 10 provides a comprehensive treatment of the enhancement of third-order nonlinear processes in periodic photonic structures and establishes the limits of the enhancement due to group velocity dispersion. Chapter 11 is dedicated to compensating group velocity dispersion in slow wave structures using gap solitons. The ability to control nonlinear processes in atomic slow light schemes is the subject of Chapter 12.

Part V, "Dynamic Structures for Storing the Light," describes photonic structures capable of dynamically changing the bandwidth and slow down factor, and thus capable of reducing deleterious signal broadening effects. It contains two chapters—Chapter 13 introduces concept of structure combining atomic and photonic resonances and Chapter 14 deals with adiabatically tuned resonators.

Finally, Part VI, "Applications," comprises four chapters. Chapter 15 deals with the severe bandwidth limitations placed by the group velocity dispersion upon the slow light in linear and nonlinear applications which are discussed in detail and a comparison is made between different schemes. Chapter 16 analyzes application of slow light to all-optical processing of signals with various modulation formats. In Chapter 17, application of slow light to packet switching is described and a comparison is made between electronic and optical schemes. Chapter 18 introduces slow light application to the analog processing for phased array antennas and other microwave photonics applications.

This book provides a snapshot of the exciting and rapidly evolving area of slow light. But we hope that it can help bring readers up to date with the most significant developments taking place in the field of slow light and stimulate readers' interest to the point where they will be enticed to make their contributions to it.

Jacob B. Khurgin
Rodney S. Tucker

REFERENCES

1. H. A. Lorentz, *Wiedem. Ann.*, 9, 641 (1880).
2. J. R. Pierce, *Bell. Syst. Tech. J.*, 29, 1 (1950).
3. L. V. Hau, S. E. Harris, Z. Dutton, and C. H. Behroozi, *Nature*, 397, 594–596 (1999).
4. D. F. Phillips, A. Fleischhauer, A. Mair, R. L. Walsworth, and M. D. Lukin, *Phys. Rev. Lett.*, 86, 783–786 (2001).

Editors

Jacob B. Khurgin has been a professor of electrical and computer engineering at Johns Hopkins University, Baltimore, Maryland, since 1988. Prior to this he was a senior member of research staff at Philips Laboratories, Briarcliff Manor, New York, where he developed various display components and systems including 3D projection TV and visible lasers pumped by electron beams. His area of expertise is in optical and electronic solid-state devices. In his 20 years at Johns Hopkins University, he had made important contributions in the fields of nonlinear optics, semiconductor optoelectronic devices, quantum-cascade lasers, optical communications, THz technology, and fundamental condensed matter physics. He has authored over 180 technical papers, 500 conference presentations, 3 book chapters, and 12 patents. He is a fellow of the Optical Society of America. Professor Khurgin has a PhD from the Polytechnic University of New York.

Rodney S. Tucker is a laureate professor in the Department of Electrical and Electronic Engineering at the University of Melbourne, Melbourne, Australia, and is the research director of the Australian Research Council Special Research Centre for Ultra-Broadband Information Networks (CUBIN). He has held positions at the University of Queensland, the University of California, Berkeley, Cornell University, Plessey Research, AT&T Bell Laboratories, Hewlett Packard Laboratories, and Agilent Technologies. He is a fellow of the Australian Academy of Science, the IEEE, the OSA, and the Australian Academy of Technological Sciences and Engineering. He received his BE and his PhD from the University of Melbourne in 1969 and 1975, respectively. He was awarded the Institution of Engineers, Australia, Sargent Medal in 1995 for his contributions to electrical engineering, and was named IEEE Lasers and Electro-Optics Society Distinguished Lecturer for the year 1995–1996. In 1997 he was awarded the Australia Prize for his contributions to telecommunications, and in 2007 he was awarded the IEEE Lasers and Electro-Optics Society's Aron Kressel Award for his contributions to high-speed semiconductor lasers.

Contributors

Mark Bashkansky
Optical Sciences Division
Naval Research Laboratory
Washington, DC

Rolf Binder
College of Optical Sciences
University of Arizona
Tucson, Arizona

Robert W. Boyd
Institute of Optics
University of Rochester
Rochester, New York

Ryan M. Camacho
Department of Applied Physics
California Institute of Technology
Pasadena, California

Philip Chak
Department of Applied Physics
California Institute of Technology
Pasadena, California

and

Department of Physics and Institute
  for Optical Sciences
University of Toronto
Toronto, Canada

Shu-Wei Chang
Department of Electrical and Computer
  Engineering
University of Illinois
Urbana, Illinois

Shun Lien Chuang
Department of Electrical and Computer
  Engineering
University of Illinois
Urbana, Illinois

Zachary Dutton
BBN Technologies
Cambridge, Massachusetts

and

Radar Division
Naval Research Laboratory
Washington, DC

Benjamin J. Eggleton
Centre for Ultrahigh-Bandwidth Devices for
  Optical Systems and School of Physics A28
University of Sydney
Sydney, New South Wales, Australia

Gadi Eisenstein
Department of Electrical Engineering
Technion
Haifa, Israel

Shanhui Fan
Ginzton Laboratory
Stanford University
Stanford, California

Alexander L. Gaeta
School of Applied and Engineering Physics
Cornell University
Ithaca, New York

Daniel J. Gauthier
Department of Physics
Duke University
Durham, North Carolina

John C. Howell
Department of Physics and Astronomy
University of Rochester
Rochester, New York

Morten Ibsen
Optoelectronics Research Centre
University of Southampton
Southampton, United Kingdom

Jacob B. Khurgin
Johns Hopkins University
Baltimore, Maryland

Thomas F. Krauss
School of Physics and Astronomy
University of St Andrews
St. Andrews, Scotland

Nai H. Kwong
College of Optical Sciences
University of Arizona
Tucson, Arizona

Ian C.M. Littler
Department of Quantum Science
School of Physical Sciences and Engineering
The Australian National University
Canberra, Australian Capital Territory
Australia

Lute Maleki
Jet Propulsion Laboratory
California Institute of Technology
Pasadena, California

Andrey B. Matsko
Jet Propulsion Laboratory
California Institute of Technology
Pasadena, California

Andrea Melloni
Politecnico di Milano
Dipartimento di Elettronica e Informazione
Milan, Italy

Joe T. Mok
Blake Dawson Patent Attorneys
Sydney, Australia

Shayan Mookherjea
University of California
San Diego, California

Francesco Morichetti
Politecnico di Milano
Dipartimento di Elettronica e Informazione
Milan, Italy

Joyce K.S. Poon
Department of Electrical Engineering and
  Department of Applied Physics
California Institute of Technology
Pasadena, California

and

Department of Electrical and Computer
  Engineering
Institute for Optical Sciences
University of Toronto
Toronto, Canada

and

Institute for Optical Sciences
University of Toronto
Toronto, Canada

Michelle L. Povinelli
Ginzton Laboratory
Stanford University
Stanford, California

Yuri Rostovtsev
Institute for Quantum Studies and Department
  of Physics
Texas A&M University
College Station, Texas

Evgeny Shumakher
Department of Electrical Engineering
Technion
Haifa, Israel

John E. Sipe
Department of Physics and Institute
  for Optical Sciences
University of Toronto
Toronto, Canada

Arthur L. Smirl
Laboratory for Photonics and Quantum
  Electronics
University of Iowa
Iowa City, Iowa

Michael Steiner
Naval Research Laboratory
Washington, DC

Contributors

C. Martijn de Sterke
Centre for Ultrahigh-Bandwidth Devices for
 Optical Systems and School of Physics A28
University of Sydney
Sydney, New South Wales, Australia

Luc Thévenaz
Institute of Electrical Engineering
Ecole Polytechnique Fédérale de Lausanne
Lausanne, Switzerland

Rodney S. Tucker
Australian Research Council Special Research
 Centre for Ultra-Broadband Information
 Networks
Department of Electrical and Electronic
 Engineering
University of Melbourne
Melbourne, Australia

Amnon Willinger
Department of Electrical Engineering
Technion
Haifa, Israel

Alan E. Willner
Department of Electrical Engineering
University of Southern California
Los Angeles, California

Zhenshan Yang
College of Optical Sciences
University of Arizona
Tucson, Arizona

and

Department of Physics and Institute
 of Optical Sciences
University of Toronto
Toronto, Canada

and

Department of Physics
Texas A&M University
College Station, Texas

Amnon Yariv
Department of Electrical Engineering and
 Department of Applied Physics
California Institute of Technology
Pasadena, California

and

Department of Electrical Engineering
California Institute of Technology
Pasadena, California

Bo Zhang
Department of Electrical Engineering
University of Southern California
Los Angeles, California

Lin Zhang
Department of Electrical Engineering
University of Southern California
Los Angeles, California

Zhaoming Zhu
Department of Physics
Duke University
Durham, North Carolina

Part I

Fundamental Physics of Slow Light in Different Media

1 Slow Light in Atomic Vapors

Ryan M. Camacho and John C. Howell

CONTENTS

1.1 Introduction..3
1.2 First Experiments in Slow Light...4
1.3 EIT 4
1.4 Two-Level Systems 8
1.5 Dispersion Management 9
1.6 Concluding Remarks..11
References 11

1.1 INTRODUCTION

While the theoretical foundations of slow light have been well-known for many years, the experimental investigation of slow light did not begin until relatively recently. From the beginning, atomic vapors have played an important role in carrying out these experiments. Many of the preliminary studies of optical group velocities were performed using atomic vapors, and atomic vapors continue to be used in both fundamental and applied slow light research.

In this chapter, we review the essential physics necessary to understand slow and fast group velocities in atomic systems, and attempt to trace the experimental observation of slow light from its origins, with a specific emphasis on work that has been done in atomic vapors. In doing so, we adopt a less common but unifying treatment, concentrating on the resonant interactions specific to each slow light medium using double-sided Feynman diagrams [1]. We take this approach for two reasons: first, it offers a pedagogical and systematic way to visualize the varied atomic resonances leading to slow light. Second, this treatment is less well-known and may provide insights not normally encountered in the usual nongraphical density matrix formalism. A more traditional approach as well as an excellent review of slow light up to 2002 has been made by Boyd and Gauthier [2] as well as Milonni [3]. Milonni also published a book in 2005 including chapters on slow, fast, and left-handed light [4].

Manifestations of slow light in atomic vapors may be broadly organized according to the type of optical resonances used to obtain the necessary dispersion. In the simplest case, slow group velocities may be obtained using a single laser in a two-level atomic system. Of recent interest has been the study of slow light in a three-level system, most commonly in the configuration of electromagnetically induced transparency (EIT). Slow light has also been observed in four-level systems, hole-burning schemes, and a variety of others, and in principle almost any optical resonance may be used. While the number of ways to achieve slow light is numerous, in this chapter we choose a few representative systems to illustrate how one may make simple predictions about any one of them, given the appropriate Hamiltonian.

1.2 FIRST EXPERIMENTS IN SLOW LIGHT

The first experimental studies of slow light were performed in the context of nonlinear optics; namely, the self-induced transparency (SIT) effect discovered by McCall and Hahn in 1967 [5]. McCall and Hahn reported an experiment in which they passed pulses generated in a ruby laser through cooled rods of ruby and observed appreciable group delay in addition to the SIT effect. That same year, Patel and Slusher demonstrated a similar effect using gaseous SF_6 as the delay medium [6]. The first study of slow light due to SIT in an atomic vapor was performed in 1968 by Bradley et al. [7], who measured time delays in potassium vapor. Interestingly, the experiment also included time-delay measurements away from SIT resonance as a control experiment. This resulted in the first measurements of slow light in the linear regime. Slusher and Gibbs then reported much more careful studies of SIT and slow light in an atomic vapor in 1972, in which a two-level transition on the D_1 line of Rb was studied [8]. Since all these studies relied on the nonlinearity generated by the pulse passing through the medium, the group delay was dependent on the width, energy, and hence, area of the pulse. In fact, some researchers reported observations of group delay as an indication of SIT, though it was soon pointed out by Courtens and Szöke [9] that this was not strictly the case.

When a transparency effect (EIT) that was linear in the input pulse was discovered by Harris and coworkers [10], it was natural that the corresponding pulse propagation dynamics was also studied in relation to the new transparency mechanism, which was taken up theoretically by Harris in 1992 [11]. The first experimental studies of linear slow light, however, were carried out much earlier by Grischkowsky [12]. Grischkowsky showed that even far away from resonance, a pulse may experience a significant reduction in group velocity and made contact with the earlier discussions of SIT by describing his results in terms of an adiabatic following effect. We will return to this class of slow light experiments after discussing slow light due to EIT.

1.3 EIT

We begin with a discussion of a three-level system in the Λ configuration (see Figure 1.1) with two fields: a weak signal beam E_s and a strong coupling beam E_c. The well-known Hamiltonian describing the atom–field interaction in the absence of damping may be written in the rotating frame as [13]

$$-\frac{\hbar}{2}\begin{bmatrix} 0 & \Omega_s & 0 \\ \Omega_s & -2\Delta_1 & \Omega_c \\ 0 & \Omega_c & -2(\Delta_1-\Delta_3) \end{bmatrix}, \quad (1.1)$$

where Ω_s and Ω_c are the Rabi frequencies induced by the signal and coupling fields, respectively, and Δ_1 and Δ_3 represent the signal and coupling field detunings from optical resonance respectively. When the two-photon detuning is zero ($\Delta_1 = \Delta_3$), we obtain eigenvalues of $\{0, \hbar/2(\Delta_1\pm\Omega_N)\}$, where we have defined a normalizing Rabi frequency $\Omega_N = \sqrt{\Omega_c^2 + \Omega_s^2}$. The vanishing energy eigenvalue corresponds to the case in which no atom field coupling exists and has an eigenvector of

$$|-\rangle = \frac{\Omega_c}{\Omega_N}|1\rangle - \frac{\Omega_s}{\Omega_N}|3\rangle. \quad (1.2)$$

When the system is in this eigenstate, no absorption takes place and hence no spontaneous emission occurs. For this reason, atoms prepared in this eigenstate are said to be in a dark state, invisible to radiation at the signal frequency. We wish to explore the dispersive properties of such a three-level system prepared in the neighborhood of this dark eigenstate, and so we must examine the atom field coupling in the vicinity of two-photon resonance and include damping. The steady state polarization nearly resonant with the signal frequency may be found by summing the polarizations

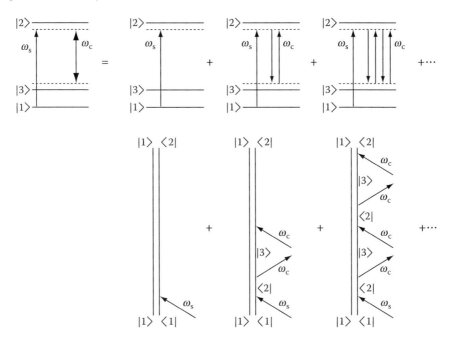

FIGURE 1.1 Energy level diagrams with corresponding Feynman diagrams illustrating how interference between various atomic absorption pathways leads to EIT.

induced by all possible excitation pathways from state $|1\rangle$ to state $|3\rangle$, conveniently represented using double-sided Feynman diagrams [1], as shown in Figure 1.1:

$$P_s = 2N\mu_{12} \frac{\Omega_s}{2} \frac{1}{\tilde{\Delta}_s} \sum_{n=0}^{\infty} r^n$$

$$= N\mu_{12} \Omega_s \frac{1}{\tilde{\Delta}_s - \frac{\Omega_c^2}{4\tilde{\Delta}_R}} \quad (1.3)$$

where the summation over $r = \Omega_c^2/(4\tilde{\Delta}_s \tilde{\Delta}_R)$ accounts for the repeated emission and absorption of a coupling photon. The quantities $\tilde{\Delta}_s = \Delta_s - i\Gamma/2$ and $\tilde{\Delta}_R = \Delta_s - \Delta_c - i\gamma$ are the complex single-photon and two-photon (Raman) detunings where Γ and γ represent the transverse excited and longitudinal ground state decay rates, respectively, N is the atomic number density, and $\Omega_j = \mathbf{E}_j \cdot \boldsymbol{\mu}_j / \hbar$ again represents the Rabi frequency induced by electric field amplitude E_j via the dipole matrix element μ_j.

We may obtain approximate expressions for the group delay and broadening by performing a series expansion of the steady state polarization resonant with the signal frequency (Equation 1.3) around zero two-photon detuning ($\Delta_R = 0$). Assuming the coupling field to be on resonance ($\Delta_c = 0$), the real and imaginary parts of the index of refraction at the signal frequency ($n_s = \sqrt{1 + \chi_s} \approx 1 + P_s/2E_s$) may be written as

$$n_s' \approx 1 + \frac{2N\mu_{12}^2}{\hbar\epsilon_0} \frac{\Delta_s}{\Omega_c^2} \quad (1.4)$$

$$n_s'' \approx \frac{2N\mu_{12}^2}{\hbar\epsilon_0} \left(\frac{\gamma}{\Omega_c^2} + \frac{2\Gamma\Delta_s^2}{\Omega_c^4} \right). \quad (1.5)$$

While these expressions are sufficient to characterize the essential delay and broadening features of an optical pulse, we take a moment to rewrite them in terms of the optical depth of the atomic vapor, an experimentally easy quantity to measure. In the absence of the coupling field, the polarization at the signal frequency is just the first term in Equation 1.3:

$$P_{s0} = \frac{N\mu_{12}\Omega_s}{2\tilde{\Delta}_s} \approx 2N\mu_{12}\Omega_s\left(\frac{\Delta_s}{\Gamma^2} + i\frac{1}{2\Gamma}\right), \quad (1.6)$$

and we may easily write the linear optical absorption coefficients $2n''\omega/c$ in the absence and presence of the coupling field, which we call α and β, respectively:

$$\alpha = \frac{N\omega\mu_{12}^2}{c\epsilon_0\hbar\Gamma}$$

$$\beta = \frac{4N\omega\mu_{12}^2\gamma}{c\epsilon_0\hbar\Omega_c^2} = \frac{4\gamma\Gamma}{\Omega_c^2}\alpha, \quad (1.7)$$

allowing us to write the real and imaginary parts of the index of refraction as

$$n'_s \approx 1 + \frac{c}{2\omega}\frac{\beta}{\gamma}\Delta_s \quad (1.8)$$

$$n''_s \approx \frac{c}{2\omega}\left[\beta + \frac{2\Gamma\beta}{\gamma\Omega_c^2}\Delta_s^2\right]. \quad (1.9)$$

A pulse which does not undergo significant distortion will travel at the group velocity given by

$$v_g = \frac{c}{n'_s + \omega_s\frac{dn'_s}{d\omega_s}} \approx \frac{c}{\omega_s\frac{dn'_s}{d\omega_s}} \quad (1.10)$$

where we have assumed that the dispersive term $\omega_s dn'_s/d\omega_s$ is much larger than the phase index n'_s. Taking the appropriate derivatives we obtain

$$t_g = \frac{L}{v_g} \approx \frac{\beta L}{2\gamma} = \frac{2\Gamma\alpha L}{\Omega_c^2}. \quad (1.11)$$

The temporal broadening of an optical pulse passing through a slow light medium originates from two sources, frequently classified as absorptive and dispersive broadening. Absorptive broadening occurs when the individual frequencies making up the pulse waveform are absorbed at different levels in the medium. This may be studied by treating the imaginary part of the index of refraction as an optical filter of the form

$$S(\Delta_s) = \exp\left[-\beta L\left(1 + \frac{2\Gamma}{\gamma\Omega_c^2}\Delta_p^2\right)\right]. \quad (1.12)$$

When the input pulse is a bandwidth limited Gaussian centered on Raman resonance, the spectrum of the output pulse is given by the product of the input pulse spectrum $A_{\text{in}}(\Delta_s)$ and the filter:

$$A_{\text{out}}(\Delta_s) = A_{\text{in}}(\Delta_s)S(\Delta_s) \propto \exp\left[-\beta L - \Delta_p^2\left(T_0^2 + \frac{2\Gamma}{\gamma\Omega_c^2}\beta L\right)\right] \quad (1.13)$$

which gives an output pulse of width

$$T_{\text{out}} = \sqrt{T_0^2 + \frac{2\Gamma}{\gamma\Omega_c^2}\beta L} = \sqrt{T_0^2 + \frac{8\Gamma^2}{\Omega_c^4}\alpha L}. \quad (1.14)$$

Slow Light in Atomic Vapors

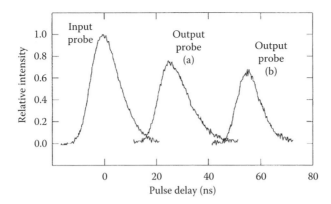

FIGURE 1.2 The first demonstration of optical pulse delay using EIT. (Reprinted from Kash, M.M. et al., *Physical Review Letters*, 82, 5229, 1999. With permission.)

Dispersive broadening may be treated by considering the difference in group delay for a pulse centered at $\Delta_s = 0$ and a pulse centered at $\Delta_s = 1/T_0$. For the case of EIT, however, and all other Lorentzian lineshapes, pulse broadening is almost entirely due to frequency-dependent absorption and the dispersive broadening may be neglected.

This and other results from the aforementioned model may be verified by comparing them to the many slow light experiments that have been done using EIT in atomic vapors. The group delay of pulses in an EIT system was studied first by Kasapi et al. [14] and then later by many others [13,15–17]. The results obtained by Kasapi et al. are shown in Figure 1.2. The experiment was performed in lead vapor, in which the authors measure an EIT linewidth 2γ of approximately 10^7 rad/s. The peak pulse transmission T may be read from the figure and used to approximate the optical depth on EIT resonance as $\beta L = -\ln(T) \approx 0.25$ for the pulse delayed by approximately 23 ns, giving a group delay of $\tau_g = \beta L/2\gamma = 25$ ns, in agreement with the data. Another noteworthy demonstration of slow light in atomic vapors using EIT was performed in 1999 by Hau et al. [18]. In their experiment they input a 2.5 µs optical pulse into a condensed cloud of sodium atoms with a spontaneous decay rate of $\Gamma = 61.3 \times 10^6$ rad/s, and calculated optical depth of $\alpha L = 63$ and $\Omega_c = 0.56\Gamma$ (Figure 1.3). When these numbers are used in Equations 1.11 and 1.14 for the pulse delay and broadening, we get a result of

$$t_g = \frac{2\Gamma\alpha L}{\Omega_c^2} = \frac{2 \times 63}{(0.56)^2 \Gamma} = 6.6 \text{ µs}$$

$$T_{\text{out}} = \sqrt{T_0^2 + \frac{8\Gamma^2}{\Omega_c^4}\alpha L} = \sqrt{(2.5 \text{ µs})^2 + \frac{8 \times 63}{(0.56)^4 \Gamma^2}} = 2.8 \text{ µs}$$

which agrees reasonably well with the experimental data reported. (The actual delay reported was 7.05 µs and the broadening, while hard to quantify from the figure, appears to be in line with the prediction.) An equivalent classical treatment of Hau et al.'s results has also been made by McDonald [19].

Shortly following the publication of Hau et al.'s results in 1999, many other experiments were reported studying slow light in coherently prepared atomic vapors. Kash et al., for example, demonstrated that very slow group velocities could be obtained without cooling the atomic sample [20], and Budker et al. reported on slow light in the closely related phenomenon of nonlinear Faraday rotation [21]. Since 1999, hundreds of experimental and theoretical papers have been published on the subject of slow and fast group velocities, with a recent emphasis on possible applications of slow light [22–28]. It should be noted that EIT, like SIT, may also be studied in the nonlinear regime, in which matched pulse solutions that have coupled slow group velocities as well as a number of other interesting properties [29,30] can be found.

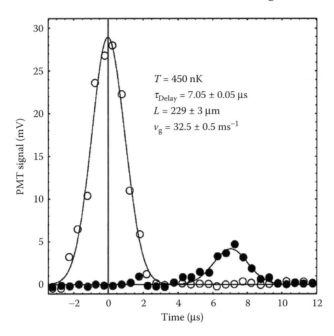

FIGURE 1.3 Delay of a 2.5 μs pulse in a low temperature cloud of sodium atoms. (Reprinted from Hau, L.V., Harris, S.E., Dutton, Z., and Behroozi, C.H., *Nature*, 397, 594, 1999. With permission.)

1.4 TWO-LEVEL SYSTEMS

While EIT provided the first dramatic reduction of group velocity, the earliest experimental studies of linear slow light were not performed using EIT, but rather using the naturally occurring resonances in atomic vapors. We may understand such experiments in the context of linear dispersion theory. For an optical field interacting with a two-level system in the linear regime, we may use Equation 1.6 to obtain an approximate group velocity:

$$n \approx 1 + \frac{c}{2\omega_s}\left(\frac{2\alpha\Delta_s}{\Gamma} + i\alpha\right) \tag{1.15}$$

yielding

$$\tau_g = \frac{L}{v_g} \approx \frac{L\omega_s \frac{dn'_s}{d\Delta_s}}{c} = \frac{\alpha L}{\Gamma} \tag{1.16}$$

The first slow light studies in this context appear to be those of Grischkowsky, whose experimental results are shown in Figure 1.4. In his experiment, he measured the difference in propagation time for left-handed (σ^-) and right-handed (σ^+) circularly polarized light pulses passing through an atomic vapor of Rb atoms. A magnetic field was used to shift the Zeeman sublevels of the relevant transitions such that the σ^- pulse was much closer to resonance than the σ^+ pulse, resulting in a greater optical depth αL for the σ^- pulse. Figure 1.4a shows an interferogram of the pulses used in parts (b) and (c) to demonstrate that the optical linewidths of the pulses are sufficiently narrow, part (b) shows the temporal profile of the linearly polarized input pulse, and part (c) shows the output pulse resolved into σ^- and σ^+ components. Each division corresponds to 5 ns in the figure. The difference in optical depth between σ^+ and σ^- pulses from the figure appears to be about $\alpha L = 1$, implying a difference in delay of $\alpha L/\Gamma = 27$ ns, in agreement with the figure (where we have used $\Gamma = 38.1$ Mrad/s for Rb). One important feature of these results is the complete separation of the pulses in time by many pulse

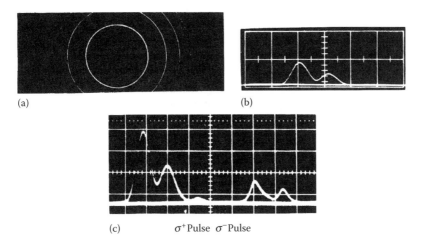

FIGURE 1.4 Relative delays of σ^+ and σ^- polarized light pulses tuned away from optical resonance. Each division represents 5 ns. (Reprinted from Grischkowsky, *Physical Review A*, 7, 2096, 1973. With permission.)

widths with minimal distortion, the first experimental demonstration of a delay-bandwidth product much greater than unity, a subject we address in the Section 1.5.

Another line of study in the linear regime was taken up by Chu and Wong [31,32] in 1982 and later by Segard and Macke in 1985 [33]. Chu and Wong measured the group velocity of a weak pulse on resonance in an alkali vapor with the purpose of demonstrating the group velocity as a distinct measurable quantity from the velocity of energy propagation. They showed that group velocity may be either greater than c or less than c in the vicinity of optical resonance. The results had been predicted by Garret and McCumber and can also be directly derived from the well-known results of Sommerfield and Brillouin [34].

1.5 DISPERSION MANAGEMENT

Many implementations of slow light require a pulse to be delayed by many times its temporal width, or equivalently maximizing the delay-bandwidth product of the pulse. To qualitatively examine how one might accomplish this, consider the dispersive term $\omega dn/d\omega$ in the denominator of the group velocity. In order to reduce the group velocity, one must make this dispersive term large, which requires a large slope of the frequency-dependent index of refraction. A steep slope requires either a large change in the nominal value of the refractive index (i.e., large dn) or a very small frequency range over which a modest change occurs (i.e., small $d\omega$). Among the more compelling reasons EIT has been useful for fundamental demonstrations of slow light is the extremely narrow resonances (<10 Hz) that can be obtained in an EIT medium, which lead to a correspondingly small $d\omega$ and hence slow group velocities. However, narrow resonances can accommodate only limited bandwidth, making a large delay-bandwidth product difficult in EIT-based systems. The few researchers who have managed delay-bandwidth products greater than unity in EIT media have done so by creating samples with very large differences in optical depth on and off resonance, thereby making the dn term large enough for multiple pulse delays. Excellent discussions regarding the maximum delay-bandwidth products attainable in EIT systems have been found in the work of Boyd et al. [35] and Matsko et al. [36], who come to different but entirely consistent conclusions. Miller [37,38] as well as Tucker et al. [23] have discussed delay-bandwidth limitations in a slightly more general context.

Alternative approaches have been suggested to overcome these delay-bandwidth limitations in atomic vapors, including channelization schemes [39,40] and ways to change the shape of the optical resonance in order to accommodate larger bandwidths [41]. The ideal resonance shape was discovered many years ago in the context of microwave physics to be a square transmission filter (in fact most radio frequency filters purchased quote a group delay among the specifications). It also happens that among the least ideal shapes is a single Lorentzian near resonance, such as those found in most EIT systems, since absorptive broadening limits the bandwidth of the input pulse to approximately the linewidth of the Lorentzian.

Far away from resonance, however, Lorentzian lineshapes have quite different delay. Working away from resonance, for example, allows one to manage second-order dispersion more effectively and increase the delay-bandwidth product for a given pulse broadening, such as that seen in the aforesaid results of Grischkowsky.

To reduce pulse broadening further, several researchers have considered the use of two resonances simultaneously. The first proposal for such a system appears to have been made by Steinberg and Chiao, who in 1994 suggested the use of two widely spaced inverted resonances to observe superluminal group velocities with little second-order dispersion [42]. They noted that there exists a point between the two resonances in which the index of refraction has no second-order term, thereby reducing dispersive broadening. In 2000, Wang et al. used two gain lines in a similar configuration to measure group velocities greater than the speed of light [43,44]. In 2003 Macke and Segard performed a careful study of such negative group velocities using double gain resonances, and also suggested that one may obtain pulse advancement by the proper spacing of two absorbing lines [45]. The first time the reciprocal arrangement to that of Steinberg and Chiao was investigated appears to have been in 2003, when Tanaka et al. used two widely spaced absorbing resonances to measure optical delays with little distortion [46]. This configuration has since been considered in a variety of theoretical and experimental studies [28,47–50].

We briefly outline the noteworthy features of this system by considering the case of two absorbing Lorentzian lines of equal width separated by a spectral distance much larger than their widths. The susceptibility of the double Lorentzian is then given by

$$\chi = \frac{N\mu^2}{\hbar\epsilon_0}\left(\frac{1}{\tilde{\Delta}_1} + \frac{1}{\tilde{\Delta}_2}\right), \quad (1.17)$$

where $\tilde{\Delta}_i$ represents the complex optical detuning $\Delta_i - i\Gamma/2$ from an optical resonance centered at ω_i.

Making the change of variables $\omega = (\omega_1 + \omega_2)/2 + \delta$ and $\omega_0 = (\omega_2 - \omega_1)/2$ and assuming that the pulse bandwidth does not exceed the spectral region between the resonances, we may write the real and imaginary parts of the refractive index as

$$n' \approx 1 + \frac{\mathcal{A}}{\omega_0^2}\delta + \frac{\mathcal{A}}{\omega_0^4}\delta^3 \quad (1.18)$$

$$n'' \approx 1 + \frac{\mathcal{A}\Gamma}{2\omega_0^2} + \frac{3\mathcal{A}\Gamma}{2\omega_0^4}\delta^2, \quad (1.19)$$

where the power series are expanded about $\delta = 0$ and $\mathcal{A} = N\mu^2/\hbar\epsilon_0$. First note the absence of a second-order term in the real part of the index of refraction, indicating the absence of second second-order group velocity dispersion. It can also be seen that $dn'/d\delta = dn''\Gamma/2$, which can be used to obtain a simple form for the group velocity. Combining this result with $\alpha_m = 2n''\omega/c$, where α_m is the optical intensity absorption coefficient of the medium at the pulse carrier frequency, one obtains an approximate group velocity Γ/α_m. This leads to a group delay of

$$\tau_g \approx \frac{\alpha_m L}{\Gamma} \quad (1.20)$$

This result is useful in the context of delay-bandwidth products since it predicts that the delay is independent of the spectral separation between the two resonances and depends only on the

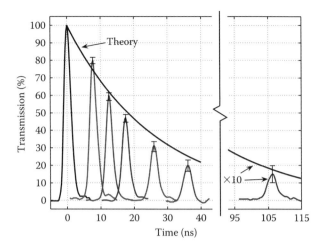

FIGURE 1.5 Multiple pulse delays with little distortion using double resonances in Rb vapor. The theory plot is taken from Equation 1.20.

absorption away from resonance and the linewidth of the resonances. Also, owing to the shape of the transparency region where delay takes place (it looks much more like a square filter than a single Lorentzian), pulse broadening is dominated by third-order dispersion rather than second-order absorption, allowing for much larger delay-bandwidth products [28].

An example of the use of two widely spaced absorbing Lorentzians to achieve large delay-bandwidth products is shown in Figure 1.5, in which optical pulses were delayed between the two hyperfine resonances in ^{85}Rb. In addition to the pulse delays, a plot of Equation 1.20 is shown, where $\Gamma = 36.1$ Mrad/s. While the delays are numerically much smaller than those achievable in EIT systems, there is, in principle, no reason why the double resonance technique could be applied to produce longer delays.

1.6 CONCLUDING REMARKS

Experiments involving slow light in atomic vapors have produced the principal results which continue to motivate slow light research. Most of the experimental and theoretical results that continue to appear in the literature may be traced to the early studies of slow light in vapors. In this chapter, we have briefly reviewed the basic physics leading to the prediction of slow group velocities in a few model systems. We have also given a brief overview of efforts that have been made to minimize group velocity dispersion in slow light systems, with an emphasis on the general principles necessary to understand the field. We have not treated all possible variations of resonances that may be used to obtain slow light in atomic systems, but have discussed a few representative systems using the same general model built on Feynman diagrams. We hope that theories for other types of systems (i.e., four-wave mixing schemes, hole burning, etc.) may be easily derived from the same framework once the basics outlined in this chapter are well understood.

REFERENCES

1. J. J. Su and I. A. Yu, *Chinese Journal of Physics* 41, 627 (2003).
2. R. W. Boyd and D. J. Gauthier, in *Progress in Optics* (Elsevier, Radarweg 29, Amsterdam 1043 NX, 2002), vol. 43, pp. 497–530.
3. P. W. Milonni, *Journal of Physics B-Atomic Molecular and Optical Physics* 35, R31 (2002).
4. P. W. Milonni, in *Fast Light, Slow Light, and Left-Handed Light* (Institute of Physics, Bristol and Philadelphia, 2005).

5. S. L. McCall and E. L. Hahn, *Physical Review Letters* 18, 908 (1967).
6. C. K. N. Patel and R. E. Slusher, *Physical Review Letters* 19, 1019 (1967).
7. D. J. Bradley, G. M. Gale, and P. D. Smith, *Nature* 225, 719 (1970).
8. R. E. Slusher and H. M. Gibbs, *Physical Review A* 5, 1634 (1972).
9. E. Courtens and A. Szöke, *Physics Letters A* 28, 296 (1968).
10. S. E. Harris, J. E. Field, and A. Imamoglu, *Physical Review Letters* 64, 1107 (1990).
11. S. E. Harris, J. E. Field, and A. Kasapi, *Physical Review A* 46, R29 (1992).
12. D. Grischkowsky, *Physical Review A* 7, 2096 (1973).
13. M. Fleischhauer, A. Imamoglu, and J. P. Marangos, *Reviews of Modern Physics* 77, 633 (2005).
14. A. Kasapi, M. Jain, G. Y. Yin, and S. E. Harris, *Physical Review Letters* 74, 2447 (1995).
15. M. Xiao, Y. Q. Li, S. Z. Jin, and J. Geabanacloche, *Physical Review Letters* 74, 666 (1995).
16. O. Schmidt, R. Wynands, Z. Hussein, and D. Meschede, *Physical Review A* 53, R27 (1996).
17. J. P. Marangos, *Journal of Modern Optics* 45, 471 (1998).
18. L. V. Hau, S. E. Harris, Z. Dutton, and C. H. Behroozi, *Nature* 397, 594 (1999).
19. K. T. McDonald, *American Journal of Physics* 68, 293 (2000).
20. M. M. Kash, V. A. Sautenkov, A. S. Zibrov, L. Hollberg, G. R. Welch, M. D. Lukin, Y. Rostovtsev, E. S. Fry, and M. O. Scully, *Physical Review Letters* 82, 5229 (1999).
21. D. Budker, D. F. Kimball, S. M. Rochester, and V. V. Yashchuk, *Physical Review Letters* 83, 1767 (1999).
22. A. B. Matsko, O. Kocharovskaya, Y. Rostovtsev, G. R. Welch, A. S. Zibrov, and M. O. Scully, in *Advances in Atomic Molecular and Optical Physics*, B. Bederson, ed. (Academic Press, New York, 2001), Vol. 46, pp. 191–242.
23. R. S. Tucker, P. C. Ku, and C. J. Chang-Hasnain, *Journal of Lightwave Technology* 23, 4046 (2005).
24. G. T. Purves, C. S. Adams, and I. G. Hughes, *Physical Review A* 74, 023805 (2006).
25. Z. Shi, R. W. Boyd, R. M. Camacho, P. K. Vudyasetu, and J. C. Howell, *Physical Review Letters* 99, 240801 (2007).
26. Z. M. Shi, R. W. Boyd, D. J. Gauthier, and C. C. Dudley, *Optics Letters* 32, 915 (2007).
27. G. S. Pati, M. Salit, K. Salit, and M. S. Shahriar, *Physical Review Letters* 99 (2007).
28. R. M. Camacho, M. V. Pack, J. C. Howell, A. Schweinsberg, and R. W. Boyd, *Physical Review Letters* 98, 153601 (2007).
29. S. E. Harris, *Physical Review Letters* 72, 52 (1994).
30. J. H. Eberly, M. L. Pons, and H. R. Haq, *Physical Review Letters* 72, 56 (1994).
31. S. Chu and S. Wong, *Physical Review Letters* 48, 738 (1982).
32. S. Chu and S. Wong, *Physical Review Letters* 49, 1293 (1982).
33. B. Segard and B. Macke, *Physics Letters A* 109, 213 (1985).
34. L. Brillouin, in *Wave Propagation and Group Velocity* (Academic Press, New York, 1960).
35. R. W. Boyd, D. J. Gauthier, A. L. Gaeta, and A. E. Willner, *Physical Review A* 71, 023801 (2005).
36. A. B. Matsko, D. V. Strekalov, and L. Maleki, *Optics Express* 13, 2210 (2005).
37. D. A. B. Miller, *Journal of the Optical Society of America B-Optical Physics* 24, A1 (2007).
38. D. A. B. Miller, *Physical Review Letters* 99, 203903 (2007).
39. Z. Dutton, M. Bashkansky, M. Steiner, and J. Reintjes, *Optics Express* 14, 4978 (2006).
40. M. Bashkansky, Z. Dutton, F. K. Fatemi, J. Reintjes, and M. Steiner, *Physical Review A* 75, 021401 (2007).
41. R. M. Camacho, M. V. Pack, J. C. Howell, and Zp, *Physical Review A* 74, 4 (2006).
42. A. M. Steinberg and R. Y. Chiao, *Physical Review A* 49, 2071 (1994).
43. L. J. Wang, A. Kuzmich, and A. Dogariu, *Nature* 406, 277 (2000).
44. K. T. McDonald, *American Journal of Physics* 69, 607 (2001).
45. B. Macke and B. Segard, *European Physical Journal D* 23, 125 (2003).
46. H. Tanaka, H. Niwa, K. Hayami, S. Furue, K. Nakayama, T. Kohmoto, M. Kunitomo, and Y. Fukuda, *Physical Review A* 68, 053801 (2003).
47. R. M. Camacho, M. V. Pack, and J. C. Howell, *Physical Review A* 73, 4 (2006).
48. B. Macke and B. Segard, *Physical Review A* 73, 043802 (2006).
49. R. M. Camacho, C. J. Broadbent, I. Ali-Khan, and J. C. Howell, *Physical Review Letters* 98, 043902 (2007).
50. Z. M. Zhu and D. J. Gauthier, *Optics Express* 14, 7238 (2006).

2 Slow and Fast Light in Semiconductors

Shu-Wei Chang and Shun Lien Chuang

CONTENTS

2.1 Introduction..13
2.2 Slow Light Based on CPO in Quantum Wells 15
2.3 Slow Light Based on CPO in Quantum Dots 21
2.4 Room-Temperature Operation of Slow Light Based on CPO in Quantum Dots 23
2.5 Fast and Slow Light Based on CPO and FWM in the Gain Regime.........................28
2.6 Slow Light Scheme Based on Spin Coherence 30
2.7 Summary 33
Acknowledgments 34
References 34

2.1 INTRODUCTION

Coherent population oscillation (CPO) is a successful method used to implement semiconductor-based slow light using the coherent pump–probe effect. The first demonstration was carried out in semiconductor multiple quantum wells (MQWs) [1–3] soon after the experimental demonstration of slow light in ruby crystal [4,5]. Later, slow light based on CPO in semiconductor QWs and quantum dots (QDs) was also successfully demonstrated at room temperature [6–10], and the slow-light regime was soon extended to the corresponding fast-light regime in semiconductor gain media [11–17] using a semiconductor optical amplifier (SOA) alone or combined with an electroabsorbtion (EA) section. The basic working principle of slow light based on most pump–probe schemes is to decrease the group velocity v_g using the sharp variation of the refractive index $\partial n'/\partial \omega$ within a narrow frequency range:

$$v_g = \frac{c}{n_g} = \frac{c}{n' + \omega \frac{\partial n'}{\partial \omega}}, \qquad (2.1)$$

where
$n_g = n' + \omega \partial n'/\partial \omega$ is the group index
n' is the real part of the refractive index n

For the CPO case, the required sharp variation of the refractive index originates from the population oscillation induced by the pump and probe signals applied to a two-level system, as indicated in Figure 2.1. The induced population oscillation generates a new polarization component and changes the susceptibility and thus the refractive index experienced by the signal. In semiconductor quantum structures, this two-level system can be the heavy-hole (HH) exciton [1–3,6,18] of QWs and QDs, or the valence and conduction ground states of QDs [18,19]. In principle, as long as the optical transition between the two states is dipole-allowed, the slow light based on CPO can be implemented.

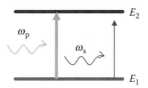

FIGURE 2.1 CPO in a simple two-level system. The photon energies of the signal and pump are close to the transition energy $E_2 - E_1$. The population beating is induced with a frequency equal to the signal–pump detuning $\omega_s - \omega_p$.

The induced population oscillation generates linear and nonlinear four-wave-mixing (FWM) polarizations which feed back to the response of the probe [2,20]. Since an extra linear polarization is induced for the probe, equivalently, the linear permittivity $\chi(\omega)$ experienced by the probe is changed. The induced population oscillation is prominent if the detuning frequency between the pump and probe is within the inverse of the population lifetime T_1^{-1} because T_1 is the timescale within which the population oscillation can follow the beating induced by the pump and probe. In semiconductors, the population lifetime T_1 is usually the radiative recombination lifetime and falls in the range of hundreds of picoseconds to one nanosecond. The induced population oscillation changes the linear refractive index in two essential respects as shown in Figure 2.2. First, in addition to the power saturation of the absorption caused by the pump, a dip with a linewidth of about T_1^{-1}, centered at the pump frequency, is present on the absorption spectrum of the probe, as indicated in Figure 2.2a. The significant population beating helps reduce the linear absorption experienced by the probe. Second, the real and imaginary parts of the relative dielectric constant $\epsilon(\omega) = \epsilon'(\omega) + i\epsilon''(\omega)$ satisfy the Kramers–Kronig (KK) relations. A dip on the imaginary part of the dielectric constant corresponds to a positive-slope variation of the real part around the same frequency range. Therefore, as shown in Figure 2.2b, after the pump is applied to the system, the probe will experience a positive-slope variation of the refractive index if its carrier frequency falls at the center of the absorption dip. This provides the necessary variation for a large group index, and thus a low group velocity.

In an absorptive medium, the nonlinear polarization does not efficiently generate the FWM signal and its effect can usually be neglected. For the gain medium (fast-light regime), whether the effect of the FWM component is negligible depends on the geometry of the pump and probe. For example, if the pump–probe scheme is in the counter-propagation configuration, the momentum

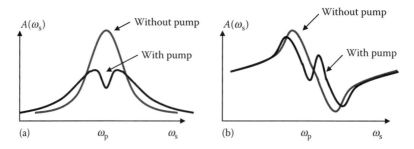

FIGURE 2.2 The variations of the absorption and real part of the refractive index of a two-level system. (a) After the pump is applied to the system, the absorption is saturated. The signal experiences the saturated absorption when the signal–pump detuning is larger compared with the inverse of the population lifetime. When the signal frequency is close to the pump frequency, the induced population beating leads to a dip on the absorption spectrum. (b) Corresponding to the absoprtion dip, a positive-slope variation of the refractive index is induced on the spectrum of the refractive index. This positive-slope variation increases the group index n_g (slowdown factor) and lowers the group velocity.

conservation required in the generation of the FWM component cannot be satisfied. In this case, the nonlinear polarization need not be considered. On the other hand, significant coupling is present in the copropagation configuration because the momentum conservation can be easily achieved. The effect of the FWM must then be taken into account [12,21,22].

Compared to slow light based on other coherent effects such as electromagnetically induced transparency (EIT) [23] and the photorefractive (PR) effect [24] or CPO using other material systems [4,5], semiconductor slow light or fast light based on CPO has less restrictions, mainly for the following reasons:

1. It is the population relaxation time that is utilized in the coherent process and not the dephasing time, which typically ranges from femtoseconds to picoseconds in semiconductors except for some limited cases which are too short to generate significant quantum coherence.
2. CPO is less affected by inhomogeneous broadening [18], which is usually inevitable and leads to a variation of the refractive index that is not as sharp [25] in semiconductor quantum structures.
3. The timescale of the population lifetime falls in the range of nanoseconds, which corresponds to a gigahertz bandwidth. The gigahertz bandwidth is much larger than slow light based on other physical mechanisms [23,24] or CPO in other material systems [4,5]. The large bandwidth is more suitable for information transmissions which require higher bandwidth.
4. Semiconductors provide easy on-chip integration with other active devices.
5. In addition to the optical pump, the bias voltage or the electrical injection current is also handy in semiconductors, which provides another degree of freedom to manipulate the optical signal.

On the other hand, semiconductor-based slow light suffers from high absorption, which prevents the scaling of the time delay with the device length. The problem of absorption can be solved by using a fast-light scheme in the gain regime since it is usually the time difference between the probe wave packet that matters [11–17,22]. There is still the problem of the limited delay-bandwidth product [26–29] which applies to most slow-light schemes. In principle, the delay-bandwidth product can be increased by using a cascade scheme [16], similar to channel dispersion [30] or wavelength division multiplexing (WDM) for semiconductor-based slow light and fast light.

In Sections 2.2 and 2.3, we will briefly review the physical mechanisms of slow light based on CPO in semiconductor QWs and QDs. The room-temperature operation of slow light based on QDs will be addressed in Section 2.4. Fast and slow lights in the gain regime are discussed in Section 2.5. Finally, a slow-light scheme based on spin coherence [31–33], which is a pump–probe scheme analogous to that of CPO, will be presented as an alternative mechanism to demonstrate slow light in semiconductors. Finally, we will summarize the semiconductor-based slow light and fast light in Section 2.7.

2.2 SLOW LIGHT BASED ON CPO IN QUANTUM WELLS

Demonstration of slow light in a semiconductor was experimentally performed in GaAs/AlGaAs QWs [1,2] at low temperature. The experimental setup is shown in Figure 2.3a. A single-mode Ti:sapphire laser acts as the pump while a tunable laser acts as the signal. The signal is modulated by a chopper for lock-in detection at a later stage. The two beams are directed at the QW sample, and the transmitted signal is detected by the lock-in amplifiers to extract the amplitude and radiofrequency (RF) phase under different pump intensities. The achieved slowdown factor was about 32,000. After the first low-temperature experiment, CPO-based slow light was pushed to room temperature [6] with a much higher pump intensity. The HH exciton was used as the two-level system for CPO, and the corresponding pump–probe scheme is shown in Figure 2.3b. The pump and signal (probe) are normally incident into QWs while the polarization of the pump $\hat{e}_p$ can be parallel or perpendicular

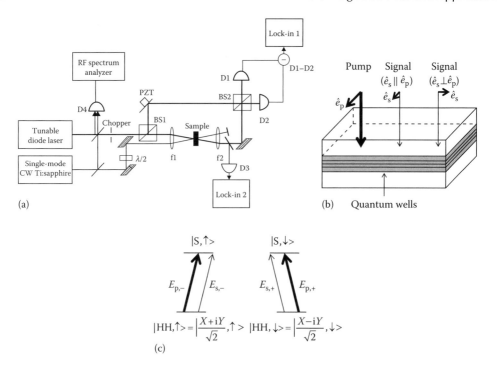

FIGURE 2.3 (a) The experimental setup of the slow-light experiment. The Ti:sapphire laser is the pump while the tunable laser acts as the signal. The phase shift and absorption are measured by two lock-in amplifiers, respectively. (From Ku, P.C., Sedgwick, F., Chang-Hasnain, C.J., Palinginis, P., Li, T., Wang, H., Chang, S.W., and Chuang, S.L., *Opt. Lett.*, 29(19), 2291, 2004. With permission.) (b) The pump–probe scheme. The signal and pump are normally incident from the top of the sample. The polarizations of the signal and pump can be parallel or orthogonal to each other. (c) The two subsystems of the CPO based on the HH exciton in QWs. The two subsystems are labeled by the respective spins and are independently excited by two circularly polarized pumps (signals), which can be decomposed from the linear polarized signal and pump. (From Chang, S.W., Chuang, S.L., Ku, P.C., Chang-Hasnain, C.J., Palinginis, P., and Wang, H., *Phys. Rev. B*, 70(23), 235333, 2004. With permission.)

to the polarization of the signal $\hat{e}_s$. Due to the selection rule of the HH exciton in [001] QWs, only the transverse electric (TE) polarization can induce its transition. There are two subsystems of the HH exciton which are labeled by their electron spin (up and down) and are dipole-active to the incident TE light, as shown in Figure 2.3c. The transitions of the two subsystems can be selectively induced by right-hand-circularly (RHC) polarized light and left-hand-circularly (LHC) polarized light, i.e., the polarization is of the form $(\hat{x} \pm i\hat{y})/\sqrt{2}$. Experimentally, only the linearly polarized pump and signal were used, which means that the two subsystems are excited simultaneously as a linearly polarized light can always be decomposed into two circularly polarized components. The experiment was carried out with normal incidence geometry. However, the waveguide geometry (SOA geometry) can also be used because the TE polarized light is allowed in a guided mode. The waveguide geometry usually leads to a smaller confinement factor compared to the normal-incidence geometry. Since the variation of the refractive index is proportional to the confinement factor, it takes a longer length to achieve the same amount of time delay for the waveguide geometry. However, since the modal absorption is also less, the light can also propagate farther. The waveguide geometry is more common in later slow-light experiments [6–14,16,17,34,35].

For QW HH excitons, not all the phenomena can be explained by a simple two-level system. However, the two-level system provides the best starting point for the basic phenomena of CPO.

Slow and Fast Light in Semiconductors

The dynamics of a simple two-level system is described by the density-matrix equation under the rotating-wave approximation:

$$\frac{\partial n}{\partial t} = -\frac{n - n^{(0)}}{T_1} - 2i[\Omega_{12}(t)\rho_{21} - \Omega_{21}(t)\rho_{12}],$$

$$\frac{\partial \rho_{21}}{\partial t} = -i\left(\omega_{21} - \frac{i}{T_2}\right)\rho_{21} - i\Omega_{21}(t)n,$$

$$n = \rho_{22} - \rho_{11},$$

$$\Omega_{21}(t) = \frac{e\mathbf{r}_{21}}{2\hbar} \cdot \left(\mathbf{E}_p e^{-i\omega_p t} + \mathbf{E}_s e^{-i\omega_s t}\right)$$

$$= \Omega_p e^{-i\omega_p t} + \Omega_s e^{-i\omega_s t},$$

(2.2)

where
- n is the difference between the population ρ_{22} of state 2 and ρ_{11} of state 1
- $n^{(0)}$ is the equilibrium value of n when neither the pump nor the signal is present
- $\rho_{21}(\rho_{12})$ is the off diagonal density matrix element
- $\Omega_{21}(t)$ is the time-dependent Rabi frequency
- $\hbar\omega_{21}$ is the energy difference between the two states
- T_2 is the dephasing time
- $e\mathbf{r}_{21}$ is the matrix element of the dipole moment
- ω_p and ω_s are the angular frequencies of the pump and signal, respectively
- Ω_p and Ω_s are the Rabi frequencies of the pump and signal, respectively

The condition $|\Omega_s| \ll |\Omega_p|$ is usually satisfied, and therefore the perturbation approximation for the signal can be used, i.e., $n = n^{(p)} + n^{(s)}$; $\rho_{21} = \rho_{21}^{(p)} + \rho_{21}^{(s)}$; $n^{(s)} \ll n^{(p)}$, $\rho_{21}^{(s)} \ll \rho_{21}^{(p)}$. The unperturbed solution of $n^{(p)}$ and $\rho_{21}^{(p)}$ due to the pump describes the saturated absorption experienced by the pump (the envelope of the absorption in the presence of the pump in Figure 2.1a):

$$n^{(p)} = \frac{n^{(0)}[1 + T_2^2(\omega_{21} - \omega_p)^2]}{1 + T_2^2(\omega_{21} - \omega_p)^2 + 4T_1 T_2 |\Omega_p|^2},$$

$$\rho_{21}^{(p)} = \tilde{\rho}_{21}^{(p)} e^{-i\omega_p t}$$

(2.3)

$$= \frac{-n^{(p)} \Omega_p e^{-i\omega_p t}}{\omega_{21} - \omega_p - \frac{i}{T_2}}.$$

The perturbed solutions for $n^{(s)}$ and $\rho_{21}^{(s)}$ not only include the phenomenon of saturated absorption but also the effect of the sharp absorption dip when the signal–pump detuning $\omega_s - \omega_p$ is small. The small population variation $n^{(s)}$ beats with the detuning frequency, which generates the linear and FWM components of $\rho_{21}^{(s)}$:

$$n^{(s)} = \tilde{n}^{(s)} e^{-i(\omega_s - \omega_p)t} + \tilde{n}^{(s)*} e^{i(\omega_s - \omega_p)t}$$

$$\rho_{21}^{(s)} = \tilde{\rho}_{21,L}^{(s)} e^{-i\omega_s t} + \tilde{\rho}_{21,FWM}^{(s)} e^{-i(2\omega_p - \omega_s)t},$$

(2.4)

where
- $\tilde{n}^{(s)}$ is the amplitude of the population beating
- $\tilde{\rho}_{21,L}^{(s)}$ and $\tilde{\rho}_{21,FWM}^{(s)}$ are the linear and FWM amplitudes of $\rho_{21}^{(s)}$ respectively

In the absorption regime, we are only interested in the linear component $\tilde{\rho}_{21,L}^{(s)}$. The perturbations $n^{(s)}$ and $\tilde{\rho}_{21,L}^{(s)}$ can be written as

$$n^{(s)} = \frac{2n^{(p)}\left(\frac{1}{\omega_{21}-\omega_p+iT_2^{-1}} - \frac{1}{\omega_{21}-\omega_s-iT_2^{-1}}\right)\Omega_p^*\Omega_s}{\omega_s - \omega_p + \frac{i}{T_1} + 2|\Omega_p|^2\left(\frac{1}{\omega_{21}-\omega_s-iT_2^{-1}} - \frac{1}{\omega_{21}-2\omega_p+\omega_s+iT_2^{-1}}\right)},$$

$$\tilde{\rho}_{21,L}^{(s)} = \frac{-n^{(p)}\Omega_s}{\omega_{21}-\omega_s-\frac{i}{T_2}}\left\{1 + \frac{2|\Omega_p|^2\left(\frac{1}{\omega_{21}-\omega_p+iT_2^{-1}} - \frac{1}{\omega_{21}-\omega_s-iT_2^{-1}}\right)}{\omega_s - \omega_p + \frac{i}{T_1} + 2|\Omega_p|^2\left(\frac{1}{\omega_{21}-\omega_s-iT_2^{-1}} - \frac{1}{\omega_{21}-2\omega_p+\omega_s+iT_2^{-1}}\right)}\right\}. \tag{2.5}$$

Equation 2.5 is most easily understood in the case $\omega_s \sim \omega_p \sim \omega_{21}$ and $|\omega_s - \omega_p| \ll T_2^{-1}$. The density matrix element $\tilde{\rho}_{21,L}^{(s)}$, which is related to the linear response of the signal, can then be approximated as

$$\tilde{\rho}_{21,L}^{(s)} \sim -in^{(p)}\Omega_s T_2\left[1 - i\frac{4T_2|\Omega_p|^2}{\omega_s - \omega_p + i\left(\frac{1}{T_1} + 4T_2|\Omega_p|^2\right)}\right]. \tag{2.6}$$

From the denominator in Equation 2.6, the detuning $\omega_s - \omega_p$ is associated with a new timescale whose inverse is $T_1^{-1} + 4T_2|\Omega_p|^2$. When the pump intensity is low (small $|\Omega_p|$), this timescale is just T_1. Under such circumstances, the response exhibits a variation characterized by the time scale T_1. Indeed, this variation corresponds to an absorption dip with linewidth of about T^{-1} and a positive-slope variation of the refractive index within the same frequency range. As mentioned earlier, the phenomenon of CPO is significant when the detuning $\omega_s - \omega_p$ is of the order of the population lifetime, which is in the nanosecond range for semiconductors. Thus, this linewidth corresponds to a gigahertz range for semiconductors. As the pump intensity becomes higher, the linewidth increases as well. The broadening of the linewidth together with the saturated absorption from the pump will finally limit the amount of slowdown factor achievable for the CPO of a simple two-level system.

Figure 2.4 shows a comparison between the experimental data and theoretical calculation [2]. Experimentally, as the pump intensity increases, the background absorption decreases as a result of the power saturation. At the same time, an absorption dip emerges from the center of the saturated background. The linewidth increases as the pump intensity increases while the depth of the dip first increases with the pump intensity but later is limited by the saturated absorption. The maximal positive slope of the corresponding variation on the refractive index, which is related to the group index n_g, or simply slowdown factor, is roughly proportional to the depth of the dip and inversely proportional to the linewidth of the dip. Therefore, as the depth of the absorption dip changes, there is a maximal slowdown factor at a given optimal pump intensity. Experimentally, the slowdown factor can be directly extracted from the measurement of the phase shift $\Delta\phi(\omega_s)$ which is approximately related to the group index (slowdown factor):

$$\Delta\phi(\omega_s) \sim \frac{2\pi fL}{c}[n_g(\omega_s) - n(\omega_s)], \tag{2.7}$$

where
  f is the modulation frequency of the signal
  L is the length of the medium which provides the effect of CPO

Slow and Fast Light in Semiconductors

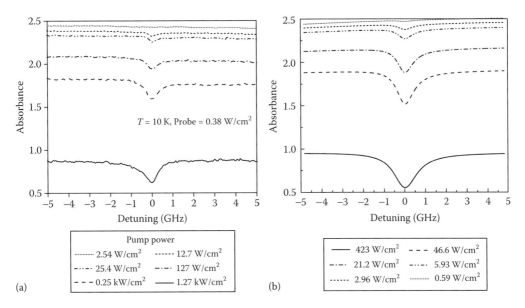

FIGURE 2.4 (a) The experimental data of the absorption experienced by the signal under different pump intensities. As the pump intensity increases, the background absorption is saturated, and a dip due to CPO emerges from the background absorption. (b) The theoretical calculation of the absorption dip, which compares well with the experimental data. (From Chang, S.W., Chuang, S.L., Ku, P.C., Chang-Hasnain, C.J., Palinginis, P., and Wang, H., *Phys. Rev. B*, 70(23), 235333, 2004. With permission.)

The experimentally measured phase shift agrees well with the theoretical calculation, as shown in Figure 2.5.

Most of the phenomena of slow light based on the QW HH exciton can be understood as the simple two-level system aforementioned. However, there are phenomena which go beyond the simple two-level system. If we treat the two spin subsystems of the QW HH excitons as just two independent two-level systems, we conclude that whether the polarizations of the pump and signal are parallel or perpendicular it shall make no difference, i.e., if we can observe absorption dip in one configuration, we can find the same phenomenon in the other. Experimentally, an absorption dip

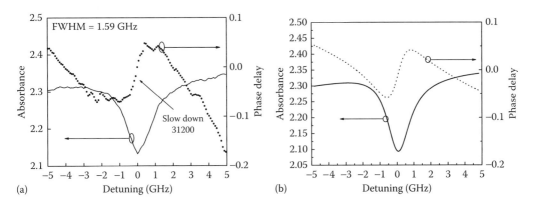

FIGURE 2.5 (a) The measured spectra of the absorption dip and phase shift. The slope at the center of the phase shift corresponds to a slowdown factor of about 32,000. (b) The corresponding theoretical calculation which compares well with the experimental data. (From Chang, S.W., Chuang, S.L., Ku, P.C., Chang-Hasnain, C.J., Palinginis, P., and Wang, H., *Phys. Rev. B*, 70(23), 235333, 2004. With permission.)

and variation of the refractive index occur only when the pump and signal polarizations are parallel, but not when the two are perpendicular. The reason for the absence of the slow-light behavior is due to the fact that the dephasing time actually depends on the populations ρ_{22} and ρ_{11} (or on n). This dependence is called excitation-induced dephasing (EID) [2,36,37] and can be approximated as

$$\frac{1}{T_2} \sim \frac{1}{T_2^{(0)}} + \gamma(n - n^{(0)}),$$

$$n = n_\uparrow + n_\downarrow,$$

(2.8)

where

γ is a proportional constant
$n_\uparrow$ and $n_\downarrow$ are the population difference of the two subsystems respectively

The two spin subsystems are coupled to each other via this additional term. It happens that only when the populations of the two subsystems oscillate in phase, which corresponds to the configuration of the parallel polarization, do the absorption dip and variation of the refractive index take place. On the other hand, if the two subsystems tend to oscillate out of phase, which is the case of the configuration of the perpendicular configuration, no slow-light phenomena are observable. Figure 2.6a shows the experimental spectra of the two configurations. There is no absorption dip for the configuration of perpendicular polarization. In Figure 2.6b, this phenomenon is calculated theoretically with a quantitative agreement that the two coupled subsystems are taken into consideration simultaneously.

Slow light based on CPO has been successfully demonstrated in QWs. In principle, other semiconductor quantum structures should also be potential candidates for slow light based on CPO. In Section 2.3, we will consider CPO based on QDs and show why slow light based on CPO can persist in semiconductor QDs which are usually fabricated with significant inhomogeneity.

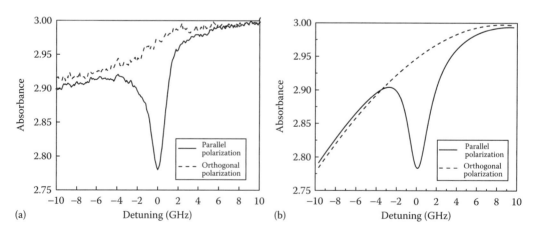

FIGURE 2.6 (a) The experimental data of the absorption experienced by the signal. The solid line is the absorption when the polarizations of the pump and signal are parallel with each other. The dashed line is the counterpart when the two polarizations are perpendicular to each other. The absence of the absorption dip for the perpendicular polarizations is due to the excitation induced dephasing and out-of-phase oscillation of the populations of the two spin subsystems. (b) The corresponding theoretical calculation. (From Chang, S.W., Chuang, S.L., Ku, P.C., Chang-Hasnain, C.J., Palinginis, P., and Wang, H., *Phys. Rev. B*, 70(23), 235333, 2004. With permission.)

2.3 SLOW LIGHT BASED ON CPO IN QUANTUM DOTS

The CPO in QDs is similar to that in QWs. Furthermore, there is three-dimensional (3D) confinement of the carriers. In principle, the 3D confinement should make room-temperature operation more plausible. Indeed, QD-based slow light was first demonstrated at room temperature [7,8] rather than at low temperature. However, one of the potential problems for most QD-based slow light schemes is the inevitable inhomogeneous broadening which can wash out the sharp absorption dip. For example, slow light based on EIT is more difficult to implement in QDs with inhomogeneous broadening [25]. Inhomogeneous broadening always exists in QDs, no matter, whether it is at low temperature or room temperature. Therefore, to demonstrate slow light using semiconductor QDs, the slow-light mechanism must be robust enough to overcome the inhomogeneous broadening. For CPO, inhomogeneous broadening degrades the performance of the QD-based slow light but does not completely eliminate the effect. The position of the absorption dip is determined by the pump frequency, ω_p and not by the resonance frequency of the transition. Although a significant inhomogeneous broadening means that only a small portion of QDs participate in CPO and contribute to slowdown the optical wave packet, the effect is not completely eliminated. In the future, if nearly identical QDs can be fabricated, slow light in QDs may be a better candidate than that in QWs or bulk.

To understand the influence of inhomogeneous broadening, let us consider the absorption spectrum illustrated in Figure 2.7. An inhomogeneous-broadened spectrum is composed of many independent homogeneous-broadened absorption peaks. If a pump is applied to this system with inhomogeneous broadening, those absorption peaks of the transitions near the pump frequency will be saturated more while those farther away are not influenced much. The nonuniform saturation leads to the wide spectral-hole burning with a linewidth of the order of the homogeneous linewidth $1/T_2$ which increases as the pump intensity becomes higher. The phenomenon of spectral-hole burning is usually observable at low temperatures for semiconductor QDs. On the other hand, carriers tend to come into equilibrium fast enough at room temperature even in the presence of the 3D confinement. The static spectral hole burning is usually washed out at room temperature. However, the effect of CPO is always present and independent of the temperature. Room-temperature CPO will be considered in Section 2.4. At low temperatures, in contrast to the big spectral hole, the amplitude of CPO is

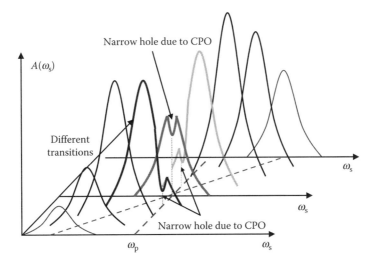

FIGURE 2.7 The inhomogeneous broadened absorption is composed of many homogeneous-broadened absorption peaks. When a pump is applied to the system, the homogeneous absorption peaks whose transition frequencies are close to that of the pump are saturated more. This nonuniform saturation leads to spectral hole burning. On the other hand, since the coherent dip of each individual homogeneous-broadened transition occurs at the same frequency (ω_p) on the absorption spectrum, the effect of CPO is not completely washed out.

significant only when the signal–pump detuning is within the inverse of the population lifetime. The coherent absorption dip (narrow hole in Figure 2.7) is always centered around the pump frequency no matter what the resonance frequency of the individual transition is. Therefore, the narrow absorption dip created is not washed out by the resonance frequency of each independent transition. However, the induced absorption dips of those transitions which are a few homogeneous linewidth $1/T_2$ away are weak as well. The weak absorption dip means that those transitions farther away from the pump frequency are not efficiently utilized. If the linewidth of the inhomogeneous broadening can be controlled below that of the homogeneous broadening, the efficiency of CPO can be greatly enhanced, i.e., it takes a lower pump intensity to induce the same amount of transparency and slowdown factor even though the peak absorption of the spectrum will be higher due to more concentrated transitions within a narrow frequency range.

Let us consider the normal incidence pump and probe again for samples with several QD layers (Figure 2.8). Figure 2.9a shows the theoretical calculation of the low-temperature absorption spectrum of QD HH-like excitons with inhomogeneous broadening for different pump intensities [18]. The spectral hole with a linewidth on the order of a terahertz is more significant as the pump intensity becomes higher. The linewidth of the spectral hole also becomes larger with an increasing pump intensity because individual absorption peaks far away from the pump frequency are finally saturated as well. This spectrum is the global absorption experienced by the signal as if the effect of CPO is absent. However, as the signal-pump detuning becomes smaller, the coherent absorption dip due to CPO, which is located at the bottom of the global spectral hole, emerges out of the absorption spectrum, as indicated in the inset of Figure 2.9a. The coherent absorption dips with a gigahertz linewidth at the bottom of the global spectral holes are shown in Figure 2.9b. The global spectral hole and the coherent absorption dip correspond to two different timescales. One is the dephasing time on the order of a femtosecond to a picosecond. The other is the population lifetime on the order of a nanosecond. Thus, the observation of two linewidths on the absorption spectrum simply means the difference in timescale.

Similar to the peak slowdown factor of QWs, there is an optimal pump intensity at which the peak slowdown factor is the maximum. The reason for the decrease of the slowdown factor at a higher pump intensity is the same as that for QWs. However, there are differences between the two cases. First, the magnitude of the peak slowdown factor differs by two orders of magnitude. The reason is that the density of active centers is small due the limited QD surface density and is not as high as that of QWs. Also, due to the finite size of QDs and low coverage ratio of the QD active layer, the transverse confinement of the normally incident light is smaller than that of QWs. These two factors

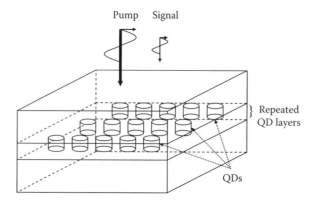

FIGURE 2.8 The configuration of the pump–probe scheme in the calculation. The pump and signal are normally incident into the active region which consists of several layers of QDs. (From Chang, S.W. and Chuang, S.L., *Phys. Rev. B*, 72(23), 235330, 2005. With permission.)

Slow and Fast Light in Semiconductors

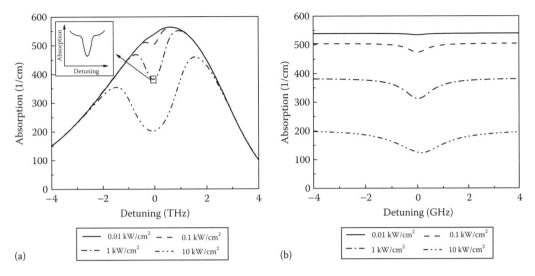

FIGURE 2.9 The calculated absorption spectra of CPO based on QDs with inhomogeneous broadening. (a) The global absorption spectrum experienced by the signal under different pump intensities. This absorption spectrum reflects the emergence of the spectral-hole burning of the inhomogeneous-broadened QDs as the pump intensity increases. The coherent absorption dip due to CPO takes place at the bottom of the spectral hole with a linewidth much narrower than that of the spectral hole, as indicated by the inset. (b) The coherent absorption dip at the bottom of the spectral hole under different pump intensities. These absorption dips are due to CPO and are much narrower than those of the incoherent spectral holes in (a). (From Chang, S.W. and Chuang, S.L., *Phys. Rev. B*, 72(23), 235330, 2005. With permission.)

result in a lower confinement factor. Second, if we use the optimal intensity as the unit to measure the intensity, the peak slowdown factor decreases much more slowly with the pump intensity in QDs with inhomogeneous broadening than its counterpart in QWs. The reason for this slow reduction is that QDs not in resonance with the pump, though less active in CPO, are difficult to saturate.

We considered CPO based on QDs at low temperature. However, for practical applications, it would be better to have a room-temperature slow-light device (e.g., an optical buffer). In Section 2.4, we discuss room-temperature CPO based on QDs.

2.4 ROOM-TEMPERATURE OPERATION OF SLOW LIGHT BASED ON CPO IN QUANTUM DOTS

Slow light based on CPO in QDs was first demonstrated experimentally at room temperature rather than low temperature [7,8]. Figure 2.10 shows the RF phase shift and small signal absorption of the CPO based slow light [8]. The clear peaks on the spectra of transmission and phase shift indicate the presence of CPO based on QDs at room temperature. Later, the electrical control of slow light based on CPO was also demonstrated [9,11,17]. Compared with the case of low temperature, the static spectral-hole burning is usually absent at room temperature. Therefore, it is a reasonable approximation to model the occupation numbers of QDs by two Fermi factors with respective quasi Fermi levels in the conduction and valence bands, taking into account the inhomogeneous broadening. In addition to the optical pump, the reverse biased voltage and forward injection current can alter the bias condition of the active region. However, the two electrical controls affect the slow light based on CPO in two distinct ways.

As shown in Figure 2.11a, in the reverse bias regime, the generated carriers of the optical pump are swept out of the QDs by the external electrical field. The effective population loss rate is determined not only by radiative recombination but also by the pullout of carriers from QDs. Thus, the linewidth

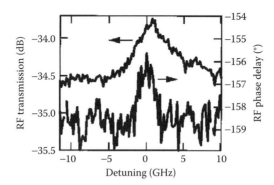

FIGURE 2.10 The experimental data of the absorption and RF phase shifts due to CPO in QDs. This experimental data shows that CPO is less affected by inhomogeneous broadening. (From Su, H. and Chuang, S.L., *Opt. Lett.*, 31(2), 271, 2006. With permission.)

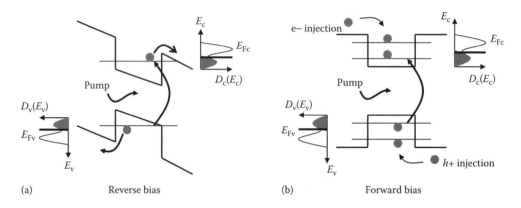

FIGURE 2.11 The electrical control of slow light in QDs. (a) The QD under reverse bias. In addition to the red shift due to the Stark effect, the effective lifetime is reduced because the applied electric field sweeps the carriers out of the QDs. (b) Under forward bias, the carriers are injected into QDs. The quasi Fermi levels of the electrons and holes are separated. The effective lifetime is fairly constant in this case, but the background absorption is saturated by the injected carriers. The saturation of the background absorption limits the depth of the absorption dip and the available slowdown factor. (From Chang, S.W., Kondratko, P.K., Su, H., and Chuang, S.L., *IEEE J. Quantum Electron.*, 43(2), 196, 2007. With permission.)

of the coherent absorption dip increases as the electrical field increases, which means that the effective lifetime is shortened. On the other hand, in the forward bias regime, the external field does not change the band profile of QDs significantly, but the forward bias current saturates the background absorption by injecting electrons and holes into QDs. The population lifetime is not significantly altered in this case, but the saturated background absorption clamps the depth of the absorption dip and reduces the available slowdown factor. This picture holds good only when the injected current is below the transparency current, that is, when population inversion is not reached. Above the transparency current, CPO and FWM in the gain regime may need to be considered simultaneously. In both the reverse and forward bias regimes, one can use electrical control to change the slowdown factor, which provides direct control over the optical buffer.

To explain how electrical control can be made to operate, it is necessary to develop a model to describe how the reverse bias and forward injection current change the phenomenon of slow light based on CPO in QDs. In the reverse bias regime, the applied electric field changes the energy levels (Stark shift) and effective lifetime of the QDs. The model is similar to that at low temperature,

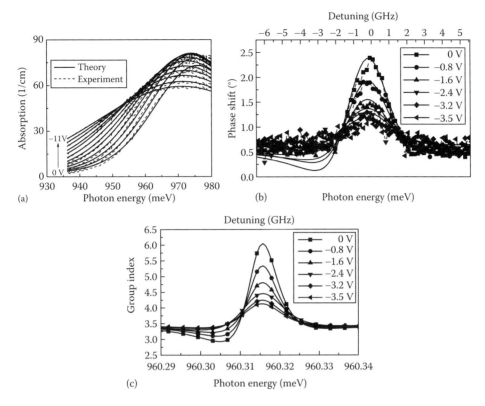

FIGURE 2.12 (a) The absorption spectra of QDs under different reverse bias voltages. The absorption is red shifted due to the Stark effect. The absorption spectrum also exhibits tunneling broadening as the reverse bias voltage increases. (b) The comparison of the experimental data and theoretical calculation of the RF phase shifts under different reverse bias voltages. The linewidth of the absorption increases as the reverse bias voltage increases, which indicates the shortening of the effective population lifetime. (c) The extracted slowdown factors corresponding to the RF phase shifts in (b). (From Chang, S.W., Kondratko, P.K., Su, H., and Chang, S.L., *IEEE J. Quantum Electron.*, 43(2), 196, 2007. With permission.)

mentioned in Section 2.4, Fermi–Dirac, except for the following conditions: (1) the carrier distributions are modeled by two distributions with two quasi Fermi levels; (2) the effective lifetime is shortened, and energy levels are shifted at each reverse bias voltage.

Figure 2.12a shows the absorption spectra of QDs under different reverse bias voltages when the optical pump is absent [18]. The sample is p-doped InGaAs QDs with five QD layers [8]. The operating wavelength is around 1.29 μm. The trend of the red shift of the absorption spectrum is similar to that of the Stark shift in QWs. With the absorption spectra at different reverse bias voltages, one can determine the necessary parameters for the simple two-level system with inhomogeneous broadening. Figure 2.12b shows the experimental data and theoretical calculations of the RF phase shift. The corresponding group index (slowdown factor) is shown in Figure 2.12c. As the reverse bias voltage increases, the magnitude of the phase shift decreases due to the broadening of the linewidth (inverse of the effective population lifetime). From the experimental data, one can calculate the corresponding slowdown factor under different reverse bias voltages. The peak slowdown factor is around 6 at zero bias. The much lower slowdown factor compared with that of the geometry of normal incidence is due to the poorer confinement factor both in the transverse and propagation directions and a lower pump intensity farther away from the input facet because of the absorption. Also, at room temperature, the dephasing time is much shorter than that at low temperature, which degrades the magnitude of CPO.

For current injection in the forward bias region, one can use the rate equations and the charge neutrality condition to relate the carrier densities n_c and n_h in the conduction and valence bands, respectively, to the injected current density J_{in}.

$$\frac{\partial n_c}{\partial t} = \frac{J_{in}}{eN_{LA}} + \left.\frac{\partial n_c}{\partial t}\right|_{loss} + \left.\frac{\partial n_c}{\partial t}\right|_{optical},$$

$$\frac{\partial n_h}{\partial t} = \frac{J_{in}}{eN_{LA}} + \left.\frac{\partial n_h}{\partial t}\right|_{loss} + \left.\frac{\partial n_h}{\partial t}\right|_{optical}, \quad (2.9)$$

$$n_c + n_A^- = n_h,$$

where
N_{LA} is the number of QD layers
n_A^- is the density of the ionized acceptors

The subscript loss means the recombination due to various recombination mechanisms while optical indicates the generation due to the presence of an optical field. At steady state, one can relate the two carrier densities to the two quasi Fermi levels and obtain the occupation numbers in the QDs. Figure 2.13a shows the experimental data and calculated results for the phase shift under different injection currents in the forward bias regime. The reduction of the phase shift is due to the saturation of the background absorption which clamps the depth of the absorption dip. Figure 2.13b shows the extracted group index (slowdown factor). Although the trend is similar to that in the reverse bias regime. the respective mechanisms of reduction are different. One can see the difference by extracting the linewidth of the RF phase and plotting it as a function of the reverse bias voltage (injection current) in the reverse bias (forward bias) regimes, as shown in Figure 2.14. The half-width-at-half-maximum (HWHM) linewidth of the RF phase shift is nearly a constant in the forward bias regime, indicating no variation for the population lifetime, while it increases as the reverse bias voltage becomes higher, reflecting the shortening of the effective population lifetime.

The above experimental data and theoretical calculations are for the p-doped samples. Therefore, before the reverse bias voltage or forward injection current is applied to the system, there are pre existing holes in the QDs, which already saturated the background absorption and therefore limited

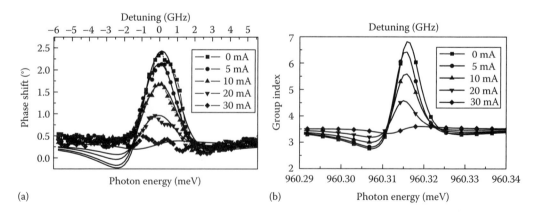

FIGURE 2.13 (a) The RF phase shifts under different forward injection currents. The reduction of the RF phase shifts are caused by the saturation of the background absoprtion which limits the depth of the absorption dips and variation of the refractive index. (b) The extracted slowdown factors corresponding to the RF phase shifts in (a). (From Chang, S.W., Kondratko, P.K., Su, H., and Chang, S.L., *IEEE J. Quantum Electron.*, 43(2), 196, 2007. With permission.)

Slow and Fast Light in Semiconductors

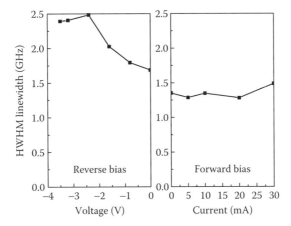

FIGURE 2.14 The extracted linewidth of the RF phase shift as a function of the reverse bias voltage and forward injection current. The linewidth stays constant in the forward bias regime, indicating a constant effective population lifetime. On the other hand, the linewidth increases with reverse bias due to the shortening of the effective lifetime. (From Chang, S.W., Kondratko, P.K., Su, H., and Chang, S.L., *IEEE J. Quantum Electron.*, 43(2), 196, 2007. With permission.)

the available absorption depth. If an intrinsic sample is used, the background absorption will increase and can sustain a larger absorption dip and higher slowdown factor. However, there is a trade off. Because of the increased background absorption, the signal and pump will be attenuated more when propagating in the QD sample. Thus, even though the larger background may increase the available slowdown factor, the required pump intensity is actually higher. Also, a larger attenuation results in a smaller signal-to-noise ratio. Figure 2.15 shows the measured slowdown factor and linewidth of the intrinsic sample. Due to a shorter population lifetime (possibly more efficient spontaneous radiative recombination) compared with the p-doped sample, the available maximum slowdown factor is only slightly better than that of the p-doped sample.

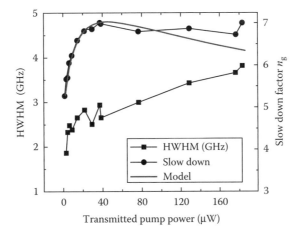

FIGURE 2.15 The slowdown factor and HWHM linewidth of the intrinsic QD sample. The intrinsic QD sample can offer a larger background absorption and may lead to a larger slowdown factor. Compared with the experimental data in Figures 2.12 and 2.13, the slowdown factor is only slightly better, mainly due to the shorter effective lifetime (larger HWHM linewidth). (From Kondratko, P.K., Chang, S.W., Su, H., and Chuang, S.L., *Appl. Phys. Lett.*, 90(25), 251108, 2007. With permission.)

In this section, we have considered CPO at room temperature. The foregoing description is still limited to the absorption regime. From the viewpoint of system applications, the attenuated signal may have unacceptably low signal-to-noise ratio. In Section 2.5 we will discuss fast light due to CPO and FWM in the gain regime. In the gain regime, although the amplified spontaneous emission is another factor limiting the signal-to-noise ratio, the restriction is less serious than that in the absorptive regime.

2.5 FAST AND SLOW LIGHT BASED ON CPO AND FWM IN THE GAIN REGIME

The immediate advantage of fast light and slow light in the gain regime is that the signal is amplified during the propagation through the device rather than attenuated [11,17]. The physics of fast light in the gain regime due to CPO is similar to that of slow light in the absorption regime [15,21]. The gain medium can be bulk, QWs or QDs as long as the background gain is high enough to sustain the necessary variation for fast light. However, there are differences between systems with distinct dimensions that can cause features in the phenomena of fast light. For example, the bulk and QW gain media usually have a significant linewidth enhancement factor while the QD gain medium has a much smaller one. As a result, while the small-signal absorption due to CPO in the QD gain medium is symmetric with respect to signal–pump detuning [11], the small-signal gain in bulk or QW gain medium has a prominent asymmetry due to the finite linewidth enhancement factor [21], as shown in Figure 2.16a. The corresponding variation in the refractive index is shown in (b). We see that on the side of negative detuning far away from the pump frequency ω_p, the positive slope of the refractive index variation is enhanced, while on the other side (positive detuning), the positive slope is reduced. Thus, if the carrier frequency of the signal is allowed to operate on the negative side of the detuning, where the slope is positive, it is possible to have a slow-light device in the gain medium, where even the small signal gain is high. On the other hand, if a fast-light device is desired, one simply operates at the asymmetric dip on the gain spectrum. Nevertheless, for both the slow-light and fast-light operation in the gain regime, CPO is indispensable from FWM, and the generated FWM signal usually cannot be separated from the signal at the output of the device. This may be one of the reasons for significant distortion of the input pulse [14].

As mentioned earlier, when the bias condition is changed, some effects originally neglected in the absorption regime can no longer be neglected. For example, below the transparency current, the effective population time is fairly constant in the forward bias regime. However, after significant

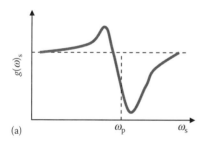 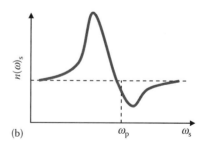

FIGURE 2.16 The small-signal spectra of gain and refractive index as a function of the signal frequency for a SOA with a finite linewidth enhancement factor. (a) The spectrum of the small-signal gain. The finite linewidth enhancement leads to an asymmetric gain dip. The gain at the negative signal–pump detuning is enhanced. (b) The spectrum of the refractive index. Corresponding to the enhanced gain at the negative detuning, there is an enhanced positive slope on the spectrum of the refractive index. This part of the refractive index can be utilized in the demonstration of the slow light. On the other hand, the negative-slope feature corresponding to the gain dip in (a) can be utilized in the demonstration of the fast light.

Slow and Fast Light in Semiconductors

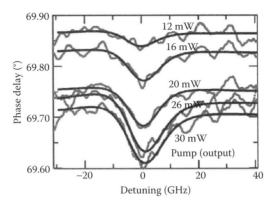

FIGURE 2.17 The RF phase shift due to fast light in the QD gain medium. Due to the significant shortening of the effective population lifetime resulting from the washout of the population grating in the wetting layer, the phase shift is much smaller than that in the absorption regime. (From Su, H. and Chuang, S.L., *Appl. Phys. Lett.*, 88(6), 061102, 2006. With permission.)

carriers are injected into the QDs, communication between carriers via the wetting layer becomes important. If the pump–probe scheme of counter propagation is used, the diffusion process in the wetting layer washes out the dynamic population grating that has resulted from CPO and the standing-wave pattern due to the pump and probe. The diffusion significantly decreases the effective lifetime to tens of picoseconds [11]. As shown in Figure 2.17, the RF phase shift in the gain regime of QDs, when compared with its counterpart in the absorption regime (Figure 2.11), is much shallower and wider (HWHM linewidth ~13 GHz). The RF phase shift due to the change of the background refractive index is even more significant than that of CPO. Although the shorter effective lifetime can increase the bandwidth of fast light, the fast-light effect is also significantly reduced.

The fast light or slow light due to CPO by itself may not be significant enough in real applications. However, if FWM is utilized, the phase shift (time delay/advance) due to slow light and fast light can be greatly increased [12,14,15,21,22]. As mentioned earlier, if the copropagation scheme is used for the pump–probe experiment, the required momentum conservation for FWM is easily satisfied, and the FWM component is amplified during the propagation through the gain medium. In this case, the effects of CPO and FWM cannot be individually identified because the two processes couple to each other during the wave propagation. The generation of linear and FWM responses and the conversion between them are not independent of each other. It is this coupling that results in a significant RF phase shift and time advance (delay) in the copropagation scheme [12–14,22]. If the signal is of the form of sinusoidal modulation, analytical expressions of phase advance and small-signal gain are available [22]. In general, for an arbitrary input pulse, since FWM in the gain regime is a nonlinear process, a numerical analysis to obtain the time delay/advance is usually required. Figure 2.18a shows the phase advance of the signal as a function of the RF modulation frequency when propagating through an SOA under different injection currents. The solid line is the theoretical calculation and agrees well with the experimental data. A larger injection current provides more carriers to the active region and leads to higher gain for both the signal and pump. At the frequency of several gigahertz, a phase advance of a few tens of degrees is induced, which is much larger than that of CPO in the absorption regime. Similar to the increase of the injected current, if the pump intensity becomes higher, the phase shift as a function of the RF modulation frequency also increases, as shown in Figure 2.18b. The above experimental data are the measured frequency response of the signal. Figure 2.18c shows the real-time data of the phase advance. This verifies that a true time advance of a few tens of degrees is achieved. One can also switch between the absorption regime (slow light) based on CPO and the gain regime based on CPO/FWM to increase the phase shift [12].

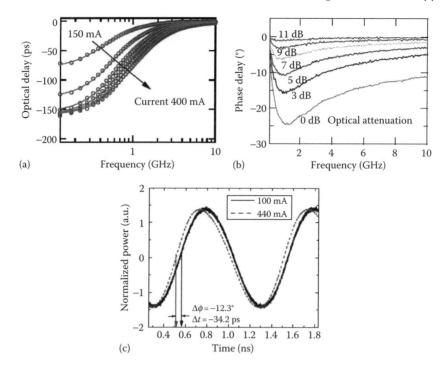

FIGURE 2.18 (a) The RF phase shift due to CPO and FWM in an SOA under different current injections. The significant increase of the RF phase shift is caused by the coupling between CPO and FWM during the nonlinear propagation of the signal through the device. (b) The RF phase shift under different pump intensities. (c) The real-time data of the RF phase shift. This experiment verifies that fast light is achieved in real time. (From Su, H., Kondratko, P.K., and Chuang, S.L., *Opt. Express*, 14(11), 4800, 2006. With permission.)

For a real application, the input signal is usually not a continuous wave (CW) carrier. Thus, it is necessary to know how a real pulse is advanced (delayed) when it propagates through a device. The experiment in Ref. [14] shows that the trend of the time delay of the pulse reflects the presence of a finite linewidth enhancement factor, as indicated in Figure 2.16b. Compared with the output with very large detuning which experiences little effect from slow or fast light, the output with the negative detuning is delayed by 0.59 ns (slow light) while that with positive detuning has a slight advance (fast light). However, the output signal corresponding to fast light is significantly distorted with a dip at the trailing edge of the pulse [14]. This indicates that distortion is an important issue in slow and fast light based on CPO and FWM in the gain medium.

So far, we have discussed slow and fast light based on CPO and FWM in both the absorption and gain regimes. In Section 2.6, we will turn to another scheme for slow light which is similar to that based on CPO. This mechanism resembles CPO in a few ways, e.g., both the linear and FWM responses are present. However, instead of the population lifetime, the timescale which is utilized will be the spin coherence time.

2.6 SLOW LIGHT SCHEME BASED ON SPIN COHERENCE

Two key requirements to demonstrate slow light are (1) a relatively long timescale which leads to sharp variations in the spectra of absorption and refractive index, and (2) a suitable pump–probe scheme to sense this long timescale. For CPO, the long timescale is the population lifetime T_1. In semiconductors, there is another long timescale which can be used in a slow-light experiment. This timescale is the electron spin coherence time, which usually ranges from a few picoseconds to one nanosecond,

Slow and Fast Light in Semiconductors

depending on the material system, doping concentration, and spin relaxation/decoherence mechanisms [38–41]. Usually, the hole spin coherence time is much shorter than the electron spin coherence time. If there is any spin coherence in semiconductors, most of it should be contributed by electron spin coherence. For [001] QW, the electron spin coherence time can be quite long (nanosecond range) in some cases, e.g., localized HH excitons [42] or simply the excitons at a low enough temperature [33]. However, in general, the electron spin coherence time in [001] QWs is still in the range of a few tens of picoseconds and is not long enough to demonstrate a significant slow-light phenomenon. One of the main reasons for the spin relaxation/decoherence in semiconductor QWs is the Dyakonov–Perel (DP) mechanism [38,39]. If this mechanism can be efficiently suppressed, the electron spin coherence time may become long. The suppression of the DP mechanism can be done in [110] QWs [38] and has been experimentally verified [43]. If we can find a suitable pump–probe scheme, this timescale may be utilized to demonstrate slow light in [110] QWs.

A particular pump–probe scheme which can sense the electron spin coherence is the double-V EIT. This pump–probe scheme utilizes the light-hole-like (LH-like) excitons whose transitions due to both TM-polarized and TE-polarized lights are allowed [31,32]. Figure 2.19a shows this pump–probe scheme based on the LH-like excitons in [110] QWs. The pump is a TE-polarized light while the signal is a TM-polarized light. Due to the band mixing which already exists at the center of the Brillouin zone for [110] QWs, the LH-like states $|\tilde{\phi}_{h1}\rangle$ and $|\tilde{\phi}_{h2}\rangle$ contain a mixture of the unperturbed Bloch states $|j, 3/2\rangle, j = \pm 3/2, \pm 1/2$. The $|\tilde{\phi}_{h1}\rangle$ state has a larger $|3/2, 1/2\rangle$ component while the $|\tilde{\phi}_{h1}\rangle$ state has a larger $|3/2, -1/2\rangle$ component. The conduction states $|\phi_c \uparrow\rangle$ and $|\phi_c \downarrow\rangle$ are identical to those in [100] QWs and contain an $|S \uparrow\rangle$ component and an $|S \downarrow\rangle$ component. The preexisting band mixing changes the selection rule of the TE-polarized light. The $|\tilde{\phi}_{h1}\rangle - |\phi_c \downarrow\rangle$ and $|\tilde{\phi}_{h2}\rangle - |\phi_c \uparrow\rangle$ transitions are no longer induced by two respective circularly polarized lights. Rather, they are individually induced by two elliptically polarized lights. The TM-polarized signal induces the $|\tilde{\phi}_{h1}\rangle - |\phi_c \uparrow\rangle$ and $|\tilde{\phi}_{h2}\rangle - |\phi_c \downarrow\rangle$ transitions while the TE-polarized pump saturates the corresponding absorptions. Figure 2.19b shows the configuration of the pump–probe scheme in real space. Since the TM-polarized signal has to be used, it is incident into the device via a waveguide geometry. On the other hand, the pump can be incident into the device by either the normal-incidence or waveguide geometry.

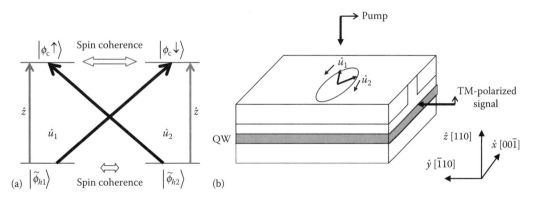

FIGURE 2.19 (a) The pump–probe scheme of double-V EIT based on spin coherence. The TE-polarized pump induces the cross transitions in the figure and the TM-polarized signal induces the vertical transitions. Due to the band mixing of [110] QWs, the two transitions due to the TE-polarized pump are induced independently by two elliptically polarized lights. The spin coherence in the conduction band plays a major role in inducing the coherent dip in the absorption spectrum. (b) The pump–probe scheme of double-V EIT. Since the signal is TM-polarized, it must be incident into the device via a waveguide geometry. On the other hand, the TE-polarized pump can be either incident into the device via the waveguide geometry or normal-incidence geometry. (From Chang, S.W., Chuang, S.L., Chang-Hasnain, C.J., and Wang, H., *J. Opt. Soc. Am. B*, 24(4), 849, 2007. With permission.)

Similar to the population beating in CPO, a tiny electron spin precession (proportional to the off diagonal density matrix element $\rho_{\uparrow\downarrow}$) with the frequency equal to the signal–pump detuning in the QW plane is induced. This precession is optically induced and its strength is proportional to the magnitude of the small signal and thus much smaller than that in the presence of an external magnetic field. Therefore, we may also think of this pump–probe scheme as one without spin precession [33]. In principle, the hole can exhibit a certain kind of precession in the "j $\sim$ |3/2, $\pm$1/2$\rangle$" spin space. However, the spin coherence time of the hole is much shorter than that of the electron, and the contribution is usually much smaller.

Most of the concepts applied to CPO are applicable to this scheme based on spin coherence. The tiny spin precession also generates linear and FWM responses to the signal. Analogous to the population lifetime, the spin coherence time is the timescale within which the spin precession can follow the beating due to pump and the signal. Thus, a coherent dip with a linewidth on the order of about the inverse of the spin coherence time is present on the absorption spectrum. Corresponding to this absorption dip, there is a sharp and positively sloped variation within the same frequency range on the spectrum of the refractive index. This positive slope can then be used to slowdown the optical wave packet. One can thus expect that most of the trends of the calculated physical quantities based on spin coherence are similar to those based on CPO. Indeed, this is the case. Interested readers can find more details in Ref. [32]. One difference between slow lights based on CPO and double-V EIT is that since a TM-polarized signal is used, the TM oscillator strength of the LH exciton is larger than the TE oscillator strength of the HH exciton. Thus, compared with the CPO case, the signal experiences a higher slowdown factor, but also higher intrinsic absorption. The higher absorption for the signal is a drawback of slow light based on double-V EIT. One can avoid the high absorption of the signal by switching the polarizations of the pump and signal. Although the signal may suffer lower absorption in this case, the slowdown factor is also reduced, and it is the pump that experiences higher absorption.

There is another drawback of the pump–probe scheme based on the LH-like exciton. Usually, the LH-like exciton lies in the HH-like continuum. The TE-polarized pump may suffer unwanted dissipation from the HH continuum states. One way to avoid this is to use tensile strain to lift the LH-like band to the top of the HH-like band [32]. With tensile strain, the LH-like excitonic absorption will have the lowest transition energy, and the unwanted dissipation from the HH continuum can be avoided. The tensile strain can be induced by using strained material or by applying uniaxial stress along the growth direction. Figure 2.20 shows the LH component ($|\pm 1/2, 3/2\rangle$) of the highest

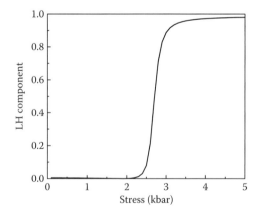

FIGURE 2.20 The LH component of the highest valence band under uniaxial stress. The uniaxial stress can induce biaxial strain, which lifts the LH-like band to the top of the HH-like band. This energy shift leads to a smaller photon energy for the LH-like exciton and avoids the unwanted dissipation of the TE-polarized pump. (From Chang, S.W., Chuang, S.L., Chang-Hasnain, C.J., and Wang, H., *J. Opt. Soc. Am. B*, 24(4), 849, 2007. With permission.)

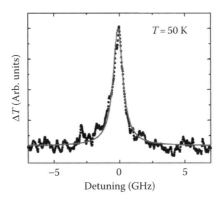

FIGURE 2.21 The transmission peak due to double-V EIT in a [001] QW. The sharp transmission peak indicates that double-V EIT is a possible candidate for the demonstration of slow light in semiconductors. (From Sarkar, S., Palinginis, P., Ku, P.C., Chang-Hasnain, C.J., Kwong, N.H., Binder, R., and Wang, H., *Phys. Rev. B*, 72(3), 035343, 2005. With permission.)

valence band at the Brillouin zone center as a function of the external stress. We can see that with a large enough stress, the highest valence band becomes LH-like. This can help reduce the extra absorption experienced by the TE-polarized pump.

Figure 2.21 shows experimental data for double-V EIT for the LH exciton in [001] QWs [33]. The experiment is carried out at 50 K. We see that a sharp peak with a linewidth of about 1 GHz is present on the transmission spectrum. The spin coherence time of these [001] QWs is actually long enough for some real applications for which a bandwidth below the gigahertz range is required. There is no strain in these QWs, therefore, the extra absorption from the HH continuum can attenuate the pump and degrade the performance of the slow-light device.

2.7 SUMMARY

We have discussed slow and fast light based on CPO in the absorption and gain regimes for semiconductor quantum structures. CPO based on HH excitons in QWs at low temperature can provide a slowdown factor of about 10^4 to 10^5. The operating temperature can also be increased to room temperature if a higher pump intensity is used. Compared with EIT and other slow-light mechanisms, CPO is also relatively immune to inhomogeneous broadening. This immunity has been verified by the slow-light experiment of QDs at room temperature. However, the low surface coverage, the low confinement factor, and inefficient usage of the QDs due to inhomogeneous broadening significantly lowers the slowdown factor.

The implementation of slow light in semiconductors also makes electrical control of the optical buffer possible. One can use both reverse bias voltages and forward injection currents to electrically control the slowdown factor. Nevertheless, slow light based on the pure CPO mechanism in the absorption regime suffers from high absorption which greatly reduces the signal-to-noise ratio. To solve this problem, CPO and FWM in the gain regime are utilized so that the signal is amplified rather than attenuated during the propagation through the whole device. The finite linewidth enhancement factor of bulk or QW SOAs enables both slow and fast light operations based on CPO. However, the nonlinear propagation process based on CPO and FWM makes the generated linear and nonlinear optical components inseparable. This inseparability is part of the reasons for the significant distortion of the optical wave packet.

Analogous to CPO, spin coherence based on the LH exciton can also be used as the mechanism to demonstrate slow light. In this case, it is the tiny spin precession rather than population beating that is utilized. However, the huge TM absorption results in a prominent absorption for either the signal or pump. High intrinsic absorption puts a restriction on the length of the device. Also, the LH-like exciton lies in the HH-like continuum and brings in unwanted dissipation. This difficulty, however, can be overcome by building tensile strain into the QW system to lower the transition energy of the LH-like exciton away from the HH-like continuum.

ACKNOWLEDGMENTS

We would like to express our gratitude to our collaborators, Professor Connie J. Chang-Hasnain, Professor Hailin Wang, Professor Pei-Cheng Ku, Jungho Kim, F. Sedgwick, P. Palinginis, Tao Li, S. Crankshaw, Hui Su, P. K. Kondratko, M. Akira, D. Nielsen, and B. Pesala, whose efforts in the research of semiconductor slow and fast light contribute to the main content of this chapter.

REFERENCES

1. P. C. Ku, F. Sedgwick, C. J. Chang-Hasnain, P. Palinginis, T. Li, H. Wang, S. W. Chang, and S. L. Chuang. Slow light in semiconductor quantum wells. *Opt. Lett.*, 29(19):2291–2293, 2004.
2. S. W. Chang, S. L. Chuang, P. C. Ku, C. J. Chang-Hasnian, P. Palinginis, and H. Wang. Slow light using excitonic population oscillation. *Phys. Rev. B*, 70(23):235333, 2004.
3. P. Palinginis, S. Crankshaw, F. Sedgwick, E. T. Kim, M. Moewe, C. J. Chang-Hasnain, H. L. Wang, and S. L. Chuang. Ultraslow light (<200 m/s) propagation in a semiconductor nanostructure. *Appl. Phys. Lett.*, 87(17):171102, 2005.
4. M. S. Bigelow, N. N. Lepeshkin, and R. W. Boyd. Superluminal and slow light propagation in a room-temperature solid. *Science*, 301:200–202, 2003.
5. M. S. Bigelow, N. N. Lepeshkin, and R. W. Boyd. Observation of ultraslow light propagation in a ruby crystal at room temperature. *Phys. Rev. Lett.*, 90(11):113903, 2003.
6. P. Palinginis, F. G. Sedgwick, S. Crankshaw, M. Moewe, and C. J. Chang-Hasnain. Room temperature slow light in a quantum-well waveguide via coherent population oscillation. *Opt. Express*, 13(24):9909–9915, 2005.
7. J. Mork, R. Kjaer, M. van der Poel, and K. Yvind. Slow light in a semiconductor waveguide at gigahertz frequencies. *Opt. Express*, 13(20):8136–8145, 2005.
8. H. Su and S. L. Chuang. Room-temperature slow light with semiconductor quantum-dot devices. *Opt. Lett.*, 31(2):271–273, 2006.
9. P. K. Kondratko, S. W. Chang, H. Su, and S. L. Chuang. Optical and electrical control of slow light in p-doped and intrinsic quantum-dot electroabsorbers. *Appl. Phys. Lett.*, 90(25):251108, 2007.
10. H. Gotoh, S. W. Chang, S. L. Chuang, H. Okamoto, and Y. Shibata. Tunable slow light of 1.3 μm region in quantum dots at room temperature. *Jpn. J. Appl. Phys., Part 1*, 46(4B):2369–2372, 2007.
11. H. Su and S. L. Chuang. Room temperature slow and fast light in quantum-dot semiconductor optical amplifiers. *Appl. Phys. Lett.*, 88(6):061102, 2006.
12. P. K. Kondratko and S. L. Chuang. Slow-to-fast light using absorption to gain switching in quantum-well semiconductor optical amplifier. *Opt. Express*, 15(16):9963–9969, 2007.
13. A. Matsudaira, D. Lee, P. K. Kondratko, D. Nielsen, and S. L. Chuang. Electrically tunable slow and fast lights in a quantum-dot semiconductor optical amplifier near 1.55 μm. *Opt. Lett.*, 32(19):2894–2896, 2007.
14. B. Pesala, Z. Y. Chen, A. V. Uskov, and C. J. Chang-Hasnain. Experimental demonstration of slow and superluminal light in semiconductor optical amplifiers. *Opt. Express*, 14(26):12968–12975, 2006.
15. A. V. Uskov and C. J. Chang-Hasnain. Slow and superluminal light in semiconductor optical amplifiers. *Electron. Lett.*, 41(16):922–924, 2005.
16. F. Ohman, K. Yvind, and J. Mork. Slow light in a semiconductor waveguide for true-time delay applications in microwave photonics. *IEEE Photon. Tech. Lett.*, 19(15):1145–1147, 2007.
17. F. Ohman, K. Yvind, and J. Mork. Voltage-controlled slow light in an integrated semiconductor structure with net gain. *Opt. Express*, 14(21):9955–9962, 2006.
18. S. W. Chang and S. L. Chuang. Slow light based on population oscillation in quantum dots with inhomogeneous broadening. *Phys. Rev. B*, 72(23):235330, 2005.
19. S. W. Chang, P. K. Kondratko, H. Su, and S. L. Chuang. Slow light based on coherent population oscillation in quantum dots at room temperature. *IEEE J. Quantum Electron.*, 43(2):196–205, 2007.

20. R. W. Boyd, M. G. Raymer, P. Narum, and D. J. Harter. Four-wave parametric interactions in a strongly driven two-level system. *Phys. Rev. A*, 24(1):411–423, 1981.
21. P. Agrawal. Population pulsations and nondegenerate four-wave mixing in semiconductor lasers and amplifiers. *J. Opt. Soc. Am. B*, 5(1):147–159, 1988.
22. H. Su, P. K. Kondratko, and S. L. Chuang. Variable optical delay using population oscillation and four-wave-mixing in semiconductor optical amplifiers. *Opt. Express*, 14(11):4800–4807, 2006.
23. L. V. Hau, S. E. Harris, Z. Dutton, and C. H. Behroozi. Light speed reduction to 17 metres per second in an ultracold atomic gas. *Nature*, 397:594, 1999.
24. E. Podivilov, B. Sturman, A. Shumelyuk, and S. Odoulov. Light pulse slowing down up to 0.025 cm/s by photorefractive two-wave coupling. *Phys. Rev. Lett.*, 91(8):083902, 2003.
25. J. Kim, S. L. Chuang, P. C. Ku, and C. J. Chang-Hasnain. Slow light using semiconductor quantum dots. *J. Phys. Cond. Matt.*, 16(35):S3727–S3735, 2004.
26. R. S. Tucker, P. C. Ku, and C. J. Chang-Hasnain. Slow-light optical buffers: Capabilities and fundamental limitations. *J. Lightwave Tech.*, 23(12):4046–4066, 2005.
27. R. S. Tucker, P. C. Ku, and C. Chang-Hasnain. Delay-bandwidth product and storage density in slow-light optical buffers. *Electron. Lett.*, 41(4):208–209, 2005.
28. A. V. Uskov, F. G. Sedgwick, and C. J. Chang-Hasnain. Delay limit of slow light in semiconductor optical amplifiers. *IEEE Photon. Tech. Lett.*, 18(6):731–733, 2006.
29. F. G. Sedgwick, C. J. Chang-Hasnain, P. C. Ku, and R. S. Tucker. Storage-bit-rate product in slow-light optical buffers. *Electron. Lett.*, 41(24):1347–1348, 2005.
30. Z. Deng, D. K. Qing, P. Hemmer, C. H. Raymond, M. S. Zubairy, and M. O. Scully. Time-bandwidth problem in room temperature slow light. *Phys. Rev. Lett.*, 96(2):023602, 2006.
31. T. Li, H. Wang, H. H. Kwong, and R. Binder. Electromagnetically induced transparency via electron spin coherence in a quantum well waveguide. *Opt. Express*, 11(24):3298–3303, 2003.
32. S. W. Chang, S. L. Chuang, C. J. Chang-Hasnain, and H. Wang. Slow light using spin coherence and v-type electromagnetically induced transparency in [110] strained quantum wells. *J. Opt. Soc. Am. B.*, 24(4):849–859, 2007.
33. S. Sarkar, P. Palinginis, P. C. Ku, C. J. Chang-Hasnain, N. H. Kwong, R. Binder, and H. Wang. Inducing electron spin coherence in GaAs quantum well waveguides: Spin coherence without spin precession. *Phys. Rev. B*, 72(3):035343, 2005.
34. F. G. Sedgwick, B. Pesala, J. Y. Lin, W. S. Ko, X. X. Zhao, and Chang-Hasnain CJ. Thz-bandwidth tunable slow light in semiconductor optical amplifiers. *Opt. Express*, 15(2):747–753, 2007.
35. M. van der Poel, J. Mork, and J. M. Hvam. Controllable delay of ultrashort pulses in a quantum dot optical amplifier. *Opt. Express*, 13(20):8032–8037, 2005.
36. H. Wang, K. Ferrio, D. G. Steel, Y. Z. Hu, R. Binder, and S. W. Koch. Transient nonlinear optical response from excitation induced dephasing in GaAs. *Phys. Rev. Lett.*, 71(8):1261, 1993.
37. H. Wang, K. B. Ferrio, D. G. Steel, P. R. Berman, Y. Z. Hu, R. Binder, and S. W. Koch. Transient four-wave-mixing line shapes: Effects of excitation-induced dephasing. *Phys. Rev. A*, 49(3):R1551, 1994.
38. M. I. Dyakonov and V. Y. Kachorovskii. Spin relaxation of two-dimensional electrons in noncentrosymmetric semiconductors. *Sov. Phys. Semicond.*, 20(1):110–112, 1986.
39. M. I. Dyakonov and V. I. Perel. Spin relaxation of conduction electrons in noncentrosymmetric semiconductors. *Sov. Phys. Solid State*, 13:3023–3026, 1972.
40. G. L. Bir, A. G. Aronove, and G. E. Pikus. Spin relaxation of electrons due to scattering by holes. *Sov. Phys. JETP*, 42:705–712, 1975.
41. R. J. Elliott. Theory of the effect of spin-orbit coupling on magnetic resonance in some semiconductors. *Phys. Rev.*, 96(2):266–279, 1954.
42. P. Palinginis and H. L. Wang. Vanishing and emerging of absorption quantum beats from electron spin coherence in GaAs quantum wells. *Phys. Rev. Lett.*, 92(3):037402, 2004.
43. Y. Ohno, R. Terauchi, T. Adachi, F. Matsukura, and H. Ohno. Spin relaxation in GaAs(110) quantum wells. *Phys. Rev. Lett.*, 83(20):4196–4199, 1999.

3 Slow Light in Optical Waveguides

Zhaoming Zhu, Daniel J. Gauthier, Alexander L. Gaeta, and Robert W. Boyd

CONTENTS

3.1	Slow Light via Stimulated Scattering	38
	3.1.1 Slow Light via SBS	38
	3.1.2 Slow Light via SRS	43
3.2	Coherent Population Oscillations	46
3.3	EIT in Hollow-Core Fibers	49
3.4	Wavelength Conversion and Dispersion	52
3.5	Conclusion	55
Acknowledgment		55
References		55

As evidenced by this book, there has been a flurry of activity over the last decade on tailoring the dispersive properties of optical materials [1]. What has captured the attention of the research community were some of the early results on creating spectral regions of large normal dispersion [2–4]. Large normal dispersion results in extremely small group velocities, where the group velocity is the approximate speed at which a pulse of light propagates through a material. We denote the group velocity by $v_g = c/n_g$, where c is the speed of light in vacuum and n_g is known as the group index. In the early experiments, described in greater detail in Chapter 1, a dilute gas of atoms is illuminated by a control or coupling beam whose frequency is tuned precisely to an optical transition of an atom. This control field modifies the absorption and dispersion properties of another atomic transition that shares a common energy level. A narrow transparency window is created on this second transition—a process known as electromagnetically induced transparency (EIT)—and within this window, v_g takes on extremely small values. Many experiments have now observed $v_g \sim 1$ m/s or less, implying $n_g > 10^8$. This result is remarkable, considering the fact that the refractive index n of a material rarely exceeds 3 in the visible part of the spectrum. What are the implications of such large group indices? What applications are enabled by this basic science discovery?

One immediate application that comes to mind is to use slow light (the situation where $v_g \ll c$) for realizing a real-time adjustable buffer for optical pulses. A buffer that is capable of delaying an entire packet of optical information can substantially increase the efficiency of routers in optical telecommunication networks [5–8], for example. The primary motive of our research over the past few years has been to develop new mechanisms for realizing slow light that operates at or near room temperature and occur in an optical waveguide. The waveguide geometry allows light–matter interactions to take place over long distances where the transverse dimension of the light is of the order of the wavelength, thereby lowering the optical power needed to create the slow-light effect. Also, a waveguide-based slow-light device can be compact and integrated with existing technologies.

The primary goal of this chapter is to review our own research on slow light in optical waveguides. In particular, we describe how slow light can be achieved by stimulated Brillouin and stimulated Raman scattering (SBS and SRS, respectively) in transparent optical fibers, by coherent population oscillations (CPO) in erbium-doped fiber amplifiers, by EIT in gas-filled hollow-core fibers, and by wavelength conversion and fiber dispersion.

3.1 SLOW LIGHT VIA STIMULATED SCATTERING

To understand how slow light can be achieved via stimulated scattering, it is important to recall the Kramers–Kronig relations [9], which relate the real and imaginary part of the complex refractive index of a causal dielectric. In particular, frequency-dependent material absorption (or gain) is necessarily associated with a frequency-dependent refractive index. Considering the definition for the group index, which is given by $n_g = n + \omega dn/d\omega$, where ω is the optical frequency, we see that the group index differs from the refractive index by the so-called dispersive term $\omega dn/d\omega$. Thus, large values of n_g are obtained when there is a substantial change in refractive index over a narrow frequency interval (making $dn/d\omega$ large), which is associated with a rapid spectral variation in the absorption of the material.

In simulated scattering processes [10], light scattering occurs as a result of highly localized changes in the dielectric constant of a medium. For sufficiently strong light fields, these changes can be induced via coupling of a material excitation to two light fields whose difference in frequencies is given by the frequency of the excitation. The excitation gives rise to a nonlinear optical coupling between the fields, which allows power to flow from one beam to another and which can give rise to absorption or amplification of a probe beam. This coupling occurs over a narrow spectral range, which gives rise to a narrow resonance that can be used to control v_g by adjusting the laser beam intensities. Important features of a stimulated scattering resonance are that it can occur at room temperature, and it is induced by a pump laser beam and hence can occur over the entire range of frequencies where the material is transparent. In our research, we have investigated the material resonances arising from both SBS and SRS, as described in Sections 3.1.1 through 3.1.5.

3.1.1 Slow Light via SBS

In the SBS process, a high-frequency acoustic wave is induced in the material via electrostriction for which the density of a material increases in regions of high optical intensity. The process of SBS can be described classically as a nonlinear interaction between the pump (at angular frequency ω_p) and a probe field (ω) through the induced acoustic wave (Ω_B) [10]. The acoustic wave in turn modulates the refractive index of the medium and scatters pump light into (out of) the probe wave when its frequency is downshifted (upshifted) by the acoustic frequency. This process leads to a strong coupling between the three waves when this resonance condition is satisfied, which results in exponential amplification (absorption) of the probe wave. Efficient SBS occurs when both energy and momentum are conserved, which is satisfied when the pump and probe waves counterpropagate.

Slow light due to SBS can be understood by studying the resonances experienced by the probe wave for the case when the medium is pumped by a continuous-wave beam. Here, we focus on the case when the frequency of a counterpropagating probe beam is near the amplifying resonance (also known as the Stokes resonance); the results are generalized straightforwardly to the case of the absorbing resonance (or anti-Stokes resonance). In the small-signal limit (i.e., pump depletion is negligible), the probe wave in the fiber (propagating in $+z$ direction) experiences an effective complex refractive index $\tilde{n}(\omega)$ given by

$$\tilde{n} = n_f - i\frac{c}{\omega}\frac{g_0 I_p}{1 - i2\delta\omega/\Gamma_B}, \tag{3.1}$$

where n_f is the modal index of the fiber mode, I_p is the pump intensity, g_0 is line-center gain factor, $\delta\omega = \omega - \omega_p + \Omega_B$, and $\Gamma_B/2\pi$ is the full width at half maximum (FWHM) linewidth of the Brillouin resonance. For a typical optical telecommunication fiber, $\Omega_B/2\pi \sim 10\,\text{GHz}$ and $\Gamma_B/2\pi \sim 30\,\text{MHz}$. The fact that the resonance linewidth is so narrow—comparable to the natural linewidth of atomic transitions used in atom-based slow light—suggests that controlling v_g in an optical fiber is achievable. From Equation 3.1, it is seen that the probe wave experiences gain and dispersion in the form of a Lorentzian-shaped resonance. The gain coefficient $g = -2(\omega/c)\text{Im}(\tilde{n})$, real refractive index $n = \text{Re}(\tilde{n})$, and group index $n_g = n + \omega(dn/d\omega)$ are given by

$$g(\omega) = \frac{g_0 I_p}{1 + 4\delta\omega^2/\Gamma_B^2}, \tag{3.2a}$$

$$n(\omega) = n_f + \frac{cg_0 I_p}{\omega} \frac{\delta\omega/\Gamma_B}{1 + 4\delta\omega^2/\Gamma_B^2}, \tag{3.2b}$$

$$n_g(\omega) = n_{fg} + \frac{cg_0 I_p}{\Gamma_B} \frac{1 - 4\delta\omega^2/\Gamma_B^2}{(1 + 4\delta\omega^2/\Gamma_B^2)^2}, \tag{3.2c}$$

respectively, where n_{fg} is the group index of the fiber mode when SBS is absent. Figure 3.1 shows the refractive index, gain, and group index for the SBS amplifying resonance. It is seen that normal dispersion near the center of the resonance leads to an increase in the group index and therefore a decrease in group velocity $v_g = c/n_g$.

For the optical data buffering application mentioned previously, one important characteristic of a slow-light device is its ability to controllably delay a pulse. For the situation where the majority of the pulse spectrum falls within the region where v_g is nearly constant, the slow-light delay (defined as the difference between the transit times with and without SBS) is given by

$$\Delta T_d = \frac{G}{\Gamma_B} \frac{1 - 4\delta\omega^2/\Gamma_B^2}{(1 + 4\delta\omega^2/\Gamma_B^2)^2}$$

$$\simeq \frac{G}{\Gamma_B}(1 - 12\delta\omega^2/\Gamma_B^2) \quad \text{when} \quad 4\delta\omega^2/\Gamma_B^2 \ll 1, \tag{3.3}$$

where $G = g_0 I_p L$ is the gain parameter whose exponential e^G is the small-signal gain, and L is the fiber length. The maximum delay occurs at the peak of the Brillouin gain ($\delta\omega = 0$) and is given simply by

$$\Delta T_d = G/\Gamma_B. \tag{3.4}$$

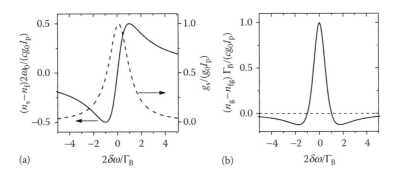

FIGURE 3.1 Large dispersion of the SBS resonance. (a) Gain (solid line) and refractive index (dashed line) of the resonance. (b) Normalized group index of the resonance. (From Zhu, Z., Gauthier, D.J., Okawachi, Y., Sharping, J.E., Gaeta, A.L., Boyd, R.W., and Willner, A.E., *J. Opt. Soc. Am. B*, 22, 2378, 2005. With permission.)

It is seen that the slow-light delay ΔT_d is tunable by adjusting the pump intensity. Equation 3.4 gives $\Delta T_d \simeq 1.2$ ns/dB for $\Gamma_B/2\pi = 30$ MHz.

Slow-light delay is always accompanied by some degree of pulse distortion due to the fact that some portion of the pulse spectrum extends to regions where there is substantial variation in the gain and group index. For example, a Gaussian-shaped pulse will be temporally broadened by a factor [12]

$$B \equiv \tau_{out}/\tau_{in} = \left[1 + (16\ln 2)G/\tau_{in}^2\Gamma_B^2\right]^{1/2}, \tag{3.5}$$

where τ_{in} and τ_{out} are the pulse widths (FWHM) of the input and output pulses, respectively, and where third and higher-order dispersion are neglected in obtaining this expression. Pulse-width broadening is mainly due to the spectral reshaping arising from the bandwidth-limited gain, which narrows the spectral width of the output pulse.

Fractional delay (the ratio of delay to the pulse width) is related to the pulse broadening by

$$\Delta T_d/\tau_{in} = \left[(B^2 - 1)G/(16\ln 2)\right]^{1/2}. \tag{3.6}$$

This result suggests that there exists a maximum achievable fractional delay in a single SBS slow-light element because G is limited due to self generation. In the self-generation process, spontaneously scattered light produced at the Stokes frequency (generated by thermal fluctuations) and near the entrance face of the fiber experiences exponential growth due to the SBS process. For sufficiently high gain (typically $G \sim 10 - 15$ for long lengths of optical fiber), the self-generated light depletes the pump beam leading to gain saturation [11]. Using a limit of $G = 15$ and a constraint of $B = 2$ gives a fractional delay limit of 2.6 for a single Lorentzian amplifying resonance. The issue of self-generation can be avoided using multiple SBS slow-light delay lines separated by attenuators as demonstrated by Song et al. [13].

Tunable slow-light delay via SBS in an optical fiber was first demonstrated independently by Song et al. [14] and Okawachi et al. [12]. In the experiment by Okawachi et al. (see Figure 3.2), a single continuous-wave narrow-linewidth tunable laser at 1550 nm was used to generate both the

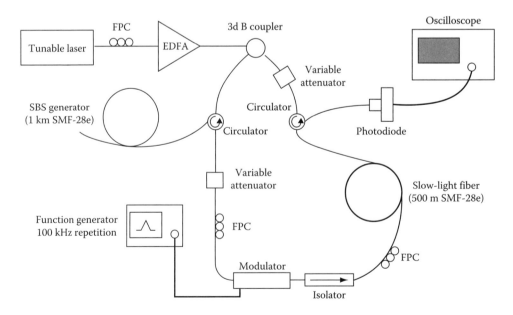

FIGURE 3.2 Experimental setup for SBS slow light. (From Okawachi, Y., Bigelow, M.S., Sharping, J.E., Zhu, Z., Schweinsberg, A., Gauthier, D.J., Boyd, R.W., and Gaeta, A.L., *Phys. Rev. Lett.*, 94, 153,902, 2005. With permission.)

Slow Light in Optical Waveguides

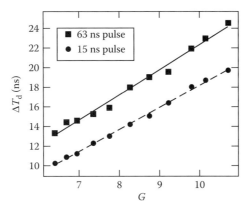

FIGURE 3.3 Demonstration of optically controllable slow-light pulse delays. Induced delay as a function of the Brillouin gain parameter G for 63 ns long (square) and 15 ns long (circle) input Stokes pulses. (From Okawachi, Y., Bigelow, M.S., Sharping, J.E., Zhu, Z., Schweinsberg, A., Gauthier, D.J., Boyd, R.W., and Gaeta, A.L., *Phys. Rev. Lett.*, 94, 153,902, 2005. With permission.)

pump wave and probe pulses. The probe wave was created from an SBS generator pumped by the light from the laser. Probe pulses counter-propagate with respect to the pump wave in a 500 m long SMF-28e fiber whereby the pulses experienced amplification and slow-light delay. The delayed output pulses were recorded by a fast detector and displayed on an oscilloscope. The experiment demonstrated that the delay can be tuned continuously by as much as 25 ns by adjusting the intensity of the pump field and that the technique can be applied to pulses as short as 15 ns. Figure 3.3 shows the measured slow-light delay as a function of the gain parameter for input pulsewidths of 63 and 15 ns. Figure 3.4 shows the delayed (solid line) and undelayed (dotted line) pulses for both pulse widths at a gain parameter of $G = 11$. A fractional slow-light delay of 1.3 was achieved for the 15 ns long input pulse with a pulse broadening of 1.4.

It is also possible to observe fast light (where $v_g > c$ or $v_g < 0$) due to SBS. In particular, anomalous dispersion occurs at the center of the absorbing resonance at the anti-Stokes frequency, which can be seen by repeating the analysis given above but with the sign of g inverted. Song et al. [14] used this effect to observe pulse advancement in an optical fiber.

Following the first demonstrations of SBS slow light in optical fibers, there has been considerable interest in exploiting the method for telecommunication applications. The appeal of the technique is that it is relatively simple, works at room temperature and at any wavelength where the material is transparent, and uses telecommunication components off the shelf.

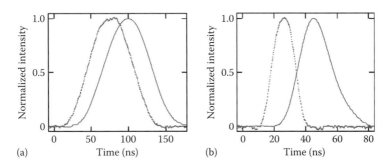

FIGURE 3.4 Temporal evolution of the Stokes pulses (with a gain parameter $G = 11$) emitted from the fiber in the absence (dotted) and presence (solid) of the pump beam for (a) 63 ns long and (b) 15 ns long input Stokes pulses. (From Okawachi, Y., Bigelow, M.S., Sharping, J.E., Zhu, Z., Schweinsberg, A., Gauthier, D.J., Boyd, R.W., and Gaeta, A.L., *Phys. Rev. Lett.*, 94, 153,902, 2005. With permission.)

One line of research has focused on reducing pulse distortion and improving system performance [15–21]. Stenner et al. were the first to consider this problem [15]. They proposed and demonstrated a general distortion management approach in which multiple SBS gain lines are used to optimize the slow-light delay under certain distortion and pump power constraints. In their experimental demonstration with two closely spaced SBS gain lines generated by a dual-frequency pump, approximately a factor of 2 increase in slow-light pulse delay was achieved as compared with the optimum single-SBS-line delay. The results suggested an effective way for reducing pulse distortion and improving SBS slow light by tailoring the gain resonance experienced by the signal pulses. Along this line, SBS slow-light efficiency (in terms of time delay per dB of gain) was also improved by customizing the gain resonance profiles using various approaches [22–24].

Another line of research has focused on broadband SBS slow light [25–31]. The width of the resonance which enables the slow-light effect limits the minimum duration of the optical pulse that can be effectively delayed without much distortion, and therefore limits the maximum data rate of the optical system. Fiber-based SBS slow light is limited to data rates less than a few tens of Mb/s due to the narrow Brillouin resonance width (~30 MHz in standard single-mode optical fibers). Herráez et al. were the first to increase the SBS slow-light bandwidth and achieved a bandwidth of about 325 MHz by broadening the spectrum of the SBS pump field. Zhu et al. extended this work to achieve an SBS slow-light bandwidth as large as 12.6 GHz, thereby supporting a data rate of over 10 Gb/s [26]. In this study, 75 ps data pulses were delayed by up to 47 ps at a gain of about 14 dB. Figure 3.5 shows the experimental results of the broadband SBS slow light. As can be seen in

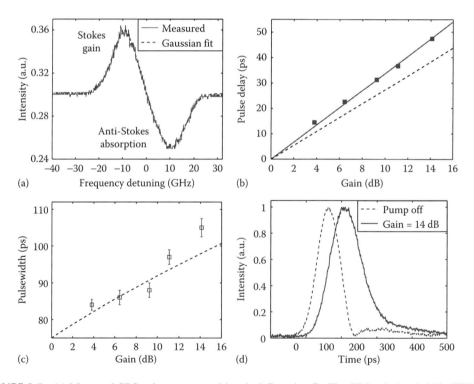

FIGURE 3.5 (a) Measured SBS gain spectrum with a dual Gaussian fit. The SBS gain bandwidth (FWHM) is found to be 12.6 GHz. Pulse delay (b) and pulse width (c) as a function of SBS gain. In (b), the solid line is the linear fit of the measured data (solid squares), and the dashed line is obtained without considering the anti-Stokes absorption. In (c), the dashed curve is the theoretical prediction. (d) Pulse waveforms at 0 dB and 14 dB SBS gain. The input data pulsewidth is ~75 ps. (From Zhu, Z., Dawes, A.M.C., Gauthier, D.J., Zhang, L., and Willner, A.E., *J. Lightw. Technol.*, 25, 201, 2007. With permission.)

Figure 3.5a, the SBS gain resonance occurring at $\omega_p - \Omega_B$ and the absorption resonance occurring at $\omega_p + \Omega_B$ are broadened to the point that they nearly overlap. In this situation, the anti-Stokes absorption tends to reduce the slow-light efficiency and limit the achievable bandwidth to about Ω_B. This limitation was recently overcome using a second broadband pump whose center frequency is higher than the first broadband pump by $2\Omega_B$ [28,29]. This second pump generates a gain resonance that compensates the absorption caused by the first pump, enabling the use of even broader pump spectra to increase the SBS slow-light bandwidth. A bandwidth of about 25 GHz was achieved using this approach [29].

Other research in SBS slow light includes decoupling gain from delay [32–34], and reducing control latency [22,35]. In SBS slow light, pulse delay is always accompanied by signal amplification. This amplification may not be desirable in situations where nearly constant signal powers are needed. To reduce the signal power variation in slow light, Zhu et al. used two widely separated anti-Stokes lines to form a slow-light element that exhibits less variation of signal amplitudes than in the single SBS resonance case [32]. In Refs. [33,34], a composite profile was created by combining a broadened absorption resonance and a narrow gain resonance. This composite profile was used to achieve slow-light delays with constant signal amplitude. Control latency in SBS slow light is determined by the smaller of the time it takes to adjust the pump laser intensity or twice the time it takes light to transit the fiber, with the latter often being a limiting factor. Therefore, use of a short length of fiber for SBS slow light can increase the speed of adjusting SBS slow light delays [22,35].

Additionally, nonsilica fibers have attracted substantial interest in the slow-light research community. For example, chalcogenide glass fibers [22,36] and tellurite glass fibers [37], which exhibit much larger SBS gain factors g_0, are being investigated for SBS slow light. A critical comparison of different fibers for SBS slow light is given in Ref. [36].

3.1.2 SLOW LIGHT VIA SRS

Slow light via SRS can also be achieved in optical fibers, but over much larger bandwidths and hence can be used with pulses of much shorter duration. The scattering arises from exciting vibrational or rotational motions in individual molecules—also known as optical phonons—as opposed to exciting sound waves as in the SBS process. Regardless of the microscopic mechanism for creating the material excitation, slow light via SRS can be similarly understood by considering the resonances induced by a pump beam propagating in the fiber.

Compared with the simple Lorentzian-shaped resonances in SBS, SRS in an optical fiber has a more complicated resonance shape, which depends on the fiber structure and material composition. Figure 3.6 shows the typical SRS spectral response function $\tilde{h}_R$ in a silica fiber [38,39], where the largest gain peak of the Stokes band occurs about 13.2 THz below the pump beam frequency. The real part of $\tilde{h}_R$ is proportional to the refractive index change due to SRS, while the imaginary part is proportional to the gain (for the Stokes band) or loss (for the anti-Stokes band). Although the gain spectrum (i.e., the imaginary part of $\tilde{h}_R$) is far from a Lorentzian shape, the real part of $\tilde{h}_R$ still has a nearly linear variation with frequency near the center of the band, suggesting that slow light is readily observable.

A model for the measured Raman response function consisting of a single damped oscillator has been proposed in Ref. [40] and is given by

$$h_R(t) = (\tau_1^2 + \tau_2^2)/(\tau_1 \tau_2^2) \exp(-t/\tau_2) \sin(t/\tau_1), \quad (3.7)$$

where $\tau_1 = 12.2$ fs and $\tau_2 = 32$ fs. Its Fourier transform $\tilde{h}_R$ is shown as thin lines in Figure 3.6. As seen in the figure, the line shape is nearly Lorentzian and thus, to good approximation, the results presented above for SBS can be applied to SRS by replacing the gain coefficient and linewidth with those for SRS. The SRS linewidth is about 9.5 THz in this model.

As a result of its broad linewidth, SRS is most useful for delaying picosecond-duration or shorter pulses. Because of the relatively small Raman gain coefficient in standard fibers, high power

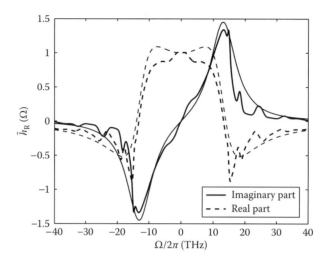

FIGURE 3.6 The real and imaginary parts of the typical Raman gain spectrum $\tilde{h}_R(\omega)$ of a silica fiber. (From Stolen, R.H., Gordon, J.P., Tomlinson, W.J., and Haus, H.A., *J. Opt. Soc. Am. B* 6, 1159, 1989; Hollenbeck, D. and Cantrell, C.D., *J. Opt. Soc. Am. B* 19, 2886, 2002.) The thin lines are obtained by fitting the experimental observations to the simple damped oscillator model. (From Blow, K.J. and Wood, D., *IEEE J. Quantum Electron.*, 25, 2665, 1989.)

continuous-wave or pulsed pump beams are usually required to obtain delays that are a sizable fraction of the input pulse width. However, the propagation of picosecond pulses under strong pumping conditions necessitates an understanding of the roles of fiber dispersion, self-phase modulation, cross-phase modulation, and group velocity mismatch and their competition with the SRS process.

Another difference between SRS and SBS is that SRS can occur for both the copropagating and counterpropagating pump–probe geometries. In the case of counterpropagating pump and probe beams, a continuous-wave pump beam is usually required so that it overlaps substantially with the probe pulse as it propagates along the fiber. However, due to the lower threshold for SBS self-generation in an optical fiber (about 2 to 3 orders of magnitude lower than that of SRS), SBS self-generation will deplete the pump beam and suppress SRS. Therefore, most SRS experiments use copropagating pump and probe beams where the pump beam is pulsed with a duration shorter than the acoustic lifetime (typically a few nanoseconds). One additional consideration for generating large SRS gain and slow-light delay is to match the group velocities of pump and probe pulses so that they do not separate temporally as they propagate through the fiber. It is possible to achieve a long walk-off length by selecting either a fiber with suitable dispersive properties or choosing a suitable pump wavelength for a given fiber.

Sharping et al. demonstrated an ultrafast all-optical controllable delay in a fiber Raman amplifier [41]. In this experiment a 430 fs pulse is delayed by 85% of its pulse width. The ability to accommodate the bandwidth of pulses shorter than 1 ps in a fiber-based system makes SRS slow-light useful for producing controllable delays in ultrahigh bandwidth telecommunication systems. Figure 3.7 shows the experimental setup where copropagating signal and pump pulses are used. Signal pulses are derived from a Ti:sapphire-laser-pumped optical parametric oscillator (OPO) and have a center wavelength of 1640 nm and a transform-limited temporal width of 430 fs. Fourier-transform spectral interferometry (FTSI) is used to measure the delay of the signal pulses since it provides the ability to accurately measure delays as small as tens of fs between pulses of very low peak power. Temporally separated reference pulses, which are used to measure the signal delay, are generated by passing the signal pulse train through an asymmetric Michelson interferometer. 500 ps long, 1535 nm pump pulses are combined synchronously and copropagate with the signal and reference pulses in a 1 km long highly nonlinear fiber (HNLF), where the signal pulse is amplified and delayed while the

Slow Light in Optical Waveguides

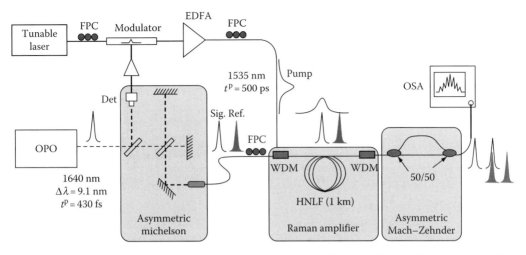

FIGURE 3.7 The experimental setup used to demonstrate slow light in a Raman fiber amplifier. (From Sharping, J.E., Okawachi, Y., and Gaeta, A.L., *Opt. Express*, 13, 6092, 2005. With permission.)

reference pulse is not amplified. The interferogram, as measured by an optical spectrum analyzer (OSA), is used to determine the slow light delay via Fourier transform.

Figure 3.8 shows the slow light results in the Raman amplifier where the signal wavelength is 1637 nm, which is very close to the peak of the Raman gain profile. The gain parameter (left axis) and signal time delay (right axis) are plotted as functions of pump peak power. It is seen that both the gain and delay vary linearly with pump power. The maximum delay achieved for the data depicted in Figure 3.8 is 140 fs, which is 40% of the transform-limited input pulse width.

Figure 3.9 shows three delayed signal pulses in the Fourier-transformed, time-domain representation, where the system is adjusted to obtain as large a delay as possible prior to each measurement. The measured delay of the pulse peak has contributions from SRS, cross-phase modulation, and wavelength shifts. For these pulses, the relative signal delay is varied from 0 to 370 ± 30 fs, which is 85% of the input pulse duration at maximum delay. The measured gain for the case of maximum delay is $G = 7$, which implies a Raman gain bandwidth of 3 THz extrapolated from $\Delta T_d = G/\Gamma_R$. It is seen that the spectral width of the pulses does not change from the zero-delay case as the delay is varied.

In addition to optical fibers, SRS slow light has also been demonstrated in a silicon-on-insulator planar waveguide [42]. Here, a group-index change of 0.15 was produced using an 8 mm long nanoscale waveguide, and controllable delays as large as 4 ps for input signal pulses as short as

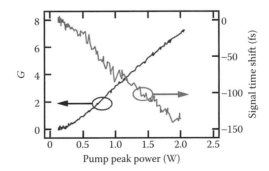

FIGURE 3.8 Signal gain and delay versus pump power in SRS slow light. (From Sharping, J.E., Okawachi, Y., and Gaeta, A.L., *Opt. Express*, 13, 6092, 2005. With permission.)

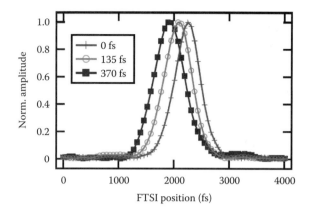

FIGURE 3.9 Amplitude of the transformed interferograms for pulse delay changes of 0 fs, 135 fs and 370 fs. (From Sharping, J.E., Okawachi, Y., and Gaeta, A.L., *Opt. Express*, 13, 6092, 2005. With permission.)

3 ps were demonstrated. This scheme represents an important step in the development of chip-scale photonics devices for telecommunication and optical signal processing.

3.2 COHERENT POPULATION OSCILLATIONS

Coherent population oscillations are a quantum effect that creates a narrow spectral hole in an absorption profile. The rapid variation of refractive index in the neighborhood of the spectral hole leads to slow or fast light (superluminal) propagation.

Coherent population oscillations occur when the ground-state population of a saturable medium oscillates at the beat frequency between a pump wave and a probe wave. The population oscillations are appreciable only for $\delta T_1 \lesssim 1$, where δ is beat frequency and T_1 is the ground state recovery time. When this condition is met, the pump wave can efficiently scatter off the temporally modulated ground-state population into the probe wave, resulting in reduced absorption of the probe wave. In the frequency domain, this leads to a narrow spectral hole in the absorption profile, and the hole has a linewidth on the order of the inverse of the excited-state lifetime.

A thorough understanding of CPO can be obtained via the density matrix equations of motion for a simple two-level system (Figure 3.10a). The shape of the probe absorption profile obtained using this formalism is given by

$$\alpha(\delta) = \frac{\alpha_0}{1+I_0}\left[1 - \frac{I_0(1+I_0)}{(T_1\delta)^2 + (1+I_0)^2}\right], \quad (3.8)$$

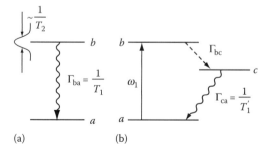

FIGURE 3.10 (a) Two-level system for observing CPO. (b) Relevant energy levels in Ruby used for CPO slow light. (From Bigelow, M.S., Lepeshkin, N.N., and Boyd, R.W., *Phys. Rev. Lett.*, 90, 113,903, 2003. With permission.)

where α_0 is the unsaturated absorption coefficient, $I_0 = \Omega^2 T_1 T_2$ is the pump intensity normalized to the saturation intensity, Ω is the Rabi frequency, and T_2 is the dipole moment dephasing time. The refractive index variation associated with the absorption feature leads to a large increase of the group index and slow light pulse propagation. The slow-light delay at line center is given by

$$\Delta T_d = \frac{\alpha_0 L T_1}{2} \frac{I_0}{(1+I_0)^3}, \qquad (3.9)$$

where L is length of the slow light medium. It is seen that there is no theoretical limit to the amount of delay achievable via CPO, since L is unbounded. In practice, however, residual absorption as well as group velocity dispersion (GVD) and spectral reshaping impose a limit to the achievable fractional delay to around 10%.

Spectral holes due to CPO were first predicted in 1967 by Schwartz and Tan [44]. The first experimental investigation of spectral holes caused by CPO observed an extremely narrow transmission window (37 Hz width) in a ruby absorption band near a wavelength of 514.5 nm [45]. Slow light using this narrow spectral feature was first demonstrated by Bigelow et al. [43], where v_g as low as 57.5 ± 0.5 m/s was observed in a 7.25 cm long ruby crystal at room temperature (the relevant energy levels are shown in Figure 3.10b). Slow light propagation was also demonstrated in an alexandrite crystal at room temperature using CPO [46], in semiconductor structures [47–50], and erbium-doped fibers [51,52].

In CPO slow light experiments, the pump and probe need not be separate beams; a single beam with a temporal modulation can experience slow-light delay. However, the modulation frequency or the pulse spectral width should be narrow enough to essentially fit within the spectral hole for the slow-light effect to be appreciable with minimum pulse distortion. This means that the slow light bandwidth is limited by the width of the spectral hole created by CPO.

In addition to an absorbing medium, CPO can also occur in an amplifying (population inverted) medium. In this case, a spectral hole is created in a gain feature, and the resulting anomalous dispersion can lead to superluminal or negative group velocities. Both slow- and fast-light effects were recently demonstrated in erbium-doped fibers where the absorption or gain can be controlled by a pump laser that creates a population inversion [51,52]. In one experiment [51], modulated or pulsed light at a wavelength of 1550 nm was delayed or advanced in 13 m long erbium-doped fiber that was pumped by a counter-propagating pump laser operating at a wavelength of 980 nm. A maximum fractional advancement of 0.12 and a maximum fractional delay of 0.09 were achieved. Figure 3.11 shows the frequency and pump power dependence of the fractional delay observed in propagation through the erbium-doped fiber, where the regime of operation can be tuned by adjusting the pump power. Figure 3.12 shows the fractional advancement as a function of pulsewidth in both the slow light and superluminal light regimes.

The superluminal light propagation with negative group velocity due to CPO has been recently observed in an erbium-doped fiber where the signal pulse appears to propagate backwards [52], which demonstrates that "backwards" propagation is a realizable physical effect.

Like other slow- and fast-light techniques discussed so far, CPO based slow- and fast-light pulse propagation also suffers from pulse distortion. In the fast light regime, pulse broadening and compression in an erbium-doped fiber amplifier (EDFA) can be described in terms of two competing mechanisms: gain recovery and spectral broadening [53]. Shin et al. observed that pulse distortion caused by these effects depends on input pulse width, pump power, and background-to-pulse power ratio. They demonstrated that pulse distortion can be minimized by a proper choice of the these parameters [53]. In their experiment, significant fractional advancement ($\sim$0.17) and minimal distortion was obtained for a background-to-pulse power ratio of $\sim$0.75, a pump power of 17.5 mW, and an input pulse duration of 10 ms.

Although the bandwidth of CPO slow or fast light in erbium-doped fibers is typically limited to the kHz frequency range, CPO slow light in semiconductors exhibits bandwidths exceeding

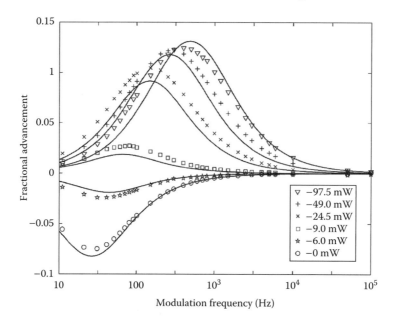

FIGURE 3.11 Frequency and pump power dependence of the fractional delay observed in propagation through erbium-doped fiber. The input was a sinusoidally modulated beam at 1550 nm with an average power of 0.8 mW. Results of the numerical model are shown as solid lines along with the experimental data points. (From Schweinsberg, A., Lepeshkin, N.N., Bigelow, M.S., Boyd, R.W., and Jarabo, S., *Europhys. Lett.*, 73, 218, 2006. With permission.)

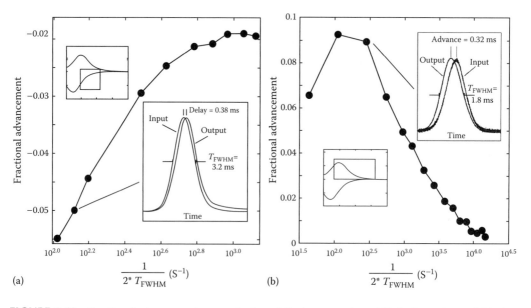

FIGURE 3.12 Fractional advancement versus the log of the inverse pulse width in the regimes of (a) slow light and (b) superluminal propagation. (From Schweinsberg, A., Lepeshkin, N.N., Bigelow, M.S., Boyd, R.W., and Jarabo, S., *Europhys. Lett.*, 73, 218, 2006. With permission.)

Slow Light in Optical Waveguides

1 GHz [47,48]. This large bandwidth, together with mature semiconductor processing techniques, makes CPO in semiconductor structures an important route to achieve chip-scale slow-light devices.

3.3 EIT IN HOLLOW-CORE FIBERS

As mentioned in the introduction to this chapter, EIT [54] is a technique that renders an atomic medium transparent over a narrow spectral range within an absorption band. It has been observed in atomic vapors [55], Bose–Einstein condensates [2], crystals [4], and semiconductor quantum wells and quantum dots [47]. Applications of EIT include ultraslow light [2,3], stored light [56,57], enhancement of nonlinear optical effects [58], and quantum information processing [59].

In its simplest form, EIT can take place in a three-level Λ system as shown schematically in Figure 3.13. A weak probe field (ω_p) is tuned near the $|1\rangle \leftrightarrow |2\rangle$ transition frequency and is used to measure the absorption spectrum of the transition, while a much stronger coupling field (ω_c) is tuned near the $|2\rangle \leftrightarrow |3\rangle$ transition frequency. The $|1\rangle \leftrightarrow |3\rangle$ transition is dipole forbidden. Quantum interference between the $|1\rangle \leftrightarrow |2\rangle$ and $|3\rangle \leftrightarrow |2\rangle$ transition amplitudes result in a cancellation of the probability amplitude for exciting state $|2\rangle$, thereby reducing the probe beam absorption. If the state $|3\rangle$ has a long lifetime, the quantum interference results in a narrow transparency window completely contained within the $|1\rangle \leftrightarrow |2\rangle$ absorption line. The rapid change in refractive index in the narrow transparency window produces an extremely low group velocity for the probe field, which leads to slow light.

Under appropriate conditions [60,61], the transparency window created by EIT is approximately Lorentzian, whereby the frequency-dependent complex absorption coefficient is given by

$$\alpha(\delta) = \alpha_0 \left(1 - \frac{f}{1 + \delta^2/\gamma^2}\right), \tag{3.10}$$

with

$$f = \frac{|\Omega_c/2|^2}{\gamma_{31}\gamma_{21} + |\Omega_c/2|^2}, \quad \gamma = \frac{|\Omega_c/2|^2}{\gamma_{21}}, \tag{3.11}$$

where Ω_c is the Rabi frequency of the strong coupling field, γ_{21} is the coherence dephasing rate of the $|1\rangle \leftrightarrow |2\rangle$ transition, and γ_{31} is the dephasing rate of the ground state coherence. The steep, linear region of the refractive index in the center of the transparency window gives rise to slow light. The slow-light delay at the center of the window is given by

$$\Delta T_d = \frac{\alpha_0 L}{2} \frac{\gamma_{21}}{\gamma_{31}\gamma_{21} + |\Omega_c/2|^2}. \tag{3.12}$$

There is no theoretical upper limit to the amount of delay that can be created by EIT as in the CPO case, while EIT has a practical advantage over CPO in that the transparency depth f can often

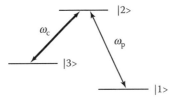

FIGURE 3.13 EIT in a three-level Λ system.

approach 1, greatly reducing residual absorption. However, spectral reshaping limits the fractional advancement to approximately

$$\Delta T_{\text{frac}} = \frac{3}{2} \frac{|\Omega_c/2|^2 \, T_{\text{FWHM}}}{\gamma_{21}}. \tag{3.13}$$

Slow light based on EIT has been demonstrated in various material systems. For example, Hau et al. observed a group velocity of 17 m/s in a Bose–Einstein condensate [2] and Turukhin et al. demonstrated the propagation of slow light with a velocity of 45 m/s through a solid crystal at a cryogenic temperature of 5 K [4].

While these results are truly impressive, there has been increased interest in achieving EIT in waveguides where coherent light–matter interactions are enhanced due to tight light confinement and long interaction lengths. For example, integrated hollow-core antiresonant reflective optical waveguides (ARROW) waveguides filled with an atomic vapor were recently proposed and investigated as a platform for chip-scale EIT [62–64]. It is also possible to observe EIT effects by placing an atomic medium outside of a waveguide. For example, Patnaik et al. proposed a fiber-based EIT slow-light scheme is which a tapered fiber is immersed in an atomic vapor, where they predict theoretically that very slow group velocities are possible [65].

Another approach is to fill a hollow-core photonic-bandgap fiber (HC-PBF) with a gas. These fibers are of great interest because the light can be confined and guided with low loss in a hollow core surrounded by a photonic crystal structure that localizes light in the core. By filling the hollow core with desirable gases, resonant optical interactions, such as EIT, can be achieved over a long interaction length. Such an HC-PBG gas cell is compact and can be integrated with existing fiber-based technologies.

EIT-based slow-light propagation has recently been demonstrated experimentally in an HC-PBF filled with acetylene [66]. Figure 3.14 shows the experimental setup for EIT slow light in a HC-PBF fiber and a scanning electron microscope image of the transverse structure of the fiber. The 1.33 m long HC-PBF has a core diameter of 12 μm and a band gap extending from 1490 to 1620 nm. The fiber ends are sealed in vacuum cells, and the fiber core is filled with 99.8% pure acetylene (C_2H_2). The energy-level scheme for the molecules used in the experiment is shown in Figure 3.15. The probe beam and the coupling beam are tuned to the R(15) and P(17) lines of $^{12}C_2H_2$ at 1517.3144 and 1535.3927 nm, respectively.

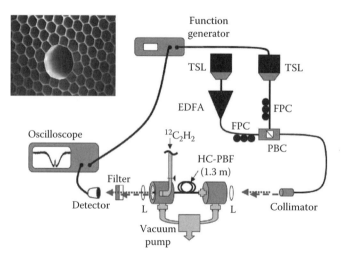

FIGURE 3.14 Experimental setup for EIT slow light in a hollow-core fiber. (From Gosh, S., Sharping, J.E., Ouzounov, D.G., and Gaeta, A.L., *Phys. Rev. Lett.*, 94, 093,902, 2005. With permission.)

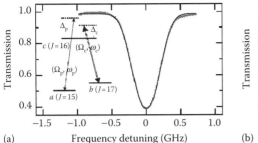
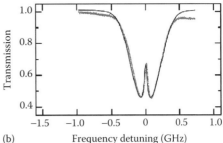

FIGURE 3.15 Measured absorption for the probe beam. (From Gosh, S., Sharping, J.E., Ouzounov, D.G., and Gaeta, A.L., *Phys. Rev. Lett.*, 94, 093,902, 2005. With permission.)

Figure 3.15a shows the 480 MHz wide Doppler-broadened absorption spectrum of probe beam in the absence of the control beam. In the presence of the control beam, a transparency window is opened. Figure 3.15b show a typical trace of the probe-field absorption in the presence of a 320 mW (measured at the output of the fiber) control beam tuned exactly to the center of the P(17) transition. The observed transparency feature in acetylene is remarkable in that the transition strength is much weaker than transition of gases commonly used in EIT.

The transparency window can be tuned by the control beam. Figure 3.16a shows the measured transparency as a function of the control power measured at the output end of the fiber together with the corresponding theoretical prediction. The measured transparency FWHM in Figure 3.16b shows a linear dependence on control power.

Slow light was demonstrated by propagating a 19 ns long probe pulse through the acetylene-filled hollow-core fiber. Figure 3.17 shows the output probe pulse with control beam on and off. The delay measured in the presence of the control beam is 800 ps. This is the first demonstration of EIT slow light at telecommunication wavelengths.

Besides acetylene, alkali atomic vapors such as rubidium have also been injected into a HC-PBF to achieve ultralow power level optical interactions [67]. Alkali atomic vapors are a favorite choice for EIT because of the relatively simple energy-level structure and large atomic absorption cross section that can result in a large optical depth for modest atomic number densities. However, due to the strong interaction of the alkali atoms with the walls of the silica fiber, making a useful HC-PBF-based alkali vapor cell has been a challenge. By coating the inner walls of the fiber core with organosilane and using light-induced atomic desorption to release alkali atoms into the core, Ghosh et al. realized a significant rubidium density in the core of an HC-PBF with an optical depth over 2000 [67]. They demonstrated EIT in such an HC-PBF Rubidium cell with a control power as low as 10 nW, which represents more than a factor of 10^7 reduction in the required control power in comparison to that needed to achieve EIT in the HC-PBF acetylene cells. These results suggest that EIT in hollow-core fibers is a viable and important route to slow light.

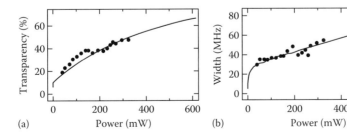

FIGURE 3.16 Transparency and bandwidth of EIT. (From Gosh, S., Sharping, J.E., Ouzounov, D.G., and Gaeta, A.L., *Phys. Rev. Lett.*, 94, 093,902, 2005. With permission.)

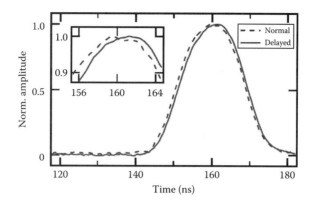

FIGURE 3.17 Observation of slow light via EIT in acetylene-filled hollow-core fiber. (From Gosh, S., Sharping, J.E., Ouzounov, D.G., and Gaeta, A.L., *Phys. Rev. Lett.*, 94, 093,902, 2005. With permission.)

3.4 WAVELENGTH CONVERSION AND DISPERSION

The slow light methods discussed so far rely on the use of either absorbing or amplifying resonances, which give rise to rapid variations in refractive index within a narrow spectral range. Dramatic slow-light delays can also be obtained by a method known as wavelength-conversion-and-dispersion that does not rely on such resonances. Instead, this method makes use of GVD of an optical waveguide and wavelength conversion techniques. In essence, the incident signal pulses are converted to another carrier frequency via a wavelength conversion device, propagated through a length of highly dispersive waveguide (with large GVD), and the delayed pulses are converted back to the original wavelength via a second wavelength conversion device. Tunable group delay of signal pulses is then obtained because of the frequency dependence of the group velocity in the dispersive waveguide. The group delay as a function of the carrier wavelength shift $\Delta\lambda$ is given by

$$\Delta T_d = LD\Delta\lambda, \tag{3.14}$$

where
  L is the length of the highly dispersive waveguide
  D is the GVD parameter of the waveguide at the incident pulse carrier frequency
  $\Delta\lambda$ is the wavelength shift of the first wavelength conversion device

This technique was first demonstrated in an optical fiber by Sharping et al. [68]. Figure 3.18 shows the schematic of the experiment setup, where wavelength conversion and reconversion are accomplished by using a fiber-based parametric amplifier. The wavelength conversion and reconversion happen in the same fiber (highly nonlinear (HNL)-dispersion-compensated fiber (DCF)) but in opposite propagation directions. The dispersive fiber (DCF) has a GVD of −74 ps/nm at the signal wavelength of 1565 nm. The signal pulses from an OPO have a pulsewidth of 10 ps and a repetition rate of 75 MHz. Tunable pulse delays in excess of 800 ps were demonstrated by varying the pump wavelength of the fiber-based parametric amplifier yielding a relative delay of 80 pulsewidths. Figure 3.19 shows the measured delay and the corresponding temporal pulse shapes. Figure 3.20 shows the optical spectra of the input and the delayed pulses. The delayed pulses have the same center wavelength as the input pulses as a result of the use of same frequency pump in the conversion and reconversion in the same fiber. The spectral shape is nearly identical down to 8 dB from the peak, at which point considerable noise is observed in the sidebands. This noise is comprised of a combination of amplified spontaneous emission from the erbium-doped fiber and parametric amplifiers.

Slow Light in Optical Waveguides 53

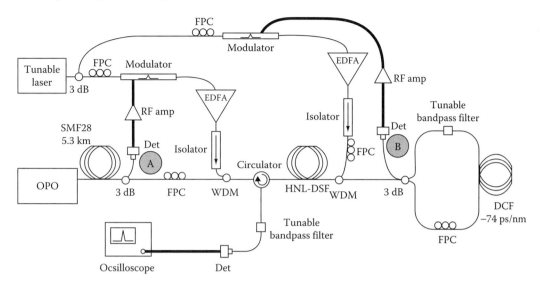

FIGURE 3.18 Experimental setup for tunable delays via wavelength conversion and dispersion. (From Sharping, J.E., Okawachi, Y., van Howe, J., Xu, C., Wang, Y., Willner, A.E., and Gaeta, A.L., *Opt. Express*, 13, 7872, 2005. With permission.)

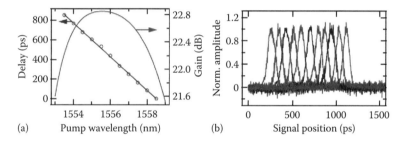

FIGURE 3.19 (a) Measured delay and calculated gain versus pump wavelength. (b) Measured pulse traces at different pump wavelengths indicated in (a). (From Sharping, J.E., Okawachi, Y., van Howe, J., Xu, C., Wang, Y., Willner, A.E., and Gaeta, A.L., *Opt. Express*, 13, 7872, 2005. With permission.)

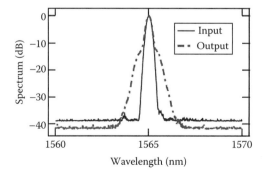

FIGURE 3.20 Spectra of input and output pulses in one implementation of the conversion and dispersion process. (From Sharping, J.E., Okawachi, Y., van Howe, J., Xu, C., Wang, Y., Willner, A.E., and Gaeta, A.L., *Opt. Express*, 13, 7872, 2005. With permission.)

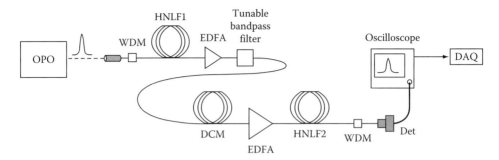

FIGURE 3.21 Experimental setup. (From Okawachi, Y., Sharping, J.E., Xu, C., and Gaeta, A.L., *Opt. Express*, 14, 12,022, 2006. With permission.)

Okawachi et al. recently simplified the fiber-based wavelength-conversion-and-dispersion technique and delayed 3.5 ps pulses up to 4.2 ns, corresponding to a fractional delay of 1200 [69]. Figure 3.21 shows the experiment setup where the wavelength conversion and reconversion are achieved with self-phase-modulation spectral broadening followed by spectral filtering. Figure 3.22 shows the measured delay as a function of the center wavelength of the tunable filter. By suitable choice of the tunable filter, the spectrum of the delayed pulse can be made identical to that of the incident pulse.

Another improved variation of the conversion–dispersion technique uses a periodically poled lithium-niobate (PPLN) waveguide for wavelength conversion [70]. The dispersion element is a DCF, and intrachannel dispersion arising from the DCF is compensated by a chirped fiber Bragg grating after the second wavelength conversion unit. The PPLN wavelength converter has the advantage of being rapidly tunable across a large bandwidth in a two-pump configuration, allowing both large delay variations and fast reconfiguration rates. In this experiment, continuous optical delay up to 44 ns is demonstrated for 10 Gb/s NRZ system, which is equal to 440 bit slots. As recently demonstrated in Ref. [71], the conversion–dispersion technique can be further improved using a silicon waveguide for the wavelength conversion stage, yielding an all-silicon integrated-circuit approach to slow light.

Compared with resonance-based slow-light techniques, the conversion–dispersion method has several advantages: (1) it has a highly controllable span of tunable delays from picosecond to nanosecond and large fractional delay, (2) it can support broad bandwidths suitable for data rates exceeding 10 Gb/s, and (3) the delayed pulses can have identical wavelength and bandwidth. Still, the reconfiguration rate is limited by the tuning speed of the filters or of the pump laser frequencies. Also, care must be taken in choosing the dispersive element and wavelength conversion range to minimize dispersive pulse broadening and maximize slow-light delay.

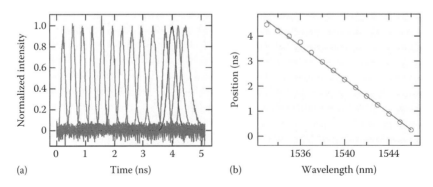

FIGURE 3.22 (a) Measured pulse traces at different pump wavelengths. (b) Measured pulse delay versus pump wavelength. (From Okawachi, Y., Sharping, J.E., Xu, C., and Gaeta, A.L., *Opt. Express*, 14, 12,022, 2006. With permission.)

3.5 CONCLUSION

Slow light is an important research area that is of fundamental interest and has great potential for applications. The use of optical waveguides is particularly attractive for application since it can be readily integrated with existing technologies. In this chapter, we have described several important developments in achieving all-optical tunable slow-light delay in optical fibers and other types of waveguides; we anticipate that some of these schemes will soon find their way into telecommunication systems, ultra high-speed diagnostic equipment, and synthetic aperture radio detection and ranging (RADAR) and laser detection and ranging (LADAR) systems.

ACKNOWLEDGMENT

We gratefully acknowledge the financial support of the DARPA DSO Slow Light program.

REFERENCES

1. R. W. Boyd and D. J. Gauthier, *Progress in Optics*, vol. 43, chap. 6, pp. 497–530 (Elsevier, Amsterdam, 2002).
2. L. V. Hau, S. E. Harris, Z. Dutton, and C. H. Behroozi, Light speed reduction to 17 meters per second in an ultracold atomic gas, *Nature* 594, 397–598 (1999).
3. M. M. Kash, V. A. Sautenkov, A. S. Zibrov, L. Hollberg, G. R. Welch, M. D. Lukin, Y. Rostovtsev, E. S. Fry, and M. O. Scully, Ultraslow group velocity and enhanced nonlinear optical effects in a coherently driven hot atomic gas, *Phys. Rev. Lett.* 82, 5229–5232 (1999).
4. A. V. Turukhin, V. S. Sudarshanam, M. S. Shahriar, J. A. Musser, B. S. Ham, and P. R. Hemmer, Observation of ultraslow and stored light pulses in a solid, *Phys. Rev. Lett.* 88, 023,602 (2002).
5. D. J. Gauthier, Slow light brings faster communication, *Phys. World* 18, 30–32 (2005).
6. D. J. Gauthier, A. L. Gaeta, and R. W. Boyd, Slow Light: from basics to future prospects, *Photon. Spectra* 44–50, March, (2006).
7. R. W. Boyd, D. J. Gauthier, and A. L. Gaeta, Applications of slow-light in telecommunications, *Opt. Photon. News* 17, 19–23 (2006).
8. E. Parra and J. R. Lowell, Toward applications of slow light technology, *Opt. Photon. News* 18, 40–45 (2007).
9. L. D. Landau and E. M. Lifshitz, *Electrodynamics of Continuous Media* (Pergamon, New York, 1960).
10. R. W. Boyd, *Nonlinear Optics*, 2nd edn. (Academic, San Diego, 2003).
11. Z. Zhu, D. J. Gauthier, Y. Okawachi, J. E. Sharping, A. L. Gaeta, R. W. Boyd, and A. E. Willner, Numerical study of all-optical slow-light delays via stimulated Brillouin scattering in an optical fiber, *J. Opt. Soc. Am. B* 22, 2378–2384 (2005).
12. Y. Okawachi, M. S. Bigelow, J. E. Sharping, Z. Zhu, A. Schweinsberg, D. J. Gauthier, R. W. Boyd, and A. L. Gaeta, Tunable all-optical delays via Brillouin slow light in an optical fiber, *Phys. Rev. Lett.* 94, 153, 902 (2005).
13. K. Y. Song, M. G. Herráez, and L. Thévenaz, Long optically controlled delays in optical fibers, *Opt. Lett.* 30, 1782–1784 (2005).
14. K. Y. Song, M. G. Herráez, and L. Thévenaz, Observation of pulse delaying and advancement in optical fibers using stimulated Brillouin scattering, *Opt. Express* 13, 82–88 (2005).
15. M. D. Stenner, M. A. Neifeld, Z. Zhu, A. M. C. Dawes, and D. J. Gauthier, Distortion management in slow-light pulse delay, *Opt. Express* 13, 9995–10,002 (2005).
16. T. Luo, L. Zhang, W. Zhang, C. Yu, and A. E. Willner, Reduction of pattern dependent distortion on data in an sbs-based slow light fiber element by detuning the channel away from the gain peak, in, *Conference on Lasers and Electro-Optics*, paper CThCC4 (2006).
17. A. Minardo, R. Bernini, and L. Zeni, Low distortion Brillouin slow light in optical fibers using AM modulation, *Opt. Express* 14, 5866–5876 (2006).
18. B. Zhang, L. Yan, I. Fazal, L. Zhang, A. E. Willner, Z. Zhu, and D. J. Gauthier, Slow light on Gbps differential-phase-shift-keying signals, *Opt. Express* 15, 1878–1883 (2007).
19. Z. Lu, Y. Dong, and Q. Li, Slow light in multi-line Brillouin gain spectrum, *Opt. Express* 15, 1871–1877 (2007).

20. Z. Shi, R. Pant, Z. Zhu, M. D. Stenner, M. A. Neifeld, D. J. Gauthier, and R. W. Boyd, Design of a tunable time-delay element using multiple gain lines for increased fractional delay with high data fidelity, *Opt. Lett.* 32, 1986–1988 (2007).
21. R. Pant, M. D. Stenner, M. A. Neifeld, Z. Shi, R. W. Boyd, and D. J. Gauthier, Maximizing the opening of eye diagrams for slow-light systems, *Appl. Opt.* 46, 6513–6519 (2007).
22. K. Y. Song, K. S. Abedin, K. Hotate, M. G. Herráez, and L. Thévenaz, Highly efficient Brillouin slow and fast light using As_2Se_3 chalcogenide fiber, *Opt. Express* 14, 5860–5865 (2006).
23. A. Zadok, A. Eyal, and M. Tur, Extended delay of broadband signals in stimulated Brillouin scattering slow light using synthesized pump chirp, *Opt. Express* 14, 8498–8505 (2006).
24. T. Schneider, R. Henker, K. U. Lauterbach, and M. Junker, Comparison of delay enhancement mechanisms for SBS-based slow light systems, *Opt. Express* 15, 9606–9613 (2007).
25. M. G. Herráez, K. Y. Song, and L. Thévenaz, Arbitrary-bandwidth Brillouin slow light in optical fibers, *Opt. Express* 14, 1395–1400 (2006).
26. Z. Zhu, A. M. C. Dawes, D. J. Gauthier, L. Zhang, and A. E. Willner, 12-GHz-bandwidth SBS slow light in optical fibers, *Optical Fiber Conference 2006*, paper PDP1 (2006).
27. Z. Zhu, A. M. C. Dawes, D. J. Gauthier, L. Zhang, and A. E. Willner, Broadband SBS slow light in an optical fiber, *J. Lightw. Technol.* 25, 201–206 (2007).
28. T. Schneider, M. Junker, and K.-U. Lauterbach, Potential ultra wide slow-light bandwidth enhancement, *Opt. Express* 14, 11,082–11,087 (2006).
29. K. Y. Song and K. Hotate, 25 GHz bandwidth Brillouin slow light in optical fibers, *Opt. Lett.* 32, 217–219 (2007).
30. L. Yi, L. Zhan, W. Hu, and Y. Xia, Delay of broadband signals using slow light in stimulated Brillouin scattering with phase-modulated pump, *IEEE Photon. Technol. Lett.* 19, 619–621 (2007).
31. L. Yi, Y. Jaouen, W. Hu, Y. Su, and S. Bigo, Improved slow-light performance of 10 Gb/s NRZ, PSBT and DPSK signals in fiber broadband SBS, *Opt. Express* 15, 16,972–16,979 (2007).
32. Z. Zhu and D. J. Gauthier, Nearly transparent SBS slow light in an optical fiber, *Opt. Express* 14, 7238–7245 (2006).
33. S. Chin, M. Gonzalez-Herraez, and L. Thévenaz, Zero-gain slow & fast light propagation in an optical fiber, *Opt. Express* 14, 10,684–10,692 (2006).
34. T. Schneider, M. Junker, and K.-U. Lauterbach, Time delay enhancement in stimulated-Brillouin scattering-based slow-light systems, *Opt. Lett.* 32, 220–222 (2007).
35. C. J. Misas, P. Petropoulos, and D. J. Richardson, Slowing of pulses to c/10 with subwatt power levels and low latency using Brillouin amplification in a Bismuth-oxide optical fiber, *J. Lightw. Technol.* 25, 216–221 (2007).
36. C. Florea, M. Bashkansky, Z. Dutton, J. Sanghera, P. Pureza, and I. Aggarwal, Stimulated Brillouin scattering in single-mode As_2S_3 and As_2Se_3 chalcogenide fibers, *Opt. Express* 14, 12,063–12,070 (2006).
37. K. S. Abedin, Stimulated Brillouin scattering in single-mode tellurite glass fiber, *Opt. Express* 14, 11,766–11,772 (2006).
38. R. H. Stolen, J. P. Gordon, W. J. Tomlinson, and H. A. Haus, Raman response function of silica-core fibers, *J. Opt. Soc. Am. B* 6, 1159–1166 (1989).
39. D. Hollenbeck and C. D. Cantrell, Multiple-vibrational-mode model for fiber-optic Raman gain spectrum and response function, *J. Opt. Soc. Am. B* 19, 2886–2892 (2002).
40. K. J. Blow and D. Wood, Theoretical description of transient stimulated Raman scattering in optical fibers, *IEEE J. Quantum Electron.* 25, 2665–2673 (1989).
41. J. E. Sharping, Y. Okawachi, and A. L. Gaeta, Wide bandwidth slow light using a Raman fiber amplifier, *Opt. Express* 13, 6092–6098 (2005).
42. Y. Okawachi, M. A. Foster, J. E. Sharping, A. L. Gaeta, Q. Xu, and M. Lipson, All-optical slow-light on a photonic chip, *Opt. Express* 14, 2317–2322 (2006).
43. M. S. Bigelow, N. N. Lepeshkin, and R. W. Boyd, Observation of ultraslow light propagation in a ruby crystal at room temperature, *Phys. Rev. Lett.* 90, 113,903 (2003).
44. S. E. Schwartz and T. Y. Tan, Wave interactions in saturable absorbers, *Appl. Phys. Lett.* 10, 4–7 (1967).
45. L. W. Hillman, R. W. Boyd, J. Krasinski, and C. R. S. Jr., Observation of a spectral hole due to population oscillations in a homogeneously broadened optical absorption line, *Opt. Commun.* 45, 416–419 (1983).
46. M. S. Bigelow, N. N. Lepeshkin, and R. W. Boyd, Superluminal and slow light propagation in a room-temperature solid, *Science* 301, 200–202 (2003).

47. P. C. Ku, F. Sedgwick, C. J. Chang-Hasnain, P. Palinginis, T. Li, H. Wang, S. W. Chang, and S. L. Chuang, Slow light in semiconductor quantum wells, *Opt. Lett.* 29, 2291–2293 (2004).
48. J. Mørk, R. Kjær, M. van der Poel, and K. Yvind, Slow light in a semiconductor waveguide at gigahertz frequencies, *Opt. Express* 13, 8136–8145 (2005).
49. P. Palinginis, S. Crankshaw, F. Sedgwick, E.-T. Kim, M. Moewe, C. J. Chang-Hasnain, H. Wang, and S.-L. Chuang, Ultraslow light (<200 m/s) propagation in a semiconductor nanostructure, *Appl. Phys. Lett.* 87, 171,102 (2005).
50. H. Su and S. L. Chuang, Room temperature slow and fast light in quantum-dot semiconductor optical amplifiers, *Appl. Phys. Lett.* 88, 61,102 (2006).
51. A. Schweinsberg, N. N. Lepeshkin, M. S. Bigelow, R. W. Boyd, and S. Jarabo, Observation of superluminal and slow light propagation in erbium-doped optical fiber, *Europhys. Lett.* 73, 218–224 (2006).
52. G. M. Gehring, A. Schweinsberg, C. Barsi, N. Kostinski, and R. W. Boyd, Observation of backwards pulse propagation through a medium with a negative group velocity, *Science* 312, 895–897 (2006).
53. H. Shin, A. Schweinsberg, G. Gehring, K. Schwertz, H. J. Chang, R. W. Boyd, Q.-H. Park, and D. J. Gauthier, Reducing pulse distortion in fast-light pulse propagation through an erbium-doped fiber amplifier, *Opt. Lett.* 32, 906–908 (2007).
54. S. E. Harris, Electromagnetically induced transparency, *Phys. Today* 50, 36–42 (1997).
55. K. J. Boller, A. Imamoglu, and S. E. Harris, Observation of electromagnetically induced transparency, *Phys. Rev. Lett.* 66, 2593–2596 (1991).
56. C. Liu, Z. Dutton, C. H. Behroozi, and L. Hau, Observation of coherent optical information storage in an atomic medium using halted light pulses, *Nature* 409, 490–493 (2001).
57. D. F. Phillips, A. Fleischhauer, A. Mair, R. L. Walsworth, and M. D. Lukin, Storage of light in atomic vapor, *Phys. Rev. Lett.* 86, 783–786 (2001).
58. H. Schmidt and A. Imamoglu, Giant Kerr nonlinearities obtained by electromagnetically induced transparency, *Opt. Lett.* 21, 1936–1938 (1996).
59. M. D. Lukin and A. Imamoğlu, Controlling photons using electromagnetically induced transparency, *Nature* 413, 273–276 (2001).
60. R. W. Boyd, D. J. Gauthier, A. L. Gaeta, and A. E. Willner, Maximum time delay achievable on propagation through a slow-light medium, *Phys. Rev. A* 71, 023,801 (2005).
61. R. W. Boyd, D. J. Gauthier, A. L. Gaeta, and A. E. Willner, Erratum: Maximum time delay achievable on propagation through a slow-light medium, *Phys. Rev. A* 72, 059,903(E) (2005).
62. D. Yin, H. Schmidt, J. Barber, and A. Hawkins, Integrated ARROW waveguides with hollow cores, *Opt. Express* 12, 2710–2715 (2004).
63. D. Yin, J. Barber, A. Hawkins, and H. Schmidt, Waveguide loss optimization in hollow-core ARROW waveguides, *Opt. Express* 13, 9331–9336 (2005).
64. W. Yang, D. B. Conkey, B. Wu, D. Yin, A. R. Hawkins, and H. Schmidt, Atomic spectroscopy on a chip, *Nat. Photon.* 1, 331–335 (2007).
65. A. K. Patnaik, J. Q. Liang, and K. Hakuta, Slow light propagation in a thin optical fiber via electromagnetically induced transparency, *Phys. Rev. A* 66, 63,808 (2002).
66. S. Ghosh, J. E. Sharping, D. G. Ouzounov, and A. L. Gaeta, Resonant optical interactions with molecules confined in photonic band-gap fibers, *Phys. Rev. Lett.* 94, 093,902 (2005).
67. S. Ghosh, A. R. Bhagwat, C. K. Renshaw, S. Goh, A. L. Gaeta, and B. J. Kirby, Low-light-level optical interactions with rubidium vapor in a photonic band-gap fiber, *Phys. Rev. Lett.* 97, 023,603 (2006).
68. J. E. Sharping, Y. Okawachi, J. van Howe, C. Xu, Y. Wang, A. E. Willner, and A. L. Gaeta, All-optical, wavelength and bandwidth preserving, pulse delay based on parametric wavelength conversion and dispersion, *Opt. Express* 13, 7872–7877 (2005).
69. Y. Okawachi, J. E. Sharping, C. Xu, and A. L. Gaeta, Large tunable optical delays via self-phase modulation and dispersion, *Opt. Express* 14, 12,022–12,027 (2006).
70. Y. Wang, C. Yu, L. S. Yan, A. E. Willner, R. Roussev, C. Langrock, M. M. Fejer, Y. Okawachi, J. E. Sharping, and A. L. Gaeta, 44-ns continuously-tunable dispersionless optical delay element using a PPLN waveguide with a two pump configuration, DCF, and a dispersion compensator, *IEEE Photonics Technol. Lett.* 19, 861–863 (2007).
71. M. A. Foster, A. C. Turener, J. E. Sharping, B. S. Schmidt, M. Lipson, and A. L. Gaeta, Broad-band optical parametric gain on a silicon photonic chip, *Nature* 441, 960–963 (2006).

4 Slow Light in Photonic Crystal Waveguides

Thomas F. Krauss

CONTENTS

4.1 Introduction..59
4.2 How Does a Photonic Crystal Generate Slow Light? 60
    4.2.1 Slow Light in 2D 63
    4.2.2 Slow Light Away from the Bandedge 63
4.3 Enhancement of Linear Interaction..65
4.4 Comparison of Cavities and Slow Light Waveguides 66
    4.4.1 Intensity Enhancement 66
    4.4.2 Bandwidth Comparison 68
    4.4.3 Comparison to Coupled Cavity Waveguides 69
    4.4.4 Implications of the FOM 69
        4.4.4.1 Nonlinear Refractive Index 69
        4.4.4.2 Linear Refractive Index 70
4.5 Losses...71
4.6 Coupling 72
4.7 Conclusions..73
References 74

4.1 INTRODUCTION

The keen interest in slow light in nanostructured dielectrics is motivated by the fact that slow light adds functionality to a material by structuring alone. Unlike the EIT-based schemes discussed elsewhere in this book, nanostructuring is wavelength-independent, i.e., it can be adjusted to any wavelength of interest within the transparency window of the material. Furthermore, it enhances the weak light–matter interaction in a material that may be of interest otherwise, such as silicon, and it adds another degree of freedom to already highly electro optic or nonlinear materials such as, chalcogenide glasses [1]. Optically linear effects such as gain, thermo optic and electro optic interaction scale with the slowdown factor, whereas nonlinear effects may scale with its square [2,3] which we shall discuss in more detail.

This chapter deals with slow light in photonic crystal waveguides based on line defects. A typical waveguide consisting of a single line of missing holes is illustrated in Figure 4.1. In comparison with single cavities which are also widely studied in the photonic crystal and nanophotonics community and offer sizeable enhancement of optically linear and nonlinear effects, slow light structures offer more bandwidth, i.e., a broader wavelength range of operation. A figure of merit (FOM) is introduced in terms of intensity enhancement vs. operational bandwidth that substantiates this claim. Bandwidth is particularly important for high data rate applications, for example, 100 Gb/s and beyond, that are

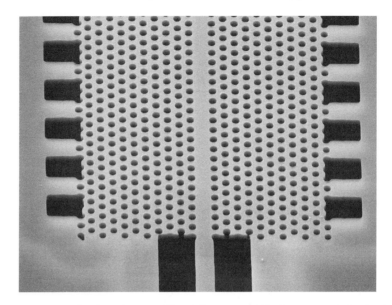

FIGURE 4.1 Micrograph of a photonic crystal line defect waveguide. The waveguide is of "W1" type, i.e., it is formed by removing a single line of holes from the perfect lattice. For 1550 nm operation, the typical lattice constant (i.e., the distance between adjacent holes) is 420 nm and the hole diameter is 250 nm. This type of generic waveguide already exhibits a slow light regime, which can be further enhanced by selectively tuning the holesize, lattice constant, or waveguide width. The waveguide is realized as an airbridge, i.e., it is formed by creating a suspended membrane in air, which maximizes the vertical refractive index contrast and thus the confinement.

elusive for electronics and may be so for some time. Furthermore, bandwidth provides robustness against environmental and technological fluctuations. While it is very difficult to achieve a specific target wavelength with a high-Q cavity, for example, which is still sensitive to temperature changes, all of these concerns are relaxed for the broad-bandwidth slow light waveguides. Therefore, devices based on slow light waveguides are a platform that can address two key issues in communications: bandwidth and switching power. The enhanced nonlinearity enables the design of low power all-optical switching and data processing device while simultaneously accommodating the large bandwidth of future ultrahigh-speed systems.

The slowdown factor S already defined elsewhere as the ratio of the phase velocity over the group velocity, $S = v_\phi/v_g$, is the measure of the slowdown and enhancement achieved by the structure, i.e., it refers to the slowdown factor to the effective index of the material. In this respect, it quantifies the effect achieved by structuring the material. Many authors use the group index n_g for the same purpose, which can be justified as it references the slowdown factor to propagation in free space. If one considers a more comprehensive FOM than slowdown factor alone and takes bandwidth and dispersion into account [4] one finds that the performance of dielectric slow light devices scales as the refractive index contrast. The index contrast plays the role of the oscillator strength in an EIT medium, and the higher the index contrast, the larger the group index that can be achieved with tolerable dispersion and bandwidth. Therefore, high refractive index structures such as photonic crystals appear particularly promising for the creation of slow light effects.

4.2 HOW DOES A PHOTONIC CRYSTAL GENERATE SLOW LIGHT?

A photonic crystal is first and foremost a grating, and most of the slow light effects can be explained from a 1D grating perspective. In fact, the coupled resonator structures discussed elsewhere in this

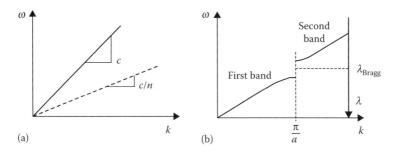

FIGURE 4.2 Dispersion diagram for (a) an optical wave propagating in free space (solid line) and a medium with constant refractive index n (dashed line), and (b) an optical wave propagating in a periodic medium. The periodicity in (b) breaks the dispersion curve into multiple bands that are labeled by increasing frequency.

book fall into the same category. Gratings and photonic crystals are conveniently described by their bandstructure. The bandstructure is an energy–wavevector (ω–k) diagram of the allowed states or modes of the crystal. The bandstructure terminology was originally developed in solid-state physics to describe the electronic properties of crystals and has proven particularly useful for semiconductors. Applying this tool to photonics has given a host of new insights, such as the development of defect waveguides and heterostructure cavities [5]; it has also provided a unified terminology for the different ways of describing the interaction of electromagnetic waves and periodic structures, such as Bloch modes and Bragg mirrors.

The simplest dispersion curve is shown in Figure 4.2a, which describes a wave propagating in a dispersion-free medium such as free space, namely a straight line of slope c, since $c = \omega/k$, which is a direct consequence of the wave equation. For a material of refractive index n, this corresponds to a slope of $c/n = \omega/k$, so the slope of the dispersion curve is inversely proportional to the refractive index. In a periodically structured medium as in Figure 4.2b, the curve is no longer a straight line but now features a discontinuity. This is due to the opening of a stopband where light is reflected. In the proximity of this stopband, the dispersion curve is no longer a straight line and one needs to distinguish between the phase velocity $v_\phi = \omega/k$ and the group velocity $v_g = d\omega/dk$, the latter denoting the local slope of the curve. For example, in Figure 4.2b, the slope of the dispersion curve for $k = \pi/a$, i.e., at the edge of the stopband, is flat, so the group velocity is zero. Slow light then refers to light that has a group velocity which is larger than zero but smaller than the phase velocity or it refers to dispersion curves with low but nonzero slope.

How does this stopband arise? Combining two materials of different refractive indices n_1 and n_2 creates reflections at the interface, according to Fresnel's equation (for normal incidence), $r = \frac{n_1 - n_2}{n_1 + n_2}$. These reflections add up in phase for a multilayer stack of equally spaced dielectric interfaces, a structure commonly known as a Bragg mirror, or a Bragg grating. The repeat unit is the lattice constant a and the wavelength range over which the reflections add up in phase is the stopband of the grating. The center wavelength of this stopband λ_{Bragg} corresponds to twice the lattice constant a, so $\lambda_{\text{Bragg}} = 2a$. This is also referred to as the Bragg condition.

In bandstructure terms, the Bragg condition occurs for $k = \pi/a$, which is the equivalent expression to $\lambda_{\text{Bragg}} = 2a$. This point has a special significance; since $G = 2\pi/a$ denotes the repeat unit cell of the grating in k-space, the boundaries of the repeat unit cell are at $k = \pm\pi/a$. According to bandstructure theory, which divides k-space into different Brillouin zones, $k = \pm\pi/a$ is then referred to as the boundary of the first Brillouin zone, with $k = \pm 2\pi/a$ the boundary of the second Brillouin zone and so on. At the Brillouin zone boundary, the dispersion curve becomes discontinuous, because the wavelengths in the stopband do not propagate as already discussed. This discontinuity is referred to as the band-edge, simply because the band suddenly stops. The wavelengths in the stopband are evanescent, so they may propagate into the grating for a short distance but are eventually reflected, hence they are not part of the allowed states of the system. The resulting dispersion curve, shown in Figure 4.2b, highlights the Brillouin zone boundary at $k = \pi/a$ and the corresponding stopband around λ_{Bragg}. Please note that the bandstructure is usually plotted in terms of frequency, but since it is more customary in photonics to specify frequencies in terms of their free-space wavelength, both are shown on the vertical axis.

The operation of a grating can now be easily divided into two separate regimes, i.e., the stopband, where light is reflected, and the passband, where light can propagate. The slow light regime lies in between, i.e., it is part of the passband but typically near the band-edge. This is the regime where the band changes slope from the initial value of c/n to zero.

How does the slow light regime actually work? We start at the band-edge and consider an incoming wave of wavevector $k = \pi/a$. This wave is reflected back, the change of direction being represented by a change of sign of the wavevector, so $k = \pi/a$ turns into $k = -\pi/a$. This phenomenon can also be explained by the Bloch theorem. The Bloch theorem, which was first developed to describe the propagation of electron waves in crystalline materials, states that waves propagating in periodic media are superpositions of multiply scattered components. Each of these components has a wavevector of $k \pm nG$ ($0 \leq n \leq \infty$) with $G = 2\pi/a$ as before. In order to explain the simple scenario discussed above, i.e., the band-edge case, we only need the first two components, so $n = 0$ and $n = 1$. The resulting wavevectors are k and $k-G$, or $k = \pi/a$ and $k = -\pi/a$, respectively, as before.

What does the corresponding Bloch mode that is created by the superposition of these two waves actually look like? A superposition of two waves of equal but opposite wavevector creates the well-known phenomenon of a standing wave, which is a wave that oscillates with a given phase velocity but has a stationary envelope, so its group velocity is zero. This explains why the slope of the dispersion curve approaches zero at the band-edge. As we move away from the band-edge, the k and $k-G$ wavevectors are no longer equal and opposite, so the resulting superposition yields a slowly moving interference pattern. This interference pattern is the slow mode and is shown in Figure 4.3b.

The slow mode is recognized by a characteristic beating pattern that arises form the mismatch between the k and $k-G$ components. As we move away from the band-edge, the mismatch increases and the period of the beating pattern becomes shorter. Additionally, the amplitude of the $k-G$ component decreases, because the multiple reflections that give rise to it are no longer in phase

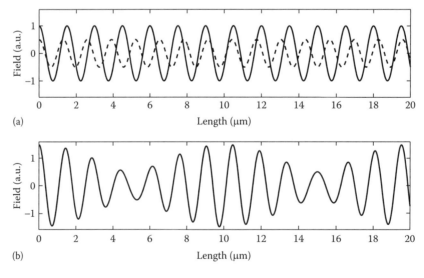

FIGURE 4.3 (a) A slow mode can be understood as the superposition of a forward (solid line, k-component) and a backward (dashed line, $k-G$ component) propagating mode. The amplitude of the $k-G$ component is lower because it is no longer in phase with the lattice. (b) The superposition of the two components yields the characteristic envelope, or beating pattern, of a slow mode.

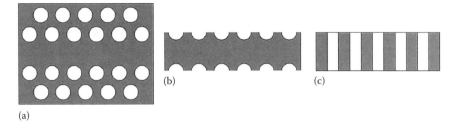

FIGURE 4.4 Similarity between a W1 photonic crystal waveguide and a 1D periodic lattice. The first line of holes adjacent to the line defect in (a) has the strongest impact on the propagation and is similar to a truncated photonic crystal waveguide (b), which can be understood as a periodic modulation of the effective index (c).

and only interfere partially. Eventually, the $k-G$ component vanishes and propagation becomes dominated by the k-component alone. The periodic material then behaves like an effective index medium of some average refractive index $n_1 < n_{\text{eff}} < n_2$. In bandstructure terms, the dispersion curve has evolved into a straight line of slope c/n_{eff}.

4.2.1 Slow Light in 2D

The above explanation is based on a simple 1D grating. By analogy, a photonic crystal waveguide operates on similar principles, as shown in Figure 4.4. The waveguide is created by removing one or more rows of holes from the perfect lattice. The strongest impact on the properties of such a waveguide is the periodic perturbation provided by the first row of holes (Figure 4.4b), which is equivalent to the effective index modulation of a 1D periodic structure (Figure 4.4c). The resulting dispersion curve is shown in Figure 4.5. Compared to Figure 4.2, the W1 mode operates in the second band, which is shown folded back into the first Brillouin zone for convenience.

4.2.2 Slow Light Away from the Bandedge

The band-edge is the most obvious place for the slow light phenomenon to occur, and it is there where most of the observations of slow light have been made. Despite this convenience, the band-edge is not the best operating point. First, the dispersion curve near the band-edge is typically parabolic, which means that the group velocity changes rapidly with frequency, so an optical pulse with finite

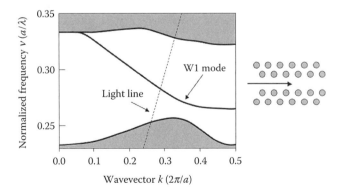

FIGURE 4.5 Dispersion curve of a W1 photonic crystal waveguide. The W1 mode and the light line, i.e., the limit of total internal reflection confinement in the vertical direction are clearly marked. The continuum of lattice modes is shown in grey.

bandwidth will disperse over a short distance [6]. Second, the band-edge presents a cut off point, where the mode turns from propagation to evanescence. Any fabrication tolerance then manifests itself as a local variation of this cut off point, so modes that may propagate as slow modes in one part of the structure may turn evanescent in another. As a result, the slow mode near the band-edge appears to be very lossy [7,8]. Both these concerns can be addressed by engineering the dispersion curve of the defect mode. Several examples of such dispersion engineering are known, for example, chirping the waveguide properties [9], changing the waveguide width [10,11] or changing the hole size [12] and position [13] of the photonic lattice adjacent to the line defect waveguide. Most of these methods (except the chirping) involve tuning the interaction between the index-guided and the bandgap-guided aspects of the W1 mode. The explanation of this interaction involves realizing that the simplistic model used before ignores the fact that a photonic crystal waveguide can confine light either by total internal reflection or by photonic bandgap effects; the modes supported can therefore be categorized as either index-guided or gap-guided, or a combination of both [10,14]. The interaction between these two types of confinement mechanism then determines the local shape of the waveguide dispersion curve. This interaction is apparent from the anticrossing between the W1 dispersion curve and the dispersion curve of the topmost lattice mode, for example, around $k = 0.35$ in Figure 4.5, as well as the fact that the W1 dispersion curve is not strictly parabolic; its shape is determined by a combination of the band-edge parabolicity and this anticrossing. An understanding of this interplay gives us a handle on shaping the dispersion curve further and allowing us to create sections of straight bands with low slope, also referred to as "flatband" slow light sections.

For example, changing the waveguide width increases the effective index of the defect mode, so its dispersion curve shifts down in the bandstructure. As the dispersion curve approaches the band of the topmost lattice mode, the interaction of the two modes distorts the shape of these bands locally. This distortion can be used as a tool to engineer the shape of the dispersion curve, especially to achieve flat sections of constant slope. As shown in Figure 4.6, such flatband slow light can be achieved for a range of parameters and group velocities. In the example, group indices of $n_g = 30, 50,$

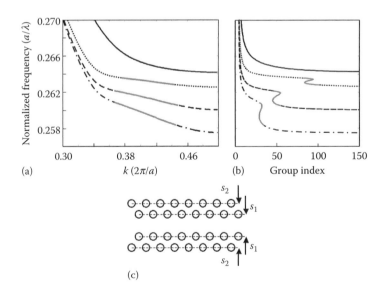

FIGURE 4.6 Illustration of the ability to engineer the group velocity over a significant range, from $n_g = 30$ to $n_g = 80$ here, as highlighted by the lighter unbroken curve sections in both (a) and (b); (a) shows the relevant dispersion curves while (b) shows the corresponding group indices on the same frequency scale. The continuous solid lines represent the case of the original W1 waveguide. This group velocity engineering is achieved by shifting the two innermost rows of holes of a W1 defect waveguide with respect to the perfect lattice, as indicated in (c) by the parameters s_1 (shift of the innermost row of holes) and s_2 (shift of second row of holes).

Slow Light in Photonic Crystal Waveguides

and 80 are demonstrated by shifting the two innermost rows of holes appropriately [13]. Note that the group indices are almost constant (within ±10% error) over approximately 20% of the Brillouin zone, which is a substantial operating range. Such sections of constant group velocity are essential for the exploitation of slow light effects, as the enhancements discussed in Section 4.3 are possible only if the optical signals do not disperse. In this respect, the demonstration of pulse compression commensurate with the group velocity reduction (a pulse was compressed 25-fold in a flatband slow light section of group index 25) [11] is significant as it highlights the fact that a constant group index can be achieved over a substantial bandwidth (here: 2.4 THz or 19 nm at 1500 nm wavelength, spanning the entire bandwidth of an fs-optical pulse). Therefore, the dispersive broadening observed in nonengineered defect waveguides [6] can indeed be overcome. Overall, it is now clear that suitably designed photonic crystal waveguides can accommodate a substantial spectral bandwidth and are therefore promising hosts for the demonstration and application of slow light effects.

4.3 ENHANCEMENT OF LINEAR INTERACTION

Most optical switching devices operate on the basis of a relative phase change of π. In a Mach–Zehnder interferometer, for instance, one requires a delay of half a wavelength in one branch with respect to the other in order to switch the device from on to off or vice versa. In principle, such a device is based on two-beam interference, with the resulting intensity I_{tot} depending on the two input intensities I_1, I_2 and their relative phase difference $\Delta\phi$.

$$I_{\text{tot}} = I_1 + I_2 + 2\sqrt{I_1 I_2} \cos \Delta\phi \qquad (4.1)$$

The phase difference $\Delta\phi$ is typically expressed in terms of the interaction length L and the difference in wavevector Δk, so $\cos \Delta\phi = \cos \Delta k L$. Δk is usually expressed as $\Delta n k_0$, with Δn, the refractive index change induced by an external effect or by the nonlinearity of the material, and k_0, the wavevector in vacuum. The slowdown factor S does not enter the discussion, so it appears as if the switching device derives no benefit from the slow light regime. There is a subtle difference, however. Since it is Δk that causes the phase difference in Equation 4.1, the Δn should be the difference in effective mode index Δn_{eff}, rather than in material refractive index Δn_{mat}. The difference is highlighted in Figure 4.7.

The dispersion curve of a given mode is described by the solid line, simplified here as consisting of a fast (steep gradient) and slow (gentle gradient) section; recall that the group velocity v_g and

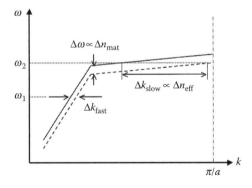

FIGURE 4.7 Explanation of the increased phase sensitivity of a slow light device; for a given index change Δn_{mat}, the dispersion curve shifts by $\Delta\omega$. In the slow light regime around ω_2, the corresponding Δk is much larger than in the fast light regime around ω_1. Therefore, a slow light device is more phase-sensitive than a fast light device.

group index n_g are given by the slope of the dispersion curve, $v_g = \frac{c}{n_g} = \frac{d\omega}{dk}$. If the material index is changed, the curve moves up or down in the diagram, as indicated by the dashed line. The effect on the mode propagating at a frequency ω is given by $\Delta k = k_0 \Delta n_{\text{eff}}$. It is clear from the figure that the Δk is much smaller for the frequency ω_1, where the mode is fast, than for the frequency ω_2, where the mode is slow. So, a slow mode experiences a larger change in effective index for a given change in material index than a fast mode, and it is obvious that the effect scales as the gradient of the dispersion curve, i.e., $\Delta n_{\text{eff}} = S \times n_{\text{mat}}$. Overall, the discussion can be summarized and the condition for switching based on Equation 4.1 can be expressed as follows:

$$\Delta \phi = \pi = \Delta k L = k_0 S \Delta n_{\text{mat}} L \tag{4.2}$$

So the slow light regime yields a large Δk for a given Δn_{mat}. An example for this effect is provided by Vlasov et al. [15]. Utilizing a photonic crystal waveguide, they demonstrate that a thermo-optically tuned Mach–Zehnder modulator requires less energy when operating in the slow light regime than it does when operating in the fast light regime. In a different embodiment, Beggs et al. [16] show a slow light enhanced directional coupler of only 5 μm length that is actuated with a refractive index change of $\Delta n = 4 \times 10^{-3}$. A comparable directional coupler of conventional design would have been 200 μm long.

4.4 COMPARISON OF CAVITIES AND SLOW LIGHT WAVEGUIDES

Cavities and slow light waveguides both enhance the electric field at the expense of bandwidth. The obvious question then is which one to use for a given application—do they both perform in a similar way or are there any differences? In order to answer this question, we compare the performance of two generic structures in terms of intensity enhancement and bandwidth, and find out whether the bandwidth penalty for a given intensity enhancement is different in the two systems. To quantify this comparison, we introduce FOM $= \frac{\Delta v_{\text{SL}}}{\Delta v_{\text{cav}}}$, with Δv_{SL} representing the useful bandwidth of the slow light structure and Δv_{cav} that of the cavity, assuming the same intensity enhancement. If FOM is larger than 1, the slow light structure offers more bandwidth, and if FOM is smaller than 1, the cavity does.

4.4.1 INTENSITY ENHANCEMENT

In a cavity, the intensity builds up because the light makes multiple round-trips; on resonance, the fraction of light lost then equals the fraction of light coupled into the cavity from the outside. To understand how much light is trapped inside the cavity, i.e., the cavity intensity I_{cav}, consider Figure 4.8a. We assume light of intensity I_0 impinging onto a loss-less Fabry–Perot cavity from the left. The cavity is made up of two mirrors with reflectivity R. Let us start inside the cavity, where light of intensity I_{cav} is propagating. At the right mirror, $(1-R)I_{\text{cav}}$ is coupled out and RI_{cav} stays inside the cavity. At the left mirror, one would then expect $R(1-R)I_{\text{cav}}$ to be coupled out. At the

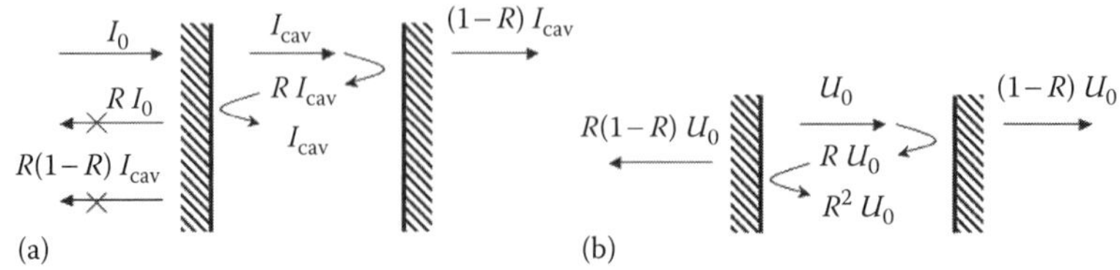

FIGURE 4.8 Sketch of the gain and loss effects at the mirrors that lead to the enhancement of intensity in an optical cavity. (a) The intensity enhancement in a cavity on resonance and (b) the dependence of the cavity quality factor on the mirror reflectivity.

Slow Light in Photonic Crystal Waveguides

same time, however, I_0 is impinging on the cavity, and RI_0 is reflected. Due to the phase change on reflection, RI_0 and $R(1-R)I_{cav}$ are π out of phase and so cancel; if these two intensities are equal, the cancellation is complete and no reflection of I_0 occurs at all. This is what is observed for a cavity on resonance. This condition is met when the two intensities are equal, $RI_0 = R(1-R)I_{cav}$, so

$$I_{cav} = \frac{I_0}{1-R} \tag{4.3}$$

The intensity inside the cavity is enhanced by $\frac{1}{1-R}$ on resonance.

In order to relate this enhancement to the cavity quality factor, we introduce the dependence of Q on the mirror reflectivity R, illustrated in Figure 4.8b. The cavity Q-factor is defined as follows:

$$Q = 2\pi \frac{\text{Energy stored}}{\text{Energy lost per cycle}} \tag{4.4}$$

Assuming that the energy stored is U_0 and that the only loss-mechanism is the nonunity mirror reflectivity, we find that the energy lost per cycle, in a single mode cavity, is given by $(1-R)U_0 + R(1-R)U_0 \approx 2(1-R)U_0$. In a multimode cavity of mode order m, the loss per cycle is reduced by m, since the mirror loss occurs only once in m cycles, so Equation 4.4 becomes

$$Q = 2\pi m \frac{U_0}{2(1-R)U_0} = \frac{m\pi}{1-R} \tag{4.5}$$

Combining Equations 4.3 and 4.5, one can easily see that $I_{cav} = I_0 \frac{Q}{m\pi}$, so the intensity in the cavity is enhanced by a factor $\frac{Q}{m\pi}$.

Slow light structures are characterized by the slowdown factor S, which is defined as the group index over the phase index, $S = \frac{n_\phi}{n_g}$. If an optical pulse enters a slow light medium, and assuming a complete transfer of energy, the front of the pulse is slowed down with respect to the tail and the pulse compresses overall. This effect is illustrated in Figure 4.9 and has been observed experimentally by Engelen et al. [11]. It is easily understood that in the absence of dispersion and nonlinearities, the shape of the incident pulse is not affected. Given also that the slowdown is not caused by the

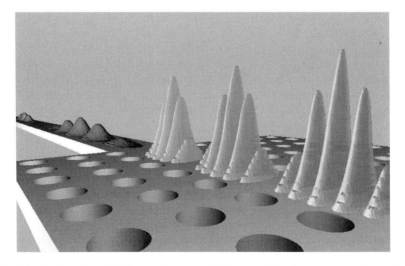

FIGURE 4.9 Sketch of a pulse of light entering a slow light waveguide; the pulse enters from the top left and is spatially compressed in the slow light regime, thereby increasing in intensity.

excitation of material resonances (as, for example, in the EIT case), the shorter pulse necessarily has to accomodate the incoming energy as a higher intensity. The pulse intensity in the slow light structure I_{SL} therefore scales as the slowdown factor, $I_{SL} = I_0 S$.

To summarize, we can write the intensity enhancement in the two types of structure as follows:

$$\frac{I_{cav}}{I_0} = \frac{Q}{m\pi} \qquad \frac{I_{SL}}{I_0} = S \qquad (4.6)$$

As we have now derived expressions for the intensity enhancement in the respective structures, we can consider the bandwidth.

4.4.2 Bandwidth Comparison

The bandwidth of a cavity is simply given by its Q-factor, i.e., by the well-known formula $Q = \frac{\nu_0}{\Delta \nu_{cav}}$, where ν_0 is the resonance frequency. Given that $\Delta \nu_{cav}$ is the full-width half maximum of the cavity, i.e., the bandwidth where the cavity transmission has dropped to 50%, we introduce a prefactor q_{cav} that takes into account the fact that the useful bandwidth may be lower than $\Delta \nu_{cav}$, for example, only between the 80% instead of the 50% transmission points, so $q_{cav} \leq 1$. Overall, we get

$$\Delta \nu_{cav} = q_{cav} \frac{\nu_0}{Q} \qquad (4.7)$$

The bandwidth of a slow light waveguide based on periodic structures can be determined via its dispersion diagram. We recall that in the dispersion diagram, the group velocity v_g is given by the slope of the dispersion curve, so $v_g = \frac{d\omega}{dk}$. Let us assume that the dispersion curve is linear in the section considered, so $\frac{d\omega}{dk} = \frac{\Delta \omega}{\Delta k}$. Introducing the group index via $v_g = \frac{c}{n_g}$, with c, the speed of light in vacuum, and converting from angular frequency, we can write

$$\Delta \nu_{SL} = \frac{1}{2\pi} \frac{c \Delta k}{n_g} \qquad (4.8)$$

The maximum Δk is given by the size of the Brillouin zone ($\Delta k = \frac{\pi}{a}$, with a being the periodicity of the system), but since the linear section of the dispersion curve does not extend across the entire Brillouin zone, we need to introduce another prefactor q_{SL} similar to the prefactor q_{cav} for cavities.

$$\Delta \nu_{SL} = q_{SL} \frac{c}{2n_g a} \qquad (4.9)$$

The operating frequency ν_0 is introduced into this equation via $\nu_0 = u\frac{c}{a}$, u being a dimensionless scaling factor ($u = \frac{a}{\lambda}$). From the aforementioned $S = \frac{n_\phi}{n_g}$, we finally obtain the following expression for the bandwidth of a slow light structure,

$$\Delta \nu_{SL} = q_{SL} \frac{\nu_0}{2uSn_\phi} \qquad (4.10)$$

The FOM between the two types of structure is given by the ratio of Equations 4.10 and 4.7,

$$\text{FOM} = \frac{\Delta \nu_{SL}}{\Delta \nu_{cav}} = \frac{q_{SL}}{q_{cav}} \frac{Q}{2uSn_\phi} \qquad (4.11)$$

Since the bandwidth comparison is meaningful only for the same intensity enhancement in both structures, we use Equation 4.6 and set $I_{cav} = I_{SL}$ to express S in terms of Q, so $S = \frac{Q}{m\pi}$. The FOM then assumes its final form,

$$\text{FOM} = \frac{q_{\text{SL}}}{q_{\text{cav}}} \frac{m\pi}{2un_\phi} \quad (4.12)$$

To appreciate the physical meaning of this FOM, let us insert some realistic values; q_{SL} can be as large as 20% [11,13] and q_{cav} can be set to 60%, so $\frac{q_{\text{SL}}}{q_{\text{cav}}} \approx \frac{1}{\pi}$. A typical operating point for a photonic crystal waveguide is around $u = 0.25 \frac{a}{\lambda}$, and the phase index is typically around $n_\phi = 2$, so the second denominator is approximately unity, $2un_\phi = 1$. This leaves the surprisingly simple expression of

$$\text{FOM} = m \quad (4.13)$$

as the final result. What does it mean? For a single mode cavity ($m = 1$), a slow light waveguide and a cavity achieve the same bandwidth for a given intensity enhancement. For a larger cavity, the FOM grows, which indicates that the slow light waveguide can accomodate more bandwidth for a given intensity enhancement. Therefore, the slow light waveguide behaves like a single mode cavity of arbitrary length.

4.4.3 COMPARISON TO COUPLED CAVITY WAVEGUIDES

It is interesting to compare the above results to coupled cavity waveguides (CCWs), also referred to as coupled resonator optical waveguides (CROWs) elsewhere in this book. Should CCWs be considered as cavities or as slow light waveguides? A single cavity supports a single photonic resonance at a given operating frequency, whereas multiple coupled cavities generate an interference of multiple photonic resonances. Since a slow mode in a photonic crystal waveguide can also be a considered an interference pattern between multiple photonic resonances, there is a clear similarity between CCWs and photonic crystal waveguides. Furthermore, a CCW is described by a transmission band that features a bandwidth and a group velocity, as is a slow light waveguide. However, in terms of the FOM, CCWs typically perform worse than photonic crystal waveguides. The FOM mainly depends on the k-range occupied by the slow light regime, as expressed by the factor q_{SL} introduced above. Since CCWs are characterized by the distance between adjacent cavities that is a multiple of the lattice constant (typically 4–10), the available k-space is reduced by the same factor. For example, in a CCW based on photonic crystal coupled heterostructure cavities [17], the dispersion curve extends from $k = 0.43 \times 2\pi/a$ to $k = 0.50 \times 2\pi/a$ and the useful slow light regime only extends over $\Delta k = 0.04 - 0.05 \times 2\pi/a$, which corresponds to a value of $q_{\text{cav}} < 0.1$; less than 10% of the k-range is useful for slow light operation. Obviously, this depends on the specific design, i.e., the more closely coupled the cavities, the larger k-range they occupy. For comparison, the best photonic crystal waveguides achieve values for q_{cav} in excess of 0.2.

4.4.4 IMPLICATIONS OF THE FOM

In order to understand the implications of this FOM, it is instructive to consider a few examples.

4.4.4.1 Nonlinear Refractive Index

All-optical effects based on nonlinear refractive index changes have been employed in a number of devices, for example, switching in a directional coupler [18] or optical bistability in a photonic crystal cavity [19]. In both cases, switching action is due to the intensity-induced refractive index change, i.e., $n = n_0 + n_2 I$, where n_0 is the linear refractive index, n_2 is the nonlinear index and I is the intensity of light. To actuate the switch, we need a path difference of $\frac{\lambda}{2}$ to build up over the interaction length L. If, for simplicity, we consider the change in material index alone and ignore the phase change argument as illustrated in Figure 4.7 and discussed in Section 4.4.4.2, we get the following comparison. Using the respective intensity enhancements from Equation 4.6 and considering that the refractive index change is $n_2 I$, we obtain

$$n_2 I_0 SL = \frac{\lambda}{2}$$

$$n_2 I_0 \frac{Q}{\pi} L = \frac{\lambda}{2}$$

If we make the interaction length m-times longer,

$$n_2 I_0 SmL = \frac{\lambda}{2}$$

$$n_2 I_0 \frac{Q}{m\pi} mL = \frac{\lambda}{2}$$

By comparing the two sets of equations above, we observe that the cavity does not benefit from an increase in length (assuming that Q is maintained, and so is the bandwidth); a longer cavity has a longer interaction length, but the resonant buildup of intensity goes down, so the two m-factors cancel; in the case of a slow light waveguide, however, a longer device leads to larger effect, because the intensity enhancement does not depend on length. Therefore, we indeed obtain an approximately m-fold stronger nonlinear interaction in a slow light waveguide, which is commensurate with the figure of merit $Fom = m$ derived above.

4.4.4.2 Linear Refractive Index

Using Equation 4.2, we can express the switching length of an interferometric device as

$$L = \frac{\lambda}{2S \Delta n} \tag{4.14}$$

A ring cavity operating on mode order m would be $nL = m\lambda$ long. If we set the two lengths identical for comparison, we obtain

$$m = \frac{n}{2 \Delta n S} \tag{4.15}$$

Let us take the example of the ring-cavity in Ref. [20], which is 12 μm in diameter and operates on a mode order of approximately 75 at $\lambda = 1570$ nm (assuming $n \approx 3$). It requires a refractive index change of $\Delta n = 10^{-4}$ to achieve switching. According to Equation 4.15, a slow light waveguide of approximately $S = 200$ would have the same length and therefore require the same switching energy.

What is the bandwidth of an $S = 200$ slow light waveguide? If we simply extrapolate the result from [11], where $\Delta \lambda = 19$ nm bandwidth was observed for a slowdown factor of $S = 12$, we obtain a value of $\Delta \lambda \approx 1$ nm for a slowdown factor of $S = 200$. In contrasts, the ring cavity in Ref. [20] exhibits a useful bandwidth of approximately $\Delta \lambda \approx 0.01-0.02$ nm, which is 50–100 times smaller than the $\Delta \lambda \approx 1$ nm of the slow light waveguide. Therefore, the reduction in bandwidth is of a similar magnitude as its mode order $m = 75$, which confirms FOM $= m$ derived above.

In practise, one would trade off the switching length/energy against bandwidth, by operating, for example, on $S = 20$, thus making the slow light waveguide longer than the cavity but gaining even more bandwidth in return.

Overall, combining the linear with the nonlinear enhancement, for example, in an all-optical Mach–Zehnder device, gives the potential of reducing the switching power by the square of the slowdown factor, as already pointed out in Refs. [2,3].

4.5 LOSSES

Propagation losses are a serious issue, especially in the slow light regime. All of the benefits and enhancements in switching and nonlinear enhancement discussed above are meaningless if in the course of slowing down, most of the light is lost. Similarly, if the loss is higher for a slow light waveguide than for an equivalent longer conventional waveguide, then the use of slow light in delay lines become less attractive. Let us examine the losses in more detail. First of all, the intrinsic losses are zero, because the waveguide modes are well defined and operate below the light line. The only losses then arise from imperfections, i.e., roughness and other deviations from the ideal structure. These losses are also referred to as extrinsic losses. Since the slow light as described in this chapter requires periodic corrugations, is there a penalty for using photonic crystal waveguides versus photonic wires? One would think that all the corrugations and surfaces present in a photonic crystal offer more opportunity for scattering than the smooth walls of a photonic wire do. In practice, however, this is not the case. While the best losses in wires reported to date are of the order 1–2 dB/cm [21,22], the best values for photonic crystal waveguides are not far behind, which are of the order 2–4 dB/cm [7,23] and may improve further.

However, once the light is slowed down, the problems arise. One of the key papers on the issue [7] suggests that losses scale as the inverse square of the group velocity, so if the light is slowed down by a factor 2, the losses go up by a factor 4. This insight was gained by studying slow light near the band-edge of a W1 waveguide and from a comparison of experimental and theoretical data. If confirmed as a general trend, this result would clearly limit the usefulness of slow light in dielectric structures. On the other hand, some experimental evidence suggests that the story is not as simple. For example, O'Faolain et al. [8] showed that the loss away from the band-edge scales as the inverse square root of the group velocity; so if the light slows down by a factor 2, the loss goes up only by a factor 1.4. Even better, in the dispersion-engineered waveguides of Li et al. [13], no change in transmission was observed for a change in group index from 5 to 30, so the light slowed down by a factor 6 without any obvious loss penalty at all. This remarkable observation is shown in Figure 4.10.

Why the discrepancy? This question is still wide open, but several observations can be made that point to limitations of the model used in Ref. [7]. The main result of the inverse square dependence of the loss on group velocity is explained with the increased density of states (which is also responsible for the enhanced light–matter interaction) in the slow light regime, as well as a backscattering component. The model is based on a perturbation approach, which assumes that the field distribution

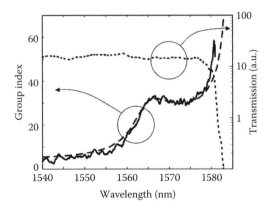

FIGURE 4.10 Transmission (dotted line) and group velocity (solid line = experiment, dashed line = model) versus wavelength for the dispersion-engineered waveguides of Li et al. [13]. There is no obvious reduction in transmission when the light slows down from c/5 to c/30.

is the same in the perfect structure as it is in the disordered one. Small deviations such as roughness and distortions in hole shape or size then act as scattering centers that radiate or backscatter the light.

1. While the perturbation approach seems justified given the small deviations present in experimental structures, with typical roughness of the order of a few nanometres and feature sizes such as hole diameter on a scale of 200–300 nm, the overall effect of the perturbation is far more dramatic than these numbers suggest. For example, the data in Ref. [7] is fitted up to a regime where the transmission is less than 10%. An effect that causes 90% or more of the light to be lost can hardly be seen as a perturbation, so the validity of the perturbative approach is questionable.
2. There have been observations (e.g., Ref. [24]) of localization due to disorder in the slow light regime. Localization requires multiple scattering, which the perturbation approach does not account for.
3. The disorder-studies by O'Faolain [8] make the point that near the band-edge, disorder causes local variations of the lattice that shift the mode cut-off. This shift leads to strong reflections that act as a significant source of loss which then dominate over other types of losses. Again, the perturbation approach does not account for such changes in mode cut-off.
4. The backscattering component, which is seen as the second factor leading to the $1/v_g^2$-dependence, may not become significant until the group velocity is very low indeed, for example, as low as c/50 or c/100, depending on roughness [25]. Therefore, its impact may have been overestimated in Ref. [7].

All these observations suggest that a more comprehensive description of the loss-dependence on group velocity is required, and that a simple fit of the data near the band-edge is not sufficient to describe the interplay between the different effects involved. The fact that there is experimental evidence for a "sweet spot" in the c/20–c/50 range shown in Figure 4.10, however, where the losses are low and have little dependence on group velocity, is very encouraging indeed. As long as one avoids excessively high slowdown factors, one can expect that the type of linear and nonlinear enhancements discussed earlier will be realized experimentally.

4.6 COUPLING

Another essential aspect of operating in the slow light regime is the ability to inject all the light into the structure; otherwise, the intensity enhancement discussed in the context of nonlinear effects cannot be realized. Can complete transfer from the fast to the slow light regime be realized without loss? It is a widely held misconception that group velocity mismatch is the key problem, which is understood in a similar way as the impedance mismatch for microwaves or Fresnel reflections in optics. A simple consideration shows that this cannot be the case; given the enormous slowdown factors achieved in ultracold gases and similar media, it would be impossible to inject light into these media if the difference in group index alone was a problem. On the other hand, there are clearly some issues, as a number of researchers have reported difficulties injecting light into the slow light regime.

1. A key issue in transitions between different types of waveguides, whether fast or slow, is the fact that the phase indices need to be well matched on both sides of the transition; otherwise, Fresnel reflections occur that scale with the phase index difference Δn. The only reason that the aforementioned ultracold gases do not offer any problems in this respect is that their phase index is close to unity, so Fresnel reflections are negligible. In photonic crystal waveguides, light is typically injected from a total internal reflection mode of similar phase index as the slow light mode, so phase matching does not tend to be the main issue either.

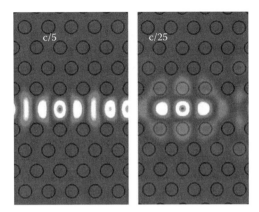

FIGURE 4.11 Mode shape of (a) fast (c/5) and (b) slow (c/25) photonic crystal waveguide modes. The difference in mode shape is part of the reason for the group velocity-dependent coupling coefficient. The characteristic envelope of the slow mode highlighted in Figure 4.3 is also recognized.

2. The next issue is mode overlap, which arises from the physical size difference between the modes in different waveguides; for example, this is the biggest problem when coupling light from an optical fiber into a nanophotonic waveguide. Coupling light into a slow mode raises this issue: As shown in Figure 4.11, a slow mode has a different shape than a fast mode, so the coupling efficiency becomes a function of group velocity. This can, at least in part, be addressed by adjusting the width of the transition and the position of the photonic crystal with respect to the termination of the waveguide, as shown by Vlasov et al. [26].
3. The main point about coupling into the slow light mode, however, is the recognition that slow light is an interference pattern between forward and backward propagating components. As shown in Figure 4.3, it is this superposition that gives rise to the characteristic envelope of the interference pattern that we refer to as the slow mode. This phenomenon intrinsically requires a photonic crystal to be present on both sides of the mode. At an interface, this is clearly not the case, as the photonic crystal is present only on one side of the mode, and the mode consists of only a forward propagating component. The slow mode interference pattern can therefore not establish immediately at the interface and it requires a transition region to build up. The most obvious solution for such a transition region is a taper that gradually increases the group index, allowing the interference pattern to establish slowly. Such a taper can be produced by slowly varying the lattice constant or the waveguide width [27,28]. A less obvious solution is the design of a multilayer type interface that combines a number of layers of different lattice constant [29]. The latter solution, however, offers the shortest transition region and can be designed to achieve unity injection for any desired group index.

Overall, it is clear that there is no intrinsic limitation to the injection efficiency, and that a number of solutions to achieve efficiencies up to 100% already exist.

4.7 CONCLUSIONS

Photonic crystal line defect waveguides provide a powerful platform for the study and exploitation of slow light effects. They belong to the class of slow light devices based on photonic resonances and share this class with individual cavities and coupled cavity waveguides. In comparison to slow light devices based on atomic resonances such as EIT, they offer a larger bandwidth of operation and more flexibility in terms of operating wavelength, which depends on the dimensions of the structure alone, mainly the lattice constant, but also the hole size and waveguide width; these parameters, when combined effectively, offer a surprising number of permutations that allow one to tailor both the slow light properties, such as bandwidth and slowdown factor, and the target wavelength, which is limited only by the transparency window of the host material. Loss and injection issues are still areas of active research, although it has now been shown that low losses can be achieved for moderate group indices in the $n_g = 20$–50 regime, and that light can be injected efficiently into such structures.

The disadvantage of the photonic crystal waveguide approach is that tuning of the slowdown factor remains a challenge because it does not come as naturally as it does to EIT-type systems, or to coupled cavities; photonic crystal line defect waveguides are therefore more suited for static delay lines, or the enhancement of optically linear and nonlinear effects.

REFERENCES

1. S. J. Madden, D.-Y. Choi, M. R. E. Lamont, V. G. Ta'eed, N. J. Baker, M. D. Pelusi, B. Luther-Davies, and B. J. Eggleton. Chalcogenide glass photonic chips. *Opt. Photon. News*, 19(1):18–23, 2008.
2. M. Soljačić, S. G. Johnson, S. Fan, M. Ibanescu, E. Ippen, and J. D. Joannopoulos. Photonic-crystal slow-light enhancement of nonlinear phase sensitivity. *J. Opt. Soc. Am. B*, 19(9):2052–2059, 2002.
3. T. F. Krauss. Slow light in photonic crystal waveguides. *J. Phys. D Appl. Phys.*, 40(9):2666–2670, 2007.
4. J. B. Khurgin. Optical buffers based on slow light in electromagnetically induced transparent media and coupled resonator structures: Comparative analysis. *J. Opt. Soc. Am. B*, 22(5):1062–1074, 2005.
5. B.-S. Song, S. Noda, T. Asano, and Y. Akahane. Ultra-high-q photonic double-heterostructure nanocavity. *Nat. Mater.*, 4:546–549, 2005.
6. R. J. P. Engelen, Y. Sugimoto, Y. Watanabe, J. P. Korterik, N. Ikeda, N. F. van Hulst, K. Asakawa, and L. Kuipers. The effect of higher-order dispersion on slow light propagation in photonic crystal waveguides. *Opt. Express*, 14(4):1658–1672, 2006.
7. E. Kuramochi, M. Notomi, S. Hughes, A. Shinya, T. Watanabe, and L. Ramunno. Disorder-induced scattering loss of line-defect waveguides in photonic crystal slabs. *Phys. Rev. B (Condens. Matter Mater. Phys.)*, 72(16):161318, 2005.
8. L. O'Faolain, T. P. White, D. O'Brien, X. Yuan, M. D. Settle, and T. F. Krauss. Dependence of extrinsic loss on group velocity in photonic crystal waveguides. *Opt. Express*, 15(20):13129–13138, 2007.
9. D. Mori, S. Kubo, H. Sasaki, and T. Baba. Experimental demonstration of wideband dispersion-compensated slow light by a chirped photonic crystal directional coupler. *Opt. Express*, 15(9):5264–5270, 2007.
10. A. Yu. Petrov and M. Eich. Zero dispersion at small group velocities in photonic crystal waveguides. *Appl. Phys. Lett.*, 85(21):4866–4868, 2004.
11. M. D. Settle, R. J. P. Engelen, M. Salib, A. Michaeli, L. Kuipers, and T. F. Krauss. Flatband slow light in photonic crystals featuring spatial pulse compression and terahertz bandwidth. *Opt. Express*, 15(1):219–226, 2007.
12. L. H. Frandsen, A. V. Lavrinenko, J. Fage-Pedersen, and P. I. Borel. Photonic crystal waveguides with semi-slow light and tailored dispersion properties. *Opt. Express*, 14(20):9444–9450, 2006.
13. J. Li, T. P. White, L. O'Faolain, A. Gomez-Iglesias, and T. F. Krauss. Systematic design of flat band slow light in photonic crystal waveguides. *Opt. Express*, 16(9):6227–6232, 2008.
14. M. Notomi, K. Yamada, A. Shinya, J. Takahashi, C. Takahashi, and I. Yokohama. Extremely large group-velocity dispersion of line-defect waveguides in photonic crystal slabs. *Phys. Rev. Lett.*, 87(25):253902, 2001.
15. Y. A. Vlasov, M. O'Boyle, H. F. Hamann, and S. J. McNab. Active control of slow light on a chip with photonic crystal waveguides. *Nature*, 438(7064):65–69, 2005.
16. D. M. Beggs, T. P. White, L. O'Faolain, and T. F. Krauss. Ultracompact and low-power optical switch based on silicon photonic crystals. *Opt. Lett.*, 33(2):147–149, 2008.
17. D. O'Brien, M. D. Settle, T. Karle, A. Michaeli, M. Salib, and T. F. Krauss. Coupled photonic crystal heterostructure nanocavities. *Opt. Express*, 15(3):1228–1233, 2007.
18. A. Villeneuve, C. C. Yang, P. G. J. Wigley, G. I. Stegeman, J. S. Aitchison, and C. N. Ironside. Ultrafast all-optical switching in semiconductor nonlinear directional couplers at half the band gap. *Appl. Phys. Lett.*, 61(2):147–149, 1992.
19. M. Notomi, A. Shinya, S. Mitsugi, G. Kira, E. Kuramochi, and T. Tanabe. Optical bistable switching action of si high-q photonic-crystal nanocavities. *Opt. Express*, 13(7):2678–2687, 2005.

20. Q. Xu, B. Schmidt, S. Pradhan, and M. Lipson. Micrometre-scale silicon electro-optic modulator. *Nature*, 435:325–327, 2005.
21. M. Gnan, S. Thoms, D. S. Macintyre, R. M. De La Rue, and M. Sorel. Fabrication of low-loss photonic wires in silicon-on-insulator using hydrogen silsesquioxane electron-beam resist. *Electron. Lett.*, 44(2):115–116, 2008.
22. F. Xia, L. Sekaric, and Y. Vlasov. Ultracompact optical buffers on a silicon chip. *Nat. Photon.*, 1(1):65–71, 2007.
23. L. O'Faolain, X. Yuan, D. McIntyre, S. Thoms, H. Chong, R. M. De La Rue, and T. F. Krauss. Low-loss propagation in photonic crystal waveguides. *Electron. Lett.*, 42(25):1454–1455, 2006.
24. J. Topolancik, B. Ilic, and F. Vollmer. Experimental observation of strong photon localization in disordered photonic crystal waveguides. *Phys. Rev. Lett.*, 99(25):253901, 2007.
25. L. C. Andreani and D. Gerace. Light-matter interaction in photonic crystal slabs. *phys. Stat. Solidi (b)*, 244(10):3528–3539, 2007.
26. Y. A. Vlasov and S. J. McNab. Coupling into the slow light mode in slab-type photonic crystal waveguides. *Opt. Lett.*, 31(1):50–52, 2006.
27. S. G. Johnson, P. Bienstman, M. A. Skorobogatiy, M. Ibanescu, E. Lidorikis, and J. D. Joannopoulos. Adiabatic theorem and continuous coupled-mode theory for efficient taper transitions in photonic crystals. *Phys. Rev. E*, 66(6):066608, 2002.
28. P. Pottier, M. Gnan, and R. M. De La Rue. Efficient coupling into slow-light photonic crystal channel guides using photonic crystal tapers. *Opt. Express*, 15(11):6569–6575, 2007.
29. J. P. Hugonin, P. Lalanne, T. P. White, and T. F. Krauss. Coupling into slow-mode photonic crystal waveguides. *Opt. Lett.*, 32(18):2638–2640, 2007.

Part II

Slow Light in Periodic Photonic Structures

5 Periodic Coupled Resonator Structures

Joyce K.S. Poon, Philip Chak, John E. Sipe, and Amnon Yariv

CONTENTS

5.1 Introduction 79
5.2 General Description 80
    5.2.1 CROW Dispersion Relation 80
    5.2.2 SCISSOR Dispersion Relation 81
    5.2.3 Comparison between CROWs and SCISSORs 83
5.3 Standing-Wave Resonators 84
    5.3.1 FP CROWs...84
    5.3.2 Two-Channel FP SCISSORs 87
5.4 Some Practical Considerations 88
    5.4.1 Finite-Size Effects 88
    5.4.2 Delay, Bandwidth, and Loss 90
5.5 Experimental Progress 92
    5.5.1 Passive Microring CROWs...92
    5.5.2 CROWs with Optical Gain 96
5.6 Conclusions..96
Acknowledgments 96
References 97

5.1 INTRODUCTION

Because of their compact, essentially chip-scale sizes, optical microresonators have been attracting considerable theoretical and experimental attention. They have applications in fields ranging from fundamental physics to telecommunications systems [1]. As optical microresonators have the capability to store light in physically small volumes, we envision that periodic coupled microresonator structures may provide a new method for controlling the group velocity of optical pulses in a compact way on a chip. To this end, two main types of structures have been proposed: coupled-resonator optical waveguides (CROWs) [2,3] and side-coupled integrated sequence of spaced optical resonators (SCISSORs) [4,5]. A CROW is a chain of resonators in which light propagates by virtue of the direct coupling between the adjacent resonators (Figure 5.1a). In contrast, a SCISSOR consists of a chain of resonators that are not directly coupled to each other, but are coupled through at least one side-coupled waveguide (Figure 5.1b). Both CROWs and SCISSORs have the potential to significantly slowdown the propagation of light.

Coupled optical resonators have already become important in nonlinear optics research as well as in telecommunication applications in recent years [4,6–8]. Systems consisting of a few coupled resonators, say $1 < N < 5$, have been proposed and demonstrated for optical filtering and modulation

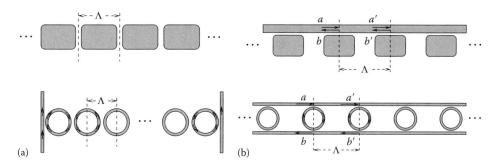

FIGURE 5.1 Schematics of a (a) CROW and (b) SCISSOR with a period of Λ. The top halves of (a) and (b) show generic implementations of a CROW and SCISSOR, respectively (the grey boxes represent generic resonators), while the bottom halves show specific implementations using ring resonators. In a CROW, the resonators are directly coupled to each other, while in a SCISSOR the coupling is mediated through waveguides. The solid arrows indicate the direction of field propagation in the waveguides.

[9–11]. CROWs and SCISSORs are large systems at the other extreme, with say $N > 10$ resonators, and can be regarded as waveguides with unique and controllable dispersion properties [2,3,6,12,13].

In this chapter, we will provide a brief overview of CROWs and SCISSORs. Interested readers should explore the references for more comprehensive details on a particular topic. For the SCISSORs, we will focus on structures where there is feedback between each unit cell as shown in Figure 5.1b since they possess more interesting dispersion relations. We shall begin with theoretical descriptions of CROWs and SCISSORs followed by a comparison between the two structures. We will then describe some interesting propagation effects in CROWs and SCISSORs consisting of single-mode standing-wave resonators. This will be followed by a discussion of some of the practical issues and trade-offs in using periodic coupled resonators to slow light. Finally, we will end with a brief review of the experimental progress in this area.

5.2 GENERAL DESCRIPTION

In this section, we will introduce the essential features of the dispersion relations of CROWs and SCISSORs. The concepts of CROWs and SCISSORs are general and applicable to many different types of resonators; thus, they can be analyzed using different types of formalisms. For this section, we have chosen to describe CROWs and SCISSORs using formalisms that are physically intuitive and yet allow us to describe the dispersive properties as generally as possible.

5.2.1 CROW DISPERSION RELATION

The dispersion relation of CROWs can be derived using a tight-binding formalism [2], transfer matrices [14], or temporal coupled-mode theory [15]. The three methods of analysis produce an identical form of the dispersion relation in the limit of weak coupling [16]. Here, we will briefly describe how the dispersion relation is derived in the tight-binding method which is applicable to arbitrary types of resonators.

In the tight-binding method, we approximate the electric field of an eigenmode $\mathbf{E}_K$ of the CROW at a frequency, ω_K, as a Bloch wave superposition of the individual resonator modes $\mathbf{E}_{\omega_0}$ [2],

$$\mathbf{E}_K(\mathbf{r}, t) = \exp(i\omega_K t) \sum_N \exp(-iNK\Lambda) \mathbf{E}_{\omega_0}(\mathbf{r} - N\Lambda\hat{\mathbf{z}}), \qquad (5.1)$$

where

the Nth resonator in the chain is centered at $z = N\Lambda$
Λ is the period
K is the Bloch wavenumber
ω_0 is the resonance frequency of an individual resonator

Substituting Equation 5.1 in Maxwell's equations and after some algebra, with the assumption symmetric to nearest neighbor coupling, we arrive at the dispersion relation of a CROW [2]:

$$\omega_K = \omega_0 \left[1 - \frac{\Delta \alpha}{2} + \kappa \cos(K\Lambda) \right], \quad (5.2)$$

where ω_0 is the resonant frequency of an individual resonator and $\Delta\alpha$ and κ are defined as

$$\Delta\alpha = \int d^3 \mathbf{r} [\epsilon(\mathbf{r}) - \epsilon_0(\mathbf{r})] \mathbf{E}_{\omega_0}(\mathbf{r}) \cdot \mathbf{E}_{\omega_0}(\mathbf{r}), \quad (5.3a)$$

$$\kappa = \int d^3 \mathbf{r} [\epsilon_0(\mathbf{r} - \Lambda\hat{\mathbf{z}}) - \epsilon(\mathbf{r} - \Lambda\hat{\mathbf{z}})] \mathbf{E}_{\omega_0}(\mathbf{r}) \cdot \mathbf{E}_{\omega_0}(\mathbf{r} - \Lambda\hat{\mathbf{z}}), \quad (5.3b)$$

where $\epsilon(\mathbf{r})$ is the dielectric coefficient of the CROW and $\epsilon_0(\mathbf{r})$ is the dielectric coefficient of an individual resonator. Therefore, the coupling parameter κ represents the overlap of the modes of two neighboring resonators and $\Delta\alpha/2$ gives the fractional self frequency shift centered at ω_0.

As evidenced by Equation 5.2, waveguiding frequency bands centered at the resonance frequencies, less the self-coupling offset, of the individual resonators arise in a CROW through the coupling between the resonators. These bands are separated in frequency by the free spectral range of the resonators. In the bandgap, an optical wave is not resonant with the structure and thus decays in the CROW. Figure 5.2 shows examples of the CROW dispersion relationship and group velocity for two values of κ. Because the group velocity, v_g, is given by $d\omega/dK$, to achieve slow light propagation, the inter-resonator coupling must be weak to attain a relatively flat propagation band in the dispersion relation.

5.2.2 SCISSOR DISPERSION RELATION

To determine the dispersion relation of a SCISSOR, transfer matrices, scattering matrices, or a generalized Hamiltonian approach can be used [17–19]. With the transfer matrices, we use the

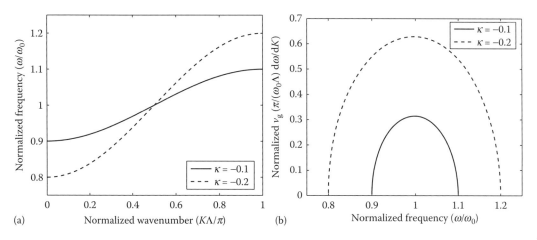

FIGURE 5.2 (a) Dispersion relation and (b) the normalized group velocity of CROWs for various values of κ and $\Delta\alpha = 0$. The normalized group velocity is defined as $\pi/(\omega_0\Lambda)d\omega/dK$.

transmission and reflection coefficients of a unit cell of the SCISSOR to propagate the fields from one unit cell to the next. The Bloch modes and dispersion relation can then be derived by imposing translation symmetry in the direction of propagation.

For a unit cell shown in Figure 5.1b, if the forward propagating and backward propagating waveguide modes are coupled to a lossless resonator mode, and the unit cell possesses mirror reflection symmetry along z, the frequency dependent transfer matrix, $T(\omega)$, for the field amplitudes labeled in Figure 5.1b is of the form [17]

$$\begin{bmatrix} a' \\ b' \end{bmatrix} = T(\omega) \begin{bmatrix} a \\ b \end{bmatrix}, \tag{5.4a}$$

$$T(\omega) = \begin{bmatrix} \exp(-i\beta\Lambda) & 0 \\ 0 & \exp(i\beta\Lambda) \end{bmatrix} \begin{bmatrix} \frac{T^2 - R^2}{T} & \frac{R}{T} \\ -\frac{R}{T} & \frac{1}{T} \end{bmatrix}, \tag{5.4b}$$

where
Λ is the period of the structure
$\beta = \omega n/c$ is the propagation constant of the waveguide
ω is the optical frequency
n is the effective index of the waveguide
c is the speed of light

We have assumed that the coupling into and out of the resonator occurs only at the point of minimum separation between the resonator and the waveguide. T and R are the transmission and reflection coefficients of the resonator. For frequencies that are in close vicinity of a resonance frequency, ω_0, T and R can be written as [19]

$$T(\omega) \approx \frac{-i\delta}{\gamma - i\delta}, \tag{5.5a}$$

$$R(\omega) \approx (-1)^q \frac{\gamma}{\gamma - i\delta}, \tag{5.5b}$$

where
$\gamma = 2\pi n \kappa_{sc}/c$
q is an integer that is determined by the symmetry of the mode, and is odd (even) for an even (odd) resonator mode
κ_{sc} is the coupling coefficient between the resonators and the waveguide at ω_0
$\delta = \omega - \omega_0(1 - \Delta\alpha')$ is the frequency detuning from the resonance frequency including the shift due to the self-coupling, $\omega_0 \Delta\alpha'$

From Equation 5.5, we see that a single resonator exhibits 100% reflection and 0% transmission when the input optical frequency is matched to the resonance frequency.

The modes and dispersion relation of an infinitely long SCISSOR can thus be found by applying the Bloch boundary conditions to Equation 5.4 and solving for the eigenvalues and eigenmodes. After some algebra, we arrive at the following dispersion relation of the SCISSOR:

$$K(\omega) = -\frac{i}{\Lambda} \ln \left\{ \left[(T_{11}(\omega) + T_{22}(\omega)) \pm \sqrt{(T_{11}(\omega) + T_{22}(\omega))^2 - 4} \right] / 2 \right\}, \tag{5.6}$$

where $T_{11}(\omega)$ and $T_{22}(\omega)$ are the appropriate elements of the matrix $T(\omega)$.

A typical dispersion relation of a SCISSOR structure is depicted in Figure 5.3. The dispersion relation exhibits two types of band gaps—the "resonator gap" that is indirect and the "Bragg gap" that is direct. By a direct gap, we mean that there is no discontinuity in the Bloch wavenumber

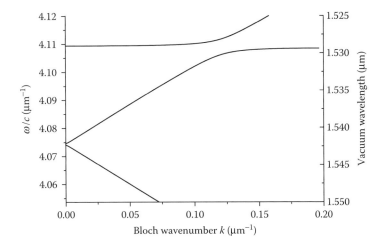

FIGURE 5.3 Typical SCISSOR dispersion relation. The parameters for an implementation consisting of ring resonators side-coupled to two waveguides are $\Lambda = 16$ μm, $n = 3.47$, a ring circumference of 26 μm, and a length-integrated coupling coefficient of $\kappa_{sc} = 0.20$. The resonator gap is near $\omega/c = 4.11$ μm^{-1} and the Bragg gap is near $\omega/c = 4.075$ μm^{-1}.

between the lower and upper band-edge frequencies. For an indirect gap, the wavevectors between the upper and lower band-edge frequencies differ by π/Λ.

The two types of bandgaps in the dispersion relation stem from two different reflection mechanisms in SCISSOR structures. For a frequency close to the Bragg frequency, the additional phase shift induced by successive unit cells of the system is an integer multiple of 2π, and a weak coupling can be enhanced via a Bragg-type process of constructive interference of reflections. This is analogous to the Bragg reflection in a 1D distributed feedback (DFB) grating structure, and leads to a gap in the dispersion relation that has the same character as a Bragg gap in DFB gratings. If the frequency of the light is close to the resonance frequency of the resonator, then the effective coupling between the two channels can become quite large, and a gap also arises. This gap is the called the "resonator gap." The differences between the two reflection mechanisms (Bragg reflection and resonant reflection) has important implications for the sorts of pulse propagation equations that can be associated with the two types of bandgaps.

5.2.3 COMPARISON BETWEEN CROWS AND SCISSORS

CROWs and SCISSORs are, in many ways, complimentary structures. In CROWs, light propagates with a purely real wavenumber at the resonance frequency of the resonators. The bandgap frequencies are not resonant with the structure. In contrast, an optical wave in a SCISSOR is reflected at the resonance frequency of the resonators and also at the Bragg frequency of the structure. On a more fundamental level, CROWs and SCISSORs differ by the number of modes in each unit cell and the way they are coupled together. The simplest CROW consists of unit cells each containing a single-mode resonator, such that each unit cell comprises one mode. For SCISSORs, the propagation bands arise from the coupling of at least three modes in each unit cell: one forward- and one backward-propagating waveguide mode which are coupled together through the resonator mode.

For slow light, we are interested in frequencies where the dispersion relation is flat. For both CROWs and SCISSORs, near the band-edge frequencies, the group velocity, v_g, approaches zero. However, the group velocity dispersion (GVD) should also be minimum $\left(\frac{\partial^2 K}{\partial \omega^2} \approx 0\right)$ to reduce pulse distortion [20]. In a CROW, as can be inferred from Figure 5.2 and Equation 5.2, the zero GVD region is at the band-center, $K\Lambda \approx \pi/2$, where the group velocity is maximum, $v_{g,\max} = |\kappa|\omega_0\Lambda$. Thus, to achieve slow light in CROWs, $|\kappa|$ should be kept weak and Λ should be kept short.

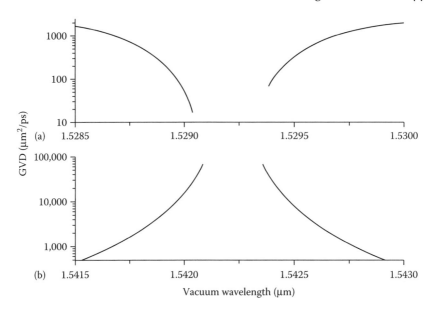

FIGURE 5.4 GVD as a function of the vacuum wavelength in the vicinity of (a) a resonator gap and (b) a Bragg gap for the system considered in Figure 5.3. The GVD is of the orders of magnitude smaller in the vicinity of a resonator gap.

For a SCISSOR, the propagation bands in the vicinities of the two types of bandgaps exhibit different GVD properties. The curvature of the dispersion relation, hence the GVD, is much smaller near a resonator gap than a Bragg gap (see, for example, Figure 5.4). The indirect nature of the resonator gap leads to a more linear dispersion relation near the resonator gap [21]. However, zero GVD is not possible at the band-edges due to the nonzero band-curvatures, so band-edge frequencies are not always desirable for linear pulse propagation. Thus, a more useful regime for slow light propagation is the flat intermediate band that results when a Bragg gap and resonator gap are tuned close to each other as in Figure 5.3. This intermediate band can support very small group velocities and can also possess very small GVD for a wide range of wavenumbers within the first Brillouin zone.

Both CROWs and SCISSORs are capable of slowing light, though the fundamental mechanisms by which the slowing is achieved are different in the two devices. Furthermore, the slow wave mechanisms and possible geometries of CROWs and SCISSORs can be combined. For example, CROWs can be side-coupled to waveguides and/or other resonators [22,23]. Combining the basic principles that underlie these structures add further complexity and richness to the types of transmission and dispersion possible in periodic coupled resonators systems.

5.3 STANDING-WAVE RESONATORS

In this section, we will describe some interesting propagation effects present in CROWs and SCISSORs based on Fabry–Perot (FP) resonators. Because FP resonators are of the standing-wave variety and support only one mode at each resonance frequency, they can exhibit considerably different properties compared to travelling wave resonators such as microrings [23,24].

5.3.1 FP CROWs

To achieve significant slowing in CROWs, and indeed in any medium, the optical delay should be achieved over as short a device length as possible in the direction of propagation. An array of evanescently coupled FP resonators may be a solution as a low index contrast slow light structure.

Periodic Coupled Resonator Structures

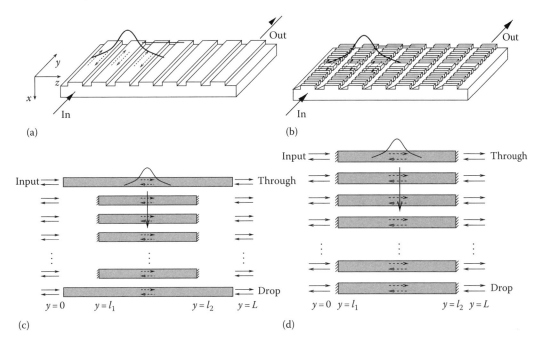

FIGURE 5.5 Schematic of (a) waveguide laser and (b) DFB laser arrays in a planar geometry as implementations of CROWs. The input/output signal can be (c) side-coupled or (d) end-coupled with respect to the array. The slanted lines represent reflectors that define each resonator. The arrows parallel to the y-axis in (c) and (d) indicate the directions of field propagation. The thicker solid arrows along z show the direction of pulse propagation.

Despite the low index contrast, a high slowing factor is obtained by decoupling the length of the device in the propagation direction from the size of the resonators. Certain implementations of these CROWs are depicted in Figure 5.5a and b.

A large slowing factor is possible, because along z, the direction of propagation, the period of the device can be short, say about 5 μm for evanescently coupled single-mode waveguides. This periodicity is similar to what is achievable in high-index contrast photonic crystal (PC), ring or disk resonators. In the y direction, propagating optical waves are resonant with the cavities. Moreover, optical gain and electronic control can be readily incorporated into the coupled waveguide array, by leveraging diode laser array techniques [25,26]. An optical signal can couple into the first array element in a side-coupled or end-coupled configuration as in Figure 5.5c and d. The output can then be out-coupled in a similar manner out of the last element of the array.

The dispersion relation of these FP-CROWs can be derived from the coupled-mode theory of conventional waveguide arrays [26] through the inclusion of the appropriate boundary conditions to describe the resonances [24], and the transmission amplitude can be calculated for a general excitation using transfer matrices [24]. From coupled-mode theory, we find the dispersion relation of an FP-CROW is

$$\omega(K) = \omega_0 - \frac{M_l c}{n} - 2\frac{\kappa_l c}{n}\cos(K\Lambda), \tag{5.7}$$

where we have assumed $n(\omega) = n(\omega_0) = n$ is a constant, and κ_l and M_l are the per unit length nearest neighbor coupling and self-coupling coefficients, respectively [26]. Since the bandwidth of a CROW is not expected to be large ($\omega/\omega_0 \ll 1$), the coupling coefficients can be assumed to be constant.

The dispersion relation described by Equation 5.7 is of the same form as the tight-binding approximate and transfer matrix results [14]. The key difference between FP and ring resonators is

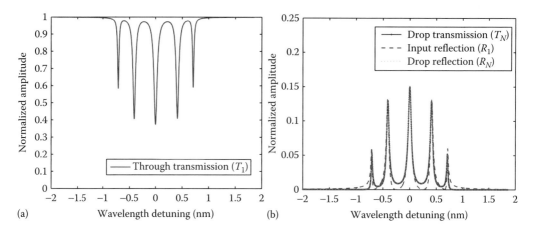

FIGURE 5.6 (a) Transmission spectrum at the through port and (b) the transmission and reflection spectra at the input and drop ports for the side-coupled array. The calculation parameters are described in the text.

that only two K vectors correspond to a particular eigen frequency for the FP resonators while there are four K vectors for the rings. This is because a ring resonator supports two degenerate modes on resonance, while an FP resonator supports only one mode on resonance.

The slowing factor at the band-center, given by the ratio of the speed of light to the maximum group velocity in the CROW, is $S = \frac{c}{v_g|_{max}} = \frac{n}{2\kappa_l \Lambda}$. Unlike coupled grating defects or ring resonators, the period Λ of the CROW is not dependent on L, the length of the resonators in the y direction. Since, for weakly coupled single-mode waveguides $\kappa_l \approx 10^{-4} - 10^{-3}$ μm^{-1} and Λ can be ~5 μm even for modest index contrast ($\Delta n/n \approx 10^{-3} - 10^{-2}$), large slowing factors of the order of a few hundred to a thousand are possible.

To determine the transmission spectra of FP-CROWs, we can describe the propagation of light $2N \times 2N$ a transfer matrix. The matrix acts on a column vector that includes the forward- and backward-propagating fields at the input side of each FP resonator. By defining the matrices for the input-side reflectors, propagation and coupling, and output-side reflectors of the structure as in Figure 5.5c and d, we can derive the transfer matrix for the system as a whole which propagates the fields from $y = 0$ to $y = L$.

The transfer matrices can account for an arbitrary input field at $y = 0$ and can be used to calculate the reflection and transmission coefficients of any resonator. However, in most cases, we are primarily interested in exciting the first element and the transmission and reflection coefficients in the first and last elements only. The transmission and reflection spectra for a CROW with five resonators with side-coupled input and output waveguides are shown in Figure 5.6. The reflectors in the calculations consist of Bragg gratings with alternating layers of thicknesses $d_H = 119$ nm and $d_L = 123$ nm, with effective indices $n_H = 3.25$ and $n_L = 3.15$, respectively. The gratings are 24 μm or 100 periods long. The waveguide sections have an effective index of 3.25 and are 50 μm long. The coupling constant is $\kappa_l = 4 \times 10^{-3}$ μm^{-1}.

By design, the standing-wave cavities support a resonance mode at a free-space wavelength of 1.551 μm. Near the resonance frequency ω_0, the transmission across cavities is increased. In contrast to CROWs consisting of travelling wave resonators (e.g., ring/disk resonators), the maximum transmission in the present situation is 25% rather than unity. This is because the standing-wave cavity has no degenerate modes at ω_0, and the fields in the cavity can decay into the two waveguides in both the forward and backward directions [27]. This limitation can be overcome by combining the output ports together, using asymmetric mirror reflectivities, or by introducing optical gain.

5.3.2 Two-Channel FP SCISSORs

SCISSORs can also consist of FP resonators as shown in Figure 5.7. Bragg gratings can be used as reflectors at the ends of each FP resonator. The unit cell of this periodic structure consists of an FP resonator side-coupled to two waveguides. This SCISSOR geometry differs from the one described in Section 5.2.2 since the single-mode cavity is coupled to two waveguides. In general, the FP resonators in the middle waveguide can be replaced with other single-mode cavities, as long as the resonators support only one mode on resonance. These FP-SCISSORs support "bright" and "dark" states accessible by tuning the relative phase of the input fields [23]. In analogy with electromagnetically induced transparency (EIT) [28], here a dark state is one in which the resonators are transparent to the optical signal even on resonance. These dark and bright states can be used to construct switchable delay lines.

To analyze the coupled waveguide structures in Figure 5.7, we can again use transfer matrices that propagate the incoming and outgoing fields along z from one unit cell to the next [23]. These matrices can be made more tractable by accounting for nearest waveguide coupling only and neglecting the coupling between the outer waveguides and the gratings. Including more general coupling between waveguides and cavities in the grating region will lead to quantitative, but not qualitative change in our results.

An example of the dispersion relation of an FP-SCISSOR is depicted in Figure 5.8. For the calculations, the gratings consist of alternating layers satisfying the quarter-wavelength condition

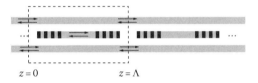

FIGURE 5.7 Schematic of standing-wave resonators side-coupled to two waveguides. The light grey regions represent the waveguiding sections and the dark regions represent the high index regions in the reflectors. The arrows indicate the direction of propagation of the optical field in each waveguide.

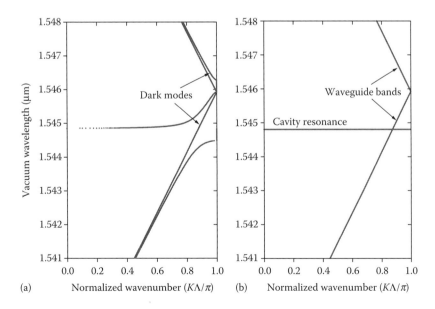

FIGURE 5.8 Dispersion relation of a two-channel FP-SCISSOR in the limit of (a) weak waveguide–cavity coupling and (b) no waveguide–cavity coupling.

at 1.55 μm for indices of 3.25 and 2.25. The cavities are 12 μm long with an index of 3.25 and a waveguide coupling strength of 8.3×10^{-3} μm^{-1}. One hundred grating periods separate successive cavities. In the figure, only the dispersion relations of non evanescent Bloch modes with real wavenumbers are shown.

The eigenmodes of the structures can be understood in terms of two different sets of modes—the dark modes that have a constant group velocity and the bright modes. The dark modes propagate unaffected by the cavities at the phase velocity of the waveguides and are represented by the linear dispersion curves. Microring SCISSORs and CROWs do not support the dark modes [14,21]. The bright mode dispersion relation is qualitatively similar to the one described in Section 5.2.2, with two types of bandgaps [21]. For the dark modes, the resonators and their feedback are transparent to the light. This effect can be understood in a similar manner as dark states in EIT and gratings [29,30]. In the dark modes, the light in the top and bottom waveguides destructively interfere such that no light is coupled into the resonator. The light in the top and bottom channels are in phase for the bright modes, but π out of phase for the dark modes. In both cases, the light in the top and bottom waveguides are equal in amplitude.

To illustrate this effect more clearly, we can adopt a simple mode-coupling model of the two waveguides and the resonators. Let A, B, C represent the envelope for the forward-going component of the wave in the top waveguide, the FP resonator, and the bottom waveguide respectively, we can write the coupled-mode equations as

$$\frac{dA}{dz} = -i\kappa B, \quad \frac{dB}{dz} = -i\kappa A - i\kappa C, \quad \frac{dC}{dz} = -i\kappa B, \tag{5.8}$$

where κ is the per-length coupling coefficient describing the waveguide–cavity coupling. The propagation constants of the eigenmodes described by Equation 5.8 are $\beta = 0, \pm\sqrt{2}\kappa$. The solution $\beta = 0$ implies that the field amplitudes are independent of z and κ. Its corresponding eigenvector is $1/\sqrt{2}\,[1\ 0\ -1]^T$, precisely representing the dark mode in which no light propagates in the resonators because of destructive interference.

Dark and bright states also exist in more general PC structures. The inset in Figure 5.9a shows a PC-SCISSOR. The waveguides and cavities of the structures are formed by line and point defects, respectively. Dispersion relations in close vicinity of the normalized resonance frequency, $\widetilde{\omega} = 0.396$, is shown in Figure 5.9a, and the corresponding field distributions are depicted in Figure 5.9b–e. The calculated fields show that the PC structures can exhibit the same characteristics as the FP-SCISSOR structures considered in Figure 5.8.

Bright and dark modes can be used to generate slow light with a switchable delay. Since these modes are distinguishable by the relative phase between the top and bottom waveguides, the delay through these devices can be adjusted only by altering the relative phase of the two input beams. By switching the relative phase between the input beams, the delay can be changed by several orders of magnitude [23].

5.4 SOME PRACTICAL CONSIDERATIONS

After describing some basic properties and propagation phenomena in the previous sections, we will now address several design issues in slowing light and making delay lines with coupled resonator structures. In particular, we will discuss the finite-size effects on the transmission of coupled resonators and the trade-offs among delay, loss, and bandwidth.

5.4.1 FINITE-SIZE EFFECTS

Finite-size effects refer to spectral features that arise when a structure is not infinitely periodic. In principle, for infinitely long CROWs and SCISSORs, the passbands in the transmission spectrum are flat as a function of wavelength. However, even in the ideal case of identical resonators, ripples

Periodic Coupled Resonator Structures

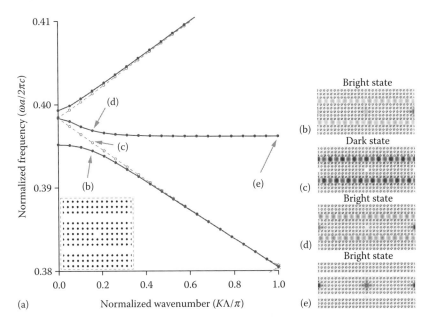

FIGURE 5.9 (a) TM dispersion of a PC SCISSOR. (b)–(e) show the fields labelled in (a). (inset) A unit cell of the PC-SCISSOR. The dark rods have an index of $n = 3.0$ and a radius of $0.2a$, where a is the lattice constant for the square lattice. The background material has $n = 1.0$.

in the transmission spectrum arise in finite resonator chains, and their depths depend on the coupling coefficients and other structural parameters. These ripples in the transmission spectrum can distort a propagating pulse and limit the bandwidths of the devices. There are filter design techniques that optimize the flatness of a transmission band by apodizing the coupling constants along a resonator chain [31–33]. Here, we shall briefly describe a more intuitive way to understand the origin of these ripples, allowing us to arrive at closed-form analytical expressions that minimize the finite length effects at and around a reference frequency, ω_{ref} [34].

As an example, let us consider a microring CROW depicted in Figure 5.10a. A typical transmission spectrum, for a finite structure consisting of $N = 10$ resonators, is shown in Figure 5.10b. For N resonators, there are N discrete transmission peaks with unity transmission. These transmission peaks arise from the resonance condition in the direction of periodicity due to the finite-size of the structure and can thus be thought of as FP fringes. This interpretation can be rigorously proven through the manipulation of transfer matrices [34,35]. Thus, to suppress the fringes at a reference frequency, ω_{ref}, we need only to slightly modify the physical parameters of the resonators at either end of the device such that the reflections cancel out the FP resonances, providing an effective antireflection (AR) coating for the structure. Practically, this AR structure can be accomplished by modifying the circumference and/or the coupling constant of the first and last ring resonators.

If both the circumference and coupling constant of the first and last rings are modified, we can find an exact expression for the ring radius and coupling coefficient for the AR structure. Defining $(L + L')/2$ as the circumference of the AR (or first/last) ring, $\theta' = \omega n L'/4c$, κ_{wg} as the modified coupling coefficients between the input/output waveguide and the ring, and $\kappa_{\text{wg}}^2 + t_{\text{wg}}^2 = 1$ for lossless coupling, we have

$$2\theta'(\omega_{\text{ref}}) = \begin{cases} \pi & \text{odd } m \\ 0 & \text{even } m \end{cases}, \quad t_{\text{wg}} = \sqrt{\eta^2 + 1} - \eta \quad (5.9)$$

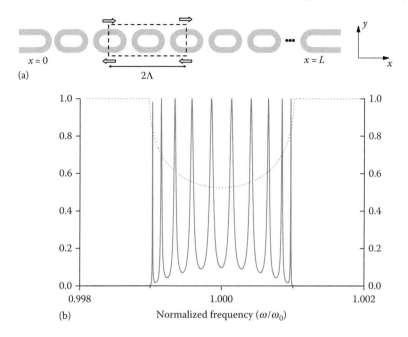

FIGURE 5.10 (a) Schematic of a coupled micro ring resonator structure. The unit cell of the structure is enclosed within the dotted line. (b) The transmission spectrum for $N = 10$ (solid line), $R_\infty(\omega)$ (dotted line). All the rings have effective index of 3.0, a circumference of 52.0 μm, and an inter-resonator coupling coefficient of $\kappa_r = 0.31$.

with

$$\eta(\omega_{\text{ref}}) = t^{-1}\sqrt{1-t^2}\sin\psi(\omega_{\text{ref}}), \quad (5.10)$$

where t is the coupler transmission coefficient, such that if κ_r is the dimensionless, length-integrated coupling coefficient between the microrings, $t^2 + \kappa_r^2 = 1$ in the case of lossless coupling.

Alternatively, the AR structure can be designed such that it differs only from a normal unit cell by its nearest neighbor coupling coefficients and possesses the same circumference as the rest of the CROW. In this case, the modified coupling coefficient between the AR ring and the input waveguide, κ'_{wg} and t'_{wg}, and the coupling coefficient between the AR ring and the next ring resonator, κ'_r and t', are given by

$$t' = t\left(1 + \kappa_r^2\right)^{-1/2}, \quad t_{\text{wg}} = \sqrt{4\eta^2 + 1} - 2\eta, \quad (5.11)$$

keeping η as defined in Equation 5.10. One can verify that the two AR structures designed according to Equations 5.9 and 5.11, respectively, lead to essentially the same transmission spectrum within the approximation $t \simeq 1$. The effectiveness of the AR structure is shown in Figure 5.11, where we have plotted the transmission spectrum of the finite structure with and without AR structure at both ends. Although the AR structure is optimized for $\omega_{\text{ref}} = \omega_0$, in practice, the transmission is improved over an appreciable range of frequencies. These results can be extended to coupled cavities that are described by the generalized Breit–Wigner formalism (e.g., photonic crystal coupled cavities).

5.4.2 Delay, Bandwidth, and Loss

Even though the structures can be engineered to improve the transmission spectra of coupled resonator devices, there remain trade-offs among the delay, bandwidth, and losses of these structures. In this section, we will highlight these issues in a CROW using the dispersion relation Equation 5.2. Similar analyzes can be carried out for SCISSORs as well as other types of periodic structures.

Periodic Coupled Resonator Structures

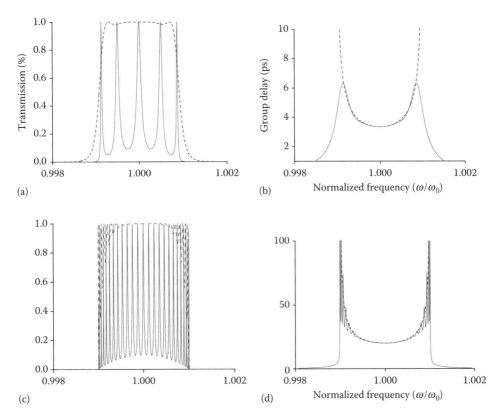

FIGURE 5.11 Transmission and group delay for finite structures with five (a, b) and 25 (c, d) cavities, using parameters as listed in Figure 5.10. Comparison of transmission for structures with (dash) or without (solid) AR modification is shown in (a), (c); Comparison of delay for a structure with AR (dash) and corresponding delay within an infinite structure (circle) is shown in (b), (d).

The trade-offs among the delay, bandwidth, and losses can be understood from the dispersion relation. An immediate consequence of Equation 5.2 is that the group velocity,

$$|v_g| \equiv \left|\frac{\partial \omega}{\partial K}\right| = |\kappa \omega_0 \Lambda \sin(K\Lambda)|, \qquad (5.12)$$

is dependent on the coupling coefficient $|\kappa|$. Since $|\kappa|$ can be controlled by the separation between adjacent resonators, we can, in principle, achieve arbitrarily large slowing down of optical pulses. However, as discussed in Section 5.2.3, the GVD is zero only at the band-center frequency of a CROW and thus a propagating pulse should be centered at or close to this frequency. Since a CROW band spans a frequency range of $\Delta\omega = 2|\kappa|\omega_0$, we define the usable bandwidth of a CROW as half of this total bandwidth centered at ω_0,

$$\Delta\omega_{\text{use}} \equiv |\kappa|\omega_0. \qquad (5.13)$$

The temporal delay of a pulse propagating through the whole length of the CROW is determined by the distance traversed in the CROW and the group velocity at $K\Lambda = \pi/2$:

$$\tau_d = \frac{N}{|\kappa|\omega_0}. \qquad (5.14)$$

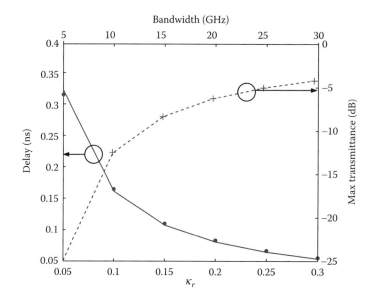

FIGURE 5.12 Trade-offs among delay, bandwidth, and loss for a CROW with 10 ring resonators. The rings have a radius of 100 μm, $n = 1.54$, and a loss of 4 dB/cm. The markers show the numerical results as computed from the transfer matrices, and the lines are calculated from Equations 5.13 through 5.15. For each value of κ_r, the coupling coefficient between the rings and the input/output waveguides was chosen to obtain a flat transmission spectrum.

Therefore, if the resonators have an intrinsic energy dissipation rate of $1/\tau_l$, the total loss accumulated in the CROW is given by

$$\alpha = \frac{\tau_d}{\tau_l} = \frac{N}{|\kappa|Q_{int}}, \qquad (5.15)$$

where Q_{int} is the intrinsic quality factor of the resonator.

Figure 5.12 summarizes the trade-offs described by Equations 5.13 through 5.15 and compares these simple, intuitive equations with numerical results obtained from transfer matrix calculations for the example of microring CROWs. To obtain a large delay using a fixed number of resonators, a weak inter-resonator coupling coefficient, κ_r, is necessary. However, as the coupling decreases, so does the bandwidth of the CROW and the overall loss of the CROW becomes more sensitive to the intrinsic losses in the individual resonator. The latter occurs because the light spends more time in a resonator before tunneling to its neighbor.

5.5 EXPERIMENTAL PROGRESS

The major challenge in realizing periodic coupled resonator devices is the fabrication of many nearly identical and relatively low-loss resonators. The tolerance on the uniformity of the resonators becomes even stricter when the resonators are weakly coupled, since the linewidth of the coupled resonators is correspondingly narrower than when they are strongly coupled. With improvements in fabrication technologies, in recent years, high-order ($N > 10$) on-chip coupled microresonator chains have been realized using microrings and photonic crystal defect cavities in silica, polymer, silicon, and compound semiconductor materials [13,36–41].

5.5.1 Passive Microring CROWs

In our research group at Caltech, we have demonstrated CROWs based on chains of high-order and weakly coupled (1% inter resonator intensity coupling coefficient) microring resonators fabricated

Periodic Coupled Resonator Structures

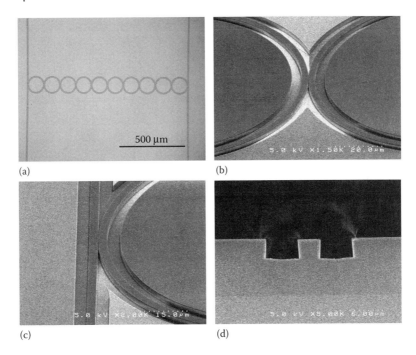

FIGURE 5.13 Optical microscope (a) and scanning electron microscope (b)–(d) images of the fabricated devices in PMMA on Cytop on silicon. (a) 10 coupled microring resonators. The ring radius is 60 μm. (b) The coupling region between two rings. (c) The coupling region between the input/output waveguide and the microring. (d) A waveguide end facet produced by cleaving.

in polymer materials [36,42]. A significant advantage of a CROW consisting of microrings rather than other types of resonators, such as spheres, disks, and photonic crystal defects, is that it possesses a clear and simple spectral response. Our experiment was the first demonstration of such high-order coupled resonators in optical polymers. Our results illustrate the viability of achieving large optical delays (>100 ps) in CROWs. Previously, Little Optics demonstrated coupled microring filters with up to 11 microring resonators in silica materials though the delay properties of the devices were not discussed [13]. Currently, the longest microring CROW demonstrated to date hails from IBM Research and consisted of 100 microring resonators fabricated in silicon-on-insulator [37].

Here, we will briefly summarize our experimental results and methods on polymer microring CROWs. The devices were fabricated directly with electron-beam lithography. Polymethyl-methacrylate (PMMA) constituted the waveguiding layer and a low-index perfluoropolymer, Cytop (Asahi Glass), was used as the lower cladding. The radius of the ring resonators was 60 μm to keep the bend loss to less than 1 dB/cm. Figure 5.13 shows several optical and scanning electron microscope pictures of the fabricated devices. The waveguides had a width of 2.9 μm and a height of 2.6 μm. The cladding regions were 4 μm wide. There was no coupling gap between the resonators and between the waveguide and first/last resonator. However, due to the radius of curvature of the rings as well as the waveguide design and index contrast, even without a coupling gap, weak coupling between the resonators was achieved.

Figure 5.14 shows the transmission spectrum of transverse electric (TE) polarized light through 10 microring resonators. The transmission spectrum does not possess spurious peaks, implying that the microring resonators were nearly identical. The propagation loss in the microring is about 15 dB/cm, equivalent to an intrinsic quality factor of 1.8×10^4. The intrinsic material losses of the polymers are several dB/cm, and the waveguiding losses were dominated by scattering from side-wall roughness.

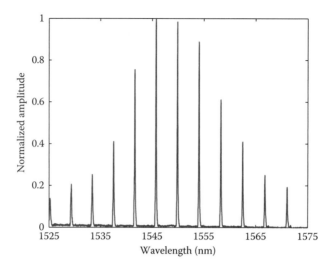

FIGURE 5.14 Drop port transmission spectrum of TE polarized light through a CROW of 10 coupled microring resonators.

We have characterized the transmission and group delay properties of CROWs with as many as 12 microring resonators. We measured the group delay using a phase-shift technique with an RF lock-in amplifier [32,42]. Figure 5.15 summarizes the transmission, group delay, phase response, and GVD for the TE polarization of a CROW consisting of 12 microrings. The narrow transmission peak, with a bandwidth at full-width half-maximum (FWHM) of about 0.13 nm (17 GHz), is indicative of weak inter-resonator coupling. The asymmetry of the transmission peak in Figure 5.15a may be due to slight polarization mixing and small deviations in the microring size and coupling in the presence of loss. Varying degrees of asymmetry were present in all of the devices measured. The inter-resonator intensity coupling coefficient from the numerical fitting of the amplitude and delay was approximately 1%.

From Figure 5.15b, we find that the group delay was 110 ± 7 ps at the transmission peak and increased to ~ 140 ps toward the edges of the transmission peak. The large group delay values greater than 200 ps are not accurate since the transmission amplitude was nearly zero at those wavelengths. We define the slowing factor, S, as the ratio of the speed of light in vacuum, c, to the speed of light modified by the inter resonator coupling, such that $S = \frac{c\tau_d}{N\Lambda}$. For the microring CROW, S is about 23 at the center of the transmission peak and about 29 at the FWHM.

The phase response was obtained by integrating the delay with respect to the frequency, and the GVD was obtained by taking the derivative of the measured group delay with respect to the wavelength. The curvatures of the theoretically calculated group delay and GVD change at the band edges due to the losses in the resonators [43]. In Figure 5.15, the GVD changes from negative to positive across the resonance peak. The high group delay and GVD at the edges of the peak may not be physical, since the transmission amplitude was low at these wavelengths. The change in the GVD and group delay curvatures at the band edges in the calculated results could not be measured, most likely because of the low transmission amplitude.

Unsurprisingly, the GVD of the CROW can be very high. The measured GVD varied from -100 to $70 \, \text{ps}/(\text{nm} \cdot \text{resonator})$ across the FWHM of the peak, with zero GVD at 1511.18 nm, near the resonance peak at 1511.15 nm. The measured GVD is significantly higher than the theoretically calculated GVD which ranges from -17 to $17 \, \text{ps}/(\text{nm} \cdot \text{resonator})$ across the FWHM of the transmission peak. The discrepancy may be a result of the deviation from the ideal scenario of identical

Periodic Coupled Resonator Structures

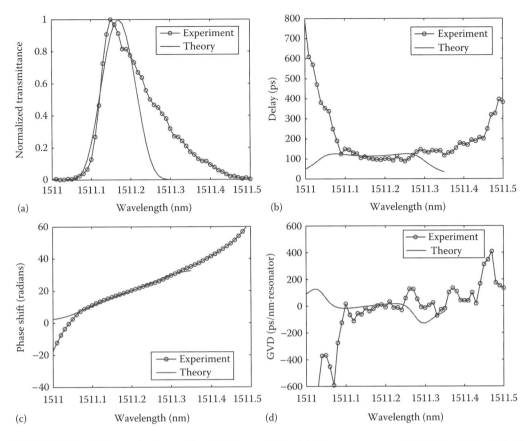

FIGURE 5.15 (a) Transmission amplitude, (b) group delay, (c) phase response, and (d) GVD of TE polarized light in a 12 microring long CROW.

resonators. Since the GVD scales as $1/v_g^3$ [42], any slight deviation of the group velocity will result in a large change in the dispersion.

Compared to other engineered waveguide structures reported to date, such as photonic crystal waveguides and fibers, the CROWs we have demonstrated possess a significantly higher GVD, even though the refractive indices of the polymer materials are relatively low. The high GVD is a consequence of the weak inter-resonator coupling. The measured GVD values of about ±100 ps/(nm · resonator) is equivalent to ±8.3 × 10⁸ ps/(nm · km), and the calculated GVD of ±17 ps/(nm · resonator) is equivalent to ±1.4 × 10⁸ ps/(nm · km). The CROWs we have presented are about 10^7 times more dispersive than conventional optical fibers, 10^6 times more dispersive than highly dispersive photonic crystal fibers [44], and approximately 100 to 1000 times more dispersive than photonic crystal waveguides [45,46]. Compared to previously reported GVD values of photonic crystal CROWs [47], the GVD of our microring CROWs is about an order of magnitude greater. With such large values of both normal and anomalous dispersion, CROWs may find applications in dispersion management and nonlinear optics [7,8,48–51].

The major drawback of passive CROWs is the loss. Assuming the input and output coupling losses were the same for the through and drop ports, the ratio between the drop port power and the difference in the "on" and "off" resonance through port powers gives an equivalent loss of 2.35 dB per resonator. Since the losses of a CROW scales linearly with the delay, the ultimate delay and number of coupled resonators that can be achieved in passive coupled resonator chains will likely be limited by the propagation loss in the resonators and not by the fabrication accuracy.

5.5.2 CROWs with Optical Gain

To overcome the limitation of optical losses, we have investigated CROWs in the form of active FP resonator arrays as described in Section 5.3.1 fabricated in InP–InGaAsP semiconductor materials [52]. The gain is supplied through the injection of electrical current.

Our results in Ref. [52] show that even though the losses can be completely compensated (since they can function as lasers), there are a number of challenges with active coupled resonators. Firstly, because ideally CROWs consist of a very large number of resonators, the fabricated device must be uniform over its footprint. This requires uniformity in the material, etching, and electrical contacts. Moreover, we found that the transmission spectrum, amplification, as well as the spontaneous emission noise depends significantly on the termination and excitation of the coupled resonators and may not necessarily mimic the properties of infinitely long structures [15].

Since the introduction of gain allows for laser oscillation, an important question is whether active periodic resonators should be operated above or below laser threshold. Subthreshold operation is simpler to understand and model, but requires highly accurate fabrication to ensure that the resonators are identical to each other. However, to suppress laser action, the input and output coupling constants as well as the inter resonator coupling strength should be large, which place a lower limit on the group velocity and net amplification attainable [15]. Operation above threshold is more complicated to analyze because locking effects may come into play but can be more interesting fundamentally. Above threshold, the CROW can lock to the input signal and also the resonators can be phase coherent with each other [25,53–57]. This can occur even when the resonators are not exactly identical. Laser action can also clamp the gain, which may help in stabilizing the operation of an amplifying coupled resonator device much like gain-clamped semiconductor optical amplifiers [58–60].

5.6 CONCLUSIONS

In summary, we have discussed two types of periodic coupled resonator structures for slow light: CROWs and SCISSORs. We have described their dispersion properties as well as several implementations using traveling-wave and standing-wave resonators. CROWs and SCISSORs are in some sense complementary structures in the way they utilize optical resonances. With improved fabrication technologies, the field of coupled microresonators has advanced rapidly in recent years. However, active devices that are readily tunable, preferably by an electrical signal, and can compensate for losses remain open avenues for exploration. The main advantages of coupled resonators for slow light over atomic or material resonances-based methods are that these devices are highly compact, chip-scale, and they offer controllable dispersion properties that are not restricted to specific resonance frequencies of the propagation medium.

ACKNOWLEDGMENTS

J. Poon wishes to thank Lin Zhu, Dr. Guy DeRose, and Professor Axel Scherer for their assistance in fabricating the devices discussed in Section 5.5 as well as the members of the Caltech Quantum Electronics group for many fruitful discussions. J. Poon and P. Chak are grateful for the financial support given by the Natural Sciences and Engineering Research Council of Canada. This work has been funded by the Defense Advanced Research Projects Agency (Slow Light Project), National Science Foundation (award no. 0438038), Hughes Research Laboratory, and Natural Sciences and Engineering Research Council of Canada.

Portions reprinted, with permission from the OSA and IEEE, from Refs. [14,16,21,23,24,34,36, 42,52]. ©2002, 2004, 2006, 2007 OSA, ©2006 IEEE.

REFERENCES

1. K. J. Vahala. Optical microcavities. *Nature*, 424(6950):839–846, 2003.
2. A. Yariv, Y. Xu, R. K. Lee, and A. Scherer. Coupled-resonator optical waveguide: A proposal and analysis. *Opt. Lett.*, 24(11):711–713, 1999.
3. N. Stefanou and A. Modinos. Impurity bands in photonic insulators. *Phys. Rev. B*, 57(19):12127–12133, 1998.
4. J. E. Heebner and R. W. Boyd. "Slow" and "fast" light in resonator-coupled waveguides. *J. Mod. Opt.*, 49(14–15):2629–2636, 2002.
5. J. E. Heebner, R. W. Boyd, and Q.-H. Park. SCISSOR solitons and other novel propagation effects in microresonator-modified waveguides. *J. Opt. Soc. Am. B*, 19(4):722–731, 2002.
6. Y. Xu, R. K. Lee, and A. Yariv. Propagation and second-harmonic generation of electromagnetic waves in a coupled-resonator optical waveguide. *J. Opt. Soc. Am. B*, 77(3):387–400, 2000.
7. D. N. Christodoulides and N. K. Efremidis. Discrete temporal solitons along a chain of nonlinear coupled microcavities embedded in photonic crystals. *Opt. Lett.*, 27(8):568–570, 2002.
8. S. Mookherjea and A. Yariv. Kerr-stabilized super-resonant modes in coupled-resonator optical waveguides. *Phys. Rev. E*, 66(4):046610, 2002.
9. C. K. Madsen. General IIR optical filter design for WDM applications using all-pass filters. *J. Lightw. Technol.*, 18(6):860–868, 2000.
10. G. Lenz, B. J. Eggleton, C. K. Madsen, and R. E. Slusher. Optical delay lines based on optical filters. *IEEE J. Quantum Elect.*, 37(4):525–532, 2001.
11. B. E. Little, S. T. Chu, W. Pan, D. Ripin, T. Kaneko, Y. Kokubun, and E. Ippen. Vertically coupled glass microring resonator channel dropping filters. *IEEE Photon. Technol. Lett.*, 11(2):215–217, 1999.
12. M. Bayindir, B. Temelkuran, and E. Ozbay. Tight-binding description of the coupled defect modes in three-dimensional photonic crystals. *Phys. Rev. Lett.*, 84(10):2140–2143, 2000.
13. B. E. Little, S. T. Chu, P. P. Absil, J. V. Hryniewicz, F. G. Johnson, F. Seiferth, D. Gill, V. Van, O. King, and M. Trakalo. Very high-order microring resonator filters for WDM applications. *IEEE Photon. Technol. Lett.*, 16(10):2263–2265, 2004.
14. J. K. S. Poon, J. Scheuer, S. Mookherjea, G. T. Paloczi, Y. Huang, and A. Yariv. Matrix analysis of microring coupled-resonator optical waveguides. *Opt. Express*, 12(1):90–103, 2004.
15. J. K. S. Poon and A. Yariv. Active coupled-resonator optical waveguides—Part I: Gain enhancement and noise. *J. Opt. Soc. Am. B*, 24(9):2378–2388, 2007.
16. J. K. S. Poon, J. Scheuer, Y. Xu, and A. Yariv. Designing coupled-resonator optical waveguide delay lines. *J. Opt. Soc. Am. B*, 21(9):1665–1673, 2004.
17. Y. Xu, Y. Li, R. K. Lee, and A. Yariv. Scattering-theory analysis of waveguide-resonator coupling. *Phys. Rev. E*, 62(5):7389–7404, 2000.
18. J. E. Heebner, P. Chak, S. Pereira, J. E. Sipe, and R. W. Boyd. Distributed and localized feedback in microresonator sequences for linear and nonlinear optics. *J. Opt. Soc. Am. B*, 21(10):1818–1832, 2004.
19. P. Chak, S. Pereira, and J. E. Sipe. Coupled-mode theory for periodic side-coupled microcavity and photonic crystal structures. *Phys. Rev. B*, 73(3):035105, 2006.
20. J. B. Khurgin. Optical buffers based on slow light in electromagnetically induced transparent media and coupled resonator structures: Comparative analysis. *J. Opt. Soc. Am. B*, 22(5):1062–1074, 2005.
21. S. Pereira, P. Chak, and J. E. Sipe. Gap-soliton switching in short microresonator structures. *J. Opt. Soc. Am. B*, 19(9):2191–2202, 2002.
22. M. F. Yanik and S. Fan. Stopping light all optically. *Phys. Rev. Lett.*, 92:083901, 2004.
23. P. Chak, J. K. S. Poon, and A. Yariv. Optical bright and dark states in side-coupled resonator structures. *Opt. Lett.*, 32(13):1785–1787, 2007.
24. J. K. S. Poon, P. Chak, J. M. Choi, and A. Yariv. Slowing light with Fabry–Perot resonator arrays. *J. Opt. Soc. Am. B*, 24(11):2763–2769, 2007.
25. D. Botez and D. R. Scifres. *Diode Laser Arrays*. Cambridge University Press, Cambridge, 1994.
26. A. Yariv. *Optical Electronics in Modern Communications*, 5th edn. Oxford University Press, New York, 1997.
27. S. Fan, P. R. Villeneuve, J. D. Joannopoulos, and H.A. Haus. Channel drop filters in photonic crystals. *Opt. Express*, 3(11):4–11, 1998.
28. L. V. Hau, S.E. Harris, Z. Dutton, and C.H. Behroozi. Light speed reduction to 17 meters per second in an ultracold atomic gas. *Nature*, 397(6720):594–598, 1999.

29. J. P. Marangos. Electromagnetically induced transparency. *J. Mod. Opt.*, 45(3):471–503, 1998.
30. E. Peral and A. Yariv. Supermodes of grating-coupled multimode waveguides and application to mode conversion between copropagating modes mediated by backward Bragg scattering. *J. Lightw. Technol.*, 17(5):942–947, 1999.
31. B. E. Little, S. T. Chu, H. A. Haus, J. Foresi, and J.-P. Laine. Microring resonator channel dropping filter. *J. Lightw. Technol.*, 15(6):998–1005, 1997.
32. C. K. Madsen and J. H. Zhao. *Optical Filter Design and Analysis: A Signal Processing Approach.* Wiley, New York, 1999.
33. A. Melloni and M. Martinelli. Synthesis of direct-coupled-resonators bandpass filters for WDM systems. *J. Lightw. Technol.*, 20(2):296–303, 2002.
34. P. Chak and J.E. Sipe. Minimizing finite-size effects in artificial resonance tunneling structures. *Opt. Lett.*, 13(17):2568–2570, 2006.
35. G. Boedecker and C. Henkel. All-frequency effective medium theory of a photonic crystal. *Opt. Express*, 11(13):1590–1595, 2003.
36. J. K. S. Poon, L. Zhu, G. A. DeRose, and A. Yariv. Transmission and group delay in microring coupled-resonator optical waveguides. *Opt. Lett.*, 31(4):456–458, 2006.
37. F. N. Xia, L. Sekaric, and Y. Vlasov. Ultracompact optical buffers on a silicon chip. *Nat. Photon.*, 1(1):65–71, 2007.
38. S. Olivier, C. Smith, M. Rattier, H. Benisty, C. Weisbuch, T. Krauss, R. Houdre, and U. Osterle. Miniband transmission in a photonic crystal waveguide coupled-resonator optical waveguide. *Opt. Lett.*, 26(13):1019–1051, 2001.
39. S. Nishikawa, S. Lan, N. Ikeda, Y. Sugimoto, H. Ishikawa, and K. Asakawa. Optical characterization of photonic crystal delay lines based on one-dimensional coupled defects. *Opt. Lett.*, 27(23):2079–2081, 2002.
40. T. D. Happ, M. Kamp, A. Forchel, J. L. Gentner, and L. Goldstein. Two-dimensional photonic crystal coupled-defect laser diode. *Appl. Phys. Lett.*, 82(1):4–6, 2003.
41. F. Pozzi, M. Sorel, Z. S. Yang, R. Iyer, P. Chak, J. E. Sipe, and J. S. Aitchison. Integrated high order filters in AlGaAs waveguides with up to eight side-coupled racetrack microresonators, in *Conference on Lasers and Electro-Optics*, p. CWK2, Optical Society of America, Washington, DC, 2006.
42. J. K. S. Poon, L. Zhu, G. A. DeRose, and A. Yariv. Polymer microring coupled-resonator optical waveguides. *J. Lightw. Technol.*, 24(4):1843–1849, 2006.
43. H. Kogelnik and C. V. Shank. Coupled-wave theory of distributed feedback lasers. *J. Appl. Phys.*, 43(5):2327–2335, 1972.
44. J. C. Knight, J. Arriaga, T. A. Birks, A. Ortigosa-Blanch, W. J. Wadsworth, and P. St. Russell. Anomalous dispersion in photonic crystal fiber. *IEEE Photon. Technol. Lett.*, 12(7):807–809, 2000.
45. M. Notomi, K. Yamada, A. Shinya, J. Takahashi, C. Takahashi, and I. Yokohama. Extremely large group-velocity dispersion of line-defect waveguides in photonic crystal slabs. *Phys. Rev. Lett.*, 87(25):253902, 2001.
46. T. Asano, K. Kiyota, D. Kumamoto, B.-S. Song, and S. Noda. Time-domain measurement of picosecond light-pulse propagation in a two-dimensional photonic crystal-slab waveguide. *Appl. Phys. Lett.*, 84(23):4690–4692, 2004.
47. T. J. Karle, Y. J. Chai, C. N. Morgan, I. H. White, and T. F. Krauss. Observation of pulse compression in photonic crystal coupled cavity waveguides. *J. Lightw. Technol.*, 22(2):514–519, 2004.
48. S. Mookherjea. Dispersion characteristics of coupled-resonator optical waveguides. *Opt. Lett.*, 30(18):2406–2408, 2005.
49. W. J. Kim, W. Kuang, and J. D. O'Brien. Dispersion characteristics of photonic crystal coupled resonator optical waveguides. *Opt. Express*, 11(25):3431–3437, 2003.
50. J. B. Khurgin. Expanding the bandwidth of slow-light photonic devices based on coupled resonators. *Opt. Lett.*, 30(5):513–515, 2005.
51. J. K. Ranka, R. S. Windeler, and A. J. Stentz. Visible continuum generation in air–silica microstructure optical fibers with anomalous dispersion at 800 nm. *Opt. Lett.*, 25(1):25–27, 2000.
52. J. K. S. Poon, L. Zhu, G. A. DeRose, J. M. Choi, and A. Yariv. Active coupled-resonator optical waveguides—Part II: Current injection InP–InGaAsP Fabry–Perot resonator arrays. *J. Opt. Soc. Am. B*, 24(9):2389–2393, 2007.
53. W. W. Chow, S. W. Koch, and M. Sargent III. *Semiconductor-Laser Physics.* Springer, Berlin, 1997.

54. H. G. Winful, S. Allen, and L. Rahman. Validity of the coupled-oscillator model for laser-array dynamics. *Opt. Lett.*, 18(21):1810–1812, 1993.
55. A. E. Siegman. *Lasers*. University Science Books, Mill Valley, 1986.
56. L. Goldberg, H. F. Taylor, J. F. Weller, and D. R. Scifres. Injection locking of coupled-stripe diode laser arrays. *Appl. Phys. Lett.*, 46(3):236–238, 1985.
57. J. P. Hohimer, A. Owyoung, and G. R. Hadley. Single-channel injection locking of a diode-laser array with a cw dye laser. *Appl. Phys. Lett.*, 47(12):1244–1246, 1985.
58. B. Bauer, F. Henry, and R. Schimpe. Gain stabilization of a semiconductor optical amplifier by distributed feedback. *IEEE Photon. Technol. Lett.*, 6(2):182–185, 1994.
59. L. F. Tiemeijer, P. J. A. Thijs, T. Dongen, J. J. M. Binsma, E. J. Jansen, and H. R. J. R. Vanhelleputte. Reduced intermodulation distortion in 1300 nm gain-clamped MQW laser amplifiers. *IEEE Photon. Technol. Lett.*, 7(3):284–286, 1995.
60. M. Bachmann, P. Doussiere, J. Y. Emery, R. NGo, F. Pommereau, L. Goldstein, G. Soulage, and A. Jourdan. Polarisation-insensitive clamped-gain SOA with integrated spot-size convertor and DBR gratings for WDM applications at 1.55 µm wavelength. *Electron. Lett.*, 32(22):2076–2078, 1996.

6 Resonator-Mediated Slow Light: Novel Structures, Applications and Tradeoffs

Andrey B. Matsko and Lute Maleki

CONTENTS

6.1 Introduction 101
6.2 Vertically Coupled Resonator Optical Waveguide as a Delay Line 102
    6.2.1 Ideal Vertically Coupled Resonator Optical Waveguide 103
    6.2.2 Absorption 105
    6.2.3 Fundamental Restrictions and Fabrication Problems 107
6.3 Interference in Resonator Chains 107
6.4 Resonator-Stabilized Oscillators 109
6.5 Slow Light in Systems with a Discrete Spectrum 110
References 115

6.1 INTRODUCTION

An active involvement of optics in information processing depends on the existence of dynamically controllable all-optical buffers. There has been a steady interest in the development of such devices and in the investigation of the fundamental principles that could lead to their development. Unlike usual optical delay lines that rely upon the phase delay of the signal light, the new generation of the controllable optical buffers is based on group delays. Phase delay $\tau_{ph} = Ln_0/c$ depends on the refractive index of the material n_0, length of the delay line L, and speed of light in the vacuum c. Group delay $\tau_g = d(\omega L n_0(\omega)/c)/d\omega$ is built upon the dispersion of the material, i.e., upon the frequency (ω) dependence of the refractive index $n_0(\omega)$.

Thousands of slow light related studies have emerged since the publication of the seminal paper of Lene Vestergaard Hau et al. [1], and the all-optical buffer activity has been started by the atomic-based slow-light experiments. Huge group delays observed in coherent atomic media has led to suggestions for applications of the media to fabricate miniature delay lines. Their advantages include small size, tunability of the delay time, and the possibility of externally controllable store/release of the optical information. Their disadvantages occur (1) due to the specific light wavelengths corresponding to appropriate atomic transitions, i.e., available wavelength ranges are very small and unique; (2) due to the residual intrinsic absorption of the atomic systems; and (3) due to the narrowness of the operational bandwidth. The delays corresponding to these linewidths are generally in the microsecond range or longer. Delays in the range of nanoseconds required in a number of practical applications cannot be easily obtained with electromagnetically induced transparency (EIT) in atomic vapors, because of the need for high power lasers for optical pumping. The high power pump lasers could interact with many atomic levels, which decreases the EIT efficiency and results in nonlinear wave mixing as well as oscillations. Moreover, it is not always possible to filter out pumping light from the signal light. Finally, parameters of EIT could depend on the probe power, as well as the drive power, in realistic atomic systems, making EIT use in optical networks difficult to implement.

There are only a few experiments where delays exceeding a full width at the half maximum of a pulse have been demonstrated [1,2]. The feasibility of the method is intensively discussed [3–5]. Large fractional delays were realized in hot atomic vapors without the involvement of low frequency atomic coherence. For instance, delays exceeding seven pulse durations with 3 dB absorption were realized in hot ^{85}Rb atomic vapors [6] when carrier frequency of the pulses was tuned to be halfway between the D_2 line hyperfine resonances. The pulses were delayed because their frequency was tuned to the transparency window limited by two strongly absorbing resonances [7]. An optical buffer based on spectral hole-burning and characterized with fractional delay on the order of three at 3 dB absorption was realized in hot rubidium vapor [8]. Again, no EIT-like effects were involved in the experiment. The disadvantage of those delays compared with EIT-based delays is in the slow tunability of the buffers.

The original idea of the application of atomic vapors for slowing light down has been expanded to various dispersive materials and structures including artificial ones. An attempt to mediate practical disadvantages of atomic group delay lines is one of the basic drives for the development. Semiconductor media, optical amplifiers, photonic crystals, and resonator structures have been proposed as potential candidates for the realization of the optical buffers.

Advantages of manmade solid-state slow light structures include: (1) simple tunability, (2) unlimited frequency selection, (3) small losses, (4) low power consumption, and (5) small size. Tuning solid-state structures can be achieved by carrier injection or electro-optically. The delay time could be changed from microseconds to nanoseconds in resonator as well as semiconductor structures. The central frequency of the transparency window of the solid-state structure is given by the structure morphology and is arbitrary. The solid-state structures could have small losses because they are based on the reflection of light, and not absorption. No optical drive is needed in solid-state structures. Therefore, they consume less power and are less noisy compared with atomic structures having the same group delay characteristics. Solid-state structures could be really tiny, while the size of the atomic systems is dictated by the size of the atomic cells that are at centimeter scale.

In this chapter, we discuss some particular applications of coupled resonators for optical buffering. We do not try to cover the entire area, but rather focus on several examples of slow light structures and discuss their properties. The examples include vertically coupled resonator optical waveguide and interfering resonator structures. We also discuss applications of the resonators for frequency stabilization.

6.2 VERTICALLY COUPLED RESONATOR OPTICAL WAVEGUIDE AS A DELAY LINE

In this section, we describe the basic properties of the vertically coupled whispering gallery mode (WGM) resonator optical delay lines [9,10], calculate group velocity as well as high-order dispersion of the delay lines, and discuss possibilities of their tuning. Recently, using a similarity between morphologies of an optical planar waveguide and an axial-symmetric solid-state WGM resonators, we demonstrated a one dimensional ring-like resonator [11,12]. A chain of such resonators, vertically coupled WGM resonator waveguide, is a novel configuration for coupled resonator optical waveguides (CROWs) and other resonator [13–17] as well as generalized optical filters-based [18–20] waveguides. Our resonator chain has a certain similarity with coil optical resonator waveguide [21], as well as with the vertically stacked multi ring resonator (VMR) [22]. The main difference is in the low contrast of our resonators. The basic features of the chain can be adjusted by changing the shape of the resonators as well as the distance between them. Unlike ordinary CROWs and VMRs, where coupled resonators are considered as lumped element objects, our structure amenable to tuning its photonic density of states similarly to such distributed systems as photonic crystals.

The vertically coupled WGM resonator waveguides belong to a new class of optical buffers having some features of optical fibers [23], open optical resonators [24,25] as well as photonic crystals [26–29]. The signature unifying these systems is that the electromagnetic field and the photon density of states can be tailored to a prescribed manner. The manipulation of the density of states can be used for the modification of linear, nonlinear, and quantum optical properties of these systems.

6.2.1 Ideal Vertically Coupled Resonator Optical Waveguide

Let us consider a resonator waveguide shown in Figure 6.1. The linearly polarized light is coupled to a mode of one of the resonators using a prism or an optical fiber coupler. The light propagates due to the coupling between adjacent resonators, occurring via evanescent field inside the rod the resonators are made of, not through the air, as is the case with conventional CROWs. To understand the difference, it is worth noting that the dimension of the evanescent field in the air outside a resonator scales as $l_e \sim \lambda/\pi(\epsilon_0 - 1)^{1/2}$, where ϵ_0 is the electric susceptibility of the resonator material and λ is the wavelength of the light in vacuum. This value is usually equal to several hundreds of nanometers. In contrast, the coupling is efficiently realized between the vertically stacked resonators that are within several microns of each other.

The propagation of light in the waveguide can be described with the usual wave equation

$$\nabla \times (\nabla \times \mathbf{E}) - k^2 \epsilon(\mathbf{r})\mathbf{E} = 0, \tag{6.1}$$

where
$k = \omega/c$ is the wave number
$\epsilon(\mathbf{r})$ is the space dependent index of refraction
$\mathbf{E}$ is the electric field of the mode
$\mathbf{r}$ is the radius vector

We assume that structure is made of a material that is lossless.

We are interested in the case of high order WGMs localized in the vicinity of the equator of the resonator. For the sake of simplicity, we consider the TE mode family and change variables in Equation 6.1 as $E = \Psi e^{\pm i\nu\phi}/\sqrt{r}$, ν is angular momentum number of the mode. We consider a low contrast structure, assuming that the resonator radius changes into $R = a - A(z)$ and $a \gg |A(z)|$. Then Equation 6.1 is transformed to

$$\frac{\partial^2 \Psi}{\partial r^2} + \frac{\partial^2 \Psi}{\partial z^2} + \left[k^2 \epsilon_0 \left(1 - 2\frac{A(z)}{a}\right) - \frac{\nu^2}{r^2}\right] \Psi = 0, \tag{6.2}$$

where expressions $\nu \gg 1$ and $\nu^2 \simeq a^2 k^2 \epsilon_0$ are used.

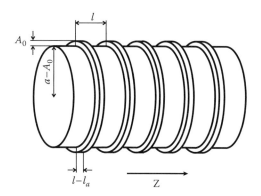

FIGURE 6.1 A chain of vertically coupled WGMRs. (From Maleki, L., Matsko, A.B., Savchenko, A.A., and Strekalov, D., *Proc. SPIE*, 6130, 61300R, 2006. With permission.)

Considering an infinite waveguide, we separate variables and introduce $\Psi = \Psi_r \Psi_z$:

$$\frac{\partial^2 \Psi_z}{\partial z^2} - 2k^2 \epsilon_0 \frac{A(z)}{a} \Psi_z = -k_z^2 \Psi_z, \qquad (6.3)$$

$$\frac{\partial^2 \Psi_r}{\partial r^2} + \left(k^2 \epsilon_0 - k_z^2 - \frac{\nu^2}{r^2} \right) \Psi_r = 0, \qquad (6.4)$$

where k_z^2 is the separation parameter.

Equations 6.3 and 6.4 can be solved analytically if $A(z)$ is a periodic function with some period l. The solution is $\Psi = \Psi_0 e^{i\beta z} \varphi(k_z z) J_\nu(k_{\nu,q} r)$, where β is a propagation constant, $J_\nu(k_{\nu,q} r)$ is a Bessel function, $\Psi_z = \exp(i\beta z) \varphi(k_z z)$ is the Bloch function [$\Psi_z(z+l) = \exp(i\beta l) \Psi_z(z)$], $k_{\nu,q}$ is the radial wave number, and q describes radial quantization of the WGMs.

Parameters β and k_z are to be derived from Equation 6.3. They stand for wave number and a counterpart of frequency of the wave propagating in the resonator chain (the frequency is determined by eigenvalues of both Equations 6.3 and 6.4). The parameters are connected by a dispersion relation which, in the simplest case of periodic grating of cylindrical WGM resonators with gaps between them, characterized by height $A(z) = A_0$ and length l_g ($l_g < l$), can be presented as

$$\cos(\beta l) = \cosh \left[l_g \sqrt{2k^2 \epsilon_0 A_0 / a - k_z^2} \right] \cos \left[k_z (l - l_g) \right]$$
$$+ \frac{k^2 \epsilon_0 A_0 / a k_z}{\sqrt{2k^2 \epsilon_0 A_0 / a - k_z^2}} \sinh \left[l_g \sqrt{2k^2 \epsilon_0 A_0 / a - k_z^2} \right] \sin \left[k_z (l - l_g) \right], \qquad (6.5)$$

where we assumed that $k(2\epsilon_0 A_0 / a)^{1/2} > k_z$ (this assumption comes from the condition that a single localized WGM resonator has at least one confined mode). As follows from Equation 6.5, k_z is determined in allowed bands separated by forbidden band gaps.

Let us characterize the frequency spectrum of the entire system. We start from expression for the frequency of WGMs derived from Equation 6.4

$$\frac{\omega^2}{c^2} \epsilon_0 = k_{\nu,q}^2 + k_z^2. \qquad (6.6)$$

In the case of small interaction between the resonators, when $k(2\epsilon_0 A_0 / a)^{1/2} \gg k_z$ and $k l_g (2\epsilon_0 A_0 / a)^{1/2} \gg 1$ we estimate from Equations 6.5 and 6.6 that each WGM mode in the resonator chain transforms into a frequency band with center

$$\frac{\omega_c}{c} \sqrt{\epsilon_0} \simeq k_{\nu,q} + \frac{1}{2k_{\nu,q}} \left[\frac{\pi m}{l - l_g} \right]^2, \qquad (6.7)$$

and width

$$\frac{\Delta \omega}{c} \sqrt{\epsilon_0} \simeq \frac{4\pi^2 m^2}{k_{\nu,q}^2 (l - l_g)^3} \sqrt{\frac{a}{2A_0}} \exp\left(-l_g k_{\nu,q} \sqrt{\frac{2A_0}{a}} \right). \qquad (6.8)$$

The spectrum of the modes due to their radial confinement, being eigenvalues of Equation 6.4, is described by

$$k_{\nu,q} \simeq \frac{1}{a} \left[\nu + \alpha_q \left(\frac{\nu}{2} \right)^{1/3} - \sqrt{\frac{\epsilon_0}{\epsilon_0 - 1}} \right], \qquad (6.9)$$

where α_q is the qth root of the Airy function, $Ai(-z)$ (q is a natural number). This is, in fact, a spectrum of high order WGMs of an infinite cylinder with radius a. We have now all the necessary elements to describe the waveguide.

One of its parameters is the group velocity, which can be evaluated using wave number β (Equation 6.5) and frequency ω (Equation 6.6). First of all, even without calculating and using the analogy with photonic band gap materials, it is clear that the group velocity of light propagating in the system, determined as $V_g = d\omega/d\beta$, approaches zero if light is tuned to the band edge. However, at that point the dispersion of the group velocity is large as well, so such a tuning is impractical. The dispersion is nearly linear in the center of the band gap, and the group velocity there can be estimated as

$$\frac{V_g\sqrt{\epsilon_0}}{c} \approx \frac{\Delta\omega\sqrt{\epsilon_0}l}{2\pi c} = \frac{2\pi m^2 l}{k_{v,q}^2 (l-l_g)^3}\sqrt{\frac{a}{2A_0}} \exp\left(-l_g k_{v,q}\sqrt{\frac{2A_0}{a}}\right). \quad (6.10)$$

It is easy to see that the group velocity exponentially decreases as the resonators are pulled apart or the depth of trenches separating them increases. The answer could be envisioned even without analytically solving the problem. Indeed, because light propagates due to coupling between adjacent resonators, the smaller the coupling, the slower the light propagation.

The vertically coupled WGMR optical delay lines have certain advantages over CROW-based optical delays [13]. The resonators are placed on a plane in a CROW. Therefore, if the group delay per resonator is τ_{gr}, the group velocity of a pulse propagating in the chain is equal to $2a/\tau_{gr}$, where a is the radius of the resonator. The delay per resonator is determined primarily by the resonator loading and can be set the same in any resonator chain. This delay also determines the frequency bandwidth of the delay line. The group velocity is equal to l/τ_{gr} for the vertically coupled resonators, if the delay per resonator is the same as in the CROW. The group velocity in the vertically coupled resonator chains is always smaller than the group velocity in CROWs for the same bandwidth of the delay line because $2a \gg l$.

6.2.2 Absorption

Could an array of vertically coupled resonators produce a larger delay compared with the delay determined by the single resonator ring down time? Such a question arises because it is known that coupled high-Q optical ring resonators enhance the performance of a single resonator which is generally characterized by the width of the resonance structure (ringdown time). Large optical group delays in chains of coupled resonators have been studied in Refs. [14,30]. It was shown recently that the Q-factor of the coupling–split modes for a system of N identical coupled resonators is greater than that of a single resonator in the chain by a factor of N, and even more, in the case of optimum coupling [31]. Stopping light all optically with a chain of interacting, tunable optical resonators was discussed in Ref. [32]. Moreover, even a couple of interacting resonators could produce unexpectedly narrow spectral lines, significantly narrower than the width of response of each resonator standing alone. This line narrowing occurs due to Fano interference effect which results in sharp asymmetric line shapes in a narrow frequency range in periodic structures and waveguide-cavity systems [33,34].

In our opinion, the results of the above mentioned studies are limited to the case of either lossless resonators or very small resonators. A chain of microresonators with quality factors restricted due to radiative decay can introduce longer group delay than the ring down time of a single resonator because radiative losses can be suppressed by interference [35–37].

The light confined in large realistic resonators experiences absorption and scattering which is independent on the configuration of the whole system and the number of the resonators. Material losses and/or losses in the mirrors limit the minimum resonances width. Furthermore, using results of previous studies [19], we conclude that our results automatically show restriction of the maximum group delay that could be achieved with optical delay lines based on resonator chains. This leads to a paradoxical conclusion that it is better to use a single optical resonator instead of a chain to realize the most efficient absolute group delay element provided by finite Q-factor resonators. On the other hand, a chain of coupled resonators can produce large fractional delays.

Let us illustrate the importance of the absorption using an example of a chain of resonators side-coupled to a waveguide. Transmission of a monochromatic electromagnetic wave of frequency ω through such a system in the case of lossless resonators can be characterized by the coefficient

$$S_{12} = \frac{\gamma_c - i(\omega - \omega_0)}{\gamma_c + i(\omega - \omega_0)}, \tag{6.11}$$

where
  S_{12} is the amplitude transmission coefficient
  γ_c is the loaded (by the resonator coupling to a waveguide) mode linewidth
  ω_0 is the resonance frequency of the mode (we assume that $|\omega - \omega_0|$ is much less than the resonator free spectral range)

For the output field envelope we find

$$E_{\text{out}}(t) = E_{\text{in}}\left(t - \frac{2\tau\mathcal{F}}{\pi}\right), \tag{6.12}$$

where $\tau = 2\pi R n_0/c$ is the round-trip time of the light in the resonator, n_0 is the refractive index of the material, R is the radius of the resonator, and $\mathcal{F}$ is the finesse determined by the coupling. We see that such a resonator results in the group delay of the pulse without phase delay of the carrier, in which respect it is similar to the EIT "slow light" phenomenon. The group delay is proportional to the light power build up in the resonator. Equation 6.12 is valid for pulses much longer than the group delay time $2\tau\mathcal{F}/\pi$. This condition is identical to the condition of the narrowness of the pulse spectrum compared to the spectral width of the resonator's mode. When N identical resonators are connected to a single waveguide, the group delay of the system is N time as much as the delay due to a single resonator [14,30]. The linewidth of the system becomes N times narrower, compared with the linewidth of a single resonator.

This conclusion is not entirely valid for a chain of resonators with absorption. The transmission coefficient should be modified in this case:

$$S_{12} = \frac{\gamma_c - \gamma - i(\omega - \omega_0)}{\gamma_c + \gamma + i(\omega - \omega_0)}, \tag{6.13}$$

where γ is the linewidth originated from intrinsic resonator losses. The average lifetime of a photon in a realistic resonator cannot exceed $(2\gamma)^{-1} = n_0(\alpha c)^{-1}$, where α is the linear loss coefficient of the resonator material. The maximum Q-factor of the resonator can then be found from

$$Q = \frac{2\pi n_0}{\alpha \lambda}. \tag{6.14}$$

The transmission is zero at the resonance and under condition of critical coupling ($\gamma_c = \gamma$).

For lossy resonators, Equation 6.12 should be modified as

$$E_{\text{out}}(t) \simeq \frac{\gamma_c - \gamma}{\gamma_c + \gamma} E_{\text{in}}\left(t - \frac{2\tau\mathcal{F}}{\pi} \frac{\gamma_c}{\gamma_c + \gamma}\right), \tag{6.15}$$

where we assume that $\gamma_c \gg \gamma$.

Transmission and dispersion for a long waveguide (waveguide length L is much greater than the resonator radius R, so that the number of the resonators is $N \approx L/2R$) coupled to many resonators could be estimated as

$$E_{\text{out}}(t) \simeq \exp\left(-\frac{\gamma L}{\gamma_c R}\right) E_{\text{in}}\left(t - \frac{2L}{\gamma_c R}\right), \tag{6.16}$$

which shows that for absorption exp(−1) maximum time delay is restricted by the properties of the material only. The linewidth of such a system is comparable with the width of an unloaded single resonator. The configuration of a large number of resonators does not lead to the efficient increase of the delay time compared with a single resonator with maximum possible Q-factor.

Let us now consider the case of the waveguide formed by vertically coupled WGM resonators. The waveguide host material possesses intrinsic losses due to scattering on the material and surface inhomogeneities, as well as absorption. The modes of the uncoupled resonators have a finite ring down time τ. Naturally, the maximum group delay in the set of coupled resonators cannot exceed this ring down time without significant absorption of the light. The minimum group velocity is then equal to $V_{g\,min} \approx l/2\pi\tau$, which corresponds to the propagation of light through a single uncoupled resonator. In the case of strong coupling between the resonators, when $\Delta\omega\tau \gg 2\pi$, light interacts with many resonators and its propagation can be studied using the formalism presented above.

6.2.3 Fundamental Restrictions and Fabrication Problems

Let us discuss fabrication issue being the same for any chain of resonators. The modes of the resonators should have approximately the same frequencies and loaded Q-factors. It is technically challenging to fabricate a pair of high-Q WGRs with coinciding optical modes. For instance, the WGRs have identical spectra if the error of their radius cutting is much less than $\Delta R = R/Q$. It is approximately 0.01 nm for a $Q = 10^8$ resonator of $2a = 0.2$ cm in diameter. Hence, it is unlikely that resonators with Q-factors exceeding a million could be efficiently coupled in a chain for the state of the art of the fabrication technology.

Even if nearly identical resonators are fabricated, their relative temperature should be stabilized better than $\Delta T = n/\kappa Q \simeq 2$ mK to avoid relative drift of the spectra, where $\kappa \simeq 10^{-5}$ K^{-1} and $n = 2.3$ are the combined thermal expansion/thermorefractive coefficient and the extraordinary index of refraction for lithium niobate, for instance. Such a thermal stabilization is challenging as well. The vertically coupled resonator chains are more compact than CROWs and, therefore, their temperature could be stabilized easier.

6.3 INTERFERENCE IN RESONATOR CHAINS

Coupled resonators can have properties mimicking those of coherent atomic media [38–44]. We here analyze the special configuration of two WGM resonators shown in Figure 6.2 [38,39] leading to sub natural (i.e., narrower than loaded), EIT-like linewidths. This scheme which utilizes linear ring resonators as WGM cavities has been widely discussed in the literature [45–47]. The existence of narrow spectral feature has been demonstrated experimentally [48] without revealing the analogy of atomic EIT and the interference effects in the resonator structure. In the case of resonators without absorption, the width of the feature may be arbitrarily narrow [38]; however, in reality, the minimum width of the resonance is again determined by the material absorption.

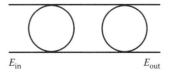

FIGURE 6.2 System of two identical WGM resonator side-coupled to two identical waveguides. (From Maleki, L., Matsko, A.B., Savchenko, A.A., and Ilchenko, V.S., *Opt. Lett.*, 29, 626, 2004. With permission.)

We find the transmission coefficient for such a configuration

$$T_P = \frac{[\gamma + i(\omega - \omega_1)][\gamma + i(\omega - \omega_2)]}{[2\gamma_c + \gamma + i(\omega - \omega_1)][2\gamma_c + \gamma + i(\omega - \omega_2)] - 4e^{i\psi}\gamma_c^2}, \quad (6.17)$$

where

- γ, γ_c, ω_1, and ω_2 are the linewidth originated from intrinsic cavity losses, linewidth due to coupling to a waveguide, and resonance frequencies of modes of the resonators, respectively
- ω is the carrier frequency (we assume that $|\omega - \omega_1|$ and $|\omega - \omega_2|$ are much less than the cavity free spectral range)
- ψ stands for the coupling phase, which may be adjusted by changing the distance between the cavities

Choosing $\exp i\psi = 1$ and assuming strong coupling regime $\gamma_c \gg |\omega_1 - \omega_2| \gg \gamma$, we see that the power transmission $|T_P|^2$ has two minima

$$|T_P|^2_{\min} \simeq \frac{\gamma^2}{4\gamma_c^2}$$

for $\omega = \omega_1$ and $\omega = \omega_2$, and a local maximum

$$|T_P|^2_{\max} \simeq \frac{(\omega_1 - \omega_2)^4}{[16\gamma\gamma_c + (\omega_1 - \omega_2)^2]^2}$$

for $\omega = \omega_0 = (\omega_1 + \omega_2)/2$. It is important to note that for $\gamma = 0$, the width of the transparency feature may be arbitrarily narrow. However, in reality, the resonance width is limited from below by a nonzero γ.

The origin of this "subnatural" structure in the transmission spectrum of the cavities is in the interference of the cavities' decay radiation. In fact, in the overcoupled regime considered here, the cavities decay primarily into the waveguides, and not into free space. Thus, there are several possible pathways for photons to be transmitted through the cavities, and the photons may interfere because they are localized in the same spatial configurations determined by the waveguides. The transmission is nearly cancelled when the light is resonant with one of the resonators' modes. However, in between the modes the interference results in a narrow transmission resonance. This phenomenon is similar to EIT originating from the interference decay, predicted theoretically in Ref. [49].

The group time delay originated from the narrow transparency resonance is approximately

$$\tau_g \simeq \frac{16\gamma_c(\omega_1 - \omega_2)^2}{[16\gamma\gamma_c + (\omega_1 - \omega_2)^2]^2} \gg \gamma_c^{-1}. \quad (6.18)$$

This delay exceeds the minimum group delay available from a single resonator.

To better understand the behavior of the system shown in Figure 6.2 let us study its eigenfrequencies which could be found from

$$[2\gamma_c + \gamma + i(\omega - \omega_1)][2\gamma_c + \gamma + i(\omega - \omega_2)] = 4\gamma_c^2, \quad (6.19)$$

where we assumed $\exp i\psi = 1$. It is easy to find roots of this equation. For the case of small frequency difference between the modes, $\gamma_c \gg |\omega_1 - \omega_2|$, we have

$$\omega_b \simeq \frac{\omega_1 + \omega_2}{2} + 4i\gamma_c, \quad (6.20)$$

$$\omega_d \simeq \frac{\omega_1 + \omega_2}{2} + i\gamma + i\frac{(\omega_1 - \omega_2)^2}{8\gamma_c}. \quad (6.21)$$

It is easy to see that there exists a normal mode (ω_d) that decays slowly compared with the mode decay of a single resonator, and a mode that decays faster (ω_b). The mode with the longer decay is responsible for the narrow spectral feature. However, the slowest decay is restricted by the cavity intrinsic Q-factor (rate γ). It means that the narrow resonance cannot be narrower than the narrowest possible resonance of a single subcritically coupled resonator.

6.4 RESONATOR-STABILIZED OSCILLATORS

Large group delays observed in resonators are useful for stabilization of microwave photonic oscillators [50]. Microwave oscillators capable of generating spectrally pure signals at high frequencies are important for a number of applications, including communications, navigation, radar, and precise tests and measurements. Conventional oscillators are based on electronic techniques that employ high quality factor resonators to achieve high spectral purity. The performance of these oscillators is nevertheless limited by the achievable Q's at room temperature, and by the sensitivity of the resonators to environmental perturbations, such as temperature and vibration. High-frequency microwave references may also be obtained by multiplication of signals generated by high-quality, but low-frequency (MHz), quartz oscillators. The noise associated with the multiplication steps, unfortunately, degrades the performance of the high-frequency signals beyond the levels required for high-end applications.

A powerful method of creating of high-purity signals is based on techniques of photonics, which are free of some of the intrinsic limitations of ultra high-frequency electronics. In particular, the opto electronic oscillator (OEO) is a photonic device that produces spectrally pure signals at many tens of GHz [51–57] limited only by the bandwidth of the modulators and detectors, which currently extends to the 100 GHz range.

The OEO is similar to a microwave oscillator in all but one key feature. In a microwave oscillator, stability is achieved through the stored microwave energy in a high-Q microwave resonator, and the oscillation frequency is locked to the frequency of the resonator. In an OEO the microwave energy, being carried as modulated light phase and/or amplitude, is stored in an optical delay line or an optical resonator. Thus, with the OEO there is no need for a microwave cavity. A long optical fiber at room temperature, for example, does the same function as a high-Q microwave resonator does at the liquid nitrogen temperature.

The OEO is a generic architecture consisting of a laser as the source of light energy. The laser radiation propagates through a modulator and an optical energy storage element, before it is converted to the electrical energy with a photodetector. The electrical signal at the output of the modulator is amplified and filtered before it is fed back to the modulator, thereby completing a feedback loop with gain, which generates sustained oscillation at a frequency determined by the filter.

Since the noise performance of an oscillator is determined by the energy storage time, or quality factor Q, the use of optical storage elements allows for the realization of extremely high Q's and thus spectrally pure signals. In particular, a long fiber delay allows realization of micro second storage times, corresponding to Q's of about a million at 10 GHz oscillation frequency. This is a high value compared with the conventional dielectric microwave cavities used in oscillators. The fiber delay line also provides for wide band frequency operation unhindered by the usual degradation of the oscillator Q with increasing frequency. Thus, spectrally pure signals at frequencies as high as 43 GHz, limited only by the modulator and detector bandwidth, have been demonstrated.

On the other hand, the long fiber delay line supports many microwave modes imposed on an optical wave. A narrow band electrical filter should be inserted into the electronic segment of the OEO feedback loop to achieve a stable single mode operation. The center frequency of this filter determines the operational frequency of the OEO. While this approach yields the desired spectrally pure high frequency signals, it nevertheless calls for an OEO configuration limited in size by the kilometers of fiber delay needed. Moreover, the long fiber delay is very sensitive to the surrounding environment so that the OEO does not produce an output with high long-term frequency accuracy and stability. The OEO is typically phase locked to a stable reference for long term stability.

The group delay of optical signal in resonant structures can be considered as a phase delay for the microwave signals impressed on the optical carrier. Hence, optical structures with large group delays can be used for stabilization of OEOs instead of long optical fiber delay lines as well as narrow band electrical filters [50]. For instance, an OEO also can be stabilized with either a high-Q optical cavity or an atomic cell. The first method allows one to choose virtually an arbitrary frequency of oscillation by tuning the cavity. The second method allows one to create a stable frequency references, if the OEO is locked to an atomic clock transition, and magnetometers, if the OEO is locked to magnetically sensitive transitions [58,59].

6.5 SLOW LIGHT IN SYSTEMS WITH A DISCRETE SPECTRUM

Group velocity is usually introduced in a class of problems where responsivity of a material can be considered as a continuous function of frequency. For instance, problems of propagation of narrowband light in systems like the atomic media belong to this class [1]. However, as was recently shown [60,61], for the broad class of systems with discreet spectra the common definition of the group velocity sometimes is not valid. An optical resonator is an example of such a system. The discreetness of the spectrum brings novel features to the notion of the group velocity defined as the velocity of a train of optical pulses. For instance, such a train can be delayed by a linear resonator much longer than the ring-down time of the resonator. Such a delay is impossible for a single pulse interacting with a linear lossless resonator, even though linear resonators as well as their chains can introduce a significant group delay [13,14,30,31].

This peculiarity arises when a conceptual transition is made from the framework of a distributed resonator to that of a lumped resonator. This is quite a common transition: while distributed resonators possess an infinite number of modes, only a finite number of modes are usually considered for their spectral studies, and often only a single mode is retained for the sake of simplicity. Such an approximation silently transforms the distributed object to a lumped one, discarding multiple phenomena, one of which is the subject of our study.

Let us outline the basic definition of the group velocity for a linear lossless medium with a continuous spectrum. A pulse of light propagates with speed equal to $V_g = c/n_g$ in such a medium, where the group index of refraction is defined as

$$n_g(\omega_0) = n(\omega_0) + \omega_0 \left.\frac{\partial n(\omega)}{\partial \omega}\right|_{\omega=\omega_0}, \qquad (6.22)$$

where

V_g is the group velocity
c is the speed of light in the vacuum
$n(\omega)$ is the frequency dependent refractive index of the material

Group index $n_g(\omega)$ can be considered frequency independent in a restricted frequency range $\delta\omega(\omega)$. The pulse of light with carrier frequency ω_0 does not change its shape while propagating in the medium if the spectral width of the pulse is much smaller than $\delta\omega(\omega_0)$.

Propagation of slow light in a dispersive medium can be characterized by the propagation of a beat note envelope of two plain monochromatic electromagnetic waves ($E_1 = \tilde{E}\exp(-i\omega_1 t + ik_1 z) + \text{c.c.}$ and $E_2 = \tilde{E}\exp(-i\omega_2 t + ik_2 z) + \text{c.c.}$) in the medium [62]. The beat note of the waves is described by $|E_1 + E_2|^2 = 2|\tilde{E}|^2[1 + \cos((\omega_1 - \omega_2)t - (k_1 - k_2)z)]$. The velocity of its propagation, $V_g = (\omega_1 - \omega_2)/(k_1 - k_2)$, corresponds to the conventional definition of the group velocity $\partial\omega/\partial k$ if $\omega_1 \to \omega_2$ and wave vector k is a continuous function of frequency ω.

Let us consider a lumped model of a ring resonator with a transfer function

$$H(\omega) = \frac{\gamma + i(\omega - \omega_0)}{\gamma - i(\omega - \omega_0)}, \qquad (6.23)$$

where γ is the full width at half maximum and ω_0 is the frequency of the resonance. A monochromatic signal with frequency ω acquires a phase shift $\arg[H(\omega)]$ when passing through the resonator. Let us assume that the length of the ring resonator is L. It is possible to write for the electric field of a monochromatic signal at the exit of the resonator the following expression

$$E(L,\omega) = E(0,\omega)e^{i\arg[H(\omega)]} = E(0,\omega)e^{ik(\omega)L}, \quad (6.24)$$

where $E(0,\omega)$ is the electric field at the entrance of the resonator. We have postulated in Equation 6.24 that propagation of light in the resonator is given by wave number k. Using Equation 6.23 we write

$$k(\omega) \equiv \frac{2}{L}\arctan\left(\frac{\omega-\omega_0}{\gamma}\right). \quad (6.25)$$

The group velocity for the resonator can be defined as

$$V_g(\omega) = \left(\frac{\partial k}{\partial \omega}\right)^{-1} = \frac{\gamma L}{2}\left[1 + \left(\frac{\omega-\omega_0}{\gamma}\right)^2\right]. \quad (6.26)$$

It is easy to see that the resonator delays the beat note of waves E_1 and E_2 by an amount of time $\tau_g(\omega_1,\omega_2) = L/V_g(\omega) \leq 2/\gamma$. The maximum delay is achieved for $\omega_1 \to \omega_2$, and, as a general rule, $\tau_g(\omega_1,\omega_2)(\omega_1-\omega_2) = L(k_1-k_2) \leq 2\pi$. As a result of the above conditions there is a general belief that the group delay introduced by a linear resonator cannot exceed the ring-down time of the resonator

$$\tau_r = \frac{\pi}{\gamma}. \quad (6.27)$$

For an optical pulse the restriction is even stronger if one assumes that the pulse propagates in the linear region of the dispersion. Let us assume that $\omega_{1,2} = \omega_0 \pm \gamma$. Then $\tau_g(\omega_1,\omega_2)(\omega_1-\omega_2) = L(k_1-k_2) = \pi$ and $\tau_g = \pi/(2\gamma) = \tau_r/2$.

The model transfer function (Equation 6.23) can be obtained from a model of a lossless distributed resonator possessing single mode family. Transfer function for such a distributed resonator is [15]

$$H(\omega) = -e^{i\varphi}\frac{1 - \sqrt{1-T}e^{-i\varphi}}{1 - \sqrt{1-T}e^{i\varphi}}, \quad (6.28)$$

where
$\varphi = 2\pi N + \Delta\varphi$ is the phase the light of frequency ω acquires during one round trip in the resonator
$2\pi N = \omega_0\tau_0$, $\tau_0 = Ln/c$ is the round trip time
$\Delta\varphi = (\omega - \omega_0)\tau_0$
T is the energy transmission coefficient of the front mirror

It is easy to see that $T/(2\tau_0) = \gamma$. Naturally Equations 6.23 and 6.28 are interchangeable (up to a constant phase shift) if $\Delta\varphi \ll 1$ and $T \ll 1$.

This conclusion is not valid for a WGM resonator or any other resonator that supports multiple modes [60,61]. The group delay in such a resonator can exceed its ring-down time significantly. This happens because a WGM resonator belongs to a class of the systems with a discrete optical spectrum. Such a resonator can be described using a lumped model within each spectral line. However, the model is not valid if the light interacts with multiple modes. The usual definition of the group velocity $\partial\omega/\partial k$ does not hold in this case. For example, expression $V_g = (\omega_1 - \omega_2)/(k_1 - k_2)$ is the only correct definition of the group velocity for the bichromatic field. A similar method should be applied to describe the propagation of a generalized optical field that has a discrete spectrum, e.g., a train of pulses, in a distributed resonator. The spectrum of the field consists of a series of arbitrarily narrowband (for an arbitrarily long train) lines enveloped by the Fourier-transform of an individual pulse. The number of "significant" spectral lines that are not too strongly suppressed by the envelope is given by essentially the duty cycle of the pulse train. This number may be just a few for a dense series of smooth (e.g., Gaussian) pulses. The group velocity of the train can be extremely small if the spectrum of the resonator with which the pulses interact is properly engineered.

We present the electric field inside the microsphere resonator as

$$E = \Psi e^{-i\omega t} + \text{c.c.}, \tag{6.29}$$

where the spatial field distribution has the general form

$$\Psi = \bar{\Psi} P_l^m(\cos\theta) J_{l+1/2}(k_{l,q} r) e^{im\phi} / \sqrt{r}, \tag{6.30}$$

θ, ϕ, and r are the spherical coordinates; indexes l, m, and q determine the spatial distribution of the field, $m = 0, 1, 2, \ldots$ and $q = 1, 2, \ldots$ are the azimuthal and radial quantization numbers, respectively, and $l = 0, 1, 2, \ldots$ is the orbital mode number. $\Psi(\theta, \phi, r)$ is the mode spatial profile, and

$$k_{l,q} \simeq \frac{1}{R}\left[l + \alpha_q \left(\frac{l}{2}\right)^{1/3}\right], \tag{6.31}$$

is the mode wave number, with $k_{l,q} = \omega n/c$, n the index of refraction of the resonator material, R the resonator radius, and α_q the qth root of the Airy function: $Ai(-\alpha_q) = 0$.

We now show that the group velocity of light propagating in the microsphere can have any desired value. We assume that a microsphere is excited with a bichromatic light by means of, e.g., a prism coupler. The frequencies of this light are resonant with two WGMs. Both modes have a nonzero electromagnetic field at the surface of the microsphere and, as a result, interfere. One can observe this interference by covering the microsphere with a fluorescent substance and taking advantage of the interaction of the evanescent field of the modes with the substance. The surface distribution of the power P scattered by the substance is described by

$$P \simeq \widetilde{P}(\theta) \left[1 + \cos(\Delta\omega_{ab} t - \Delta m_{ab} \phi)\right], \tag{6.32}$$

where $\widetilde{P}(\theta)$ is a normalization function, $\Delta\omega_{ab} = 2\pi \Delta\nu_{ab}$, and $\Delta m_{ab} = l_a - l_b$ when the modes belong to the fundamental azimuthal family. According to Equation 6.32, Δm_{ab} is the number of maxima of the interference pattern on the surface of the resonator. The pattern moves with a velocity $V_g = R\Delta\omega_{ab}/\Delta m_{ab}$ along the surface of the microsphere. This is the velocity of the beat note of the bichromatic light propagating in the resonator, i.e., the group velocity. It's value is equal to $1.4 \times 10^{-2} c$ for the selected WGMs $((a)$ and $(b))$. The value of the group velocity is much smaller when the frequency difference between the modes is small. The group velocity goes to zero if $\omega_a - \omega_b \to 0$, i.e., in this case the light is stopped inside the resonator. Finally, the group velocity is negative when $\Delta m_{ab} < 0$, or superluminal when $R/\Delta m_{ab}$ is large enough.

We have discussed the propagation of the light field (E) inside the resonator. Let us now find how the group velocity inside the resonator compares to the effective group velocity for the light that has passed through the resonator (group delay of the field E_{out} with respect to E_{in}). Consider a resonator with two fiber couplers. Assuming that the resonator is lossless and that the entrance/exit coupling efficiencies are identical, we infer that each harmonic of the bichromatic light that travels through the resonator experiences a phase shift:

$$E_{\text{out } a} = E_{\text{in } a} \exp(i\pi m_a), \quad (6.33)$$

$$E_{\text{out } b} = E_{\text{in } b} \exp(i\pi m_b). \quad (6.34)$$

The beat note of the harmonics acquires a phase shift $\pi(m_a - m_b)$, which corresponds to $\pi(l_a - l_b)$ for the main mode sequence. Keeping in mind that the light has traveled a distance πR, we find the group velocity

$$V_g = \frac{R \Delta \omega_{ab}}{l_a - l_b}. \quad (6.35)$$

The velocity can be either subluminal or superluminal depending on the WGMs that the light interacts with. The group delay $\tau_g = \pi(l_a - l_b)/\Delta\omega_{ab}$ does not depend on the spectral width of the modes, and consequently their ring-down times, thus proving the assertion made above.

It is easy to find now the value of product

$$\tau_g \Delta \nu_{ab} = \frac{l_a - l_b}{2}, \quad (6.36)$$

where $2\pi \Delta \nu_{ab} = \Delta \omega_{ab}$. This value can easily be bigger than 1. The fraction group delay is twice bigger for the resonator with a single prism coupler

$$\tau_g \Delta \nu_{ab} = l_a - l_b. \quad (6.37)$$

We have considered the possibility of the delay of a beat note of two and three monochromatic waves. In principle, it is possible to engineer the shape of the WGM resonator (not necessary a sphere) as well as the evanescent field coupler in such a way that a beat note of more than two waves could be delayed. It is possible to construct a WGM resonator based delay line for a train of optical pulses, as well.

To confirm this theoretical prediction we have performed an experiment with a WGM resonator to demonstrate the case of stopped light, $\Delta\omega_{ab} = 0$. The resonator, a microsphere with radius $R = 150$ μm, is fabricated with optical grade fused silica obtained from a multimode fiber. A taper is manually pulled out from the fiber using a hydrogen–oxygen micro torch. The thin end of the taper having approximately 50 μm in diameter is gradually heated in a hydrogen flame until the sphere of the required size appears.

We use a single unmodulated 635 nm diode laser to demonstrate zero group velocity of light that should be seen as a stationary interference pattern generated on the surface of the microsphere by running monochromatic waves. The light is sent into the resonator with an angle-cut fiber attached to a micromanipulator. The frequency of the laser is swept in a 5 GHz frequency span at the rate of 20 scans per second. The output of the fiber coupler is directed to an optical detector (Thorlabs Det110) and the signal from the detector is observed on an oscilloscope screen. The setup allows exciting different WGMs and selecting them by frequency. The coupling efficiency is controlled by the gap between the coupler and the surface of the silica resonator. Dry resonators made in this way have an intrinsic quality factor $Q \approx 10^9$ limited by the dust contamination of their surface. Nanometer scale dust particles attach to the resonator surface resulting in an easily recognizable surface glow.

The fiber coupler and the resonator are placed inside a fluidic mini-cell in the focal plane of a microscope for visualization of the pattern. Visualization at 670 nm is made using 1 μmol solution of fluorescent dye Cy5 in methanol. Elastically scattered radiation at 635 nm is blocked by a thin film notch filter installed between the cell and the microscope. Spectral measurements are conducted while visually observing the fluorescence of the dissolved dye at the surface of the resonator.

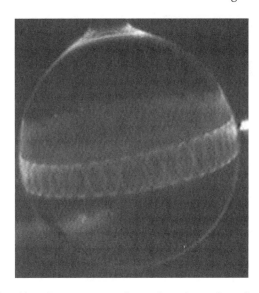

FIGURE 6.3 An example of interference pattern observed on the surface of a 300 μm-diameter optically pumped microsphere. Note the static interference pattern that exists despite the traveling-wave (clockwise) excitation. Light coupling is achieved with angle-cleaved fibers.

Absorption of light by the dye solution leads to a reduction of the Q-factor to $Q \approx 10^6$, which effectively eliminates the residual frequency difference of nearly degenerate WGMs belonging to different mode families. As a result, a monochromatic light can simultaneously interact with several WGMs. The reduction of the Q-factor also ensures the absence of Rayleigh scattering on the resonator surface, preventing light reflection inside the resonator.

The interference of different groups of modes resulted in different stationary fluorescence patterns (see, e.g., Figure 6.3 as well as patterns presented in Ref. [61]). The observed patterns are created by the interference of more than two spatially overlapping WGMs frequencies, which are within their spectral widths. The stationarity of these patterns created by the *running* wave confirms the theoretical prediction. In particular, the experiment allows observing the case of stopped light. It worth noting, that the stopped light does not carry any information because it represents a beat-note of several frequency degenerate running monochromatic waves.

Our observations are not in contradiction to the well known fact that modes belonging to the same optical resonator are orthogonal. Put in mathematical terms, given the wave equation and the particular boundary conditions of the resonator shape one obtains a set of eigenvalues and eigenvectors which by definition have zero overlap in time and space. Physically this means that all resonator modes must differ by either frequency or spatial distribution. This works perfectly well for WGM resonators, and our observation does not contradict it. We are able to observe the interference pattern because modes overlap in space due to their finite bandwidth, so a monochromatic light source simultaneously excites several modes.

Physically, the WGM resonator has nothing to do with the slow group velocity of light. This is intuitively obvious from the fact that the group delay does not depend on the quality factor of the resonator. The same kind of delay can be realized if one sends each harmonic of the pulse train into an optical waveguide of selected length (see Figure 6.4). The length of each waveguide should be bigger than the length of the previous one by a given value. The higher frequency corresponds to the longer waveguide. Such a selection maintains normal group velocity dispersion and results in the delay of the pulse train. Though the scheme in Figure 6.4 is obvious and well known, its realization is not always technically feasible, especially if the pulses have low repetition rate. An advantage of

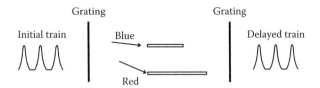

FIGURE 6.4 An equivalent scheme for the group delay line based on the WGM resonator. (From Matsko, A.B., Savchenko, A.A., Ilchenkov, V.S., Strekalov, D., and Maleki, L., *SPIE Proc.*, 6452, 64520P, 2007. With permission.)

the WGM resonator is in its ability to support multiple modes suitable for the realization of the delay. Another advantage is in a possibility of the direct visualization of the slow light propagation.

REFERENCES

1. L. V. Hau, S. E. Harris, Z. Dutton, and C. H. Behroozi, Light speed reduction to 17 metres per second in an ultracold atomic gas, *Nature* 397, 594–598 (1999).
2. A. Kasapi, M. Jain, G. Y. Yin, and S. E. Harris, Electromagnetically induced transparency: propagation dynamics, *Phys. Rev. Lett.* 74, 2447–2450 (1995).
3. R. W. Boyd, D. J. Gauthier, A. L. Gaeta, and A. E. Willner, Maximum time delay achievable on propagation through a slow-light medium, *Phys. Rev. A* 71, 023801 (2005).
4. A. B. Matsko, D. V. Strekalov, and L. Maleki, On the dynamic range of optical delay lines based on coherent atomic media, *Opt. Express* 13, 2210–2223 (2005).
5. J. B. Khurgin, Optical buffers based on slow light in electromagnetically induced transparent media and coupled resonator structures: comparative analysis, *J. Opt. Soc. Am. B* 22, 1062–1074 (2005).
6. R. M. Camacho, M. V. Pack, and J. C. Howell, Low-distortion slow light using two absorption resonances, *Phys. Rev. A* 73, 063812 (2006).
7. B. Macke and B. Segard, Propagation of light-pulses at a negative group-velocity, *Eur. Phys. J. D* 23, 125–141 (2003).
8. R. M. Camacho, M. V. Pack, and J. C. Howell, Low-distortion slow light using two absorption resonances, *Phys. Rev. A* 74, 033801 (2006).
9. A. B. Matsko, A. A. Savchenkov, and L. Maleki, Vertically coupled whispering-gallery-mode resonator waveguide, *Opt. Lett.* 30, 3066–3068 (2005).
10. L. Maleki, A. B. Matsko, A. A. Savchenkov, and D. Strekalov, Slow light in vertically coupled whispering gallery mode resonators, *Proc. SPIE* 6130, 61300R (2006).
11. A. A. Savchenkov, I. S. Grudinin, A. B. Matsko, D. Strekalov, M. Mohageg, V. S. Ilchenko, and L. Maleki, Morphology-dependent photonic circuit elements, *Opt. Lett.* 31, 1313–1315 (2006).
12. I. S. Grudinin, A. Savchenkov, A. B. Matsko, D. Strekalov, V. Ilchenko and L. Maleki, Ultra high Q crystalline microcavities, *Opt. Commun.* 265, 33–38 (2006).
13. A. Yariv, Y. Xu, R. K. Lee, and A. Scherer, Coupled-resonator optical waveguide: a proposal and analysis, *Opt. Lett.* 24, 711–713 (1999).
14. A. Melloni, F. Morichetti, and M. Martinelli, Linear and nonlinear pulse propagation in coupled resonator slow-wave optical structures, *Opt. Quantum Electron.* 35, 365–379 (2003).
15. J. E. Heebner, P. Chak, S. Pereira, J. E. Sipe, and R. W. Boyd, Distributed and localized feedback in microresonator sequences for linear and nonlinear optics, *J. Opt. Soc. Am. B* 21, 1818–1832 (2004).
16. J. K. S. Poon, L. Zhu, G. A. DeRose, and A. Yariv, Transmission and group delay of microring coupled-resonator optical waveguides, *Opt. Lett.* 31, 456–458 (2006).
17. J. K. S. Poon, L. Zhu, G. A. DeRose, and A. Yariv, Polymer microring coupled-resonator optical waveguides, *J. Lightw. Technol.* 24, 1843–1849 (2006).
18. G. Lenz, B. J. Eggleton, C. R. Giles, C. K. Madsen, and R. E. Slusher, Dispersive properties of optical filters for WDM systems, *IEEE J. Quantum Electron.* 34 1390–1402 (1998).
19. G. Lenz, B. J. Eggleton, C. K. Madsen, and R. E. Slusher, Optical delay lines based on optical filters, *IEEE J. Quantum Electron.* 37 525–532 (2001).

20. F. N. Xia, L. Sekaric, and Y. Vlasov, Ultracompact optical buffers on a silicon chip, *Nat. Photon.* 1, 65–71 (2007).
21. M. Sumetsky, Uniform coil optical resonator and waveguide: transmission spectrum, eigenmodes, and dispersion relation, *Opt. Express* 13, 4331–4340 (2005).
22. M. Sumetsky, Vertically-stacked multi-ring resonator, *Opt. Express* 13, 6354–6375 (2005).
23. L. Tong, R. R. Gattass, J. B. Ashcom, S. He, J. Lou, M. Shen, I. Maxwell, and E. Mazur, Subwavelength-diameter silica wires for low-loss optical wave guiding, *Nature* 426, 816–819 (2003).
24. K. J. Vahala, Optical microcavities, *Nature* 424, 839–846 (2003).
25. A. B. Matsko and V. S. Ilchenko, Optical resonators with whispering gallery modes I: basics, *J. Sel. Top. Quantum Electron.* 12, 3–14 (2006).
26. E. Yablonovitch, Inhibited spontaneous emission in solid-state physics and electronics, *Phys. Rev. Lett.* 58, 2059–2062 (1987).
27. J. B. Khurgin, Light slowing down in Moiré fiber gratings and its implications for nonlinear optics, *Phys. Rev. A* 62, 013821 (2000).
28. V. N. Astratov, R. M. Stevenson, I. S. Culshaw, D. M. Whittaker, M. S. Skolnick, T. F. Krauss, and R. M. De La Rue, Heavy photon dispersions in photonic crystal waveguides, *Appl. Phys. Lett.* 77, 178–180 (2000).
29. R. Binder, Z. S. Yang, N. H. Kwong, D. T. Nguyen, and A. L. Smirl, Light pulse delay in semiconductor quantum well Bragg structures, *Phys. Stat. Solidi B* 243, 2379–2383 (2006).
30. J. E. Heebner, R. W. Boyd, and Q. H. Park, Slow light, induced dispersion, enhanced nonlinearity, and optical solitons in a resonator-array waveguide, *Phys. Rev. E* 65, 036619 (2002).
31. D. D. Smith, H. Chang, and K. A. Fuller, Whispering-gallery mode splitting in coupled microresonators, *J. Opt. Soc. Am. B* 20, 1967–1974 (2003).
32. M. F. Yanik and S. Fan, Stopping light all optically, *Phys. Rev. Lett.* 92, 083901 (2004).
33. S. Fan, Sharp asymmetric line shapes in side-coupled waveguide-cavity systems, *Appl. Phys. Lett.* 80, 908–910 (2002).
34. S. Fan, W. Suh, and J. D. Joannopoulos, Temporal coupled-mode theory for the Fano resonance in optical resonators, *J. Opt. Soc. Am. A* 20, 569–572 (2003).
35. E. I. Smotrova, A. I. Nosich, T. M. Benson, and P. Sewell, Threshold reduction in a cyclic photonic molecule laser composed of identical microdisks with whispering-gallery modes, *Opt. Lett.* 31, 921–923 (2006).
36. M. L. Povinelli and S. H. Fan, Radiation loss of coupled-resonator waveguides in photonic-crystal slabs, *Appl. Phys. Lett.* 89, 191114 (2006).
37. D. P. Fussell and M. M. Dignam, Engineering the quality factors of coupled-cavity modes in photonic crystal slabs, *Appl. Phys. Lett.* 90, 183121 (2007).
38. L. Maleki, A. B. Matsko, A. A. Savchenkov, and V. S. Ilchenko, Tunable delay line with interacting whispering-gallery-mode resonators, *Opt. Lett.* 29, 626–628 (2004).
39. A. B. Matsko, A. A. Savchenkov, D. Strekalov, V. S. Ilchenko, and L. Maleki, Interference effects in lossy resonator chains, *J. Mod. Opt.* 51, 2515–2522 (2004).
40. M. F. Yanik, W. Suh, Z. Wang, and S. H. Fan, Stopping light in a waveguide with an all-optical analog of electromagnetically induced transparency, *Phys. Rev. Lett.* 93, 233903 (2004).
41. A. Naweed, G. Farca, S. I. Shopova, and A. T. Rosenberger, Induced transparency and absorption in coupled whispering-gallery microresonators, *Phys. Rev. A* 71, 043804 (2005).
42. Y. P. Rakovich, J. J. Boland, and J. F. Donegan, Tunable photon lifetime in photonic molecules: a concept for delaying an optical signal, *Opt. Lett.* 30, 2775–2777 (2005).
43. Q. F. Xu, S. Sandhu, M. L. Povinelli, J. Shakya, S. H. Fan, and M. Lipson, Experimental realization of an on-chip all-optical analogue to electromagnetically induced transparency, *Phys. Rev. Lett.* 96, 123901 (2006).
44. Q. Xu, J. Shakya, and M. Lipson, Direct measurement of tunable optical delays on chip analogue to electromagnetically induced transparency, *Opt. Express* 14, 6463–6468 (2006).
45. B. E. Little, S. T. Chu, H. A. Haus, J. Foresi, and J. P. Laine, Microring resonator channel dropping filters, *J. Lightw. Technol.* 15, 998–1005 (1997).
46. P. Urquhart, Compound optical-fiber-based resonators, *J. Opt. Soc. Am. A* 5, 803–812 (1988).
47. K. Oda, N. Takato, and H. Toba, A wide-FSR waveguide double-ring resonator for optical FDM transmission systems, *J. Lightw. Technol.* 9, 728–736 (1991).
48. S. T. Chu, B. E. Little, W. Pan, T. Kaneko, and Y. Kukubun, Second-order filter response from parallel coupled glass microring resonators, *IEEE Photon. Technol. Lett.* 11, 1426–1428 (1999).

49. A. Imamoglu, Interference of radiatively broadened resonances, *Phys. Rev. A* 40, 2835–2838 (1989).
50. D. Strekalov, D. Aveline, N. Yu, R. Thompson, A. B. Matsko, and L. Maleki, Stabilizing an optoelectronic microwave oscillator with photonic filters, *J. Lightw. Technol.* 21, 3052–3061 (2003).
51. X. S. Yao and L. Maleki, Optoelectronic microwave oscillator, *J. Opt. Soc. Am. B* 13, 1725–1735 (1996).
52. Y. Ji, X. S. Yao, and L. Maleki, Compact optoelectronic oscillator with ultralow phase noise performance, *Electron. Lett.* 35, 1554–1555 (1999).
53. T. Davidson, P. Goldgeier, G. Eisenstein, and M. Orenstein, High spectral purity CW oscillation and pulse generation in optoelectronic microwave oscillator, *Electron. Lett.* 35, 1260–1261 (1999).
54. S. Romisch, J. Kitching, E. Ferre-Pikal, L. Hollberg, and F. L. Walls, Performance evaluation of an optoelectronic oscillator, *IEEE Trans. Ultraconics Ferroelectrics Freq. Control* 47, 1159–1165 (2000).
55. X. S. Yao and L. Maleki, Multiloop optoelectronic oscillator, *IEEE J. Quantum Electron.* 36, 79–84 (2000).
56. S. Poinsot, H. Porte, J. P. Goedgebuer, W. T. Rhodes, and B. Boussert, Continuous radio-frequency tuning of an optoelectronic oscillator with dispersive feedback, *Opt. Lett.* 27, 1300–1302 (2002).
57. D. H. Chang, H. R. Fetterman, H. Erlig, H. Zhang, M. C. Oh, C. Zhang, and W. H. Steier, 39-GHz optoelectronic oscillator using broad-band polymer electrooptic modulator, *IEEE Photon. Technol. Lett.* 14, 191–193 (2002); ibid 14, 579 (2002).
58. A. B. Matsko, D. Strekalov, and L. Maleki, Magnetometer based on the opto-electronic microwave oscillator, *Opt. Commun.* 247, 141–148 (2005).
59. D. Strekalov, A. B. Matsko, N. Yu, A. A. Savchenkov, and L. Maleki, Application of vertical cavity surface emitting laser in self-oscillating atomic clocks, *J. Mod. Opt.* 53, 2469–2484 (2006).
60. A. B. Matsko, A. A. Savchenkov, V. S. Ilchenko, D. Strekalov, and L. Maleki, The maximum group delay in a resonator: an unconventional approach, *SPIE Proc.* 6452, 64520P (2007).
61. A. A. Savchenkov, A. B. Matsko, V. S. Ilchenko, D. Strekalov, and L. Maleki, Direct observation of stopped light in a whispering-gallery-mode microresonator, *Phys. Rev. A* 76, 023816 (2007).
62. M. M. Kash, V. A. Sautenkov, A. S. Zibrov, L. Hollberg, G. R. Welch, M. D. Lukin, Y. Rostovtsev, E. S. Fry, and M. O. Scully, Ultraslow group velocity and enhanced nonlinear optical effects in a coherently driven hot atomic gas, *Phys. Rev. Lett.* 82, 5229–5232 (1999).

7 Disordered Optical Slow-Wave Structures: What Is the Velocity of Slow Light?

Shayan Mookherjea

CONTENTS

7.1 Introduction: The Tight-Binding Optical Waveguide 120
    7.1.1 Slow-Wave Dispersion Relationship 120
    7.1.2 Optical Signal Processing: The Next Generation? 121
    7.1.3 Puzzling Question: The Velocity of Disordered Light 122
    7.1.4 Care Needed When Using "Bandsolver" Simulation Tools 123
    7.1.5 Density of States 123
7.2 Formalism 124
7.3 Spectrum of the Solutions: General Principles 125
    7.3.1 Experimental Determination of M and the Coupling Coefficients 125
7.4 Special Forms of the Coupling Matrix 126
    7.4.1 Quick Method to Calculate the Density of States $\rho(\omega)$ 127
7.5 Models of Disorder and the Calculation of $\rho(\omega)$ 128
    7.5.1 Randomness in the Coupling Coefficients 128
    7.5.2 Randomness in the Diagonal Terms 130
7.6 Velocity of Slow Light in Disordered Structures 131
7.7 Localization of Fields 135
7.8 Summary 140
Acknowledgments 140
Appendix A: Two Pendulum Bobs Coupled by a Spring 140
Appendix B: Derivation of Equation 7.6 and the Coupling Coefficient κ 141
    Directional Waveguide Couplers 142
References 143

Optical slow-wave structures such as coupled photonic crystal cavities, coupled microrings, coupled quantum wells, etc. are promising candidates for slow light devices because of the nature of their dispersion relationship which follows the tight-binding model well-known in solid-state physics. In a perfectly uniform structure, the slope of the dispersion curve, which defines the group velocity, is zero at the edges of the waveguide band, leading to slow light and enhanced light–matter interaction. Although the effects of group velocity dispersion (and higher-order dispersion) can limit the performance of slow light structures [1–4], the effects of dispersion can, in principle, be compensated by a variety of mechanisms, including some of which are used commonly in optical fiber communications.

A more fundamental limit is imposed by the effects of disorder which may arise either from roughness, or more generally, from a number of issues related to fabrication or self-assembly imperfections. All the known methods of fabricating optical slow-wave structures show structural imperfections on at least the nanometer length scale. Although the resultant effect of such small disorder may be thought to be very weak, we show that in fact, the consequences on the slowing down of light can be severe. Here, we derive and discuss a model of the effects of weak disorder on the velocity of slow light in such structures, which also applies to studies of ballistic transport in any weakly disordered tight-binding lattice.

7.1 INTRODUCTION: THE TIGHT-BINDING OPTICAL WAVEGUIDE

It is rare to find (or better still, invent) a physical waveguiding mechanism that allows a significant degree of control over how light propagates—over its velocity, or its dispersion, for example. Optical slow-wave structures such as coupled photonic crystal cavities, coupled microrings, coupled Fabry–Perot cavities, etc. are "engineered dispersion" waveguides which can be lithographically patterned or self-assembled on length scales comparable to the optical wavelength. Our hope is that they lead to the invention of photonic chip-scale devices with unprecedented control over light propagation.

Optical slow-wave structures consist of a chain or network of repeated unit cells in which light propagates by hopping from one unit cell to its nearest neighbors. Each unit cell could consist, e.g., of a microring resonator, a defect resonator in a photonic crystal, or a Fabry–Perot cavity. This underlying physical principle of nearest-neighbor photon hopping can be used to derive an analytical description of the waveguide dispersion, of pulse propagation, and of various nonlinear effects such as second-harmonic generation or the optical Kerr effect.

The tight-binding coupled harmonic oscillator model used in these calculations is important not only in optics, but also in many other branches of physics. It is one of the most fundamental and widely applied physical models of propagation, used extensively in solid-state physics to describe electron propagation and phonon vibration in crystalline lattices. As there are many analogies between the Schrödinger equation and the electromagnetic wave equation [5,6], it is not surprising that the tight-binding theory should also have applications in optical structures, particularly those related to lattices or photonic crystals. Indeed, recent theoretical and experimental research has developed many aspects of the tight-binding model for photonics [7–15].

7.1.1 Slow-Wave Dispersion Relationship

In the tight-binding method applied to a chain of unit cells in a one-dimensional (1D) lattice, indexed by $n = 1, 2, \ldots, N$ and with intersite distance R, an excitation is described as a set of time-dependent coefficients, $\{a_n\}$, which represent the oscillation amplitudes of the individual field modes at the lattice sites. The evolution of the amplitudes are given by

$$i\frac{da_n}{dt} + \Omega a_n + \Omega \kappa (a_{n-1} + a_{n+1}) = 0, \tag{7.1}$$

where Ω and κ are the self-coupling and the (dimensionless) nearest-neighbor coupling coefficients, respectively (see Table 7.1). Assuming periodic boundary conditions,* we guess a solution of the form

$$a_n(t) = \frac{1}{\sqrt{N}} e^{i(\omega t - nk_m R)}, \tag{7.2}$$

---

* In assuming periodic boundary conditions, we rely on our physical intuition that the "bulk" optical properties within the waveguide are unaffected by the exact nature of the boundaries. In chains consisting of only a few unit cells (say, less than 10), this is obviously not true [16]. Chains of resonators with gain may also be sensitive to boundary conditions [17].

TABLE 7.1
Typical Values of Microresonator Coupling Coefficients

	$|\kappa|$	Reference
Microspheres		
2–5 µm polystyrene	$2.8 - 3.5 \times 10^{-3}$	[47]
4.2 µm polystyrene	1.3×10^{-2}	[74]
2D photonic crystal defects (one missing hole, H1, even symmetry)		
1 row between defects	4.7×10^{-2}	[28]
2 rows between defects	5.4×10^{-3}	[19]
3 rows between defects	3.7×10^{-3}	[75]
	1.3×10^{-3}	[19]
Microrings		
60 µm radius polymer (poly methyl methacrylate [PMMA])	1.2×10^{-1}	[9]
6.5–9 µm radius Si on SiO_2	$0.22 - 0.34$	[14]
Fabry–Perot		
Silicon superlattice	1.23×10^{-2}	[76]

where $k_m = m \, 2\pi/(NR)$. Substituting Equation 7.2 into Equation 7.1, we obtain

$$\frac{\omega_m}{\Omega} = 1 + 2\kappa \, \cos(k_m R), \quad m = 0, 1, \ldots, N-1 \quad (7.3)$$

are the N normalized eigenfrequencies of the chain. In the limit of large N, Equation 7.3 defines ω as a continuous function of k, which, in the first Brillouin zone, takes values from $-\pi/R$ to π/R. Analogous dispersion relationships in 2D and 3D are shown in Figure 7.1. As $k \to 0, \pm\pi/R$, the group velocity $v_g = d\omega/dk \to 0$, i.e., slow light is expected at the edges of the band. Slow light may also be observed at band-center, with a slowing factor that depends on the magnitude of the coupling coefficient: The time taken to propagate from input to output for light of frequency $\omega = \Omega$ (at band-center) is $\tau_{\min} = N/(2\Omega|\kappa|)$.

The eigenvectors corresponding to the eigenfrequencies given by Equation 7.3 are known as the normal modes of the structure, and allow us to calculate the Green function and the density of states per site [18].

If there exists significant coupling between next-to-nearest neighbor unit cells, in addition to the nearest-neighbor coupling considered above, then the dispersion relationship is modified as shown in [19]. Close spacing between adjacent unit cells, or coupling to slab modes or radiation modes may give rise to inter-resonator coupling over distances exceeding the nearest-neighbor spacing.

7.1.2 Optical Signal Processing: The Next Generation?

Photonic devices which contain several optical resonators and involve interactions between them are becoming important in optical signal processing [1,15,20–24] and may also be described by the tight-binding or nearest-neighbor coupling model. Many potential applications of coupled-resonator waveguides and filters, e.g., for delay lines, optical digital filters, and optical buffers, will benefit from cascading a larger number of resonators. For example,

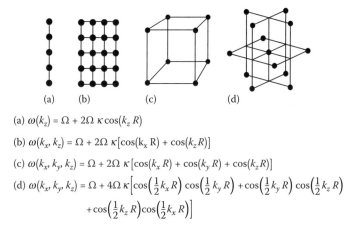

(a) $\omega(k_z) = \Omega + 2\Omega\,\kappa\cos(k_z R)$

(b) $\omega(k_x, k_z) = \Omega + 2\Omega\,\kappa[\cos(k_x R) + \cos(k_z R)]$

(c) $\omega(k_x, k_y, k_z) = \Omega + 2\Omega\,\kappa[\cos(k_x R) + \cos(k_y R) + \cos(k_z R)]$

(d) $\omega(k_x, k_y, k_z) = \Omega + 4\Omega\,\kappa\left[\cos\left(\tfrac{1}{2}k_x R\right)\cos\left(\tfrac{1}{2}k_y R\right) + \cos\left(\tfrac{1}{2}k_y R\right)\cos\left(\tfrac{1}{2}k_z R\right) + \cos\left(\tfrac{1}{2}k_z R\right)\cos\left(\tfrac{1}{2}k_x R\right)\right]$

FIGURE 7.1 Schematics of some optical slow-wave coupled-resonator structures where each black circle corresponds to a single unit cell, consisting of a resonator, coupled to its nearest-neighbors indicated by lines. Unit cells may be laid out (a) along straight lines, or in (b) 2D or (c) 3D simple cubic arrays, or in more complicated arrangements such as (d) the face-centered cubic lattice. In each case, the distance between neighboring unit cells is taken as R, and the dispersion relationships are indicated, assuming that the fields in the unit cells are s-like, i.e., real, nondegenerate and only dependent on $|\mathbf{r}|$, without angular variations.

1. The time taken by a pulse to traverse a coupled-resonator delay line scales linearly with N, the number of resonators [1].
2. In the coupled-resonator implementation of a digital filter, the average group delay scales linearly with N, as does the nonlinear sensitivity, which measures the nonlinear phase change per unit intensity from the optical Kerr effect [25].
3. The number of resonators needed to store N_{st} bits in a coupled-resonator "optical buffer" scales as $(N_{st})^{3/2}$ (taking into the account the effects of higher-order dispersion) [3].

Such optical structures are described by the tight-binding model, and in this chapter, they will generally be called coupled-resonator optical waveguides (CROWs) [26]. The fundamental mechanism of light propagation in CROWs is different than in conventional waveguides: A CROW is formed by placing optical resonators in a linear (or 2D/3D) array, so as to guide light from one end of this chain to the other by photon hopping between adjacent resonators, i.e., the spatio-temporal overlap of the electromagnetic fields of the resonators [8,27]. Usually 10–100 coupled resonators are sufficient to experimentally map out the tight-binding dispersion relationship [28,29].

However, it is unlikely that either the eigenfrequency of the resonators or the inter-resonator coupling coefficient will be exactly identical throughout a large ensemble. The study of disorder is important in any practical optical device, but it is especially important for a waveguiding structure such as the coupled-resonator chain, which relies on the constructive interference of the pairwise interactions of many repeated unit cells.

7.1.3 Puzzling Question: The Velocity of Disordered Light

According to the tight-binding dispersion relation, Equation 7.3, slow light should be observed at the edges of the waveguiding band. In fact, $v_g \equiv d\omega/dk \to 0$ as $k \to \pm\pi/R$, i.e., the velocity of light should go to zero—a reduction in v_g over six orders of magnitude compared to its typical value at band-center—but experimental observations (cited later) measure a band-edge reduction of v_g of only one or two orders of magnitude. This large disagreement merits explanation, since optical and electron microscope images suggest good control over the fabrication process with few nonuniformities in the structure, and sidewall roughness not exceeding a few nanometers [14,30], leaving little room for future improvement. Moreover, band-gap characteristics of periodic dielectric slow-wave structures are largely preserved in the presence of weak disorder [31,32].

This raises the fundamental question: in a weakly perturbed tight-binding lattice such that ballistic excitations can still propagate from one end to the other of the chain, what is the velocity of such excitations, and how does it scale with disorder?

7.1.4 Care Needed When Using "Bandsolver" Simulation Tools

Optical slow-wave structures, including photonic crystal waveguides and CROWs are frequently studied using plane-wave or eigenfunction expansion [33] or finite-difference time-domain method [19] algorithms. One may consider adding random perturbations to the dielectric distribution and calculating the new bandstructure. Such a calculation, using a plane-wave expansion method, is shown in Figure 7.2. Even if the effects of disorder can be seen in the spatial distribution of the field, the dispersion relationship still predicts zero velocity of light at the band-edges.

This is wrong: the error comes from the assumption of the periodic boundary conditions used to restrict the calculation region to a finite-size unit cell, which always leads to the band bending to zero slope at the edges of the dispersion relationship, and thus a false prediction of zero velocity.

7.1.5 Density of States

A deeper understanding of what happens at the band edge of an $\omega - k$ dispersion relationship comes from a new framework for analyzing optical slow-wave structures, which yields, as the main quantities of interest, $\rho(\omega)$ and $\rho(k)$, the densities of states in ω and k spaces, i.e., the number of

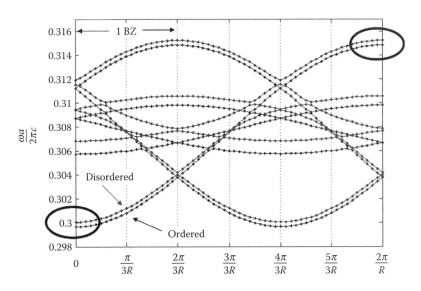

FIGURE 7.2 The dispersion relationship, in the repeated zone scheme, for a slow-wave structure consisting of coupled defect resonators in a photonic crystal slab ($n_{slab} = 2.65$), with lattice periodicity $a = 477$ nm, hexagonal arrangement of air holes, and inter-resonator spacing $R = 1.65$ μm. A super cell of three unit cells was used to calculate the dispersion using a plane-wave expansion method; hence, the unfolded band extends across three Brillouin zones (BZ). In the disordered case (blue lines), 5 nm of root-mean-squared (r.m.s.) positional inaccuracy was incorporated, compared to the ordered case (black lines). At the edges of the slow-wave band, the slope $d\omega/dk$ still goes to zero, because of enforcing periodic boundary conditions at the edges of the supercell.

eigenmodes per unit volume and per unit optical frequency or wavenumber, detuning from the center of the waveguiding band.*

It is shown by numerically calculating the eigenfrequencies of weakly disordered structures of a modestly finite length, that these eigenfrequencies lie on a curve for which an analytical expression is known. Thereafter, one can use the analytical expression for $\rho(\omega)$ to obtain answers about the velocity of propagation and localization in disordered lattices as if one knows $\rho(\omega)$ with greater accuracy. The disordered $\rho(k)$ is obtained from first-order perturbation theory, consistent with the assumption of weak disorder.

Although the density of states is obtained from a time-domain formalism, which is the natural approach for studying resonances, lasing, and other collective oscillatory phenomena in coupled resonators, there are other approaches one can use to calculate $\rho(\omega)$. A spatial transfer matrix approach [34] is often used to investigate transmission in 1D random chains, and to study localization phenomena [35,36]. There is a vast literature on this subject in condensed matter physics related to calculation of the Lyapunov exponents and Anderson localization [18,37–40].

7.2 FORMALISM

The coupling of the nearest-neighbor unit cells in an optical slow-wave waveguide is usually well described by coupled-mode theory, which is a well-known perturbative description of the induced interactions between nominally orthogonal modes of resonators. Louisell's formalism of two pendula coupled by a spring [41] describes the salient features of this model (see Appendix A). Other useful references are the descriptions of coupled optical resonators in the time domain [42–44]. Similar equations are used in the quantum mechanics of various spin-1/2 systems, Bloch equations, calculation of the Rabi frequency, etc. [45] as well as optical analogues of quantum-mechanical two-level systems [46,47].

Assuming that the slow-wave structure consists of N unit cells, the field in each unit cell is characterized by m eigenmodes. A column vector (state vector) lists these field components,

$$\mathbf{u} = \begin{pmatrix} \mathbf{u}_1 \\ \mathbf{u}_2 \\ \vdots \\ \mathbf{u}_N \end{pmatrix}, \tag{7.4}$$

where $\mathbf{u}_l = (a_l \; b_l \; c_l \; \ldots \; m_l)^T$ describes the m eigenmodes of the lth unit cell. Physically, these scalar coefficients appear in the description of the electric field,

$$\mathbf{E}(\mathbf{r}, t) = \sum_{l=1}^{N} \frac{1}{2} \left[\hat{\mathbf{e}}_{a_l}(\mathbf{r}) \, a_l(t) \, e^{i\omega_{a_l} t} + \hat{\mathbf{e}}_{b_l}(\mathbf{r}) \, b_l(t) \, e^{i\omega_{b_l} t} + \cdots + \hat{\mathbf{e}}_{m_l}(\mathbf{r}) \, m_l(t) \, e^{i\omega_{m_l} t} \right] + \text{c.c.} \tag{7.5}$$

as the slowly varying complex amplitudes which multiply the normalized modes of the unit cell, where $\hat{\mathbf{e}}_{a_l}(\mathbf{r})$ is a vector field with unit magnitude and describes the spatial profile of one particular mode and ω_{a_l} is its eigenfrequency, etc. We will assume that the eigenfrequencies are degenerate ($\omega_{a_l} = \omega_{b_l} = \cdots = \omega_{m_l}$) or close enough in value to permit adiabatic coupling between the modes within the framework of linear optics. In a slow-wave structure consisting of cascaded microring resonators, e.g., each unit cell consists of a single ring resonator. Aside from the longitudinal mode spectrum imposed by the periodic boundary conditions along the direction of wave propagation,

---

* In fact, one can guess that $\rho(\omega)$ and $\rho(k)$ will be needed, since the definition of group velocity, $v_g \equiv d\omega/dk$ involves a ratio of the changes in ω and in k around a particular $\omega - k$ point, and such changes are connected, of course, to the density of points (states) along the ω and k axes.

we count two modes at each frequency because of the two-fold degeneracy in the direction of field circulation (clockwise or counter-clockwise).

As derived in Appendix B, the time evolution of **u** is determined by a matrix M,

$$i\frac{d}{dt}\mathbf{u} = M\mathbf{u}, \qquad (7.6)$$

where M has the following structure,

$$M = \begin{pmatrix} \Omega_1 & C_{12} & 0 & 0 & \cdots \\ C_{21} & \Omega_2 & C_{23} & 0 & \cdots \\ 0 & C_{32} & \Omega_3 & C_{34} & \cdots \\ 0 & 0 & C_{43} & \Omega_4 & \cdots \\ \vdots & \vdots & \vdots & \vdots & \end{pmatrix}, \qquad (7.7)$$

with each of the Ω_l's and C_{kl}'s themselves being $(m \times m)$ square matrices. The diagonal elements of Ω_l are the self-coupling coefficients of the lth unit cell, the off-diagonal elements of Ω_l represent the coupling of the internal modes of a single unit cell (e.g., coupling between the clockwise and counter-clockwise modes of a ring resonator because of Rayleigh scattering from the slightly uneven sidewalls [48]) and C_{kl} represent the coupling coefficients between adjacent unit cells.

7.3 SPECTRUM OF THE SOLUTIONS: GENERAL PRINCIPLES

Eigensolutions of Equation 7.6 have the time-evolution behavior $\mathbf{u} \sim \exp(i\omega t)$ and, therefore, are found by solving the determinantal equation in the variable ω,

$$|M - \omega I| = 0, \qquad (7.8)$$

where I is the unit matrix. Algorithms to calculate the eigenvalues of large matrices (with complex entries) are described in textbooks [49]; some specific cases of practical importance are discussed in Section 7.4.

M has a total of $N \times m$ eigenvalues, not all of which may be distinct. We will label the set of eigenvalues by $\{\omega_1, \omega_2, \ldots\}$ and the eigenvectors by $\{\mathbf{v}_1, \mathbf{v}_2, \ldots\}$. Assuming that the unit cells are nominally identical ($\omega_{a_l} = \omega_{b_l} = \cdots = \omega_{m_l}$), real solutions to Equation 7.8 correspond to collective resonances of the system which have a fixed temporal phase relationship between the individual unit cells, and are sometimes called super-modes. (This phase relationship may be stabilized in principle by nonlinear effects [50] or by gain [51]; neither of these additional complications are discussed here.)

7.3.1 EXPERIMENTAL DETERMINATION OF M AND THE COUPLING COEFFICIENTS

How can one experimentally estimate the elements of the matrix M, i.e., the coupling coefficients? If we measure the eigenfrequencies and the corresponding eigenvectors (i.e., excitation amplitudes of the unit cells) by a combination of imaging and spectroscopic techniques [13,52], then we may construct M as follows.

If we assume that the matrix M has distinct eigenvalues, i.e., $\omega_i \neq \omega_j$ for $i \neq j$, or is Hermitian (even if the eigenvalues are not distinct), then

$$M = T \operatorname{diag}\{\omega_1, \omega_2, \ldots\} T^{-1} \qquad (7.9)$$

where T is the matrix constructed from columns of the (linearly independent) eigenvectors, $T = [\mathbf{v}_1 \mathbf{v}_2 \cdots]$ [53].

It may seem restrictive to be forced to assume that the eigenvalues of M are distinct, but in practice, there is no great penalty associated with this assumption. Clearly, the elements of the matrix M can be known only to within some limited precision, and they can also vary with changing environmental conditions, e.g., fluctuating temperature, mechanical vibrations, applied electric or magnetic fields, etc. It can be proved that the eigenvalues of M depend continuously on the elements of the matrix [53]. There are two consequences of this observation.

First, the distinct-eigenvalue M calculated by Equation 7.9 is indistinguishable from the real M, which may have multiple eigenfrequencies, because of the following Lemma:

LEMMA 7.1

Let M be a square matrix with multiple eigenvalues. Then, there is a matrix $\widetilde{M}$ such that $\widetilde{M}$ has distinct eigenvalues, such that for any $\epsilon > 0$, $|\widetilde{m}_{ij} - m_{ij}| < \epsilon$.
Proof See Ref. [53].

Second, for Hermitian matrices, we can use Weyl's inequalities to estimate the error in the eigenvalues from this imprecision in knowing the exact values of M.

LEMMA 7.2

Let $\widetilde{M}$ represent the matrix near to M such that $\widetilde{M} = M + P$ where P is the error matrix with elements p_{ij}. Then the error in estimating the eigenvalue ω_i of M is bounded,

$$|\delta \omega_i| \leq \max_i \left(\sum_j |p_{ij}| \right). \tag{7.10}$$

Proof See Ref. [53].

7.4 SPECIAL FORMS OF THE COUPLING MATRIX

In order to understand the spectral characteristics of large ensembles of coupled resonators, we first consider two special forms of the coupling matrix M.

EXAMPLE 7.1

$$M_I = \begin{pmatrix} \omega_1 & \kappa_1 & 0 & 0 & \cdots \\ \kappa_1^* & \omega_2 & \kappa_2 & 0 & \cdots \\ 0 & \kappa_2^* & \omega_3 & \kappa_3 & \cdots \\ 0 & 0 & \kappa_3^* & \omega_4 & \cdots \\ \vdots & \vdots & \vdots & \vdots & \end{pmatrix} \tag{7.11}$$

where the ω's and κ's are numbers rather than matrices is a tridiagonal Hermitian matrix describing the 1D nearest-neighbor-coupling model restricted to a single nondegenerate field mode in each unit cell.

7.4.1 QUICK METHOD TO CALCULATE THE DENSITY OF STATES $\rho(\omega)$

The characteristic polynomial $\phi(\omega) = \det(M_I - \omega I)$ can be found according to the following recursive procedure:

$$\begin{aligned}
\phi_0(\omega) &= 1, \\
\phi_1(\omega) &= \omega_1 - \omega, \\
\phi_n(\omega) &= (\omega_n - \omega)\phi_{n-1} - |\kappa_{n-1}|^2 \phi_{n-2}, \quad (n \geq 2)
\end{aligned} \quad (7.12)$$

where we continue the calculation until we reach $N = \dim(M_I)$. The zeros of the characteristic polynomial correspond, of course, to the eigenvalues of M_I.

The sequence of polynomials $\phi_0(\omega), \phi_1(\omega), \ldots, \phi_N(\omega)$ forms a Sturm sequence, which have the following property:

LEMMA 7.3

Let $\phi_0(\omega), \phi_1(\omega), \ldots, \phi_n(\omega)$ be a Sturm sequence on the interval (a, b). Let $\sigma(\omega)$ be the number of sign changes in the ordered array of numbers $\phi_0(\omega), \phi_1(\omega), \ldots, \phi_N(\omega)$. Assuming that $\phi_N(a) \neq 0$ and $\phi_N(b) \neq 0$, the number of zeros of the function ϕ_N in the interval (a, b) equals $\sigma(b) - \sigma(a)$.
Proof See Ref. [53].

Assuming that there are N unit cells in the waveguide, the number of zeros of ϕ_N in the interval (ω_k, ω_{k_1}) approximates $\rho(\omega_k)$. Therefore, to calculate $\rho(\omega)$:

1. Divide the ω axis into segments demarcated by $\{\omega_1, \omega_2, \ldots\}$.
2. At each ω_k, use Equation 7.12 to calculate the Sturm sequence of polynomials $\phi_0(\omega_k), \ldots, \phi_N(\omega_k)$.
3. Count the number of sign changes in that list of numbers, which defines $\sigma(\omega_k)$.
4. Except for a normalizing factor, $\rho(\omega_k/2 + \omega_{k+1}/2) = \sigma(\omega_{k+1}) - \sigma(\omega_k)$.

EXAMPLE 7.2

$$M_{II} = \begin{pmatrix} \Omega & C_{12} & C_{13} & C_{14} & \cdots \\ 0 & \Omega & C_{23} & C_{24} & \cdots \\ 0 & 0 & \Omega & C_{34} & \cdots \\ 0 & 0 & 0 & \Omega & \cdots \\ \vdots & \vdots & \vdots & \vdots & \end{pmatrix} \quad (7.13)$$

is a block-triangular non-Hermitian matrix which represents a system with unidirectional coupling and is important in studies of side-coupled waveguides and the side-coupled integrated spaced sequence of resonators (SCISSOR) waveguide geometry [54]. The characteristic polynomial which determines the eigenvalues of M_{II} is

$$\phi(\omega) = [\det(\Omega - \omega I)]^n \quad (7.14)$$

where $n = \dim(M_{II})/\dim(\Omega)$. The eigenvalues do not depend on any of the elements of C, i.e., the inter-resonator coupling coefficients—they simply do not appear in the characteristic polynomial (however, see Appendix A for a caveat). Therefore, such a system of coupled resonators does not suffer from mode-splitting, and the spectrum of eigenvalues is insensitive to variations in the coupling coefficients. This agrees with the comment made by Heebner et al. [55] that the "optical properties (of the SCISSOR waveguide) are independent of whether or not all the spacings between neighboring resonators are the same."

The structure of M_{II} is sufficient to avoid mode-splitting, but is not necessary. Another example is shown in Ref. [56], where mode-splitting occurs through some coupling coefficients, and not through some others.

A non-Hermitian structure for the coupling matrix M can also be achieved by using a grating-assisted directional coupler with gain or loss in the coupling region results in asymmetric coupling between two waveguides [57]. Slow-wave structures with gain are analyzed in Ref. [58]. A complex index perturbation in a multimode waveguide can act as a unidirectional coupler, e.g., between the transverse electric (TE) and transverse magnetic (TM) modes of the waveguide [59]. One area of applications of such components is to realize higher-order optical filters and add–drop components without the disadvantages of mode-splitting.

7.5 MODELS OF DISORDER AND THE CALCULATION OF $\rho(\omega)$

7.5.1 RANDOMNESS IN THE COUPLING COEFFICIENTS

In large collections of microresonators, it is unlikely that the parameters describing matrix M in Equation 7.7 will be identical throughout the ensemble. (Explicit expressions for the coupling coefficients between waveguides in terms of the transverse field profiles and the refractive index distribution may be found in Appendix B.)

Following the general classification of Ziman [38], we consider two types of disorders: (1) isotopic disorder, with randomness in the diagonal elements of M, and (2) interaction disorder, characterized by variations in the off-diagonal coupling terms. We will first discuss the latter type, since the off-diagonal terms correspond to the interaction energy of the coupled-resonator model, and hence relate to the fundamental issue of how resonators couple.

If the matrix M is tri-diagonal, the density of states can easily be obtained as discussed in Section 7.4.1. In the general case, we obtain the density of states from the list of eigenvalues, using Monte-Carlo techniques: we generate a histogram of the eigenvalues by dividing the ω axis into bins and counting how many eigenvalues fall within each bin. Invariably, any one particular calculation will also generate some noise in the results. Therefore, we sum up the results of a large number of independent calculations ($N_{trials} = 256$ seems more than sufficient) and take a slice through this histogram, cutting off those bins which have registered less than $\sqrt{N_{trials}}$ counts. (The results seem insensitive to the value of this threshold.) The density of states $\hat{\rho}(\omega)$ is normalized so that the area under the curve $\int d\omega \, \hat{\rho}(\omega) = 2$, which we assume is the number of modes in a single resonator (e.g., clockwise (cw) and counterclockwise (ccw) circulating modes of a microring).

If $\hat{\rho}(\omega)$ is further divided by NR, we obtain the number of states per unit frequency per unit volume (which $= 1/(\pi c)$ for classical waves in 1D, and has units of s/m).

In the following examples, we calculate the eigenspectrum for an ensemble of 256, 512, or 1024 resonators, each with two degenerate (equi frequency) modes, e.g., cw and ccw circulating in the case of ring, disk, or sphere resonators. The coupling matrix has the form

$$M = \begin{pmatrix} \Omega & C_{12} & 0 & 0 & \cdots \\ C_{21} & \Omega & C_{23} & 0 & \cdots \\ 0 & C_{32} & \Omega & C_{34} & \cdots \\ 0 & 0 & C_{43} & \Omega & \cdots \\ \vdots & \vdots & \vdots & \vdots & \end{pmatrix}, \quad (7.15)$$

where

$$\Omega = \begin{pmatrix} \Delta\omega_1 & \sigma_l \\ \sigma_l & \Delta\omega_2 \end{pmatrix}, \quad C_{l+1,l} = \begin{pmatrix} 0 & \kappa^{(1)}_{l+1,l} \\ \kappa^{(2)}_{l+1,l} & 0 \end{pmatrix},$$

$$C_{l,l+1} = \begin{pmatrix} 0 & \kappa^{(2)}_{l+1,l} \\ \kappa^{(1)}_{l+1,l} & 0 \end{pmatrix}. \tag{7.16}$$

The form of $C_{l+1,l}$ and $C_{l,l+1}$ arise from the fact that in a chain of coupled resonators, the cw-circulating mode of one resonator couples to the ccw-circulating modes of its nearest neighbors, and vice versa. We will ignore the self-coupling terms $\Delta\omega_1 = \Delta\omega_2 = 0$, since a constant value has no effect on the spectrum, and the effect of random variations in these terms is discussed later. The coupling coefficients $\kappa^{(1)}_{l+1,l}$, $\kappa^{(2)}_{l+1,l}$, and σ_l are independent, identically distributed real* random variables chosen from a uniform random distribution. Whereas the κ's represent the inter-resonator coupling coefficients, the σ's represent the disorder-induced coupling between nominally orthogonal modes within a single resonator [48,60].

Figure 7.3 was calculated for 256 coupled resonators forming the slow-wave structure and shows that, in the presence of disorder in the coupling coefficients, the density of states no longer exhibits a singularity at the band-edge. Instead, it extends smoothly outwards, with an exponential decay as shown in the inset to Figure 7.3. In this calculation, it was assumed that κ is chosen from the uniform random distribution over the interval 10 ± 2.5 Mrad/s.

Based on the relationship between the density of states and the group velocity (see Section 7.6), we conclude that the absence of a singularity at the band-edge implies that the group velocity no longer goes to zero. Physically, the zero-group-velocity mode in the ordered model is a very special resonance, formed by the static, in-phase excitation of the chain of resonators; randomness in the coupling coefficients destroys this precise balance of the relative phase shifts between neighboring resonators, and can easily lead to a propagation of the excitation along the chain.

To provide a theoretical expression for $\rho(\omega)$, which accurately models the lineshapes shown in Figures 7.3 and 7.4, we use a mathematical expression obtained by Smith [61] for the $X - Y$ model with random coupling constants (see also [62]). Using the approach of Dyson [63], Smith evaluated analytically an asymptotically accurate expression for the density of states of a disordered 1D lattice with nearest-neighbor coupling—see Equations 4.6, 4.8, and 4.16 in chapter 4 in Smith [61], which we repeat here. In terms of the normalized frequency detuning $y = \omega/\omega_{\text{edge}}$,

$$\rho(y) \approx \frac{1}{2\pi} \left\{ \left(1 - \frac{y^2}{4}\right)^{-1/2} + \frac{1}{4N} \left(1 - \frac{y^2}{4}\right)^{-3/2} \right\}, \quad \text{for} \quad |y| < 2 - \delta, \tag{7.17a}$$

$$\rho(y) \approx 0.18 N^{1/3} \left[1 - 0.53 \left\{ N^{2/3} \left(\frac{y^2}{4} - 1\right) \right\}^2 \right], \quad \text{for} \quad |y| \in (2 - \delta, 2 + \delta), \tag{7.17b}$$

$$\rho(y) \approx \frac{y}{2\pi \sinh\varphi} \{2N \varphi(\cosh\varphi - 1) + \varphi - 1\}$$
$$\times \exp[-\varphi - 2N(\sinh\varphi - \varphi)], \quad \text{for} \quad |y| > 2 + \delta, \tag{7.17c}$$

with the following definitions,

$$\varphi \equiv \cosh^{-1}\left(\frac{y^2}{2} - 1\right), \quad \text{and} \quad \delta \equiv N^{-2/3}.$$

---

* In the practically relevant case of the tridiagonal Hermitian matrix (see Section 7.4), it is unnecessary to consider complex variations in κ since, according to Equation 7.12, only $|\kappa|^2$, which is real, participates in the calculation of the characteristic polynomial.

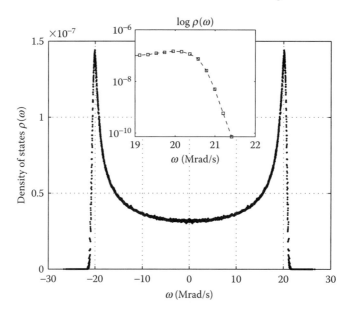

FIGURE 7.3 Density of states $\rho(\omega)$ for the disordered case, where ω is measured in Mrad/s from zero detuning. Compared to the ordered case, the band-edge is no longer a singularity: the inset shows that the tail of $\rho(\omega)$ reaches a definite peak value, depending on the degree of disorder, and is exponential (linear on a logarithmic scale) into the tail. The density of states away from the band-edges is relatively unaffected. (Reproduced from Mookherjea, S., *J. Opt. Soc. Am. B*, 23, 1137, 2006. With permission.)

N is a parameter (usually $N \gg 1$) that represents the r.m.s. variation in the coupling coefficients, whose value we choose such that lineshape described by Equations 7.17a–c best fits the numerically calculated $\rho(\omega)$ distribution such as shown in Figures 7.3 and 7.4. At the end of Section 7.6, we will write down the expression relating N to the r.m.s. variation in the coupling coefficient, $\delta\kappa$, and its average value, κ.

7.5.2 Randomness in the Diagonal Terms

In Section 7.5.1, we discussed the calculation of $\rho(\omega)$ in the presence of disorder in the coupling coefficients which occur in the off-diagonal components of the matrix M. Randomness in the diagonal terms corresponds to variations from one resonator to another in the self-coupling coefficients, as indicated by Equation 7.B.3.

We recall Gerschgorin's theorem,

LEMMA 7.4

Each eigenvalue ω of the matrix $M = \{m_{ij}\}$ lies in at least one of the closed disks

$$|\omega - m_{ii}| \leq \sum_{j \neq i} |m_{ij}|.$$

Proof See Ref. [53].

Thus, a non zero value for $\Delta\omega$ has at least the effect of shifting the center of the disks to $\pm\Delta\omega$ even if the radii of the disks are unchanged. Random perturbations in the diagonal elements of M translate

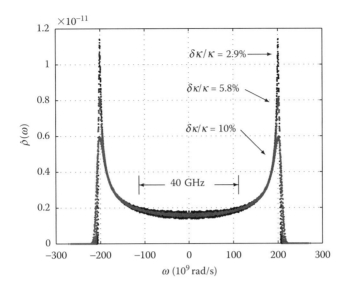

FIGURE 7.4 The density of states, $\hat{\rho}(\omega)$ obtained from a numerical calculation of the eigenvalues of the matrix M in Equation 7.6. Indicated values of $\delta\kappa/\kappa$ represent the standard deviation in the coupling coefficients. The horizontal axis is the radian frequency detuning from band-center. (Reproduced from Mookherjea, S. and Oh, A., *Opt. Lett.*, 32, 289, 2007. With permission.)

the various disks randomly to the left or to the right, which blurs out any peaks or valleys that are visible in the ordered spectrum. Similar to the previously discussed case, the density of states no longer exhibits a singularity at the band-edge, and the velocity of light cannot go to zero.

A serious problem may occur if the variations in the eigenfrequency of the individual resonators are large. Such a situation can occur when very high-Q resonators are coupled weakly and the coupling integral is not robust to variations in the surrounding environment. A study of transmission characteristics has been carried out [32], showing that if the structural variations cause the eigenfrequencies of the resonators to vary by more than the original coupling bandwidth, i.e., $\Omega\kappa$ in Equation 7.1, then the transmission is significantly degraded.

As Equation 7.B.3 shows, such variations, e.g., in $\omega_{b_l} - \omega_{a_l}$, generally result in the elements of the matrix M becoming time-varying. The form of Equation 7.6 remains unchanged; however, its solutions can no longer be described in terms of the linear theory of modes as developed in this chapter. In fact, a variety of other modes, originally ignored due to phase-matching considerations, may be excited in such a structure.

7.6 VELOCITY OF SLOW LIGHT IN DISORDERED STRUCTURES

Having obtained the distribution of eigenmodes for a weakly perturbed tight-binding lattice by solving Equation 7.8, we calculate the density of states, $\rho(\omega)$, which is defined such that $\rho(\omega)\,d\omega$ is the number of eigenstates in a small neighborhood of frequencies around ω. Figure 7.4 shows the normalized density of states, $\hat{\rho}(\omega)$ (normalization: $\int d\omega \hat{\rho}(\omega) = 2$) corresponding to a tight-binding slow-wave structure consisting of $N = 512$ resonators, each with two modes ($m = 2$) which could model, e.g., the cw and ccw modes of a microring. The inter-resonator coupling coefficients are chosen from the uniform (or alternatively, the Gaussian) random distribution, with the definition,

$$\hat{\kappa} \sim \kappa + \delta\kappa\, U[-1/2, 1/2], \quad \text{(Uniform)} \tag{7.18a}$$

$$\text{or,} \quad \hat{\kappa} \sim \kappa + \delta\kappa\, G[0; 1], \quad \text{(Gaussian)} \tag{7.18b}$$

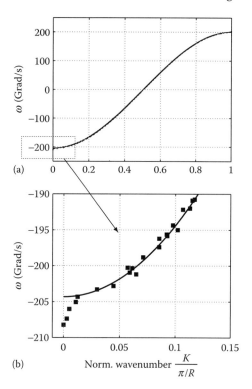

FIGURE 7.5 The dispersion relationship of a weakly disordered optical tight-binding lattice for $\delta\kappa/\kappa = 3\%$ in the uniform random distribution is compared to the theoretical result (straight line). (a) The shape of the ideal dispersion curve, indicated by the solid line, is reproduced over most of the range by the data points, except near the band-edge, where as shown by (b), the slope in the disordered structure is not exactly zero, and a tail extends into the band-gap region. (Reproduced from Mookherjea, S. and Oh, A., *Opt. Lett.*, 32, 289, 2007. With permission.)

where $\kappa = 100 \times 10^9$ rad/s, $U[-1/2, 1/2]$ is the uniform random distribution between $-1/2$ and $+1/2$ and $G[0; 1]$ is the Gaussian random distribution with mean 0 and variance 1. 1024 Monte-Carlo simulations are performed for each value of $\delta\kappa/\kappa$. (Some of the values of $\delta\kappa/\kappa$ chosen to generate this figure are intentionally rather high, to clearly show the change of the shape of the density of states function.)

Figure 7.5 shows the numerically calculated dispersion relationship for weak disorder $\delta\kappa/\kappa = 3\%$ in the uniform random distribution. The dispersion relationship for the uniform tight-binding lattice and the disordered structure are practically coincident except near the band-edges, as shown in Figure 7.5b. We observe no isolated islands of disordered states beyond the tail of the spectrum, in agreement with Dean's earlier studies of glass-like lattices [64].

Since $\rho(\omega)\,d\omega = \rho(k)\,dk$, our knowledge of $\rho(\omega)$ and $\rho(k)$ determines the group velocity, $v_g \equiv d\omega/dk = \rho(k)/\rho(\omega)$. At band-center, $\rho(\omega)$ takes its minimum value, and using $R = 10$ μm as an example for coupled microrings,

$$[v_g]_{max} = \frac{1}{[\hat{\rho}(\omega)]_{min}} \frac{1}{\pi/10 \text{ μm}} = 1.0 \times 10^6 \text{ m/s}. \tag{7.19}$$

This value agrees with that obtained directly from the dispersion relationship, $[v_g]_{max} = (\delta\omega/2) \times R = 1 \times 10^6$ m/s, where $2\,\delta\omega$ is the full-width of the pass-band ($2\,\delta\omega = 400 \times 10^9$ from Figure 7.4).

Note that approximately one order-of-magnitude reduction in $[v_g]_{max}$ may be achieved by using coupled photonic crystal defect resonators instead of microrings.

Now we turn our attention to the velocity of light in the disordered structure. Bearing in mind the conditions for its use as an optical delay line, we require the structure in which light can be mostly transmitted through in an "elastic" phase-coherent manner, i.e., in the ballistic transport regime. Indeed, both simulations [32,36] and experimental observations in the microwave regime [31] suggest that transmission is largely preserved in the presence of weak disorder, but the effect of disorder on the slow velocity of light was not directly measured or studied.

Change in $\rho(k)$: In the absence of disorder, the density of states in k-space is such that a total of $N \times m$ states are evenly distributed in the first Brillouin zone between $k = -\pi/R$ and $k = \pi/R$, i.e., $\rho_{ideal}(k) = Nm/(2\pi/R)$, where R is the inter-resonator spacing. In the presence of disorder, we represent by $\phi(k)$ the (average) phase shift induced over length $L = NR$ at wavenumber k, so that

$$\phi(k) \sim \sum_{k'} (k - k')L \times (\text{number of states at } k') \times (\text{probability of transition: } k' \to k)$$

$$= \int dk' \, (k - k')L \, \rho_{ideal}(k') \, W(k, k'), \tag{7.20}$$

where, from elementary scattering theory, $W(k, k') = |\langle \Psi_k | \delta U(\mathbf{r}) | \Psi_{k'} \rangle|^2$, in terms of the spatial profile of the disorder, $\delta U(\mathbf{r})$, and the eigenmode Ψ as described in Equation 7.31.

From the resonance condition $kL + \phi(k) = 2m\pi$ for some integer m, we obtain,

$$\rho(k) = \rho_{ideal}(k) \left(1 + \frac{1}{L} \frac{d\phi}{dk}\right). \tag{7.21}$$

Change in $\rho(\omega)$: This was discussed in detail in Section 7.5. To summarize, in the absence of disorder, the density of states $\rho(\omega)$ exhibits a divergence at the band edge, $\rho(\omega) \propto (\omega_{edge} - \omega)^{-1/2} \to \infty$ as $\omega \to \omega_{edge}$ and $v_g \to 0$. This is no longer true in the presence of disorder, as shown in Figure 7.4. As the strength of disorder increases, $\rho(\omega)$ and the dispersion relationship (Figure 7.5) extend beyond precisely ± 200 Grad/s. According to the photonic sum rule [65], the density of states is redistributed within the available spectrum such that the integral over the density of states is constant. Thus, the peak value of $\rho(\omega)$ is reduced.

Thus, after the previous steps, we obtain the equation describing the group velocity in a weakly disordered optical slow-wave structure:

$$v_g = \frac{1}{\hat{\rho}(\omega)} \frac{1}{2\pi/R} \left(1 + \frac{1}{NR} \frac{d\phi}{dk}\right). \tag{7.22}$$

Figures 7.4 and 7.5 show that the group velocity in the tight-binding lattice is quite insensitive to weak disorder at band-center and therefore, $[v_g]_{max}$ is approximately equal to the value given in Equation 7.19. The strongest effects of disorder are felt at the band-edges, and we assume in this paper, for simplicity, that $d\phi/dk$ can be neglected compared to the large changes in $\rho(\omega)$ in this narrow range of ω.

We introduce a parameter called the band-edge slowing factor, defined as

$$S_{be} \equiv \frac{v_g \text{ at band-center}}{v_g \text{ at band-edge}}, \tag{7.23}$$

and in Figure 7.6, we plot S_{be} versus $(\delta\kappa/\kappa)^{-1}$. The numerically calculated data points lie almost exactly on a straight line fit (on a log–log scale),

$$\log_{10}(S_{be}) = 0.644 \log_{10}\left(\frac{\kappa}{\delta\kappa}\right) + 0.272, \quad \text{(Uniform)} \tag{7.24a}$$

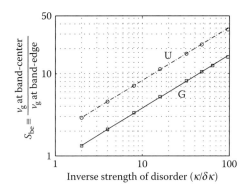

FIGURE 7.6 The band-edge slowing factor obtained from numerical calculations of $\rho(\omega)$. The horizontal axis is the inverse strength of disorder, $(\delta\kappa/\kappa)^{-1}$, varying from 50% to 1%. U and G refer to the uniform and Gaussian random distributions in Equations 7.18a–b. Equations 7.24a and b describe the straight-line best-fit to the data points. The vertical distance between the lines is approximately $(2/3)\log_{10}\sqrt{12}$, to equate the variances of the two distributions. (Reproduced from Mookherjea, S. and Oh, A., *Opt. Lett.*, 32, 289, 2007. With permission.)

$$\log_{10}(S_{be}) = 0.648 \log_{10}\left(\frac{\kappa}{\delta\kappa}\right) + 0.281 - 0.648 \log_{10}\sqrt{12}, \quad \text{(Gaussian)} \quad (7.24b)$$

The last factor on the right hand side of Equation 7.24b represents the scaling of $\delta\kappa/\kappa$ to equate the variances in the two distributions Equations 7.18a and b. (The same value of $\delta\kappa/\kappa$ implies a different normalized variance in the two distributions.)

To compare these observations with theoretical calculations, we refer to the investigations of Dyson [63] and Smith [61], calculating analytically the frequency spectrum of a disordered chain of coupled masses connected by elastic springs, and of a linear chain of magnetic spins in an external field, respectively. The randomness in the coupling coefficients was modeled as a generalized Poisson distribution, which permits explicit analytical evaluation of $\hat{\rho}(\omega)$ (see Equations 4.6, 4.8, and 4.16 in Smith [61]), and leads to

$$\log_{10}(S_{be})|_{\text{theory}} = 0.667 \log_{10}\left(\frac{\kappa}{\delta\kappa}\right) + 0.313 \quad (7.25)$$

In view of the differences between a uniform random distribution and the generalized Poisson distribution, the agreement between Equations 7.24a and b and Equation 7.25 is quite satisfactory, yielding the inference, $S_{be} \sim (\kappa/\delta\kappa)^{2/3}$.

The value of $\delta\kappa/\kappa$ can be connected to the r.m.s. roughness or positional inaccuracy in nanometers either by performing a series of Monte-Carlo simulations, or by theoretical arguments, as shown in Appendix B. As an order-of-magnitude estimate, see Equation 7.B.10,

$$\delta\kappa/\kappa \approx 1\% \text{ per nm of roughness or inaccuracy} \quad (7.26)$$

and therefore,

$$\frac{v_g \text{ at band-center}}{v_g \text{ at band-edge}} = \left(\frac{100}{\text{r.m.s. roughness in nm}}\right)^{2/3}. \quad (7.27)$$

Thus, for variations in the inter-resonator coupling coefficients in the range of 1% to 10%, or r.m.s. positional accuracy in the range of 1–10 nm, which covers the range of practical interest with current fabrication technology, slow light at the band-edge is expected to be only 10 to 30 times slower than at the band-center.

This agrees with the experimentally measured values of this parameter, which are $S_{be} \approx 7$ (optical regime: [9]) and $S_{be} \approx 6.5$ (microwave regime [28]).*

Finally, we can calculate the slowing factor of a weakly disordered slow light structure. In an undisordered structure, at band-center ($\omega = \Omega$) the time taken to propagate from input to output is $\tau_{min} = N/(\Omega \kappa)$, and the group velocity v_g at band-center $= 2\kappa \Omega R$ (with Ω expressed in hertz rather than radians per second). In the presence of disorder, we can use Equation 7.25 and thereby calculate that the slowing factor is

$$S \equiv \frac{c}{v_g \text{ at band-edge}} = \frac{\lambda}{R} \frac{1}{(\delta \kappa^2 \cdot \kappa)^{1/3}}. \qquad (7.28)$$

Equation 7.28 shows that $S \to \infty$ (since v_g at band-edge $\to 0$) if $\delta \kappa \to 0$. But, for a typical structure, if $\kappa = 10^{-2}$, $\delta \kappa = 5\%$ of κ, and $R = 10\lambda$, then $S = 74$, a much more modest slowing factor. Experimental observations also indicate $S \approx 10 - 100$ [9,14,28].

Equation 7.28 suggests that the slowing factor is enhanced if R (the distance between unit cells) is comparable to λ and therefore suggests the use of 100 nm–1 µm length scales rather than significantly larger structures as effective delay lines. (The theoretical validity of the nearest-neighbor coupling model is questionable if $\lambda \ll R$.) However, $\delta \kappa$ tends to be significantly larger at the smaller length scales, since surface roughness and other fabrication imperfections play a more significant role. The sweet spot of slow-light operation lies somewhere in the middle between the micro- and nanoregimes, but should move towards smaller length scales as fabrication technology improves.

We can also establish the connection between $\delta \kappa$ and the parameter N which appears in Equations 7.17a–c,

$$S_{be} \equiv \frac{\rho(\omega_{peak}, \text{ i.e., } y = 2)}{\rho(\omega_{band\text{-}center}, \text{ i.e., } y = 0)} \approx 1.131 \frac{N^{4/3}}{N + 0.25}. \qquad (7.29)$$

Using Equation 7.25 to write S_{be} in terms of $\delta \kappa / \kappa$, and assuming that $N \gg 1$ (typical values are 100–1000), we find

$$N \approx \left(0.408 \frac{\delta \kappa}{\kappa}\right)^{-2}. \qquad (7.30)$$

In words, the disorder-less limit, $\delta \kappa \to 0$, corresponds to taking $N \to \infty$ in Equations 7.17a–c.

7.7 LOCALIZATION OF FIELDS

Having discussed the distribution of eigenvalues (frequencies) of weakly disordered lattices and their usefulness in predicting the slow velocity of light, the spatial distribution of the excitations remains to be discussed. Recall that eigensolutions of Equation 7.6 have the time-evolution behavior $\mathbf{u} \sim \exp(i\omega t)$ and, therefore, are found by solving the determinantal equation $|M - \omega I| = 0$, where I is the unit matrix.

When there is no disorder, the corresponding spatial part of each eigenmode is, in the tight-binding approximation,

$$\langle z | \Psi_k \rangle \equiv \sum_{n=1}^{N} e^{-inkR} \mathbf{E}_{\text{single}}(\mathbf{r} - nR\hat{\mathbf{z}}) \qquad (7.31)$$

---

* In Ref. [28], the authors derived Figure 7.4 (inset) by fitting the data points by a sum-of-cosines, which artificially forces the slope at the band-edge to zero, even if this is not indicated by the measured data points. We extracted $S_{be} \approx 6.5$ from the data points (open circles) shown in Figure 7.4.

i.e., a Bloch sum over the single-resonator fields $\mathbf{E}_{\text{single}}(\mathbf{r})$. Ψ_k satisfies the generalized eigenvalue equation

$$\nabla \times \nabla \times |\Psi_k\rangle = \epsilon_{\text{wg}} E_k |\Psi_k\rangle, \tag{7.32}$$

where $E_k \equiv (\omega_k/c)^2$. We write the disorder as $\epsilon_{\text{wg}}(\mathbf{r}) \to \epsilon_{\text{wg}}(\mathbf{r}) + \delta\epsilon_{\text{wg}}(\mathbf{r})$, and the new modes as $|\psi_k\rangle$, which satisfy

$$\nabla \times \nabla \times |\psi_k\rangle - \delta\epsilon_{\text{wg}} E_k |\psi_k\rangle = \epsilon_{\text{wg}} E_k |\psi_k\rangle. \tag{7.33}$$

Defining the operator $H \equiv \nabla \times (\nabla \times) - \delta\epsilon_{\text{wg}} E_k$, the solution can be written as

$$|\psi_k\rangle = |\Psi_k\rangle - (\epsilon_{\text{wg}} - H)^{-1} \delta\epsilon_{\text{wg}} E_k |\Psi_k\rangle, \tag{7.34}$$

which may be verified by multiplying the left and right hand sides by $(\epsilon_{\text{wg}} - H)$. In terms of Green's function for the disordered structure, $G(z, z'; E_k)$, the solution may be written as

$$\psi_k(\mathbf{r}) = \Psi_k(\mathbf{r}) - \frac{\omega_k^2}{c^2} \sum_{\mathbf{r}'} G(\mathbf{r}, \mathbf{r}'; E_k) \delta\epsilon_{\text{wg}}(\mathbf{r}') \Psi_k(\mathbf{r}'). \tag{7.35}$$

$G(\mathbf{r}, \mathbf{r}'; E_k)$ is not known exactly, of course, but may be found quasi analytically from various well-known theories such as the average t-matrix approximation (ATA), or the coherent potential approximation (CPA) [18,39]. However, these methods do not work well in the region of our interest—the band-edge, where the density of states is dominated by contributions from resonance or localized eigenstates, as the numerical results suggest.

Compared to the 2D or 3D cases (see Figure 7.1), calculation of G in the 1D case is the most difficult, since for E_k within the passband of the structure, reflected or transmitted waves from one scattering center propagate with constant amplitude throughout the linear chain and are invariably scattered again by the other impurities. Thus, multiple scattering terms cannot be ignored, which leads to very complicated expressions even with only two defects in an otherwise perfect lattice (see [18]). In practical structures, the presence of loss and finite coherence length of photons ameliorates the situation slightly. To obtain a very rough idea of the behavior of Equation 7.35, we approximate $G(z, z'; E_k)$ by $G_0(z, z'; E_k)$ for $z' \neq z$, where G_0 is Green's function for the unperturbed structure [18], and, for $z' = z$, by $-\pi\rho(z, E)$, in terms of the local density of states, calculated numerically as before, which actually does not depend on z. According to Equation 7.35, the fields $\psi_k(\mathbf{r})$ which are most perturbed are those for which $\rho(E_k)$ is the most significantly affected (see Figure 7.4)—i.e., the band-edge states. This is shown in Figure 7.7.

The eigenvectors of the disordered coupling matrix may also be obtained numerically: Figure 7.8 shows the eigenmodes for the three highest-magnitude eigenvalues (near the upper band-edge, $\omega_{\text{edge}} = +200$ Grad/s) with $\delta\kappa/\kappa = 3\%$ variation in the coupling coefficients, the same as used to calculate Figure 7.5. The field distributions in the weakly disordered case are distorted versions of the modes calculated in the ordered case. In this example, the disordered fields are extended, i.e., the localization length extends beyond the length of the structure (256 unit cells), and are thus capable of transporting energy from one end of the chain to the other. In other words, these modes can contribute to the overall transmission coefficient at the appropriate optical frequency.

This observation leads to a working definition of weak disorder: we call the disorder weak if the intersite coupling potential fluctuations are smaller in magnitude than the energy level spacing of the ordered lattice. Since the energy level spacing (i.e., the separation between successive eigenvalues) becomes smaller as N, the number of unit cells, increases, the disorder-induced mixing of Bloch states is weak, and the eigenmodes can spatially span the extent of the waveguide, as long as N is not too large [66]. In the so-called thermodynamic limit $N \to \infty$, all the eigenmodes are localized.

These qualitative arguments can be verified using the numerically obtained density of modes, $\rho(\omega)$, or alternatively, the theoretical approximations to the curve, Equations 7.17a–c. We consider the

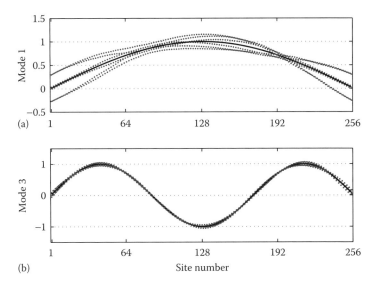

FIGURE 7.7 Modal field amplitudes at the resonator sites for a slow-wave structure consisting of 256 coupled unit cells, calculated using Equation 7.35. Excitation amplitudes are shown for (a) the mode nearest the band-edge, and (b) the third mode counting inwards from the band-edge. The black line is the distribution of amplitudes in the ideal (ordered) case, and the blue dots are calculated in the disordered case, with 1% r.m.s. variation in $\epsilon(\mathbf{r})$ from one resonator site to the next. A few different Monte-Carlo iterations are superimposed, showing that the variation in the excitation amplitudes is much larger in the band-edge mode (a), compared to a mode (b) further inside the band.

localization length $\gamma^{-1}(\omega)$, which describes the number of unit cells spanned by a particular localized field distribution whose eigenvalue is ω. In other words, $\gamma^{-1}R$ tells us what is the propagation distance for light of frequency ω.

The Herbert–Jones–Thouless formula [37,67] obtains the relationship between $\gamma(\omega)$ and $\rho(\omega)$,

$$\gamma(\omega) = \int_{-\infty}^{\infty} d\omega' \, \rho(\omega') \log|\omega - \omega'| \qquad (7.36)$$

where we have normalized the coupling coefficients to unity-mean (to eliminate an additive term, $\log(\kappa)$), in the above expression. Since $\rho(\omega)$ drops off rapidly beyond $\pm \omega_{\text{edge}}$, a numerical evaluation of the above integral converges rapidly.

Figure 7.9 shows the localization length $\gamma^{-1}(\omega)$ calculated using Equations 7.17a–c, as well as obtained directly from the r.m.s. width of the disordered field distributions. Figures 7.10 and 7.11 show the spatial profile of states closest to the band-edge. A sequence of 1024 unit cells is assumed to be coupled, with $\delta\kappa/\kappa = 1\%$ in the case of Figure 7.10, and $\delta\kappa/\kappa = 6\%$ in the case of Figure 7.11. In the latter case, a greater number of band-edge states are clearly localized. However, as shown in Figure 7.9, the localization length for all but the very last few states greatly exceeds the device lengths in today's coupled-resonator optical slow-wave structures. With better fabrication technology, as well as a deeper understanding of band-edge optical physics, longer structures may be fabricated and studied in the near future. (Recently, one-dimensional optical localization has been observed experimentally in coupled-resonator, coupled-waveguide, and photonic crystal waveguides [68–70].)

To summarize this section, the theory predicts localization of all eigenmodes in the limit $N \to \infty$; however, it is quite appropriate to assume that the fields may extend over the entire lattice for finite

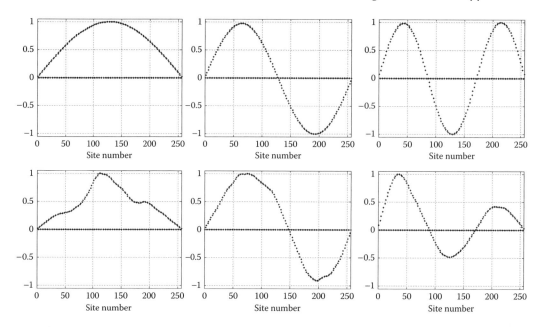

FIGURE 7.8 Field amplitudes at the resonator sites for a slow-wave structure consisting of 256 coupled unit cells, calculated directly by finding the eigenvectors of the coupling matrix. The top row shows the eigenmodes in the perfectly ordered case for the three highest-magnitude eigenvalues (near the band-edge), and the bottom row shows representative random field distributions for the weakly disordered case ($\delta\kappa/\kappa = 3\%$ variation in the coupling coefficients). The eigenvalue for the left-most mode on the bottom row ($\omega = 200.012$ Grad/s) lies beyond the band-edge of the ordered chain; nevertheless, the field is spatially extended, not localized, in this finite-length chain. Figures 7.10 and 7.11 show band-edge localization in longer chains of unit cells.

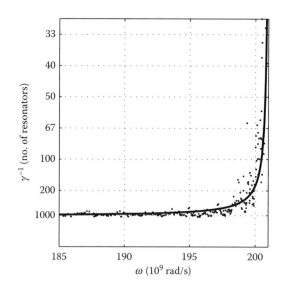

FIGURE 7.9 Localization length (width of the field distribution measured in number of resonator sites) for disorder defined by $\delta\kappa/\kappa = 6\%$ in Equation 7.18a for frequencies near the band-edge. The solid line is obtained from the theoretical expression for the density of states, and the dots are obtained from the numerically calculated field distributions. $\gamma^{-1} \times R$ defines the propagation distance for light of frequency ω, with a maximum value limited by the number of unit cells considered in the model ($N = 1024$), and thus corresponding to the entire length of the structure.

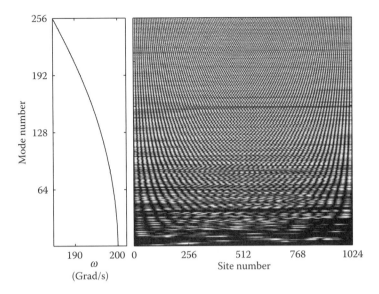

FIGURE 7.10 Band-edge localization of fields: the vertical axis indexes the mode number, starting from the band-edge mode at the bottom up to one-eighth of the band. The left figure shows the corresponding eigenfrequency. The right figure shows the eigenmode, i.e., the spatial field pattern as would be imaged by a camera. Modes near the bottom of the vertical axis correspond to band-edge states and are localized, whereas modes closer to the center of the band are extended. In these calculations, $N = 1024$ resonators, each with two modes were coupled with coupling disorder $\delta\kappa/\kappa = 1/100 = 1\%$.

and practically relevant lengths of coupled optical resonators (say, $N \leq 100$) in the weakly disordered case (say, $\delta\kappa/\kappa \leq 1/10$). It seems justified to assume that end-to-end quasi ballistic propagation exists in the weakly disordered slow-wave structure, and therefore investigate the velocity of such excitations, as we did in the earlier section.

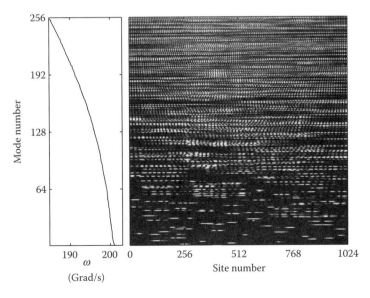

FIGURE 7.11 Same as Figure 7.10, except that the coupling disorder is larger, with $\delta\kappa/\kappa = 1/16 \approx 6\%$. Localized states are found in a wider range of ω near the band-edge.

7.8 SUMMARY

We have discussed optical slow-wave structures based on a density of states formalism, so that we can answer the question of what is the velocity of slow light in a weakly disordered slow-wave structure.

In particular, we have analyzed the spectrum of the optical tight-binding waveguide consisting of a large number of resonators coupled in a 1D chain configuration, with disorder in the coupling coefficients, as well as in the individual eigenfrequencies of the unit cells. We have presented a fast way to calculate the density of states in the nearest-neighbor nondegenerate case, and discussed how the density of states can be calculated in more general cases.

Unlike in the ordered case, the disordered model does not exhibit a singularity in the density of states at the band-edge. Consequently, the zero-group-velocity mode predicted for the ordered chain is not sustained, and applications of coupled resonator devices which rely on this feature are not robust to variations in the coupling coefficients.

As discussed in Section 7.7, for small values of the disorder and in lattices of modest size, consisting of up to 100 resonators (addressing the current lithographic fabrication or self-assembly capabilities of coupled resonators), the field distributions are mostly extended from input to output ends of the chain; a few localized modes can be seen at the very edge of the band. Extended field distributions contribute to wave propagation at the appropriate frequency from one end of the lattice to the other, although the velocity of propagation of such excitations was hitherto unknown. The relationship between this velocity and the degree of disorder was studied in Section 7.6.

The dispersion relationship of a uniform structure predicts that $v_g \to 0$ at the band-edge, i.e., the band-edge slowing factor S_{be} is infinitely large. Thus, S_{be} can serve as a figure-of-merit for how close an optical slow-wave structure is, compared to the ideal. We have argued that, for variations in the inter-resonator coupling coefficients in the range of 1% to 10%, which covers the range of practical interest with current fabrication technology, velocities at the band-edge is expected to be only 10 to 30 times slower than at band-center, as predicted by Equation 7.27. This theoretical calculation is in agreement with experimental observations made by researchers in both the microwave and optical regimes.

ACKNOWLEDGMENTS

The author is grateful to Andrew Oh for assistance with numerical simulations. This work was funded in part with support from the National Science Foundation (Dr. R. Hui and Dr. L. Goldberg). The San Diego Supercomputing Center (SDSC) provided computational resources, and the Academic Internship Program at UCSD supported the participation of Andrew Oh.

APPENDIX A: TWO PENDULUM BOBS COUPLED BY A SPRING

The close relationship between the equations of motion of pendulum bobs coupled by springs and the evolution equations of more general forms of coupled resonator systems has been discussed by Louisell [41]. Consider two identical pendula, each of mass m, coupled by a weightless spring with spring constant k. The displacements from the rest position of the pendula are labeled by x_1 and x_2, and their velocities by v_1 and v_2. From these quantities, the normal mode amplitudes are defined,

$$a_1 = \frac{\sqrt{m}}{2}(v_1 + i\omega_1 x_1),$$
$$a_2 = \frac{\sqrt{m}}{2}(v_1 + i\omega_2 x_2), \quad (7.A.1)$$

which obey the following evolution equation,

$$\frac{d}{dt}\begin{pmatrix} a_1 \\ a_2 \\ a_1^* \\ a_2^* \end{pmatrix} = C \begin{pmatrix} a_1 \\ a_2 \\ a_1^* \\ a_2^* \end{pmatrix}, \qquad (7.A.2)$$

where the elements of $C = \{c_{kl}\}$, called the mode coupling coefficients, are

$$\begin{aligned} c_{11} &= -c_{33} = i\omega_1 \left(1 + \frac{k}{2m\omega_1^2}\right), \\ c_{22} &= -c_{44} = i\omega_1 \left(1 + \frac{k}{2m\omega_2^2}\right), \\ c_{13} &= c_{21} = -c_{23} = -c_{31} = c_{41} = -c_{43} = -i\frac{k}{2m\omega_1}, \\ c_{12} &= -c_{14} = c_{24} = c_{32} = -c_{34} = -c_{42} = -i\frac{k}{2m\omega_2}. \end{aligned} \qquad (7.A.3)$$

The total energy of the two coupled pendula is

$$\begin{aligned} E &= |a_1|^2 + |a_1^*|^2 + |a_2|^2 + |a_2^*|^2 \\ &\quad - \frac{k}{2m}\left(\frac{a_1 - a_1^*}{\omega_1} - \frac{a_2 - a_2^*}{\omega_2}\right)^2. \end{aligned} \qquad (7.A.4)$$

The first four terms are nonzero even when the pendula are decoupled ($k = 0$) and represent the potential energy of the two pendula. The last term defines the potential energy associated with the coupling mechanism, and arises from the term $V = (k/2)(x_1 - x_2)^2$. When $x_1 = x_2$, and the bobs are at rest, the spring is not under tension, and this term is zero. As $|x_1 - x_2|$ increases, the tension increases and so does the stored potential energy in the spring.

The pendula are said to be weakly coupled when the last term in Equation 7.A.4 is smaller than the sum of the other terms. Mathematically, the weak-coupling condition is

$$\frac{k}{2m} \ll \omega_{1,2}^2. \qquad (7.A.5)$$

and if this is satisfied, the on-diagonal coupling coefficients defined in Equation 7.A.3 become independent of k.

Similar observations apply to the diagonal coupling coefficients of Equation 7.7 and to its particular form chosen in Equation 7.13. When appropriate, the weak-coupling assumption allows us to ignore, to leading order, the perturbative effect of γ in the diagonal elements of the matrix Ω.

APPENDIX B: DERIVATION OF EQUATION 7.6 AND THE COUPLING COEFFICIENT κ

In a guided-wave structure, the electromagnetic field $\mathbf{E}(\mathbf{r}, t)$ described by Equation 7.5 obeys the wave equation,

$$\nabla^2 \mathbf{E}(\mathbf{r}, t) - \mu\epsilon(\mathbf{r})\frac{\partial^2}{\partial t^2}\mathbf{E}(\mathbf{r}, t) = \mu \frac{\partial^2}{\partial t^2}\mathbf{P}(\mathbf{r}, t) \qquad (7.B.1)$$

where $\mathbf{P}(\mathbf{r}, t)$ is the perturbation polarization which represents any deviation of the polarization from that of the unperturbed waveguide.

In the absence of coupling between the various modes of the resonators, the elements of **u** are time-invariant and are completely determined by the initial conditions, $\mathbf{E}(\mathbf{r}, t_0)$. We write the (linear) perturbation polarization $\mathbf{P}(\mathbf{r}, t) = \Delta\epsilon(\mathbf{r}, t)\mathbf{E}(\mathbf{r}, t)$. From the wave equation, we obtain

$$\epsilon(\mathbf{r}) \sum_{l=1}^{N} \omega_{a_l} \left[i \frac{da_l}{dt} \right] e^{i\omega_{a_l} t} \hat{\mathbf{e}}_{a_l}(\mathbf{r}) + \cdots + \text{c.c.}$$

$$\approx \Delta\epsilon(\mathbf{r}, t) \sum_{l=1}^{N} \frac{1}{2} \left[\omega_{a_l}^2 \hat{\mathbf{e}}_{a_l}(\mathbf{r}) a_l(t) e^{i\omega_{a_l} t} + \cdots + \text{c.c.} \right] \quad (7.B.2)$$

assuming that $a_l, b_l, \ldots$ and $\Delta\epsilon$ vary slowly compared to the optical carrier frequencies $\omega_{a_l}, \omega_{b_l}, \ldots$.

The structure of Equation 7.B.2 suggests that, by projecting out various components using the orthogonality of eigenmodes, we can obtain the evolution equations for the coefficients. For example,

$$i\frac{d}{dt} a_l(t) = \frac{\omega_{a_l}}{2} \langle \hat{\mathbf{e}}_{a_l} | \Delta\epsilon | \hat{\mathbf{e}}_{a_l} \rangle a_l$$

$$+ \frac{\omega_{b_l}^2}{2\omega_{a_l}} \langle \hat{\mathbf{e}}_{a_l} | \Delta\epsilon | \hat{\mathbf{e}}_{b_l} \rangle e^{i(\omega_{b_l} - \omega_{a_l})t} b_l + \cdots \quad (7.B.3)$$

using the standard bra-ket shorthand,

$$\langle \hat{\mathbf{e}}_{a_l} | \epsilon | \hat{\mathbf{e}}_{a_l} \rangle \equiv \int d^3\mathbf{r}\, \epsilon(\mathbf{r}) \left| \hat{\mathbf{e}}_{a_l}(\mathbf{r}) \right|^2 = 1, \text{ etc.} \quad (7.B.4)$$

Collecting all these equations, we can form a matrix equation for $\dot{\mathbf{u}}_l = (\dot{a}_l\, \dot{b}_l\, \dot{c}_l\, \ldots\, \dot{m}_l)^T$, where $\dot{a}_l(t) \equiv da_l/dt$, etc. Assuming that only adjacent resonators (nearest neighbors) couple to one another, we can eliminate those coupling terms where the spatial overlap between $\hat{\mathbf{e}}_{a_l}$ and $\hat{\mathbf{e}}_{b_l}$ is small. Furthermore, if two frequencies, say ω_{a_l} and ω_{b_l} for the lth resonator, are considerably different in value, e.g., $|\omega_{a_l} - \omega_{b_l}| = O(\omega_{a_l})$, then time-averaging of Equation 7.B.3 over a few optical cycles—which does not change the slowly varying left-hand side—shows that those two modes do not couple, since the coupling coefficient time averages to zero. With these simplifications, Equation 7.6 is obtained.

Assuming, as we do in this chapter, that $\Delta\epsilon$ is not just slowly varying but independent of time, and that $\omega_{a_l} = \omega_{b_l} = \cdots = \omega_{m_l}$, we observe that the coefficients on the right-hand side of Equation 7.B.3 and consequently, the elements of the matrix M in Equation 7.6 are independent of time. This allows the solution of Equation 7.6 to be written as the linear superposition of eigenmodes, as investigated in Section 7.3. The more general case, where the right-hand side of Equation 7.B.3 is not independent of time, is considerably harder to solve, and no general theory exists.

DIRECTIONAL WAVEGUIDE COUPLERS

For the case of microring or racetrack resonators which are coupled by waveguide directional couplers (two identical single-mode waveguides in close-proximity to each other), the coupling coefficients can be explicitly written in terms of the various geometric dimensions and dielectric coefficients.

$$\kappa = -i\sin(\tilde{\kappa}L)\, e^{-i\phi}, \quad (7.B.5)$$

where L is the length of the directional coupler and $\phi = (\tilde{M} + \beta)L$ with the definitions

$$\tilde{\kappa} = \frac{\omega\epsilon_0}{4} \int d\mathbf{r}_\perp \left(n_c^2 - n_1^2\right) \mathbf{E}^{(1)}(\mathbf{r}_\perp) \cdot \mathbf{E}^{(2)}(\mathbf{r}_\perp)$$

$$\tilde{M} = \frac{\omega\epsilon_0}{4} \int d\mathbf{r}_\perp \left(n_c^2 - n_1^2\right) \mathbf{E}^{(1)}(\mathbf{r}_\perp) \cdot \mathbf{E}^{(1)}(\mathbf{r}_\perp) \quad (7.B.6)$$

in terms of the modal profiles $\mathbf{E}^{(1)}(\mathbf{r}_\perp)$ and $\mathbf{E}^{(2)}(\mathbf{r}_\perp)$ for the two waveguides and the geometry of the refractive index distributions. In Cartesian coordinates, $n_{1,2}(x,y)$ are the cross-sectional refractive index distributions for each of the two waveguides considered separately (assuming the other waveguide is not present nearby), which were used to calculate $\mathbf{E}^{(1,2)}(x,y)$ and $\beta = 2\pi n_{\text{eff}}/\lambda$, and $n_c(x,y)$ is the cross-sectional refractive index distribution in the coupling region, when the two waveguides are close to each other.

If simple expressions can be derived from the eigenvalue problem to describe the transverse modes (e.g., see [71]), one can write down expressions for κ in terms of the physical parameters describing the structure, such as the width w of the waveguide, the various refractive indices (effective index n_{eff}), the inter-waveguide separation s, the wavelength of light λ, and the length of the coupler L,

$$\kappa = -i \sin\left[\frac{2h^2 p\, e^{-ps}}{\beta w (h^2 + p^2)} L\right] e^{-i\phi} \tag{7.B.7}$$

where $\beta = 2\pi n_{\text{eff}}/\lambda$, and h and p are obtained from the eigenvalue problem

$$(pd)^2 + (hd)^2 = (k_0 d)^2 (n_{\text{core}}^2 - n_{\text{clad}}^2),$$
$$pd = hd \tan(hd), \tag{7.B.8}$$

as the inverse length scales associated with the field inside and outside the guiding core of the waveguide, i.e., outside the waveguide core, the field amplitude decays as $\exp(-px)$ and within the guiding layer, the field amplitude behaves as $\cos(hx)$.

Assuming that κ is small, we can replace $\sin(\theta) \approx \theta$ in Equation 7.B.7. Assuming a perturbation in the inter-waveguide spacing $s \rightarrow s + \delta s$, where $p\,\delta s \ll 1$, we find that

$$\left|\frac{\delta \kappa}{\kappa}\right| \approx p\,\delta s. \tag{7.B.9}$$

Therefore, the fractional change of κ if $\delta s = 1$ nm is given simply by p.

If we are within the single-mode limit, then Equation 7.B.8 yields the condition $p \in \{k_0/\sqrt{2}, k_0\} \times \sqrt{n_{\text{core}}^2 - n_{\text{clad}}^2}$. A typical value of $\delta\kappa/\kappa$ can be calculated assuming $n_{\text{core}}^2 = 10$, $n_{\text{clad}}^2 = 1$, $k_0 = 2\pi/1.5$ μm, and $\delta s = 1$ nm,

$$\left|\frac{\delta\kappa}{\kappa}\right| = 0.89\,\% - 1.26\,\% \text{ per nm}. \tag{7.B.10}$$

This derivation of $\delta\kappa/\kappa$ reflects the basic fact that fields outside a waveguide or resonator decay exponentially with distance, at a spatial rate that is proportional to the square root of the difference of the relative dielectric constants of the core and cladding regions, and inversely proportional to the wavelength. Thus, even if it is approximate, Equation 7.B.10 gives a useful order-of-magnitude estimate for a wide variety of dielectric resonator coupling mechanisms.

REFERENCES

1. J. K. S. Poon, J. Scheuer, Y. Xu, and A. Yariv. Designing coupled-resonator optical waveguide delay lines. *J. Opt. Soc. Am. B*, 21(9):1665–1673, 2004.
2. W. J. Kim, W. Kuang, and J. D. O'Brien. Dispersion characteristics of photonic crystal coupled resonator optical waveguides. *Opt. Express*, 11:3431–3437, 2003.
3. J. B. Khurgin. Optical buffers based on slow light in electromagnetically induced transparent media and coupled resonator structures: Comparative analysis. *J. Opt. Soc. Am. B*, 22:1062–1074, 2005.
4. S. Mookherjea, D. S. Cohen, and A. Yariv. Nonlinear dispersion in a coupled-resonator optical waveguide. *Opt. Lett.*, 27:933–935, 2002.

5. E. Burstein and C. Weisbuch, (Eds.). *Confined Electrons and Photons: New Physics and Applications*, NATO Advanced Science Institute Series. Plenum Press, New York, 1993.
6. J. D. Joannopoulos, R. D. Meade, and J. N. Winn. *Photonic Crystals*. Princeton University Press, Princeton, 1995.
7. S. Mookherjea and A. Yariv. Optical pulse propagation in the tight-binding approximation. *Opt. Express*, 9(2):91–96, 2001.
8. G. Gutroff, M. Bayer, J. P. Reithmaier, A. Forchel, P. A. Knipp, and T. L. Reinecke. Photonic defect states in chains of coupled microresonators. *Phys. Rev. B*, 64:155313, 2001.
9. J. K. S. Poon, L. Zhu, G. DeRose, and A. Yariv. Transmission and group delay of microring coupled-resonator optical waveguides. *Opt. Lett.*, 31:456–458, 2006.
10. T. D. Happ, M. Kamp, A. Forchel, J.-L. Gentner, and L. Goldstein. Two-dimensional photonic crystal coupled-defect laser diode. *Appl. Phys. Lett.*, 82:4–6, 2003.
11. C. M. de Sterke. Superstructure gratings in the tight-binding approximation. *Phys. Rev. E*, 57(3):3502–3509, 1998.
12. V. N. Astratov, J. P. Franchak, and S. P. Ashili. Optical coupling and transport phenomena in chains of spherical dielectric microresonators with size disorder. *Appl. Phys. Lett.*, 85:5508–5510, 2004.
13. V. Zhuk, D. V. Regelman, D. Gershoni, M. Bayer, J. P. Reithmaier, A. Forchel, P. A. Knipp, and T. L. Reinecke. Near-field mapping of the electromagnetic field in confined photon geometries. *Phys. Rev. B*, 66:115302, 2002.
14. F. Xia, L. Sekaric, and Y. Vlasov. Ultracompact optical buffers on a silicon chip. *Nat. Photon.*, 1:65–71, 2007.
15. A. Melloni, F. Morichetti, and M. Martinelli. Optical slow wave structures. *Opt. Photon. News*, 14:44–48, 2003.
16. Y.-H. Ye, J. Ding, D.-Y. Jeong, I. C. Khoo, and Q. M. Zhang. Finite-size effects on one-dimensional coupled-resonator optical waveguides. *Phys. Rev. E*, 69:056604, 2004.
17. J. K. S. Poon and A. Yariv. Active coupled-resonator optical waveguides—part I: Gain enhancement and noise. *J. Opt. Soc. Am. B*, 24(9):2378–2388, 2007.
18. E. N. Economou. *Green's Functions in Quantum Physic*. Springer, Berlin, 3rd edn, 2006.
19. Y. Xu, R. K. Lee, and A. Yariv. Propagation and second-harmonic generation of electromagnetic waves in a coupled-resonator optical waveguide. *J. Opt. Soc. Am. B*, 17(3):387–400, 2000.
20. J. V. Hryniewicz, P. P. Absil, B. E. Little, R. A. Wilson, and P.-T. Ho. Higher order filter response in coupled microring resonators. *IEEE Photon. Technol. Lett.*, 12:320–322, 2000.
21. B. Liu, A. Shakouri, and J. E. Bowers. Wide tunable double ring coupled lasers. *IEEE Photon. Technol. Lett.*, 14:600–602, 2002.
22. D. Rabus, M. Hamacher, H. Heidrich, and U. Troppenz. High-Q channel dropping filters using ring resonators with integrated SOAs. *IEEE Photon. Technol. Lett.*, 14:1442–1444, 2002.
23. R. Iliew, U. Peschel, C. Etrich, and F. Lederer. Light propagation via coupled defects in photonic crystals. In *Conference on Lasers and Electro-Optics, OSA Technical Digest, Postconference Edition*, Vol. 73 of *OSA TOPS*, pp. 191–192, OSA, Washington, DC, 2002.
24. M. T. Hill, H. J. S. Dorren, T. de Vries, X. J. M. Leijtens, J. H. den Besten, B. Smalbrugge, Y.-S. Oei, H. Binsma, G.-D. Khoe, and M. K. Smit. A fast low-power optical memory based on coupled micro-ring lasers. *Nature*, 432:206–209, 2004.
25. Y. Chen, G. Pasrija, B. Farhang-Boroujeny, and S. Blair. Engineering the nonlinear phase shift with multistage autoregressive moving-average optical filters. *Appl. Opt.*, 44:2564–2574, 2005.
26. A. Yariv, Y. Xu, R. K. Lee, and A. Scherer. Coupled-resonator optical waveguide: A proposal and analysis. *Opt. Lett.*, 24(11):711–713, 1999.
27. N. Stefanou and A. Modinos. Impurity bands in photonic insulators. *Phys. Rev. B*, 57(19):12127–12133, 1998.
28. M. Bayindir, B. Temelkuran, and E. Ozbay. Tight-binding description of the coupled defect modes in three-dimensional photonic crystals. *Phys. Rev. Lett.*, 84(10):2140–2143, 2000.
29. J. K. S. Poon, J. Scheuer, S. Mookherjea, G. T. Paloczi, Y. Huang, and A. Yariv. Matrix analysis of microring coupled-resonator optical waveguides. *Opt. Express*, 12:90, 2004.
30. D. O'Brien, M. D. Settle, T. Karle, A. Michaeli, M. Salib, and T. F. Krauss. Coupled photonic crystal heterostructure nanocavities. *Opt. Express*, 15:1228–1233, 2007.

31. M. Bayindir, E. Cubukcu, I. Bulu, T. Tut, E. Ozbay, and C. Soukoulis. Photonic band gaps, defect characteristics, and waveguiding in two-dimensional disordered dielectric and metallic photonic crystals. *Phys. Rev. B*, 64:195113, 2001.
32. B. Z. Steinberg, A. Boag, and R. Lisitsin. Sensitivity analysis of narrowband photonic crystal filters and waveguides to structure variations and inaccuracy. *J. Opt. Soc. Am. A*, 20:138–146, 2003.
33. S. G. Johnson, M. L. Povinelli, M. Sojacic, S. Jacobs, and J. D. Joannopoulos. Roughness losses and volume–current methods in photonic-crystal waveguides. *Appl. Phys. B*, 81:283–293, 2005.
34. J. B. Pendry and A. MacKinnon. Calculation of photon dispersion relations. *Phys. Rev. Lett.*, 69:2772–2775, 1992.
35. C. Barnes, T. Wei-chao, and J. B. Pendry. The localization length and density of states of 1D disordered systems. *J. Phys. Condens. Matter*, 3:5297–5305, 1991.
36. H. Matsuoka and R. Grobe. Effect of eigenmodes on the optical transmission through one-dimensional random media. *Phys. Rev. E*, 71:046606, 2005.
37. D. J. Thouless. A relation between the density of states and range of localization for one dimensional random systems. *J. Phys. C*, 5:77–81, 1972.
38. J. M. Ziman. *Models of Disorder*. Cambridge University Press, Cambridge, 1979.
39. A. Gonis. *Green Functions for Ordered and Disordered Systems*. North-Holland, Amsterdam, 1992.
40. P. Sheng. *Introduction to Wave Scattering, Localization and Mesoscopic Phenomena*, 2nd edn. Springer, Berlin, 2006.
41. W. H. Louisell. *Coupled Mode and Parametric Electronics*. John Wiley & Sons, New York, 1960.
42. H. A. Haus. *Electromagnetic Noise and Quantum Optical Measurements*. Springer, Berlin, 2000.
43. B. E. Little, S. T. Chu, H. A. Haus, J. Foresi, and J. P. Laine. Microring resonator channel dropping filters. *J. Lightwave Technol.*, 15:998–1005, 1997.
44. U. Peschel, A. L. Reynolds, B. Arredondo, F. Lederer, P. J. Roberts, T. F. Krauss, and P. J. I. de Maagt. Transmission and reflection analysis of functional coupled cavity components. *IEEE J. Quantum Electron.*, 38:830–836, 2002.
45. C. Cohen-Tannoudji, B. Diu, and F. Laloë. *Quantum Mechanics*. John Wiley & Sons, New York, 1977.
46. R. J. C. Spreeuw and J. P. Woerdman. Optical atoms. In E. Wolf, (Ed.), *Progress in Optics Vol. XXXI*, pp. 263–319. North-Holland, Amsterdam, 1993.
47. T. Mukaiyama, K. Takeda, H. Miyazaki, Y. Jimba, and M. Kuwata-Gonokami. Tight-binding photonic molecule modes of resonant bispheres. *Phys. Rev. Lett.*, 82(23):4623–4626, 1999.
48. D. S. Weiss, V. Sandoghdar, J. Hare, V. Lefevre-Seguin, J.-M. Raimond, and S. Haroche. Splitting of high-Q Mie modes induced by light backscattering in silica microspheres. *Opt. Lett.*, 20:1835–1837, 1995.
49. W. H. Press, S. A. Teukolsky, W. T. Vetterling, and B. P. Flannery. *Numerical Recipes in Fortran 77*. Cambridge University Press, Cambridge, 1996.
50. S. Mookherjea and A. Yariv. Kerr-stabilized super-resonant modes in coupled-resonator optical waveguides. *Phys. Rev. E*, 66:046610, 2002.
51. D. Botez. Monolithic phase-locked semiconductor laser arrays. In D. Botez and D. R. Scifres, (Eds.), *Diode Laser Arrays*, pp. 1–71. Cambridge University Press, Cambridge, 1994.
52. B. M. Möller, M. V. Artemyev, and U. Woggon. Coupled-resonator optical waveguides doped with nanocrystals. *Opt. Lett.*, 30:2116–2118, 2005.
53. J. N. Franklin. *Matrix Theory*. Dover, New York, 2000.
54. J. E. Heebner, R. W. Boyd, and Q.-H. Park. SCISSOR solitons and other novel propagation effects in microresonator-modified waveguides. *J. Opt. Soc. Am. B*, 19:722–731, 2004.
55. J. E. Heebner, P. Chak, S. Pereira, J. E. Sipe, and R. W. Boyd. Distributed and localized feedback in microresonator sequences for linear and nonlinear optics. *J. Opt. Soc. Am. B*, 21:1818–1832, 2004.
56. S. Mookherjea. Mode cycling in microring optical resonators. *Opt. Lett.*, 30:2751–2753, 2005.
57. M. Kulishov, J. M. Laniel, N. Bélanger, and D. V. Plant. Trapping light in a ring resonator using a grating-assisted coupler with asymmetric transmission. *Opt. Express*, 13:3567–3578, 2005.
58. S. Mookherjea. Using gain to tune the dispersion relationship of coupled-resonator optical waveguides. *IEEE Photon. Technol. Lett.*, 18:715–717, 2006.
59. M. Greenberg and M. Orenstein. Unidirectional complex grating assisted couplers. *Opt. Express*, 12:4013–4018, 2004.
60. T. J. Kippenberg, S. M. Spillane, and K. J. Vahala. Mode coupling in traveling-wave resonators. *Opt. Lett.*, 27:1669–1671, 2002.

61. E. R. Smith. One-dimensional $x-y$ model with random coupling constants. I. Thermodynamics. *J. Phys. C*, 3:1419–1432, 1970.
62. I. M. Lifshits, S. A. Gredeskul, and L. A. Pastur. *Introduction to the Theory of Disordered Systems*. John Wiley & Sons, New York, 1988.
63. F. J. Dyson. The dynamics of a disordered linear chain. *Phys. Rev.*, 92:1331–1338, 1953.
64. P. Dean. Vibrations of glass-like disordered chains. *Proc. Phys. Soc.*, 84:727–744, 1964.
65. S. M. Barnett and R. Loudon. Sum rule for modified spontaneous emission rates. *Phys. Rev. Lett.*, 77:2444–2446, 1996.
66. F. Dominiguez-Adame and V. A. Malyshev. A simple approach to Anderson localization in one-dimensional disordered lattices. *Am. J. Phys.*, 72(2):226–230, 2004.
67. D. C. Herbert and R. Jones. Localized states in disordered systems. *J. Phys. C Solid State Phys.*, 4:1145–61, 1971.
68. S. Mookherjea, J. S. Park, S.-H. Yang, and P. R. Bandaru. Localization in silicon nanophotonic slow-light waveguides. *Nature Photonics*, 2:90–93, 2008.
69. J. Topolancik, B Ilic, and F. Vollmer. Experimental observation of strong photon localization in disordered photonic crystal waveguides. *Phys. Rev. Lett.*, 99:253901, 2007.
70. Y. Lahini, A. Avidan, F. Pozzi, M. Sorel, R. Morandotti, D. N. Christodoulides, and Y. Silberberg. Anderson localization and nonlinearity in one-dimensional disordered photonic lattices. *Phys. Rev. Lett.*, 100:013906, 2008.
71. A. Yariv. *Optical Electronics in Modern Communications*, 5th edn. Oxford, New York, 1997.
72. S. Mookherjea. Spectral characteristics of coupled resonators. *J. Opt. Soc. Am. B*, 23:1137–1145, 2006.
73. S. Mookherjea and A. Oh. Effect of disorder on slow light velocity in optical slow-wave structures. *Opt. Lett.*, 32:289–291, 2007.
74. Y. Hara, T. Mukaiyama, K. Takeda, and M. Kuwata-Gonokami, Heavy photon states in photonic chains of resonantly coupled cavities with supermonodispersive microspheres, *Phys. Rev. Lett.*, 94:203905, 2005.
75. M. F. Yanik and S. Fan, Stopping light all optically, *Phys. Rev. Lett.*, 92:083901, 2004.
76. M. Ghulinyan, M. Galli, C. Toninelli, J. Bertolotti, S. Gottardo, F. Marabelli, D. S. Wiersma, L. Pavesi, and L. C. Andreani, Wide-band transmission of nondistorted slow waves in one-dimensional optical superlattices, *Appl. Phys. Lett.*, 88:241103, 2006.

Part III

Slow Light in Fibers

8 Slow and Fast Light Propagation in Narrow Band Raman-Assisted Fiber Parametric Amplifiers

Gadi Eisenstein, Evgeny Shumakher, and Amnon Willinger

CONTENTS

8.1	Introduction	149
8.2	Theoretical Model	151
	8.2.1 SRS-Assisted OPA in Isotropic Fibers	151
	8.2.2 SRS-Assisted OPA in a Birefringent Fiber	153
	8.2.2.1 Statistical Analysis of Gain and Delay Limits	156
	8.2.3 Averaged PMD Model of Fiber NB-OPA	159
	8.2.4 Effect of Longitudinal Variations of Propagation Parameters	161
	8.2.4.1 Uniqueness and Spatial Resolution	161
	8.2.4.2 Estimation Procedure	162
	8.2.4.3 Results	162
8.3	Experimental Results	164
Appendix A: Group Delay Calculations in Isotropic Fibers		167
Appendix B: Gain Calculations for a Birefringent Fiber		168
Appendix C: Group Delay Calculations in Birefringent Fibers		169
References		171

8.1 INTRODUCTION

Optical parametric amplification (OPA), employing optical fibers, is a well-known nonlinear process that has been explored extensively for the past few decades [1–6]. A conventional fiber OPA relies on a pump propagating in the anomalous dispersion regime that yields a broad band gain spectrum which is symmetric around the pump wavelength.

A parametric gain spectrum of a completely different nature is obtained when the pump propagates in the normal dispersion regime (where β_2 is positive) of a fiber whose β_4 parameter is negative. Phase matching conditions are satisfied in this case in two narrow spectral regions both of which are widely detuned from the pump wavelength [7,8]. A large narrow band optical parametric gain, with an accompanying large change in the group index, is obtained in each of these regions. On the short

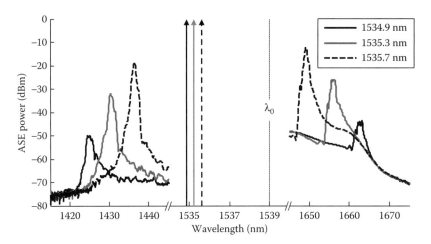

FIGURE 8.1 Measured NB-OPA ASE spectra in a 1 km DSF for different pump wavelengths and a constant pump power of 5 W.

wavelength side of the pump, the group refractive index increases whereas on the long wavelength side, it decreases. Slow and fast light propagation is naturally possible in the two respective spectral windows [9].

Exemplary spontaneous emission (which is proportional to gain) spectra in a narrow band optical parametric amplifier (NB-OPA), measured for a 1 km long dispersion-shifted fiber (DSF) are shown in Figure 8.1. The particular fiber has a zero dispersion wavelength λ_0 of 1539 nm and the pump power was 5 W. As the pump wavelength is detuned away from λ_0, the gain spectra shift and their peak levels reduce. The asymmetry between the short and long wavelength gain regions stems from stimulated Raman scattering (SRS) which is significant due to the large detuning between the pump and the gain region.

Slow light propagation in an NB-OPA was first demonstrated in Ref. [9]. Figure 8.2 describes pump power dependence of the delay experienced by a 70 ps wide pulse in a 1 km DSF.

The exemplary measured data in Figures 8.1 and 8.2 demonstrate the potential of NB-OPAs as wide band slow and fast light media with a large tuning range. However, a careful examination

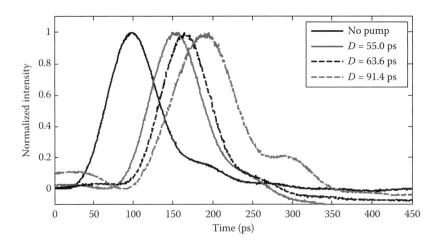

FIGURE 8.2 Measured pump power dependent delay of 70 ps wide pulses in 1 km long DSF. (From Dahan, D. and Eisenstein, G., *Opt. Express*, 13, 6234, 2005.)

reveals many nonobvious details such as gain spectrum asymmetries and distortions experienced by the delayed pulses. These have a profound effect on the performance of slow and fast light systems and therefore it is crucial to model the NB-OPA accurately.

SRS has two contributions to the parametric gain. The imaginary part of the Raman susceptibility reduces the overall gain in the short wavelength (slow light) region while increasing it in the fast light region. At the same time, the real part of the Raman susceptibility contributes to the phase matching conditions [10] further affecting the two gain spectra. The concept of SRS contribution to phase matching originates from early work by Bloembergen and Shen [11]. It was previously analyzed and demonstrated experimentally [8,12,13] for nonlinear optical fibers but its influence on NB-OPAs is particularly large due to the sensitivity to operational conditions of this gain mechanism.

The efficiency of parametric processes and of SRS depends on the mutual state of polarization (SOP) of the interacting fields [6,14,15]. Random birefringence modifies the SOP of the propagating fields and therefore the efficiency of the distributed interaction [16,17].

Fiber imperfections that bring about longitudinal variations of linear and nonlinear propagation parameters also influence the distributed nonlinear processes. The longitudinal parameter distributions can be extracted with a high spatial resolution using an optimization procedure that makes use of measured NB-OPA gain spectra [18]. These distributions should be included in the detailed propagation model in order to increase its accuracy further.

Precise modeling requires therefore the inclusion of the coupling of parametric processes to SRS in the presence of random fiber birefringence and longitudinal variations of linear and nonlinear propagation parameters. This chapter describes a comprehensive model for slow and fast light propagation in a Raman-assisted NB-OPA. A separate subsection outlines the parameter extraction procedure which has important implications not related to slow and fast light propagation. Experimental results of measured gain spectra as well as of delayed high speed data signals are also described. The experimental results are compared with modeling predictions showing superb fits, highlighting the viability of comprehensive modeling of slow and fast light propagation.

8.2 THEORETICAL MODEL

This section addresses the key issues governing slow and fast light propagation in a Raman-assisted NB-OPA. Section 8.2.1 describes, using a scalar model, the effect of the full complex Raman susceptibility on the gain and delay spectra. The role of random birefringence is added in Section 8.2.2. This necessitates a complex stochastic model which yields statistical characteristics of the NB-OPA performance. Section 8.2.3 outlines a computationally efficient approach to estimate the performance of an NB-OPA where the longitudinal variations of the birefringence are averaged in a manner that accounts for every possible realization. This enables to compute the characteristics of an average fiber. Section 8.2.4 describes the longitudinal parameter extraction procedure and the effect of propagation parameter variation on the gain and delay spectra of the NB-OPA.

8.2.1 SRS-Assisted OPA in Isotropic Fibers

The coupling of SRS and parametric amplification is a fundamental property of any nonlinear medium first formulated by Bloembergen in 1964 [11] and studied in optical fibers both theoretically [8] and experimentally [10,13]. For a perfectly isotropic fiber, the model describing SRS-assisted OPA is scalar.

The nonlinear susceptibility P_{NL} has two components

$$P_{NL} = \varepsilon_0 \int \int \int \tilde{\chi}^{(3)}(t_1, t_2, t_3) E(t-t_1) \cdot E(t-t_2) E(t-t_3) \, dt_1 dt_2 dt_3$$

$$= \varepsilon_0 \left(\chi_K |E(t)|^2 + \int_{-\infty}^{t} \tilde{\chi}_R(t-\tau) |E(\tau)|^2 \, d\tau \right) E(t) \tag{8.1}$$

The first is due to the instantaneous Kerr effect χ_K from which it is common to define the well-known nonlinear coefficient γ_K. The second contribution to P_{NL} is $\tilde{\chi}_R$, which results from the delayed SRS process, where the well-known (and measurable) Raman gain coefficient is related to the imaginary part of the SRS susceptibility $g_R(\omega) = -(4\gamma_K/3\chi_K)\,\text{Im}\,[\chi_R(\omega)]$ with $\chi_R(\omega) = \int_{-\infty}^{\infty} \tilde{\chi}_R(t)e^{j\omega t}dt$ being the Fourier transform (FT) of the time response $\tilde{\chi}_R(t)$.

The combined effects of four wave mixing (FWM) and SRS are described by three coupled equations for the envelopes of the pump, signal, and idler ($A_q = A_p, A_s, A_i$), where each field is formulated as $E_q = F(x, y) \cdot A_q(z)\, e^{j\beta_q z - j\omega_q t}$. $F(x,y)$ is the transverse field distribution of the propagating mode. Considering a lossless fiber, an undepleted pump and continuous wave (CW) fields, the equation governing propagation of the pump envelope is

$$\frac{\partial A_p}{\partial z} = j\gamma_K \left[1 + \frac{2\chi_R(0)}{3\chi_K}\right] |A_p|^2 A_p \equiv j\gamma |A_p|^2 A_p, \tag{8.2}$$

where the Raman susceptibility is evaluated at $\omega = 0$. The total nonlinear coefficient γ contains two contributions, one from the Kerr effect and the other from the real part of χ_R. The fractional contribution due to SRS is denoted by f which in all silica fibers equals 0.18 [12]. This means that in cases where SRS plays a measurable role, the pure Kerr nonlinearity (γ_K) differs from that which is extracted in self-phase modulation (SPM) experiments $\gamma_K = (1-f)\gamma$.

A simple manipulation of Equation 8.2 leads to an analytical expression for f

$$f = \frac{\gamma - \gamma_K}{\gamma} = \frac{2}{3}\frac{\chi_R(0)}{\chi_K}\frac{\gamma_K}{\gamma}. \tag{8.3}$$

The propagation of the signal and idler is governed by the following coupled equations, which include the contribution to phase matching of the real part of the Raman susceptibility

$$\begin{aligned}\frac{\partial A_s}{\partial z} &= j\gamma \left[(1-f) + 1 + f\underline{\chi}_R\right] P_p A_s + j\gamma \left[(1-f) + f\underline{\chi}_R\right] A_p^2 A_i^* e^{-j\Delta\beta\cdot z} \\ \frac{\partial A_i}{\partial z} &= j\gamma \left[(1-f) + 1 + f\underline{\chi}_R^*\right] P_p A_i + j\gamma \left[(1-f) + f\underline{\chi}_R^*\right] A_p^2 A_s^* e^{-j\Delta\beta\cdot z},\end{aligned} \tag{8.4}$$

where $\underline{\chi}_R = \underline{\chi}_R(\Omega_s)$ is the normalized values of the Raman susceptibility defined using the symmetry of the Raman response $\underline{\chi}_R(\Omega_s) = \left[\underline{\chi}_R(\Omega_i)\right]^* \equiv \underline{\chi}_R$ at a given frequency shift $\Omega_s = \omega_s - \omega_p = -\Omega_i$. The resulting FWM process contains the memory properties of SRS.

Invoking the manipulation $A_{s,i} = B_{s,i} \exp\{j(2\gamma P_p - \Delta\beta/2)z\}$, where $\Delta\beta = \beta_s + \beta_i - 2\beta_p$ is the phase mismatch, it is possible to describe the evolution of $\underline{b} = [B_s\ B_i^*]^T$ in the form of an ordinary differential equation with constant coefficients

$$\partial_z \underline{b} = j \begin{bmatrix} \left(\eta_{KR} + \frac{\Delta\beta}{2}\right) & \eta_{KR} \\ -\eta_{KR} & -\left(\eta_{KR} + \frac{\Delta\beta}{2}\right) \end{bmatrix} \underline{b}, \tag{8.5}$$

where $\eta_{KR} = \gamma \left[(1-f) + f\underline{\chi}_R\right] P_p$ is the Kerr–Raman coefficient.

The solution is similar in form to that of conventional FWM without SRS [5] with the gain coefficient, g, obeying the characteristic equation $g^2 + (\eta_{KR} + (\Delta\beta/2))^2 - \eta_{KR}^2 = 0$. The total accumulated gain experienced by the signal is

$$G = \left|\frac{B_s}{B_{s,0}}\right|^2 = \left|\cosh(gz) + j\frac{\eta_{KR} + (\Delta\beta/2)}{g}\sinh(gz)\right|^2. \tag{8.6}$$

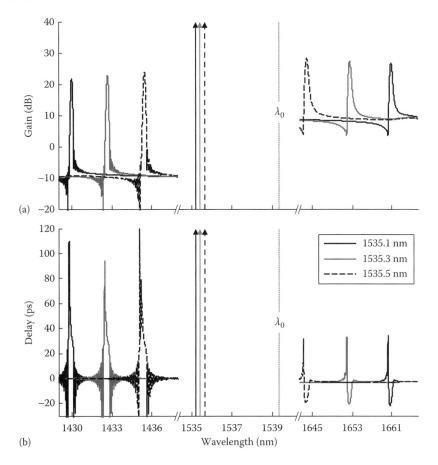

FIGURE 8.3 Calculated gain (a) and delay (b) spectra using the scalar model of SRS-assisted OPA in a 1 km DSF.

Note that it is possible to obtain an analytical expression for the gain (Equation 8.6) only for the case of lossless fiber.

Calculated gain and delay spectra for different pump wavelengths are shown in Figure 8.3. The calculation assumes a 1 km DSF with $\lambda_0 = 1539$ nm, $\beta_3 = -5.617 \times 10^{-4}$ ps^3/km, $\beta_4 = 0.12315$ ps^4/km, a measured nonlinear coefficient of $\gamma = 2.3$ W^{-1}/km and a pump power of 2 W. The delay spectra were calculated based on the movement of the center of gravity of the propagating power envelope. The technique is described in Appendix A.

The gain spectra in Figure 8.3a reveal the main features of the SRS-assisted NB-OPA. As the pump wavelength detuning from λ_0 increases, the gain level reduces and the bandwidth narrows. The asymmetry between the short and long wavelength spectra is a direct consequence of SRS. At the short wavelength spectra, the background gain is negative since SRS induces loss whereas in the long wavelength spectra, the narrow spectra lie on top of the positive Raman gain. A different asymmetry occurs in the delay spectra as SRS increases the delay (at short wavelengths) but reduces the advancement in the fast light region.

8.2.2 SRS-Assisted OPA in a Birefringent Fiber

Propagation in a birefringent fiber requires modeling using vector formalism [19] since the components of each propagating field are not restricted to one fiber axis. The field envelopes are described hereon using the Jones notation for the SOP

$$|A\rangle \equiv \begin{bmatrix} A_x & A_y \end{bmatrix}^T \Leftrightarrow \langle A| \equiv \begin{bmatrix} A_x^* & A_y^* \end{bmatrix}. \tag{8.7}$$

Linear propagation in a birefringent fiber is described by

$$\partial_z |A\rangle = \left[\left(-\frac{\alpha}{2} + j\sum_{n=2}^{\infty} \frac{\beta_{(\omega_0)}^{(n)}}{n!}(j\partial_t)^n\right)\underline{\underline{I}} - \frac{j}{2}\omega_0\left(\vec{\beta}\cdot\vec{\sigma}\right)\right]|A\rangle \equiv \underline{\underline{L}}_0[j\partial_t]|A\rangle, \tag{8.8}$$

where

$\underline{\underline{L}}_0[\omega]$ is a linear operator
ω_0 is the carrier frequency
$\underline{\underline{I}}$ is the 2 × 2 unity matrix
α is the fiber loss coefficient
$\beta_{(\omega_0)}^{(n)}$ is the nth order linear dispersion

The fiber birefringence is denoted by a three-dimensional vector $\vec{\beta}$ (of units s/m) in Stokes space. Pauli's spin matrices, σ_i, are used to transform the two-dimensional Jones notation to the three-dimensional Stokes space [19]. The transformation is described in Appendix B.

$$\vec{\beta}\cdot\vec{\sigma} = \begin{bmatrix} \beta_1 & \beta_2 - j\beta_3 \\ \beta_2 + j\beta_3 & -\beta_1 \end{bmatrix} \tag{8.9}$$

The coupling of the components causes changes in the SOP which for a time-varying envelope results in polarization mode dispersion (PMD). PMD is commonly quantified by a parameter D_p [16] measured in units of ps/$\sqrt{km}$.

The nonlinear polarization which is induced by two fields has one contribution from the instantaneous Kerr effect (χ_K) and a second due to the delayed SRS process ($\widetilde{\chi}_R$). The nonlinear polarization ($\vec{P}_{NL}$) is similar in form to Equation 8.1, but the electric field $\vec{E}$ and the polarization $\vec{P}_{NL}$ are now two-dimensional vectors

$$\vec{P}_{NL} = \varepsilon_0 \chi_K \left(\vec{E}\cdot\vec{E}\right)\vec{E} + \varepsilon_0 \left(\int_{-\infty}^{t} \widetilde{\chi}_R(t-\tau)\left[\vec{E}(\tau)\cdot\vec{E}(\tau)\right]d\tau\right)\vec{E}(t). \tag{8.10}$$

Equation 8.10 considers contributions to the Raman susceptibility, $\widetilde{\chi}_R$, resulting only from copolarized waves since cross polarized waves yield a Raman gain which is reduced by at least two orders of magnitude [20].

Adopting the notation $\vec{E}_q = F(x,y) \cdot \left|A_q(z)\right\rangle e^{j\beta_q z - j\omega_q t}$, the signal and idler field envelopes are further manipulated according to $|A_{s,i}\rangle = |\underline{A}_{s,i}\rangle \exp\{-j\Delta\beta_{s,i}z\}$, where $\Delta\beta_{s,i} = \beta_{s,i} - \beta_p$. For the CW case under the undepleted pump approximation, the propagation of the pump, signal, and idler waves is described by three coupled equations

$$\partial_z |A_p\rangle = \underline{\underline{L}}_p |A_p\rangle + \underline{\underline{S}}_p |A_p\rangle \tag{8.11}$$

$$\partial_z |A_s\rangle = \underline{\underline{L}}_s |A_s\rangle + j\Delta\beta_s |A_s\rangle + \left(\underline{\underline{X}}_p + \underline{\underline{R}}_p\right)|A_s\rangle + \underline{\underline{F}}_s |A_i^*\rangle \tag{8.12}$$

$$\partial_z |A_i\rangle = \underline{\underline{L}}_i |A_i\rangle + j\Delta\beta_i |A_i\rangle + \left(\underline{\underline{X}}_p + \underline{\underline{R}}_i\right)|A_i\rangle + \underline{\underline{F}}_i |A_s^*\rangle \tag{8.13}$$

The linear operator $L_{p,s,i}$ (L_0 in Equation 8.8) is used for the CW case with the argument $\omega = 0$. The other operators are matrices denoting different nonlinear effects: SPM, cross-phase modulation (XPM), FWM, and SRS, respectively:

$$\underline{\underline{S}}_p = j\frac{\gamma_K}{3}\left(2\langle A_p|A_p\rangle \underline{\underline{I}} + |A_p^*\rangle\langle A_p^*|\right) + 2j\frac{\gamma_K}{3}\frac{\chi_R(0)}{\chi_K}\langle A_p|A_p\rangle \quad (8.14)$$

$$\underline{\underline{X}}_p = 2j\frac{\gamma_K}{3}\left(\langle A_p|A_p\rangle \underline{\underline{I}} + |A_p\rangle\langle A_p| + |A_p^*\rangle\langle A_p^*|\right) \quad (8.15)$$

$$\underline{\underline{F}}_{s,i} = j\frac{\gamma_K}{3}\left(\langle A_p^*|A_p\rangle \underline{\underline{I}} + 2|A_p\rangle\langle A_p^*|\right) + j\frac{2\gamma_K}{3\chi_K}\chi_R(\Omega_{s,i})|A_p\rangle\langle A_p^*| \quad (8.16)$$

$$\underline{\underline{R}}_{s,i} = j\frac{2\gamma_K}{3\chi_K}\left(\chi_R(0)\langle A_p|A_p\rangle \underline{\underline{I}} + \chi_R(\Omega_{s,i})|A_p\rangle\langle A_p|\right) \quad (8.17)$$

Equations 8.11 through 8.13 describe the complete effect of nonlinearities in the presence of birefringence which in practical fibers has random longitudinal variations. Direct solutions are computationally cumbersome but can be simplified by averaging over the SOP [15]. The averaging process is justified since the correlation length of the residual birefringence (a few meters) is significantly shorter than the characteristic nonlinear length of the fiber ($\sim$1 km). This means that changes of the SOP caused by anisotropic nonlinear processes are negligible compared with those caused by linear birefringence.

The averaging is performed in the rotated pump frame using the transition matrix

$$\underline{\underline{U}} = \cos\left(\frac{\Delta\varphi}{2}\right)\underline{\underline{I}} - j\left(\hat{\beta}\cdot\vec{\sigma}\right)\sin\left(\frac{\Delta\varphi}{2}\right) \quad (8.18)$$

The phase shift is $\Delta\varphi = \omega_p|\vec{\beta}|z$ and the envelopes of the waves are $|A\rangle = \underline{\underline{U}}|X\rangle$, where $|X\rangle$ is the envelope in the rotated pump frame.

The outcome of the averaging process is that on average, an observer in the rotated frame of the pump observes the pump experiencing no birefringence and only polarization independent SPM, with a reduced nonlinear coefficient. The governing equation of the pump averaged Jones vector, in its rotated frame, is thus given by

$$\partial_z|X_p\rangle = -\frac{\alpha}{2}|X_p\rangle + j\gamma_K\left[\frac{8}{9} + \frac{2}{3}\frac{\chi_R(0)}{\chi_K}\right]P_p|X_p\rangle \equiv -\frac{\alpha}{2}|X_p\rangle + j\gamma P_p|X_p\rangle \quad (8.19)$$

where $P_p = \langle A_p|A_p\rangle = |A_{p,x}|^2 + |A_{p,y}|^2$ is the pump power.

The same averaging process is implemented to derive coupled averaged equations for the signal and idler propagation, also in the rotated pump frame. Invoking the manipulation $|X_{s,i}\rangle = |Y_{s,i}\rangle \exp\{j\gamma P_p \cdot L_{\text{eff}}(z)\}$ yields

$$\partial_z|Y_s\rangle = \left(-\frac{\alpha}{2} + j\Delta\beta_s\right)|Y_s\rangle - \frac{j}{2}\Omega_s\left(\vec{b}\cdot\vec{\sigma}\right)|Y_s\rangle + j\gamma q_s\chi_{s,p}^*|p\rangle + j\gamma q_s\chi_{i,p}|p\rangle \quad (8.20)$$

$$\partial_z|Y_i\rangle = \left(-\frac{\alpha}{2} + j\Delta\beta_i\right)|Y_i\rangle - \frac{j}{2}\Omega_i\left(\vec{b}\cdot\vec{\sigma}\right)|Y_i\rangle + j\gamma q_i\chi_{i,p}^*|p\rangle + j\gamma q_i\chi_{s,p}|p\rangle \quad (8.21)$$

where $q_{s,i} \triangleq 1 - f + f\chi_R(\Omega_{s,i})$, and the effective length is $L_{\text{eff}}(z) = (1 - e^{-\alpha z})/\alpha$ for finite losses and $L_{\text{eff}}(z) = z$ in the lossless case.

Rapid variations in the SOP reduce the Kerr effect by a factor of 8/9. Rearranging Equation 8.19 leads to $\gamma_K = \frac{9}{8}(1-f)\gamma$ (where f in the vector model is the same as in the scalar case, see Section 8.2.1). This means that the pure Kerr nonlinearity differs from that which is extracted in experiments performed using birefringent fibers, where both SRS and PMD play a measurable role.

The second term in Equations 8.20 and 8.21 describes the birefringence induced changes in the SOP of the waves which depend on the detuning from the pump and on $\vec{b}$, a modified Stokes vector of the birefringence, as viewed from the rotating pump frame. The third and fourth terms in Equations 8.20 and 8.21 describe XPM, FWM, and SRS which depend on the relative SOPs of the three respective waves: $\chi_{s,p} = \langle Y_s | p \rangle$ and $\chi_{i,p} = \langle Y_i | p \rangle$, where $|p\rangle$ is the SOP of the pump $|X_p\rangle = |p\rangle \exp\{j\gamma P_p \cdot L_{\text{eff}}(z)\}$.

Averaging over the rapid changes of the SOP (which are due to the linear birefringence) does not eliminate the stochastic nature of the propagation because there are also slow SOP variations. These are due to nonlinear polarization rotations and to a relative linear birefringence that stems from the extremely large detuning between the interacting fields. Since the nature of the propagation remains stochastic, a deterministic evaluation of the OPA characteristics is not possible. Solving the propagation model described in Equations 8.20 and 8.21 yields only the characteristics of a specific (random) fiber realization. Repeating the solution provides statistical properties from which the OPA performance can be estimated. Each solution is obtained by dividing the fiber into small segments of constant birefringence, whose values are randomly generated (within a well-defined statistical set). Sufficient accuracy is obtained for segment lengths of 1–2 m and 10^4–10^5 random fiber realizations. The exemplary calculations described hereon are for the $\sim$1 km DSF used in the scalar case, Section 8.2.1. The calculations assume $\alpha = 0$ since this lossless case simplifies the numerical calculations. Random birefringence is characterized by the PMD parameter with values of 0.05, 0.1, and 0.15 ps/$\sqrt{\text{km}}$.

8.2.2.1 Statistical Analysis of Gain and Delay Limits

The OPA gain limits in a birefringent fiber are the maximum and minimum gain values, at a given wavelength, which are obtained for a given fiber realization by scanning all possible relative SOPs at the input of the fiber. The gain limits are computed from the expression for the overall OPA gain given in Appendix B.

Figure 8.4 describes the normalized empirical probability distribution functions (PDFs) of the peak gain levels for the three different PMD values. Figure 8.4a, which depicts the maximum gain, reveals that for an increasing PMD value, not only does the range of attainable gain shift towards lower values (as expected from lower parametric efficiency) but the distribution also narrows. The PDFs of the minimum gain peak point is shown in Figure 8.4b. For all the examined PMD values,

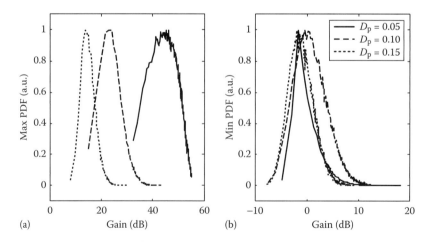

FIGURE 8.4 Normalized empirical PDFs of (a) maximum and (b) minimum gain for D_p values of 0.05, 0.1, and 0.15 $\sqrt{\text{km}}$.

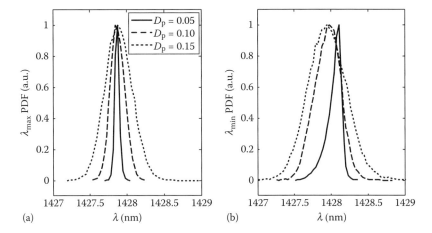

FIGURE 8.5 Normalized empirical PDFs of the wavelength for which gain is (a) maximum and (b) minimum for D_p values of 0.05, 0.1, and 0.15 ps/$\sqrt{\text{km}}$.

the PDFs are centered near transparency and their standard deviations increase moderately with the PMD values.

PMD also causes variations in the wavelength where phase matching conditions are satisfied. This is presented in Figure 8.5, which shows the PDFs of the wavelength for which the maximum and minimum gain spectra peak. In both the cases, the deviation of the peak wavelength increases for a larger PMD.

The input relative polarization angles between the pump and the signal, required to achieve maximum and minimum gain, are presented in Figure 8.6a and b, respectively. For a small PMD value, the coherent nature of the parametric process exhibits itself by tending towards coaligned SOPs for maximum efficiency and cross-aligned SOPs for minimum efficiency. The PDFs change for larger PMD values, as the optimum in both the minimum and the maximum gains shifts toward 45° and the distributions become wider.

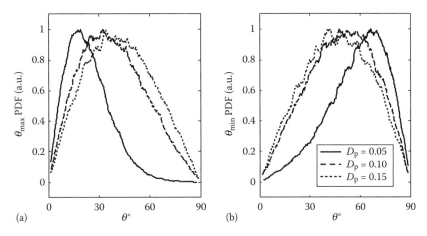

FIGURE 8.6 Normalized empirical PDFs of the relative input polarization angle for which the gain is (a) maximum and (b) minimum for D_p values of 0.05, 0.1, and 0.15 ps/$\sqrt{\text{km}}$.

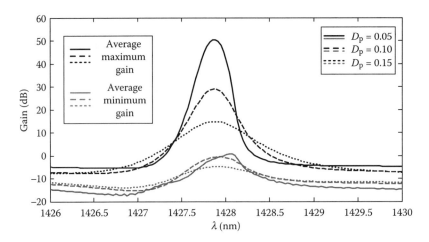

FIGURE 8.7 Maximum (black) and minimum (grey) averaged gain limit spectra for D_p values of 0.05, 0.1, and 0.15 ps/$\sqrt{km}$.

The optimum input relative phase is also obtained from the solution. It turns out to be uniformly distributed for all PMD values.

A given fiber has a particular birefringence distribution. Averaging among the results obtained from many random fiber realizations yields mean values of the gain limits from which the overall performance of the OPA can be deduced. The mean gain limit spectra are described in Figure 8.7.

Large gain levels are attainable for small PMD values but this is accompanied by a large sensitivity to the input SOP. The range of minimum average gain values for the different PMD values is substantially smaller, consistent with Figure 8.4.

Similar conjectures are valid for the mean delay spectra shown in Figure 8.8. These delay values are calculated for the specific input SOPs that match the minimum and maximum gain limits, respectively. The procedure of delay calculations in the vector model is detailed in Appendix C. While these are not necessarily the delay limits, they indicate the average delay performance of the OPA in the presence of birefringence. An increase of the PMD does not decrease the delay significantly

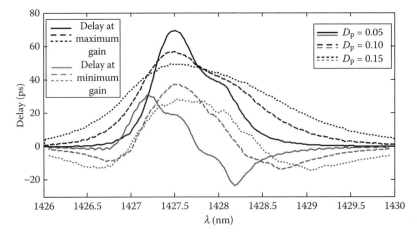

FIGURE 8.8 Average delay spectra at the maximum (black) and minimum (grey) gain limits for D_p values of 0.05, 0.1, and 0.15 ps/$\sqrt{km}$.

8.2.3 Averaged PMD Model of Fiber NB-OPA

The performance of the NB-OPA in the presence of random birefringence was described in Section 8.2.2. This section outlines a different approach in which the OPA performance is estimated by averaging the birefringence longitudinal variations in a manner that accounts for every possible realization [15]. This approach has many computational advantages since averaging over the random longitudinal distribution of birefringence results in a set of deterministic (rather than stochastic) differential equations which are easier to solve. The outcome is an estimate for the performance of an average fiber with a given D_p.

The Ito and Stratonovich interpretations of stochastic integration [21] are used to formulate the longitudinal averaging over the random birefringence vector $\vec{b}$ (Equations 8.20 and 8.21). $\vec{b}$ is assumed to be a three-dimensional white Gaussian noise source with zero mean and a δ-function correlation [22].

$$\overline{\vec{b}(z)} = \underline{0} \tag{8.22}$$

$$\overline{\vec{b}(z_1) \cdot \vec{b}^{\mathrm{T}}(z_2)} = \frac{1}{3}\frac{3\pi}{8}D_p^2 \underline{I}\delta(z_1 - z_2) \tag{8.23}$$

A set of six coupled equations describing the propagation of the power of the waves and their various mutual interactions is developed in order to arrive at an average gain spectrum

$$\partial_z \left(P_p \overline{P_s}\right) = (-2\alpha + \eta_R)\, P_p \overline{P_s} + \eta_R \overline{V_s} + \eta_{\mathrm{NL}} \overline{U_r} + \eta_R \overline{U_i} \tag{8.24}$$

$$\partial_z \left(P_p \overline{P_i}\right) = (-2\alpha - \eta_R)\, P_p \overline{P_i} - \eta_R \overline{V_i} + \eta_{\mathrm{NL}} \overline{U_r} - \eta_R \overline{U_i} \tag{8.25}$$

$$\partial_z \overline{U_r} = -\left(2\alpha + \frac{1}{2}\eta_d\right) \overline{U_r} + \kappa \overline{U_i} + \eta_{\mathrm{NL}} \left(P_p \overline{P_s} + P_p \overline{P_i}\right) + \eta_{\mathrm{NL}} \left(\overline{V_s} + \overline{V_i}\right) \tag{8.26}$$

$$\partial_z \overline{U_i} = -\left(2\alpha + \frac{1}{2}\eta_d\right) \overline{U_i} - \kappa \overline{U_r} - \eta_R \left(P_p \overline{P_s} + P_p \overline{P_i}\right) - \eta_R \left(\overline{V_s} - \overline{V_i}\right) \tag{8.27}$$

$$\partial_z \overline{V_s} = -\left(2\alpha + \eta_d + \eta_R\right) \overline{V_s} + \eta_R P_p \overline{P_s} + \eta_{\mathrm{NL}} \overline{U_r} + \eta_R \overline{U_i} \tag{8.28}$$

$$\partial_z \overline{V_i} = -\left(2\alpha + \eta_d - \eta_R\right) \overline{V_i} - \eta_R P_p \overline{P_i} + \eta_{\mathrm{NL}} \overline{U_r} - \eta_R \overline{U_i} \tag{8.29}$$

In addition to the two powers, the variables in Equations 8.24 through 8.29 describe the mutual interactions between the waves: $V_s = \vec{p} \cdot \vec{y}_s$, $V_i = \vec{p} \cdot \vec{y}_i$, and $U = U_r + jU_i = 2j\chi_{s,p}\chi_{i,p}$, where $\vec{y}_{s,i} = \langle Y_{s,i}|\vec{\sigma}|Y_{s,i}\rangle$ are the Stokes vectors of the signal and the idler and $\vec{p} = \langle X_p|\vec{\sigma}|X_p\rangle$ is the Stokes vector of the pump. The upper bars in Equations 8.24 through 8.29 denote the appropriate mean values. For simplicity, the effective nonlinear parameter γ_e is defined together with the well-known Raman gain and phase coefficients

$$\gamma_e \overset{\Delta}{=} \frac{8}{9}\gamma_K = (1-f)\gamma \tag{8.30}$$

$$\gamma f \underline{\chi}_R(\Omega_s) \equiv \psi_R^s - j\frac{g_R^s}{2} \tag{8.31}$$

Together with the PMD parameter D_p, these coefficients determine the propagation parameters in Equations 8.24 through 8.29 that describe the contributions of PMD, Raman gain, and nonlinear phase shift.

$$\eta_d = \frac{1}{3}\frac{3\pi}{8}\Omega_s^2 D_p^2 \tag{8.32}$$

$$\eta_R = \frac{g_R^s}{2} P_p \tag{8.33}$$

$$\eta_{NL} = \left(\gamma_e + \psi_R^s\right) P_p \tag{8.34}$$

Finally the net phase mismatch among the three interacting waves is

$$\kappa = \beta_s + \beta_i - 2\beta_p + 2\left(\gamma_e + \psi_R^s\right) P_p \tag{8.35}$$

In a lossy fiber, the pump power varies according to $\partial_z P_p = -\alpha P_p$ and Equations 8.24 through 8.29 can only be solved numerically. For zero losses, these equations become simple ordinary differential equations with constant coefficients that can be solved analytically

$$\partial_z \underline{r} = \underline{\underline{M}}\, \underline{r} \tag{8.36}$$

where a vector variable $\underline{r}$ and the matrix $\underline{\underline{M}}$ are defined as

$$\underline{r} = \left[P_p \overline{P}_s, P_p \overline{P}_i, \overline{U}_r, \overline{U}_i, \overline{V}_s, \overline{V}_i\right]^T \tag{8.37}$$

$$\underline{\underline{M}} = \begin{bmatrix} \eta_R & 0 & \eta_{NL} & \eta_R & \eta_R & 0 \\ 0 & -\eta_R & \eta_{NL} & -\eta_R & 0 & -\eta_R \\ \eta_{NL} & \eta_{NL} & -\tfrac{1}{2}\eta_d & \kappa & \eta_{NL} & \eta_{NL} \\ -\eta_R & \eta_R & -\kappa & -\tfrac{1}{2}\eta_d & -\eta_R & \eta_R \\ \eta_R & 0 & \eta_{NL} & -\eta_R & \eta_R - \eta_d & 0 \\ 0 & -\eta_R & \eta_{NL} & -\eta_R & 0 & -\eta_R - \eta_d \end{bmatrix} \tag{8.38}$$

A comparison between the average gain spectra in Section 8.2.2 and the solution of the averaged set of Equation 8.36 is depicted in Figure 8.9 for a lossless fiber. The latter considers parallel pump and signal input SOPs while the former describes optimal input SOP, as discussed in Section 8.2.2. Three PMD values are compared with the pump power and wavelength being 5 W and 1534.95 nm,

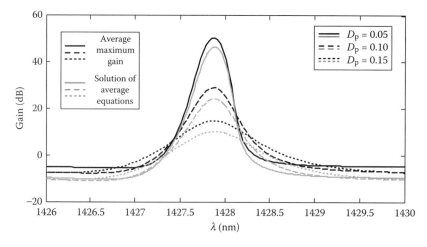

FIGURE 8.9 Gain spectra for D_p values of 0.05 ps/$\sqrt{\text{km}}$ (solid), 0.1 ps/$\sqrt{\text{km}}$ (dashed), and 0.15 ps/$\sqrt{\text{km}}$ (dotted).

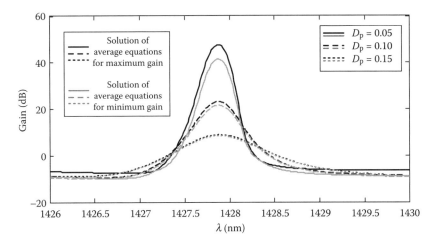

FIGURE 8.10 Maximum and minimum gain spectra for the averaged equations.

respectively. The black lines depict the average maximum attainable gain and the grey lines signify the solution of the averaged equations for parallel input SOPs.

For all three cases, the two solutions are close. PMD has a large effect on the gain with the peak for the largest PMD value being some 40 dB lower than the corresponding peak gain for the lowest PMD.

Since the averaging was performed over every possible realization of the birefringence distribution, this model represents a hypothetical average fiber. A close examination of the values in the variable vector (Equation 8.37) reveals that only the relative SOP input angle (embedded in V_s) is important and that the optimum gain is achieved with copolarized input waves. This is naturally not true for a specific realization of the birefringence distribution. The calculated average gain spectra are shown in Figure 8.10. The sensitivity to the input SOP is not maintained as the difference between the maximum and minimum gain limits almost vanishes.

Nevertheless, the spectral form of the gain of the average fiber (for coaligned polarizations) resembles that of the average maximum gain spectra in Figure 8.9. Thus, the average equations avail a tool for the estimation of the performance of the OPA in near optimum conditions using significantly simplified calculations.

8.2.4 Effect of Longitudinal Variations of Propagation Parameters

Imperfections in fiber manufacturing cause longitudinal variations of all fiber properties [23]. Since NB-OPAs characteristics are extremely sensitive to the spatial distribution of linear and nonlinear propagation parameters, they can be used as an efficient tool to extract parameter distributions with high accuracy. Moreover, the inclusion of these spatial variations improves the predictability of the NB-OPA characteristics.

Most known parameter extraction techniques [23,24] determine only the variations of λ_0 and do not offer spatial resolutions of the order of singles meters which are crucial for NB-OPAs. Since NB-OPAs operate at large signal-pump detunings of 100 nm or more, it is required to estimate, in addition to λ_0, the variations of the entire dispersion function as well as the nonlinear propagation parameters such as the effective mode area (A_{eff}).

8.2.4.1 Uniqueness and Spatial Resolution

Estimating the propagation parameter distributions from measured OPA efficiency spectra is a complex nonlinear inverse problem. The uniqueness of a possible solution is hard to guarantee and it is extremely difficult to obtain a bound on the attainable spatial resolution. This is especially so if the problem is formulated in terms of a cumbersome mathematical framework as the one outlined in Section 8.2.2. A significantly simpler formulation [23] can be used to overcome this problem. The OPA efficiency of a lossless medium is expressed in terms of the amplified output intensity

$$I(\lambda_p) \propto \left| \int_0^L dz \, \exp\left\{ i \int_0^z \Delta\beta_{\lambda_p}(z')dz' \right\} \right|^2 \tag{8.39}$$

Expanding $\Delta\beta_{\lambda_p}$ allows to manipulate Equation 8.39 into an expression which resembles the squared absolute value of the FT of some phase only function $\exp\{i\varphi(z)\}$. Knowing $\exp\{i\varphi(z)\}$ leads to an easy determination of the parameter distribution.

The problem amounts then to finding the function $\varphi(z)$ which requires performing an inverse FT while only the absolute values of the transform (λ_p-dependent parametric gain) are given. Uniqueness of such an inversion is not guaranteed except in very specific cases of an analytic $\varphi(z)$ function and a slightly lossy medium [25,26].

Obtaining a resolution bound is very challenging in the case of a nonlinear inversion problem. However, the bound can be estimated rather well by considering a similar linear problem and its corresponding inversion.

The FT formulation is valid only for small excursions from λ_p since it requires the argument of the transform kernel (exponent) to be a linear function of the transform variable (λ_p). From the argument of FT resolution, the estimation is sensitive to such Δz spans that cause the most distant argument values (belonging to the most spectrally remote pump wavelengths) to attain a difference of 2π. The large detuning of NB-OPAs enables, potentially, extremely high resolutions of the order of single meters.

8.2.4.2 Estimation Procedure

The estimation procedure assumes that every parameter $a(z)$ undergoes small perturbations of the form $a(z) = a_0 + \Delta a(z)$, with a_0 being the mean value and $\Delta a(z)$ a deterministic deviation along the fiber. The deviation of every propagation parameter is assumed to be a continuous function of z. $\Delta a(z)$ can therefore be spanned on a finite domain (the fiber of length L) by an orthogonal functional basis [23]. Postulating that only long range variations exist, the series can be truncated. Good results with a relatively small number of spanning coefficients are attainable with a basis of Chebyshev functions.

Having a set of spanning coefficients, it is possible to calculate an average performance of a fiber (having a known value of D_p). The procedure concentrates on finding an optimum set of coefficients for each parameter that produces the closest match to a measured parametric gain spectrum.

The parametric efficiency is influenced by dispersion parameters as well as by spatially inhomogeneous nonlinear parameters. The procedure at hand does not assume constant nonlinearity along the fiber, as do other known techniques [23,24] and therefore allows estimating the nonlinear parameter distribution. The optimizations process estimates the distributions of three linear parameters $\Delta\lambda_0(z)$, $\beta_3(z)$, and $\beta_4(z)$ and of $A_{\text{eff}}(z)$ which describes perturbations of the nonlinear parameters.

The estimation procedure is formulated as an optimization problem which requires a method suitable for problems of high dimensionality. One such method is a genetic optimization algorithm called particle swarm optimization [27], which covers efficiently the solution space within a small number of iterations, each requiring few solutions.

8.2.4.3 Results

As an example we describe extracted parameter distributions for DSFs with lengths of 200 m and 1 km. Figure 8.11a shows in open circles a measured parametric gain spectrum of the 200 m long

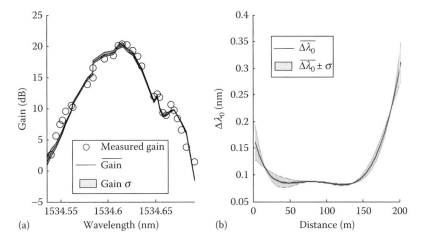

FIGURE 8.11 (a) Measured and calculated parametric gain; (b) estimated λ_0 distribution.

DSF. A spectrum of maximum obtainable gain values is used in the present case which allows employing the averaged set of Equations 8.24 through 8.29. The algorithm starts from a random distribution and is run for a certain number of iterations. The entire process is repeated for a predetermined number of times to test convergence.

Using this spectrum as the input to the estimation procedure, four parameter distributions are calculated. One of these, $\Delta\lambda_0(z)$ is shown in Figure 8.11b together with the standard deviation of convergence. The solid lines in Figure 8.11a describe the calculated gain spectra considering all the four parameter distributions. The average fit discrepancy is very small testifying to the accuracy of the extraction procedure.

A second example, for a 1 km long DSF, is shown in Figure 8.12. The measured and calculated parametric gain and the four estimated mean parameter distributions $\lambda_0(z)$, $\beta_3(z)$, $\beta_4(z)$, and $A_{\text{eff}}(z)$ are shown in Figure 8.12a and b, respectively.

Using the extracted parameters of the 1 km DSF (Figure 8.12) alters all calculated NB-OPA characteristics. For example, instead of the PDFs of Figures 8.4 and 8.5, the normalized distributions

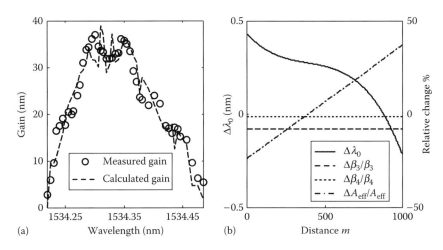

FIGURE 8.12 (a) Measured parametric gain; (b) parameter distributions for a 1 km long DSF.

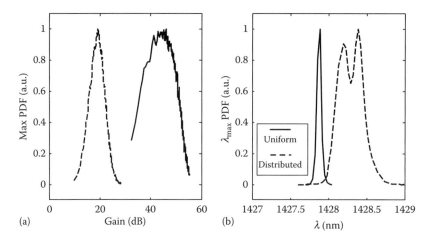

FIGURE 8.13 Normalized empirical PDFs of (a) gain (b) wavelength where phase matching conditions are satisfied for a 1 km long DSF with a D_p value of 0.05 ps/$\sqrt{km}$.

of maximum gain and that of the wavelength, where maximum phase matching takes place, become as shown in Figure 8.13 for the case of $D_p = 0.05$ ps/$\sqrt{km}$.

The effect of distributed propagation parameters is the same as an increased PMD value. Namely, the distribution narrows the range of possible gain values and reduces the mean gain value. At the same time, it broadens significantly the distribution of wavelengths ensuring maximum phase matching.

8.3 EXPERIMENTAL RESULTS

This section describes several experimental results which prove the broad band nature of NB-OPA slow light media and the capabilities to delay high bit rate digital data. The experimental results are compared to calculations and show superb fits which highlight the capabilities of the comprehensive model presented in Section 8.2 and the need to accurately predict the performance of slow and fast light systems.

Various aspects of the NB-OPA slow light medium were characterized using the experimental setup described in Figure 8.14.

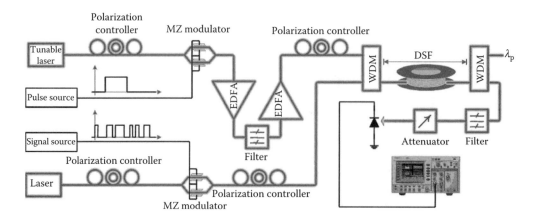

FIGURE 8.14 NB-OPA experimental setup.

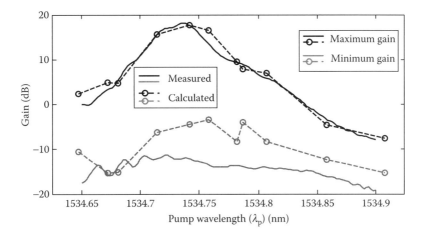

FIGURE 8.15 Measured and calculated (averaged) polarization dependent gain.

The high pump power used in the experiments is achieved by amplification of a low duty cycle pulsed source. The pump and signal are combined and are copropagated in the DSF. The signal is filtered at the output and characterized using a fast photodetector and a sampling oscilloscope. The portion of the signal stream which coincides with the pump pulse and experiences delay is easily identified.

First shown are the measurements of minimum and maximum attainable gain spectra in a 1 km long DSF. In Figure 8.15, these results are compared to the theoretical predictions based on the model presented in Section 8.2 using the estimated parameter distribution of the fiber described in Section 8.2.4. The PMD parameter was taken to be $D_p = 0.06$ ps/$\sqrt{km}$ and the pump power was 5 W.

The measured spectra were obtained by scanning, at each pump wavelength, all possible input SOPs and recording the maximum and minimum obtained gain values. The measured and calculated maximum gain spectra fit superbly while the minimum gain spectra differ. The discrepancy is due to the fact that the parameter distribution used in the calculations relies only on the maximum gain spectrum.

Propagation of single pulses was studied by delaying single 70 ps pulses in a 200 m long DSF. Figure 8.16 compares linear transmission (with the pump off) with the delayed pulses under two

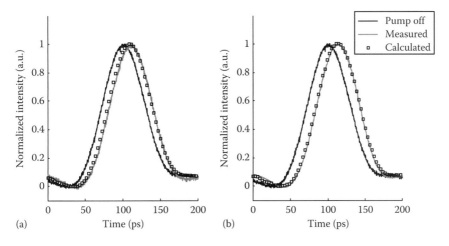

FIGURE 8.16 Delay of 70 ps pulses in a 200 m long DSF (a) $G = 10.3$ dB and (b) $G = 38.5$ dB.

OPA gain levels. For a gain of 10.3 dB, the delay is 8.4 ps whereas for a gain of 38.5 dB, the delay increases to 12.3 ps.

The delayed pulses were reproduced theoretically by solving the propagation equation of the total field envelope while including extracted parameter distributions. The calculations were done employing the split-step FT (SSFT) algorithm [28]. The delay in this example is moderate due to the short fiber but the calculated predictions, shown in open squares, fit the measurements perfectly.

Figure 8.17 describes the performance for a 10 Gbit/s digital data in the nonreturn to zero (NRZ) modulation format. A time-domain trace (Figure 8.17a) and the corresponding eye pattern (Figure 8.17b), both measured at the output of an NB-OPA based on a 2 km long DSF, are shown. The linearly propagating signal is shown as a reference, yielding a delay of approximately 130 ps with very minor distortions.

Linear and delayed propagations of a portion of the NRZ stream in a 1 km DSF are shown in Figure 8.18a. The measured delay in this case was 55 ps. The simulated delayed pulse train (computed using the SSFT method) is shown in open squares and fits every detail of the measured pulses very well.

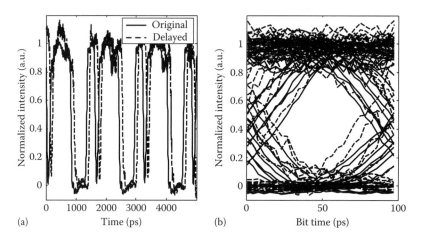

FIGURE 8.17 (a) Original and delayed time-domain traces of a 10 Gbit/s NRZ stream and (b) corresponding eye patterns.

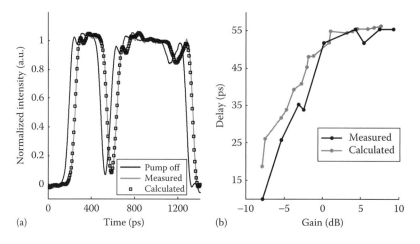

FIGURE 8.18 (a) Part of a 10 Gbit/s NRZ stream and (b) measured delay versus parametric gain.

Since all slow light systems introduce signal distortion, it is important to define the delay values accurately. For simple pulses, the shift of the center of gravity, described in Appendixes A and C suffice. For a random data stream however, where distortions due to pattern effects may occur, a more precise measure is needed. A useful delay measure is the temporal shift of the optimum sampling point which yields the largest eye opening—a criterion widely used in standard communications [29].

Measured and simulated delay of the 10 Gbit/s data stream are plotted in Figure 8.18b versus parametric gain. The delay range shown is 10–55 ps. The agreement between theoretical prediction and experiments is very good for large gain values. In the negative gain regime, the measurements are noisy and an accurate estimation of the optimum sampling point requires averaging over many bits which is difficult due to the limited statistical ensemble. The delay reaches a maximum value of ~55 ps beyond which it is constant. Larger gain values cause signal distortion where the delay of the pulse leading edge increases but the temporal shift of the trailing edge remains unchanged. Consequently, the best sampling point does not change and the same level of delay is maintained.

APPENDIX A: GROUP DELAY CALCULATIONS IN ISOTROPIC FIBERS

The dependence of the group delay on the phase response is formulated here for the scalar case. It serves to describe any scalar transmission system, and as the foundation for the delay calculation of the more complex vector case, described in Appendix C.

The basic delay definition stems from the center of gravity of the temporal shape of the field envelope $A(z,t)$ which describes the moment t_z in which $A(z,t)$ crosses an arbitrary point z along the fiber

$$t_z = \frac{\int t P(t) dt}{\int P(t) dt} = \frac{\int t |A(t)|^2 dt}{\int |A(t)|^2 dt} = -\frac{j \int A^*(\omega) (\partial A(\omega)/\partial \omega) d\omega}{\int |A(\omega)|^2 d\omega} \quad (8.A.1)$$

The last equality is due to Parseval's theorem, where the FT of the pulse envelope is denoted by $A(\omega) = \int_{-\infty}^{\infty} A(t) e^{j\omega t} dt$. Appropriately, the delay experienced by the pulse traveling along the fiber is

$$\tau = -\frac{j \int A_{out}^* \partial_\omega A_{out} d\omega}{\int |A_{out}|^2 d\omega} + \frac{j \int A_{in}^* \partial_\omega A_{in} d\omega}{\int |A_{in}|^2 d\omega} \quad (8.A.2)$$

The input and the output envelopes are related by a linear time invariant (LTI) transmission function, $A_{out}(t) = h(t) * A_{in}(t)$, whose frequency response is $H(\omega) = \sqrt{G(\omega)} \cdot e^{j\Phi(\omega)}$.

The above integrals in Equation 8.A.2 can be rewritten as

$$\int A_{out}^* \partial_\omega A_{out} d\omega = \int \left[G A_{in}^* \partial_\omega A_{in} + \left(\frac{1}{2} \partial_\omega G + j G \partial_\omega \Phi \right) |A_{in}|^2 \right] d\omega \quad (8.A.3)$$

$$\int |A_{out}|^2 d\omega = \int G |A_{in}|^2 d\omega \quad (8.A.4)$$

The nominators in (8.A.2) are all pure real and therefore

$$j \int A_{in}^* \partial_\omega A_{in} d\omega = -\int \text{Im} \left[A_{in}^* \partial_\omega A_{in} \right] d\omega \quad (8.A.5)$$

$$j \int A_{out}^* \partial_\omega A_{out} d\omega = -\int \left(G \, \text{Im} \left[A_{in}^* \partial_\omega A_{in} \right] + G \partial_\omega \Phi |A_{in}|^2 \right) d\omega \quad (8.A.6)$$

Substituting Equations 8.A.5 and 8.A.6 into Equation 8.A.2 yields the delay expression. For a known input spectrum and particular gain and phase responses, the delay is

$$\tau = \frac{\int \left(G \, \text{Im} \left[A_{in}^* \partial_\omega A_{in} \right] + G \partial_\omega \Phi |A_{in}|^2 \right) d\omega}{\int G |A_{in}|^2 d\omega} - \frac{\int \text{Im} \left[A_{in}^* \partial_\omega A_{in} \right] d\omega}{\int |A_{in}|^2 d\omega} \quad (8.A.7)$$

Since the assumption of an LTI transmission fiber holds in many cases, (8.A.7) holds for any input spectrum. Simplification is possible for very narrow band input spectra, centered around a carrier frequency ω_c. These approximate CW waves for which the gain and phase responses are defined. Integration over a narrow frequency span of the input spectrum can be approximated by integration over a δ-function, $\delta(\omega - \omega_c)$ which produces an approximate form of the integrals in Equation 8.A.7.

$$\int G |A_{in}|^2 \, d\omega \simeq G(\omega_c) \int |A_{in}|^2 \, d\omega \tag{8.A.8}$$

$$\int \left(G \, \text{Im} \left[A_{in}^* \partial_\omega A_{in} \right] + G \partial_\omega \Phi |A_{in}|^2 \right) d\omega \simeq G(\omega_c) \int \text{Im} \left[A_{in}^* \partial_\omega A_{in} \right] d\omega$$
$$+ G(\omega_c) \left. \frac{\partial \Phi}{\partial \omega} \right|_{\omega_c} \int |A_{in}|^2 \, d\omega \tag{8.A.9}$$

Substituting in Equation 8.A.7, and canceling out equal and negative terms, leads to a simple (and intuitive) expression for the delay of a narrow band pulse

$$\tau = \left. \frac{\partial \Phi}{\partial \omega} \right|_{\omega_c} \tag{8.A.10}$$

APPENDIX B: GAIN CALCULATIONS FOR A BIREFRINGENT FIBER

A vector model for the input and output wave envelopes is examined where the propagation along the fiber is described in the frequency domain by a 2×2 transmission matrix $\underline{\underline{H}}(\omega)$, so that $|A_{out}\rangle = \underline{\underline{H}} |A_{in}\rangle$. Using the expressions for the input and output powers, the gain is

$$G = \frac{P_{out}}{P_{in}} = \frac{\langle A_{out} | A_{out} \rangle}{\langle A_{in} | A_{in} \rangle} = \frac{\langle A_{in} | \underline{\underline{H}}^\dagger \underline{\underline{H}} | A_{in} \rangle}{\langle A_{in} | A_{in} \rangle} \tag{8.B.1}$$

Generally $\underline{\underline{H}}$ is a complex matrix, but $\underline{\underline{H}}^\dagger \underline{\underline{H}}$ is a Hermitian matrix, that can be expressed [19] in the form of

$$\underline{\underline{H}}^\dagger \underline{\underline{H}} = G_0 \underline{\underline{I}} + \vec{h} \cdot \vec{\sigma} \tag{8.B.2}$$

where
  G_0 is a real number
  $\underline{\underline{I}}$ is the 2×2 unity matrix
  $\vec{h}$ is a real three-dimensional vector in Stokes space

Using Pauli's spin matrices σ_i, the two-dimensional Jones notation is transformed into the three-dimensional Stokes space

$$\sigma_1 = \begin{bmatrix} 1 & 0 \\ 0 & -1 \end{bmatrix} \quad \sigma_2 = \begin{bmatrix} 0 & 1 \\ 1 & 0 \end{bmatrix} \quad \sigma_3 = \begin{bmatrix} 0 & -j \\ j & 0 \end{bmatrix} \tag{8.B.3}$$

$$\vec{a} = \langle A | \vec{\sigma} | A \rangle = (a_1, a_2, a_3) \Leftrightarrow a_i = \langle A | \sigma_i | A \rangle \tag{8.B.4}$$

$$\vec{a} \cdot \vec{\sigma} = \begin{bmatrix} a_1 & a_2 - ja_3 \\ a_2 + ja_3 & -a_1 \end{bmatrix} \tag{8.B.5}$$

Employing this notation, it is possible to describe circular two-dimensional polarizations using real three-dimensional vectors in Stokes space. Moreover, it can be shown that for any complex vector $|A_{in}\rangle$

$$\left\langle A_{in} | \vec{h} \cdot \vec{\sigma} | A_{in} \right\rangle = \vec{h} \cdot \vec{a}_{in} \tag{8.B.6}$$

Thus for unity input power, the gain for any input SOP is

$$G = \left\langle A_{\text{in}} | \underline{\underline{H}}^\dagger \underline{\underline{H}} | A_{\text{in}} \right\rangle = G_0 + \vec{h} \cdot \vec{a}_{\text{in}} \tag{8.B.7}$$

A more detailed investigation of the properties of Hermitian and spin matrices [19] leads to the minimum and maximum obtainable gain values. Given a three-dimensional Stokes vector $\vec{a}$, a Hermitian matrix in the form of Equation B.5 has real and opposite eigenvalues, with appropriate orthogonal eigenvectors. The positive eigenvalue equals $|\vec{a}|$ and its eigenvector is parallel to the appropriate two-dimensional Jones vector $|A\rangle$

$$(\vec{a} \cdot \vec{\sigma}) |A\rangle = |\vec{a}| \cdot |A\rangle \tag{8.B.8}$$

Applying these properties to Equation 8.B.2, allows any two orthonormal eigenvectors of $\left(\vec{h} \cdot \vec{\sigma}\right)$ to span any input SOP by

$$|A_{\text{in}}\rangle = c_+ |H_+\rangle + c_- |H_-\rangle \tag{8.B.9}$$

The eigenvector $|H_+\rangle$ matches the positive eigenvalue $|\vec{h}|$, and for unity power input we have $|c_+|^2 + |c_-|^2 = 1$. Substituting Equation 8.B.9 in Equation 8.B.6 yields

$$\vec{h} \cdot \vec{a}_{\text{in}} = \left(|c_+|^2 - |c_-|^2\right) |\vec{h}| \tag{8.B.10}$$

The maximum and minimum obtainable gain values, and the corresponding input SOPs are

$$\begin{aligned} G_{\text{max}} &= G_0 + |\vec{h}| \Leftrightarrow |A_{\text{max}}\rangle \propto |H_+\rangle \\ G_{\text{min}} &= G_0 - |\vec{h}| \Leftrightarrow |A_{\text{min}}\rangle \propto |H_-\rangle \end{aligned} \tag{8.B.11}$$

APPENDIX C: GROUP DELAY CALCULATIONS IN BIREFRINGENT FIBERS

Calculation of the group delay in a birefringent fiber starts from the definition of the temporal center of gravity

$$t_z = \frac{\int t P(t) \, dt}{\int P(t) \, dt} = \frac{\int t \langle A(t) | A(t) \rangle \, dt}{\int \langle A(t) | A(t) \rangle \, dt} = -\frac{j \int \langle A(\omega) | \frac{\partial}{\partial \omega} | A(\omega) \rangle \, d\omega}{\int \langle A(\omega) | A(\omega) \rangle \, d\omega} \tag{8.C.1}$$

Similar to the scalar case, the FT of the pulse envelope and Parseval's theorem are employed to yield the delay

$$\tau \triangleq \frac{\int t \langle A_{\text{out}} | A_{\text{out}} \rangle \, dt}{\int \langle A_{\text{out}} | A_{\text{out}} \rangle \, dt} - \frac{\int t \langle A_{\text{in}} | A_{\text{in}} \rangle \, dt}{\int \langle A_{\text{in}} | A_{\text{in}} \rangle \, dt} \tag{8.C.2}$$

Assuming that the spectral shape of the input pulse is common to both components of the input field leads to

$$|A_{\text{in}}(\omega)\rangle = \tilde{A}(\omega) |a_{\text{in}}\rangle \tag{8.C.3}$$

The input spectrum is given by the scalar function $\tilde{A}(\omega)$ and the input SOP is given by $|a_{\text{in}}\rangle$, which is frequency independent. Continuing in a similar manner to Appendix A and considering that the integrals in Equation 8.C.2 must be real, results, using Equation 8.C.3, in

$$\int \langle A_{\text{in}} | A_{\text{in}} \rangle \, dt = \frac{1}{2\pi} \int \left| \tilde{A} \right|^2 d\omega \tag{8.C.4}$$

$$\int t \langle A_{\text{in}} | A_{\text{in}} \rangle \, dt = \frac{1}{2\pi} \int \text{Im}\left[\tilde{A}^* \partial_\omega \tilde{A}\right] d\omega \tag{8.C.5}$$

In order to examine the output pulse, the propagation model outlined in Appendix B is exploited with a transmission matrix $\underline{\underline{H}}(\omega)$.

The appropriate integrals in Equation 8.C.2 are evaluated as

$$\int \langle A_{\text{out}} | A_{\text{out}} \rangle \, dt = \frac{1}{2\pi} \int \langle A_{\text{in}} | \underline{\underline{H}}^\dagger \underline{\underline{H}} | A_{\text{in}} \rangle \, d\omega \tag{8.C.6}$$

$$\int t \, \langle A_{\text{out}} | A_{\text{out}} \rangle \, dt = \frac{-j}{2\pi} \int \left(\langle A_{\text{in}} | \underline{\underline{H}}^\dagger \partial_\omega \underline{\underline{H}} | A_{\text{in}} \rangle + \langle A_{\text{in}} | \underline{\underline{H}}^\dagger \underline{\underline{H}} \partial_\omega | A_{\text{in}} \rangle \right) d\omega \tag{8.C.7}$$

Substituting Equation 8.C.3 in Equations 8.C.6 and 8.C.7 while retaining the notation for the gain as in Equation 8.B.7, the expressions in the integrands become

$$\langle A_{\text{in}} | \underline{\underline{H}}^\dagger \underline{\underline{H}} | A_{\text{in}} \rangle = |\tilde{A}|^2 \left(G_0 + \vec{h} \cdot \vec{a}_{\text{in}} \right) \tag{8.C.8}$$

$$\langle A_{\text{in}} | \underline{\underline{H}}^\dagger \underline{\underline{H}} \partial_\omega | A_{\text{in}} \rangle = \tilde{A}^* \partial_\omega \tilde{A} \cdot \left(G_0 + \vec{h} \cdot \vec{a}_{\text{in}} \right) \tag{8.C.9}$$

$$\langle A_{\text{in}} | \underline{\underline{H}}^\dagger \partial_\omega \underline{\underline{H}} | A_{\text{in}} \rangle = |\tilde{A}|^2 \langle a_{\text{in}} | \underline{\underline{H}}^\dagger \partial_\omega \underline{\underline{H}} | a_{\text{in}} \rangle \tag{8.C.10}$$

where $\vec{a}_{\text{in}} = \langle a_{\text{in}} | \vec{\sigma} | a_{\text{in}} \rangle$ is the Stokes vector matching the SOP of the input wave $|a_{\text{in}}\rangle$. Examining only the imaginary part of the integrands, Equation 8.C.7 transforms into

$$\int t \, \langle A_{\text{out}} | A_{\text{out}} \rangle \, dt = \frac{1}{2\pi} \int \text{Im} \left[|\tilde{A}|^2 \langle a_{\text{in}} | \underline{\underline{H}}^\dagger \partial_\omega \underline{\underline{H}} | a_{\text{in}} \rangle \right] d\omega$$

$$+ \frac{1}{2\pi} \int \text{Im} \left[\tilde{A}^* \partial_\omega \tilde{A} \cdot \left(G_0 + \vec{h} \cdot \vec{a}_{\text{in}} \right) \right] d\omega \tag{8.C.11}$$

The expressions formulated above are valid for any input pulse spectrum, and the delay of the center of gravity is given by

$$\tau \simeq \frac{\int \text{Im} \left[\tilde{A}^* \partial_\omega \tilde{A} \cdot \left(G_0 + \vec{h} \cdot \vec{a}_{\text{in}} \right) \right] d\omega}{\int |\tilde{A}|^2 \left(G_0 + \vec{h} \cdot \vec{a}_{\text{in}} \right) d\omega} + \frac{\int \text{Im} \left[|\tilde{A}|^2 \langle a_{\text{in}} | \underline{\underline{H}}^\dagger \partial_\omega \underline{\underline{H}} | a_{\text{in}} \rangle \right] d\omega}{\int |\tilde{A}|^2 \left(G_0 + \vec{h} \cdot \vec{a}_{\text{in}} \right) d\omega} - \frac{\int \text{Im} \left[\tilde{A}^* \partial_\omega \tilde{A} \right] d\omega}{\int |\tilde{A}|^2 d\omega} \tag{8.C.12}$$

Equation 8.C.12 can be simplified for a narrow input spectral width centered around a carrier frequency ω_c. As in Appendix A, the integrals are approximated as being sampled by a δ-function at ω_c so that Equation 8.C.12 is simplified to

$$\tau \simeq \frac{\text{Im} \left[\langle a_{\text{in}} | \underline{\underline{H}}^\dagger \partial_\omega \underline{\underline{H}} | a_{\text{in}} \rangle \big|_{\omega_c} \right]}{\left(G_0 + \vec{h} \cdot \vec{a}_{\text{in}} \right) \big|_{\omega_c}} \tag{8.C.13}$$

The nominator in Equation 8.C.13 is simplified by expanding the imaginary part and defining the following matrix

$$\underline{\underline{M}} \triangleq \frac{1}{2j} \left(\underline{\underline{H}}^\dagger \partial_\omega \underline{\underline{H}} - \left(\partial_\omega \underline{\underline{H}} \right)^\dagger \underline{\underline{H}} \right) \tag{8.C.14}$$

By definition, $\underline{\underline{M}}$ is Hermitian and thus can be expressed using a real scalar and a three-dimensional real vector

$$\underline{\underline{M}} = M_0 \underline{\underline{I}} + \vec{m} \cdot \vec{\sigma} \tag{8.C.15}$$

This leads to a simple expression for the delay in a birefringent fiber which holds true for any narrow band input pulse spectrum. As expected, the delay depends on input SOP and of course on the carrier frequency

$$\tau \simeq \frac{M_0 + \vec{m} \cdot \vec{a}_{in}}{G_0 + \vec{h} \cdot \vec{a}_{in}} \qquad (8.C.16)$$

REFERENCES

1. A. Hasegawa and W. F. Brinkman, Tunable coherent IR and FIR sources utilizing modulational instability, *IEEE J. Quantum Electron.*, 16(7), 694–697, 1980.
2. R. H. Stolen and J. E. Bjorkholm, Parametric amplification and frequency conversion in optical fibers, *IEEE J. Quantum Electron.*, 18(7), 1062–1072, 1982.
3. F. S. Yang, M. E. Marhic, and L. G. Kazovsky, CW fiber optical parametric amplifier with net gain and wavelength conversion efficiency >1, *Electron. Lett.*, 32(25), 2336–2338, 1996.
4. J. Hansryd, P. A. Andrekson, M. Westlund, L. Lie, and P. O. Hedekvist, Fiber-based optical parametric amplifiers and their applications, *IEEE Sel. Top. Quantum Electron. Lett.*, 8(3), 506–520, 2002.
5. G. P. Agrawal, *Nonlinear Fiber Optics*, 2nd edn., chap. 10. Academic, San Diego, CA, 1995.
6. K. Inoue, Polarization effect on four-wave mixing efficiency in a single-mode fiber, *IEEE J. Quantum Electron.*, (28), 883–894, 1992.
7. M. E. Marhic, K. Y. K. Wong, and L. G. Kazovsky, Wide-band tuning of the gain spectra of one-pump fiber optical parametric amplifiers, *IEEE. J. Sel. Top. Quantum Electron.*, 10(5), 1133–1141, 2004.
8. E. Golovchenko, P. V. Mamyshev, A. N. Pilipetskii, and E. M. Dianov, Mutual influence of the parametric effects and stimulated Raman scattering in optical fibers, *IEEE J. Quantum Electron.*, 26(10), 1815–1820, 1990.
9. D. Dahan and G. Eisenstein, Tunable all optical delay via slow and fast light propagation in a Raman assisted fiber optical parametric amplifier: A route to all optical buffering, *Opt. Express*, 13, 6234–6249, 2005.
10. F. Vanholsbeeck, P. Emplit, and S. Coen, Complete experimental characterization of the influence of parametric four-wave mixing on stimulated Raman gain, *Opt. Lett.*, 28(20), 1960–1962, 2003.
11. N. Bloembergen and Y. R. Shen, Coupling between vibrations and light waves in Raman laser media, *Phys. Rev. Lett.*, 12(18), 504–507, 1964.
12. R. Stolen, J. P. Gordon, W. J. Tomlinson, and H. A. Haus, Raman response function of silica-core fibers, *J. Opt. Soc. Am. B*, 6(6), 1159–1166, 1989.
13. A. Hsieh, G. Wong, S. Murdoch, S. Coen, F. Vanholsbeeck, R. Leonhardt, and J. D. Harvey, Combined effect of Raman and parametric gain on single-pump parametric amplifiers, *Opt. Express*, 15(13), 8104–8114, 2007.
14. C. McKinstrie, H. Kogelnik, R. Jopson, S. Radic, and A. Kanaev, Four-wave mixing in fibers with random birefringence, *Opt. Express*, 12(10), 2033–2055, 2004.
15. Q. Lin and G. P. Agrawal, Vector theory of stimulated Raman scattering and its application to fiber-based Raman amplifiers, *J. Opt. Soc. Am. B*, 20(8), 1616–1631, 2003.
16. Q. Lin and G. P. Agrawal, Effects of polarization-mode dispersion on fiber-based parametric amplification and wavelength conversion, *Opt. Lett.*, 29(10), 1114–1116, 2004.
17. A. Galtarossa, L. Palmieri, M. Santagiustina, and L. Ursini, Polarized backward Raman amplification in randomly birefringent fibers, *J. Lightwave Technol.*, 24(11), 4055–4063, 2006.
18. E. Shumakher, A. Willinger, R. Blit, D. Dahan, and G. Eisenstein, High resolution extraction of fiber propagation parameters for accurate modeling of slow light systems based on narrow band optical parametric amplification, in *Proc. OFC*, Anaheim, 2007, paper OTuC2.
19. J. P. Gordon and H. Kogelnik, PMD fundamentals: Polarization mode dispersion in optical fibers, *Proc. Natl. Acad. Sci. USA*, 97(9), 4541–4550, 2000.
20. D. J. Dougherty, F. X. Kartner, H. A. Haus, and E. P. Ippen, Measurement of the Raman gain spectrum of optical fibers, *Opt. Lett.*, 20(1), 31–33, 1995.
21. C. W. Gardiner, *Handbook of Stochastic Methods*. Springer-Verlag, Berlin, 1990.
22. M. Karlsson and J. Brentel, Autocorrelation function of the polarization-mode dispersion vector, *Opt. Lett.*, 24(14), 939–941, 1999.

23. I. Brener, P. P. Mitra, D. D. Lee, D. J. Thomson, and D. L. Philen, High-resolution zero-dispersion wavelength mapping in single-mode fiber, *Opt. Lett.*, 23(19), 1520–1522, 1998.
24. A. Mussot, E. Lantz, A. Durecu-Legrand, C. Simonneau, D. Bayart, T. Sylvestre, and H. Maillotte, Zero-dispersion wavelength mapping in short single-mode optical fibers using parametric amplification, *Photon. Technol. Lett.*, 18(1), 22–24, 2006.
25. A. M. J. Huiser and H. A. Ferwerda, On the problem of phase retrieval in electron microscopy from image and diffraction pattern: II. On the uniqueness and stability, *Optik (Stuttgart)*, 46, 407–420, 1976.
26. A. M. J. Huiser, P. van Toorn, and H. A. Ferwerda, On the problem of phase retrieval in electron microscopy from image and diffraction pattern: III. The development of an algorithm, *Optik (Stuttgart)*, 47, 1–8, 1977.
27. J. Kennedy and R. Eberhart, Particle swarm optimization, in P*roc. of the IEEE Int. Conf. on Neural Networks*, Piscataway, NJ, pp. 1942–1948, 1995.
28. O. V. Sinkin, R. Holzlöhner, J. Zweck, and C. R. Menyuk, Optimization of the split-step Fourier method in modeling optical-fiber communications systems, *J. Lightwave Technol.*, 21(1), 61–68, 2003.
29. G. P. Agrawal, *Fiber-Optic Communication Systems*, 3rd edn. Wiley, New York, 2002.

9 Slow and Fast Light Using Stimulated Brillouin Scattering: A Highly Flexible Approach

Luc Thévenaz

CONTENTS

9.1 Monochromatic Pump 175
9.2 Modulated Pump 181
9.3 Multiple Pumps 186
References 190

The predominant place occupied by optical fibers in modern photonic systems has steadily stimulated research to realize slow and fast light devices directly in this close-to-perfect transmission line. This potentially offers the key advantage of a seamless and flexible integration in most optical transmission systems. The main obstacle for such a realization is related to the highly disordered amorphous nature of the silica constituting the optical fibers, prohibiting the use of narrowband atomic transitions. The most efficient approach to create a narrowband gain or loss in optical fibers remains the exploitation of an optical interaction requiring a strict phase matching condition to be satisfied. This can be realized using the nonlinear optical response of the material that offers the possibility to transfer the energy from one optical wave to another lightwave. For instance, a resonant coupling over a narrow frequency range is observed in parametric interactions, which are conditioned to the generation of a third idler wave for the fulfillment of a phase matching condition. The optical wave that benefits from the energy transfer will actually experience a linear gain and therefore be subject to light slowing, while the depleted wave undergoes a linear loss and will experience a fast light effect.

Among all parametric processes observed, silica-stimulated Brillouin scattering (SBS) turns out to be the most efficient. In its most simple configuration the coupling can be realized between two optical waves exclusively propagating in opposite directions in a single mode fiber, through the stimulation by electrostriction of a longitudinal acoustic wave that plays the role of the idler wave in the interaction (Boyd, 2003). This stimulation is efficient only if the two optical waves show a frequency difference resulting in an interference beating resonant with an acoustic wave. This acoustic wave in turn induces a dynamic Bragg grating in the fiber core that diffracts the light from the higher frequency wave back into the lower frequency wave. A schematic description of the SBS parametric process is shown in Figure 9.1.

As a result of the slow acoustic velocity ($\sim$5800 m/s) compared with the speed of light and the long acoustic lifetime of silica ($\sim$10 ns), a very strict phase matching condition must be satisfied, giving rise to an efficient conversion to the acoustic wave only if the frequency difference between the

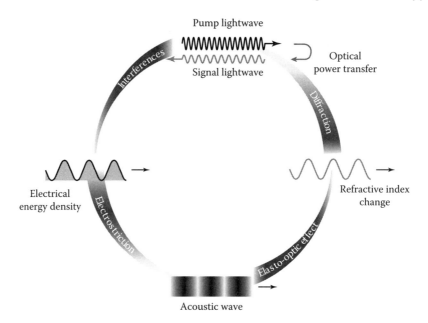

FIGURE 9.1 Schematic description of SBS as a parametric process coupling the pump and signal lightwaves through the intercession of an idler acoustic wave. The final energy transfer is possible through the successive realization of four effects (counterclockwise succession in the diagram).

optical waves is precisely set to a value known as the Brillouin frequency shift ν_B that is essentially due to the acoustic velocity and the light wavelength in the fiber. This results in a very narrowband resonant coupling that translates into narrowband gain or loss for the interacting lightwaves. In optical fibers, when represented as a function of the frequency difference between the two contrapropagating lightwaves, this gain or loss spectral distribution faithfully follows a Lorentzian distribution, centered at the Brillouin frequency shift ν_B (10–11 GHz at $\lambda = 1550$ nm) and with a 30 MHz full width at half maximum (FWHM) width (Niklès et al., 1997). In practice, a very efficient narrowband amplification or attenuation can be created in any plain silica single-mode fiber by propagating an intense monochromatic optical wave called pump in the fiber. A signal wave propagating in the opposite direction experiences a gain if its frequency is downshifted by ν_B with respect to the pump, or a loss if it is upshifted by ν_B. The narrowband nature of the interaction makes the signal propagate in a slow light regime if there is a gain, and in fast light if there is a loss (Boyd and Gauthier, 2002). The system is highly flexible, since it can be operated at any wavelength and in nearly all types of single-mode fibers. The only practical difficulty resides in the precise and stable spectral positioning of the signal that must be accurately frequency-shifted with respect to the pump, in the megahertz range.

Effective experimental solutions have been found to overcome this difficulty and the first experimental demonstrations of slow and fast light in optical fibers were performed using SBS (Okawachi et al., 2005; Song et al., 2005a). This approach demonstrated that efficient delays can be realized and turned out to be an excellent platform to test the validity of the theoretical models describing slow and fast light, regarding the perfect Lorentzian distribution of the spectral resonance and the well-controlled experimental conditions that can be implemented. In particular, the relationship between gain and delay, T can be explicitly established in the case of a Lorentzian resonance (see Chapter 3 of this book):

$$T = \frac{G}{2\pi \Delta \nu} \quad \text{with} \quad G = g_o I_p L_{\text{eff}} \tag{9.1}$$

where
- $\Delta \nu$ is the half width at half maximum of the Lorentzian distribution (Hz)
- g_o is the peak value of the Brillouin linear gain (m/W)
- L_{eff} is the effective length of fiber (m)
- I_p is the intensity of the pump (W/m²)

Using the standard value of 30 MHz for $\Delta \nu$, a delay of 1 ns/dB gain can be calculated from Equation 9.1. This value was confirmed in the first experimental demonstration (Song et al., 2005). This relation also shows that the delay depends linearly on the pump power in a nondepleted regime: this property is a clear advantage for applications and can even be crucial for the processing of analog signals. Of course the amount of delay cannot be extended without a limit as there are practical limitations to the maximal gain that will be discussed later.

While narrowbanding, SBS became an interesting tool to demonstrate the feasibility of slow and fast light in optical fibers. However, SBS delay lines are impracticable for implementation in high data rate systems if the bandwidth is narrow. Recent works have demonstrated that a unique property of SBS—the possibility to superpose many spectral resonances and thus synthesize nearly all possible gain spectral distributions—allows the experimental conditions to be adapted to almost all types of signal and has offered the possibility to implement unprecedented slow and fast light systems, such as delay lines handling multi Gbit/s data rate, with optimized gain spectral distribution and no change in signal amplitude.

The different slow and fast light implementations of SBS are described here in order of increasing complexity, from the early demonstration using a single monochromatic pump through the more complex gain spectra generated using a modulated pump to finally the most advanced systems combining gain and loss spectra generated by multiple pumps.

9.1 MONOCHROMATIC PUMP

Equation 9.1 shows that the delay T induced by the narrowband resonance depends only on the amount of total gain G and identical delays are obtained for a constant product $I_p \times L_{\text{eff}}$. In a standard SMF28 fiber, a 30 dB gain using the natural amplification through SBS requires a pump power of 22 mW for an effective length of 1 km and can even be reduced below 1 mW for very long fibers such as those showing the maximal effective length $L_{\text{eff}} = \alpha^{-1}$ fixed by the linear attenuation α.

This immediately shows that important gains can be reached at moderate pump powers if kilometer-long fibers are used, provided that the spectral width of pump and signal are well contained in the $\sim$25 MHz spectral span of the Brillouin resonance. The natural linewidth of a single frequency semiconductor laser easily fulfills this condition when it is operated in continuous wave (CW) emission. As far as the signal is concerned its bandwidth is required to be much smaller than 25 MHz, corresponding to pulse trains showing a width well exceeding 40 ns.

The major experimental difficulty in demonstrating slow light using SBS is to ensure that the pump and signal optical waves are exactly separated by the $\sim$11 GHz Brillouin shift ν_B, with the additional requirement that this difference must be kept within a megahertz stability to maintain the signal at the center of the SBS narrowband resonance. This causes practical difficulty in using distinct free-running semiconductor lasers to generate the pump and signal waves launched at the opposite fiber ends. The first two experimental demonstrations of slow and fast light in optical fibers, based on SBS using a CW pumping, were realized independently and thus used different techniques to generate pump and signal waves with a stable frequency difference.

The first reported experimental demonstration was realized at Ecole Polytechnique Fédérale in Lausanne (EPFL), Switzerland (Song et al., 2005), and was based on a technique using the modulation sidebands produced from a single laser source by a Mach–Zehnder guided-wave intensity electro-optic modulator (EOM). The operating point of the EOM was set, so that the carrier is suppressed and a bifrequency signal is produced only in the presence of the upper and lower modulation sidebands.

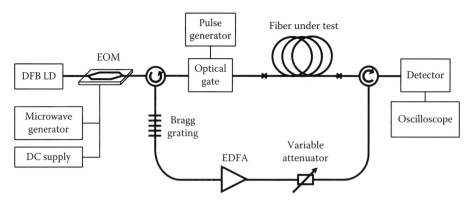

FIGURE 9.2 Experimental setup of the EPFL experiment to demonstrate slow and fast light using SBS in optical fibers. Pump and signal waves are generated from a single distributed feedback (DFB) laser source by modulating the CW laser light by a microwave generator. The two interacting lightwaves are separated in different fiber channels using the fiber Bragg grating. (From Song, K.-Y., González Herráez, M., and Thévenaz, L., *Opt. Express*, 13(1), 82, 2005. With permission.)

The frequency separation between these two modulation sidebands is equal to twice the frequency of the sine electrical signal applied on the EOM electrodes. So the pump and signal waves can be obtained by modulating the EOM using a microwave signal at exactly half of the Brillouin shift v_B. These optical waves can then be separated and directed into separate fibers using a narrowband filter such as a fiber Bragg grating. This technique, invented by the EPFL team in the 1990s, is currently extensively used in many configurations of Brillouin distributed fiber sensor (Niklès et al., 1997). Since the light for both pump and signal waves comes from the same laser source the advantage of the technique is to offer a perfect stability for their frequency difference, which is fixed by the microwave generator and can be perfectly adjusted to make the signal wave exactly match the center of the Brillouin resonance. The scheme of the experiment is shown in Figure 9.2, in which a fiber Bragg grating connected to a circulator makes the separation of the pumping and signal channels possible. The EPFL team could obtain Brillouin gains up to 30 dB in an 11.8 km standard fiber using a moderate pump power and delays of 0.97 ns/dB of gain were experimentally observed for a 100 ns pulse train, very close to the predicted result of 1 ns/dB. In addition, the experimental configuration could be readily modified to easily observe fast light in a Brillouin loss regime. This was simply achieved by slightly shifting the laser frequency, so that the other sideband is reflected by the fiber Bragg grating. The role of each sideband is swapped, the lower frequency sideband taking the role of pump wave.

Figure 9.3 shows the measured delays obtained in slow and fast light regimes, for two types of fibers. Pulses of 100 ns were continuously delayed from −8 ns in a fast light regime to 32 ns in a slow light propagation. It must be pointed out that a dispersion-shifted fiber (DSF) shows a smaller delay per dB gain as a result of the broader linewidth of the Brillouin resonance, as stated in Equation 9.1. But in a DSF each dB of amplification (or loss) requires less pump power, as this type of fiber normally presents a smaller core area, so that a higher intensity is obtained for the same pump power. The same range of delays was thus obtained in a twice shorter fiber, accordingly.

In the same year, slow light delays were also reported as the results of an independent work at Cornell University, NY, USA, using a different approach for the generation of the two interacting lightwaves (Okawachi et al., 2005). In this case a single laser source is also used for the generation of pump and signal, but the higher frequency wave is obtained directly from the laser while the lower frequency wave is produced by launching the laser light in a distinct segment of fiber and collecting the amplified spontaneous Brillouin emission. This process generates a monochromatic wave downshifted in frequency by the exact Brillouin shift of the fiber, though showing a finite

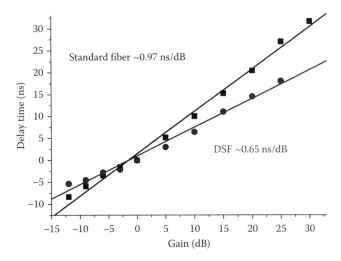

FIGURE 9.3 Delay time of the pulse as a function of the Brillouin gain using the configuration shown in Figure 9.2 for a standard single mode fiber (square) and a dispersion-shifted fiber (DSF) (circle). In a gain situation the pulse is delayed while it is accelerated in a loss configuration. (From Song, K.-Y., González Herráez, M., and Thévenaz, L., *Opt. Express*, 13(1), 82, 2005. With permission.)

linewidth of nearly 10 MHz. The power conversion from the incoming light to this amplified scattered wave is complete once a critical input power is reached. For a several kilometer-long fiber this critical power is in the milliwatt range and an intense backscattered wave can be obtained and used to generate the signal wave. This approach offers the advantage of a reduced number of devices and equipments to generate the two interacting waves, easily separated in distinct fibers with no filtering. This simplicity in turn requires that the fibers used for the generation of the signal wave and for the delaying section are identical and in similar conditioning (temperature, strain) to secure the same Brillouin shift v_B. The amplified spontaneous emission can also show power and frequency fluctuations that may impair the signal purity.

The experimental configuration developed by the Cornell team is shown in Figure 9.4. The delay of a 63 ns pulse could be varied from 14 to 25 ns, as shown in Figure 9.5. The largest delay was obtained for a 48 dB Brillouin gain that gives a slope factor of 0.52 ns of delay per dB of gain. Using a model presented in their paper, the authors calculated an effective linewidth for the Brillouin resonance of 70 MHz, much broader than the standard value of 25 MHz. Among the possible causes are poor homogeneity of the delaying fiber, resulting in a possible mismatch between the signal frequency and the Brillouin resonance, and a spectral broadening of the signal wave as a result of the amplified spontaneous Brillouin emission. This team also carried out an interesting experiment using a shorter 15 ns pulse, as shown in Figure 9.5, for which they could obtain for the first time in fibers a delay exceeding one pulse width, more exactly a fractional delay of 1.3. This delay exceeds the limit of approximately 1 for a distortion-free delay using slow light and as expected from the predictions the results showed a substantial distortion. The technique was demonstrated only in the slow light case, but can certainly also be implemented for fast light generation by using the amplified spontaneous Brillouin emission as pump wave.

The maximum delay that can be reached is fixed by the gain that would sufficiently amplify thermally scattered photons to make their power comparable to that of the pump wave. A simple model shows that this situation is reached for a total gain of 91 dB (or e^{21}) (Smith, 1972). It means that delays up to nearly 90 ns can be theoretically generated using SBS. In real situations, spurious reflections show that gains above 40 dB are difficult to achieve without the presence of a strong amplified spontaneous signal screening the pulse train. The situation can be circumvented by inserting

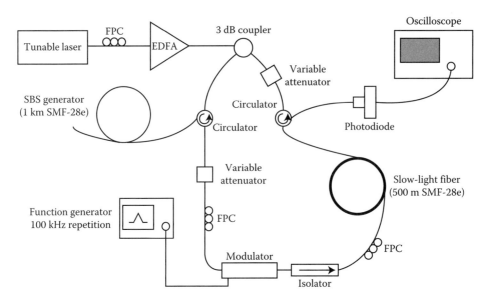

FIGURE 9.4 Experimental setup of the Cornell experiment to demonstrate slow and fast light using SBS in optical fibers. Pump and signal waves are generated from a single laser source using a 1 km segment of fiber as SBS generator. (From Okawachi, Y. et al., *Phys. Rev. Lett.*, 94, 153902, 2005. With permission.)

spectrally neutral attenuators between cascaded delaying fibers. This way the attenuators cancel the gain experienced by the signal in the delaying fiber segments, but have no delaying effect as expected from their broadband response. The delays therefore accumulate in each fiber segment while the signal and the amplified spontaneous wave amplitudes are maintained at a reasonable level thanks to the inserted attenuators. This scheme was practically demonstrated using four cascaded delaying segments, each of them generating a maximum of 40 ns delay associated with a 40 dB gain (Song et al., 2005c). A total delay of 152 ns for a 40 ns input pulse was experimentally realized, corresponding to a fractional delay of 3.6, as shown in Figure 9.6. Nevertheless, as theoretically predicted, the pulse experienced a very substantial pulse broadening by a factor of 2.4 for the maximal delay.

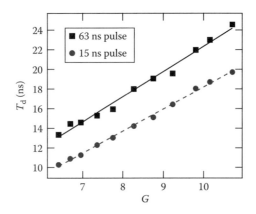

FIGURE 9.5 Induced delay as a function of the Brillouin gain parameter G for 63 ns long (square) and 15 ns long (circle) input Stokes pulses using the Cornell experiment sketched in Figure 9.4. (From Okawachi, Y. et al., *Phys. Rev. Lett.*, 94, 153902, 2005. With permission.)

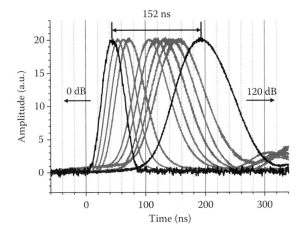

FIGURE 9.6 Temporal evolution of 40 ns pulses after propagation through four cascaded fibers separated by spectrally neutral attenuators to maintain the signal amplitude at a practical level. The gain is continuously varied from 0 to 120 dB (equivalent nonattenuated value) to reach the record 152 ns delay in optical fibers. For such a large fractional delay (3.6) the pulse experiences an important broadening. (From Song, K.-Y., González Herráez, M., and Thévenaz, L., *Opt. Express*, 30(14), 1782, 2005. With permission.)

Even though impressive delays were obtained in these pioneering experiments, they were achieved in kilometer-long fibers and these delays represent a minor fraction of the transit time along the fiber. The corresponding group index change ΔN_g is in the 10^{-3} to 10^{-2} range and the modification of the absolute group velocity is only marginal, far from the amazing results obtained in atomic media. For different gains realized by SBS, Figure 9.7 shows the group index change as a function of the fiber length along which the gain is actually realized. This figure shows that for a 30 dB gain an index change of the order of unity is possible in a ~4 m long fiber, but such a gain can only

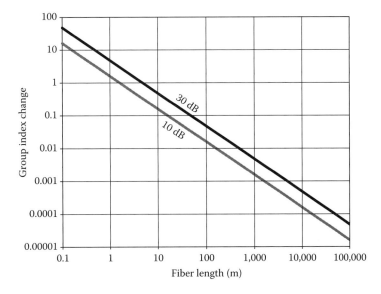

FIGURE 9.7 Induced group index change as a function of the fiber length over which the SBS gain is realized, for two typical values of gain in standard single-mode fibers. A large gain can be maintained in a shortened fiber by proportionally increasing the pump power.

be reached with a pump power of about 10 W. This was experimentally realized by the team at the EPFL, Switzerland, and they demonstrated in a 2 m fiber the slowing of the signal down to a velocity of 71,000 km/s, corresponding to a group index of 4.26 (González Herráez et al., 2005). The high pump power was obtained by using long pump pulses that were synchronized to the signal pulses to produce a uniform pumping of the delaying fiber during the signal propagation. The team basically used the same experimental configuration as in their original demonstration using long fibers.

Using the same setup, the EPFL team could drastically accelerate the signal in a fast light configuration using the Brillouin loss process. They could reach the regime of superluminal propagation, namely a group velocity exceeding the vacuum light velocity. The maximal advancement attained 14.4 ns, to be compared to the 10 ns of normal propagation time in the 2 m fiber. This is a situation of negative group velocity and literally it means that the main feature of the signal exits the fiber before entering it. This can actually be visualized in Figure 9.8 where it can be clearly seen that the peak of the pulse is simultaneously present at the input and the output of the fiber. This situation is of course possible only at the expense of a major reshaping of the pulse leading edge to satisfy all the principles given by causality and relativity for the transfer of information. These results were the first demonstration that a negative group velocity can actually be achieved in optical fibers.

The ability to realize delays in a short fiber is not only a scientific curiosity, but has an important practical impact since the reconfiguration time for the delays is based on the transit time through the delaying fiber. The total gain is mostly obtained by the product of pump power and fiber length, and therefore shortening of the fiber requires an increase in the pump power in the same proportion to maintain the same gain and thus the same delay. Using a pump power of many watts limits the practicability and a substantial research effort has been taken to develop fibers in materials showing a larger natural Brillouin gain. The first progress was reported by the University of Southampton, England, using bismuth-oxide optical fibers (Jáuregui Misas et al., 2007). The improvement was important, as they could obtain a fivefold reduction of the group velocity in a 2 m fiber using just 400 mW of CW pump power. A higher efficiency was even obtained by using As_2Se_3 chalcogenide fibers for which a 46 ns delay was realized in a 5 m fiber using only 60 mW of pump power (Song

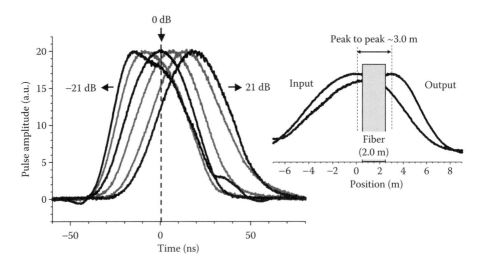

FIGURE 9.8 Temporal evolution of pulses for different SBS gain/loss induced in a 2 m fiber, resulting in a group index varying from −0.7 to 4.26. The inset compares the respective positions on a distance scale of the pulse propagating in normal conditions (0 dB gain) and of the most advanced pulse (12 dB loss), demonstrating that in this case the pulse peak exits the fiber before entering it. This situation corresponds to a negative group velocity. The fiber length is represented by the shaded area. (From González Herráez, M., Song, K.-Y., and Thévenaz, L., *Appl. Phys. Lett.*, 87, 081113, 2005. With permission.)

et al., 2006). The authors introduced a simple figure of merit to compare the delaying potential of different types of fiber and this comparison shows that the chalcogenide fiber has a figure of merit four times greater than the bismuth-oxide fiber and 110 times greater than a standard silica fiber. This figure of merit can be simply interpreted this way: 1 m of chalcogenide fiber shows the same delay as 4 m of bismuth-oxide fiber or 110 m of standard silica fiber using the same pump power. Of course using such fibers in exotic glasses is relevant only for short delaying segments since they normally show a linear attenuation far exceeding that of silica fiber.

9.2 MODULATED PUMP

A very interesting feature of SBS is the possibility to describe it, under some general assumptions, as a linear system from the point of view of the signal. The interaction of the pump and signal fields, represented by their complex amplitudes field A_p and A_s, respectively, with the acoustic material density wave of complex amplitude ρ is governed by a set of three coupled differential equations in time t and position z (Agrawal, 2006):

$$\frac{\partial A_p}{\partial z} - \frac{n}{c}\frac{\partial A_p}{\partial t} = \frac{\alpha}{2}A_p + i\kappa\rho A_s$$
$$\frac{\partial A_s}{\partial z} + \frac{n}{c}\frac{\partial A_s}{\partial t} = -\frac{\alpha}{2}A_s - i\kappa\rho^* A_p \quad (9.2)$$
$$\frac{\partial \rho}{\partial t} + \frac{\Gamma}{2}\rho = -i\Lambda A_p A_s^*$$

where
 n is the index of refraction
 c is the vacuum light velocity
 α is the linear attenuation of light
 Γ is the acoustic decay rate

The Brillouin coupling coefficients κ and Λ represent the photoelastic and electrostrictive effects, respectively.

If any change in the amplitudes is effective on a timescale much larger than the acoustic decay time, this steady state can be described by setting all explicit derivatives with respect to t to zero, so that the density ρ is given by

$$\rho = -2i\frac{\Lambda}{\Gamma}A_p A_s^* \quad (9.3)$$

This approximation simply assumes that the acoustic wave adapts instantaneously to any change in the optical fields. Assuming moreover that the pump field is much stronger than the signal and its amplitude is not significantly modified by the interaction (nondepletion regime), the equation governing the signal amplitude in Equation 9.2 can be written after inserting Equation 9.3:

$$\frac{\partial A_s}{\partial z} = \frac{1}{2}\left[g_B|A_p|^2 - \alpha\right]A_s \quad (9.4)$$

where $g_B = 4\kappa\Lambda/\Gamma$ is the usual Brillouin gain coefficient and shows a Lorentzian distribution peak when the frequency difference between pump and signal is equal to the Brillouin shift ν_B. Equation 9.4 is a simple linear transformation of the signal amplitude A_s if the pump can be assumed unchanged with position z. In practice, it means that the transformation of each frequency component of the signal can be calculated separately as a monochromatic wave and then recombined to obtain the global signal transformation. Equation 9.4 also shows that the phase of the pump has no impact on the signal transformation, the sinusoidal index perturbation induced by the acoustic wave instantaneously adapts to any phase change between the pump and the signal.

If the pump is modulated, its complex amplitude A_p can be expanded into its distinct frequency components:

$$A_p = A_{po} \int_{-\infty}^{+\infty} F(f) df \quad \text{with} \quad \int_{-\infty}^{+\infty} |F(f)|^2 df = 1 \tag{9.5}$$

The spectral decomposition identifies the modulated pump to a set of monochromatic waves that each generates its own acoustic wave while beating with the signal wave. As a result of their oscillatory nature, the cross beating terms between the pump spectral components can be neglected since they contribute inefficiently to the signal growth process in the right-hand term of Equation 9.2. The combined effect of each pump spectral component can be represented by an effective Brillouin gain spectral distribution g_B^{eff} calculated from the convolution of the intrinsic Brillouin gain g_B with the pump power spectrum, so that Equation 9.4 can be simply expressed for a modulated pump as

$$\frac{\partial A_s}{\partial z} = \frac{1}{2} \left[g_B^{\text{eff}} |A_{po}|^2 - \alpha \right] A_s \quad \text{with} \quad g_B^{\text{eff}}(f) = g_B(f) \otimes |F(f)|^2 \tag{9.6}$$

where the symbol $\otimes$ conventionally denotes the convolution operator.

This result has a tremendous impact on applications for signal transformation through SBS. It means that the gain spectral distribution can be drastically modified and shaped by the pump power spectrum. This is depicted in Figure 9.9, showing that the result of the convolution is to smooth the pump power spectrum. Actually the spectral distribution of the effective Brillouin gain g_B^{eff} is essentially given by the pump power spectrum when it is much broader than the natural Brillouin resonance ($\sim$25 MHz at $\lambda = 1550$ nm). It must be pointed out again that the pump phase information does not impact on the effective gain spectral distribution, including any phase difference between the different modulation lines.

The interest of a modified SBS gain spectrum has soon been identified after the pioneering experiments using SBS delaying and a first demonstration was realized using the simplest possible polychromatic spectrum made of just two-frequency components. The result of the convolution is to produce a double resonance and if the peak separation is comparable to the natural Brillouin linewidth, the overlapping of the resonances generates a reverse index slope and fast light can be realized in gain regime (Song et al., 2005b). This situation is depicted in Figure 9.10, showing the additional advantage that the delaying effect is produced with a much reduced signal amplification. It turns out that the latter feature has a limited advantage, as a large gain is still present in the center of the two resonance peaks that may amplify the spontaneous emission and limits the maximum pump power to generate the delays. This cannot be solved by producing slow light in the loss regime as gain resonance peaks are generated symmetrically in the Stokes band.

But this demonstration clearly shows that new schemes can be designed for slow light using the unique feature of pump modulation. In the particular case of the two-frequency pump, the delay can

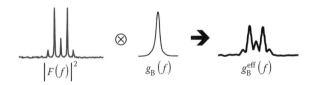

FIGURE 9.9 The effective Brillouin gain spectrum g_B^{eff} is given by the convolution of the pump power spectral distribution $|F(f)|^2$ and the intrinsic Brillouin gain spectrum g_B, offering the possibility to synthesize tailored gain spectral distribution.

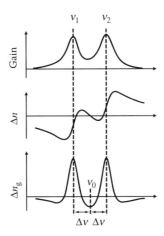

FIGURE 9.10 SBS gain, refractive index change, and group index change induced by a two-frequency pump source. The overlapping of the two resonances creates an inversion of the refractive index slope, making possible the realization of fast light in gain regime. (From Song, K.-Y., González Herráez M., and Thévenaz, L., *Opt. Express*, 13(24), 97585, 2005. With permission.)

be continuously tuned by changing the frequency separation between the two spectral lines instead of varying the pump power, even moving from a slow light to a fast light propagation. Modification of the actual spectral shape of the resonance also contributes to the optimization of the delaying effect. It turns out that for a similar spectral width a rectangular spectral distribution gives rise to a steeper index slope than a bell-shaped distribution. In addition, it can also significantly improve the distortion experienced by the signal. Under this scope, an interesting study was carried out by a team in Naples, Italy, in which they evaluated and measured the impact on the slow light response of a three-tone pump spectrum, as produced by an intensity modulator (carrier + two sidebands) (Minardo et al., 2006). By varying the amplitude and the frequency separation between evenly spaced frequency components they could show that a flattened spectrum maximizes the efficiency of delaying effect.

It must be pointed out that similar effects can be obtained entirely passively by appending segments of fibers showing different Brillouin frequency shifts v_B as a result for instance of different core doping concentrations. In this case, the global Brillouin gain of the concatenated fibers manifests as the spectral superposition of the discrete intrinsic Brillouin gain of each segment, fully identical to the effective gain spectrum produced by a modulated pump spectrum containing discrete frequency components (Chin et al., 2007).

All these experiments using a moderate number of discrete lines in the pump spectrum do not significantly break the limitation due to the narrowband nature of SBS, so that this type of slow light was prematurely classified as irrelevant for multi Gbit/s transmissions. An important step was made when it was experimentally demonstrated that the effective Brillouin gain spectrum can be continuously broadened using a randomly modulated pump source (González Herráez et al., 2006). A broadened smooth SBS gain spectrum up to 325 MHz FWHM was generated by directly modulating the current of the pump laser diode using a pseudorandom bit generator. Pulse as short as 2.7 ns could be delayed in a similar way using the natural Brillouin resonance, as shown in Figure 9.11, but with a linewidth of the gain resonance that can be potentially and arbitrarily extended.

This extension of the bandwidth has opened a wide field of applications, from microwave analog signals to Gbits/s optical transmissions. It has definitely removed a deadlock previously considered impossible to overcome. Nevertheless it must be pointed out that this bandwidth extension requires to raise the pump power proportionally to the amount of spectral broadening to maintain the same fractional delay, and even to the square of the relative spectral broadening to reach the same absolute

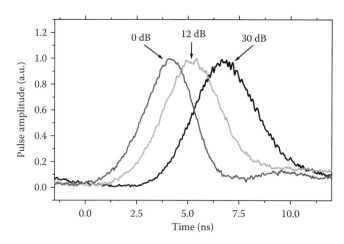

FIGURE 9.11 Time evolution of 2.7 ns pulses for different SBS gain generated with a randomly modulated pump showing an effective linewidth of 325 MHz. (From González Herráez, M., Song, K.-Y., and Thévenaz, L., *Opt. Express*, 14(4), 1395, 2006. With permission.)

delay. This comes from the fact that on one hand the peak effective SBS gain is decreased proportionally to the broadening as a result of the convolution, and on the other hand the slow light delay is inversely proportional to resonance linewidth, as shown in Equation 9.1.

The realization of broadband delays using SBS was thoroughly studied at Duke University by broadening the pump spectrum using an electrical noise source superposed on the injection current of the pump laser diode (Zhu et al., 2007). The SBS gain spectrum could be broadened up to the point where the Stokes and anti-Stokes bands start to overlap and mutually neutralize, as shown in Figure 9.12. This effect prevents any further extension of the gain linewidth and can be seen as the limit to the practical broadening. This corresponds to an equivalent gain bandwidth of 12.6 GHz which can be considered the maximal bandwidth that can be obtained using a single broadened pump. As presented in the next section this limit can actually be overcome by using multiple pumps. Figure 9.13 shows the delaying of 75 ps pulses, making this type of delay line compatible with 10 Gbit/s data rate.

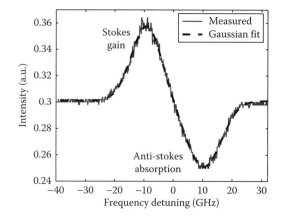

FIGURE 9.12 Extreme broadening of the SBS gain spectrum obtained using a pump diode laser driven by a noise source, showing the overlapping of the gain and loss spectra. (From Zhu, Z., Dawes, A.M.C., Gauthier, D.J., Zhang, L., and Willner, A.E., *J. Lightwave Technol.*, 25(1), 201 2007. With permission.)

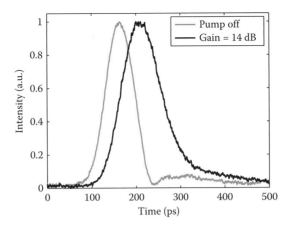

FIGURE 9.13 Time evolution of 75 ps pulses delayed by the effective SBS gain spectrum shown in Figure 9.12 (equivalent linewidth 12 GHz). (From Zhu, Z., Dawes, A.M.C., Gauthier, D.J., Zhang, L., and Willner, A.E., *J. Lightwave Technol.*, 25(1), 201, 2007. With permission.)

The effective SBS gain spectra obtained using the pseudorandom bit generator or the noise source present a typical bell-shaped distribution and are not ideal for the phase gradient through the gain resonance and the distortion (Pant et al., 2008). An interesting implementation was realized at Tel Aviv University in Israel, in which the gain spectral distribution was shaped to maximize the phase gradient and thus the delaying strength (Zadok et al., 2006). The spectrum of a pumping semiconductor laser was tailored using a combination of a deterministic, periodic current modulation together with a small random component to eventually obtain a continuous effective SBS gain spectrum with sharp edges, as shown in Figure 9.14. For equal pump powers and gain bandwidths, such a tailored gain spectrum introduces 30%–40% longer delays than standard Lorentzian resonance.

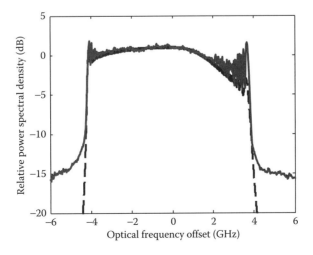

FIGURE 9.14 Optimized pump spectrum with sharp edges to enhance the delaying effect, obtained by a synthesized pump chirp based on the combination of pseudorandom and deterministic electrical driving signals. (From Zadok, A., Eyal, A., and Tur, M., *Opt. Express*, 14(19), 8498, 2006. With permission.)

9.3 MULTIPLE PUMPS

SBS offers an additional degree of freedom through the superposition of gain and loss spectral distributions. This can be realized by using pumps separated by twice the Brillouin frequency shift ν_B (~21–22 GHz at $\lambda = 1550$ nm), in which case the Brillouin loss spectrum of the lower frequency pump 1 superposes on the Brillouin gain spectrum of the higher frequency pump 2, as shown in Figure 9.15. This possibility offers new functionalities through the complete or partial overlapping of gain and loss spectral distributions showing identical or different broadenings. In the case of two pumps as depicted in Figure 9.15 and for a signal frequency in the overlapping spectral region, Equation 9.6 is simply transformed into:

$$\frac{\partial A_s}{\partial z} = \frac{1}{2} \left[g_B^{\text{eff2}} |A_{\text{po2}}|^2 - g_B^{\text{eff1}} |A_{\text{po1}}|^2 - \alpha \right] A_s \quad (9.7)$$

where the indices 1 and 2 indicate the effective gains and powers of the lower and higher frequency pumps, respectively.

This feature was exploited at Tokyo University to further extend the bandwidth of an SBS delay line over the 12 GHz limit given by the broadening of a single pump laser (Song and Hotate et al., 2007). In that case, the two pumps were exactly separated by twice the Brillouin frequency shift ν_B and were identically broadened, so that the loss spectrum of pump 1 is perfectly cancelled by the gain spectrum of pump 2 at any frequency. This way the gain spectrum of pump 1 is no longer compensated by its own loss spectrum and the effective gain broadening can be extended till the loss spectrum of pump 2 starts to overlap the gain spectrum of pump 1. The situation is depicted in Figure 9.16 and it shows that the limit of the effective gain linewidth can actually be doubled to reach nearly 25 GHz. Thirty-seven picosecond pulses were delayed up to 10.9 ps and the effective gain spectral distribution was measured to be approximately Gaussian with a estimated linewidth of 27 GHz, as shown in Figure 9.17.

It must be pointed out that this scheme can be further extended by adding another broadened pump separated from pump 2 by twice the Brillouin shift ν_B to compensate the loss spectrum of pump 2. This makes possible the extension of the bandwidth by another 12 GHz to get close to the 40 GHz limit needed for the compatibility with ultrahigh data rate optical communications. The bandwidth

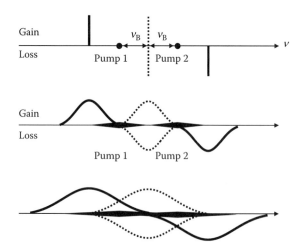

FIGURE 9.15 Gain and loss spectra produced by two pumps separated by twice the Brillouin shift ν_B, for gradual broadening of the pump spectrum. The gain and loss spectra in the middle of the covered spectral range mutually cancel, so that the actual bandwidth can be doubled. (From Song, K.-Y. and Hotate, K., *Opt. Lett.*, 32(3), 217, 2007. With permission.)

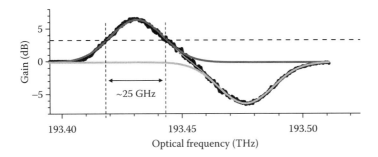

FIGURE 9.16 Measured SBS amplification produced by two distinct pump sources separated by twice the Brillouin frequency shift v_B. The loss and gain spectra of individual pumps perfectly compensate in the middle of the spectral range, so that the amplification bandwidth can be extended to $2v_B$ as depicted in Figure 9.15. (From Song, K.-Y. and Hotate, K., *Opt. Lett.*, 32(3), 217, 2007. With permission.)

can thus be increased by v_B steps by incrementing the number of pumps. This raises a serious practical difficulty as such a scheme would require a massive broadening of the pump's emission spectrum, with the immediate consequence that the power of each individual pump must also be increased in the same proportion to maintain the fractional delaying strength.

The superposition of gain and loss spectra of different spectral widths also offers the possibility to achieve slow and fast light in a fully transparent regime. This situation is similar to ideal electromagnetically induced transparency, as a narrowband gain can open a transparency window in a broadband loss spectrum.

Let $G_{g,l}$ and $\Delta v_{g,l}$ be the peak gain/loss value and linewidth of the superposed gain (g) and loss (l) spectra, respectively. As a result of the linear transformation experienced by the signal as given by Equation 9.4 and using Equation 9.1, the total delay and gain experienced by a signal spectrally placed at the center of the superposed resonance will be

$$T = \frac{G_g}{2\pi \Delta v_g} - \frac{G_l}{2\pi \Delta v_l} \qquad G = G_g - G_l \qquad (9.8)$$

If $G_g = G_l$ and $\Delta v_g \neq \Delta v_l$, then:

$$T = \frac{G_{g,l}}{2\pi}\left(\frac{1}{\Delta v_g} - \frac{1}{\Delta v_l}\right) \neq 0 \qquad G = 0 \qquad (9.9)$$

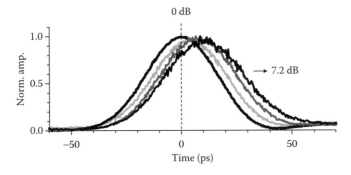

FIGURE 9.17 Time evolution of 37 ps pulses delayed by the effective SBS gain spectrum shown in Figure 9.16 (equivalent linewidth 25 GHz). The maximum delay is 10.9 ps for a 7.2 dB gain. (From Song, K.-Y. and Hotate, K., *Opt. Lett.*, 32(3), 217, 2007.)

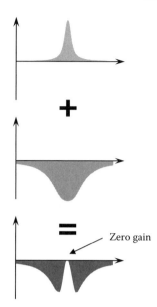

FIGURE 9.18 Superposition of gain and loss resonances with different linewidths results in a spectral transmission showing a narrow fully transparent spectral window that creates slow light propagation with zero gain.

so that a delay or an advancement can be generated in the total absence of loss or gain. This situation is depicted in Figure 9.18 in the spectral domain and clearly shows that the gain compensation maintains the sharp spectral transitions necessary to induce a slow and fast light propagation.

This situation was experimentally demonstrated by the EPFL team in Switzerland, in both slow and fast light regimes (Chin et al., 2006). The principle of the experimental setup is shown in Figure 9.19, in which the narrowband gain resonance was produced by a monochromatic pump and the broadband loss by a spectrally broadened distinct pump laser. The powers of the two pump lasers are adjusted so that their peak SBS gain or loss exactly compensates, as shown in Figure 9.20. The delaying effect is then obtained by varying in the same proportion the powers of the two pumps, using a broadband variable attenuator for instance. Figure 9.21 shows that delays similar to those obtained

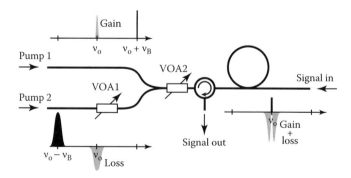

FIGURE 9.19 Principle of the experimental configuration to generate zero-gain spectral resonances, in which two distinct optical pumps showing different spectral width were used to produce spectrally superposed Brillouin gain and loss with different linewidth (VOA, variable optical attenuator). (From Chin, S., Gonzalez-Herraez, M., and Thévenaz, L., *Opt. Express*, 14(22), 10684, 2006. With permission.)

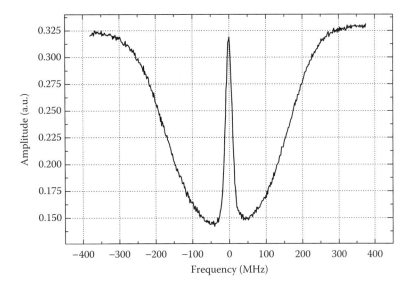

FIGURE 9.20 Measured spectral transmission after propagation in a fiber where SBS gain and loss curves of different linewidth are spectrally superposed, showing that the transparency condition is experimentally realized for the central frequency. (From Chin, S., Gonzalez-Herraez, M., and Thévenaz, L., *Opt. Express*, 14(22), 10684, 2006. With permission.)

using a single monochromatic pump are achieved, but with less than 1 dB variations in the signal intensity throughout the 12 ns delay range. Signal advancement using fast light was also observed by simply swapping the spectral positions of the pumps. This result has an important practical impact, as any change in the signal intensity remains an unwanted side effect of the delaying process.

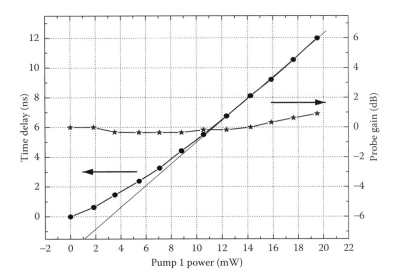

FIGURE 9.21 Induced delays (dots) and intensity change (stars) as a function of the narrow linewidth pump power for a signal propagating in the middle of the spectral feature shown in Figure 9.20, showing that efficient delays can be induced with a minor signal intensity change. (From Chin, S., Gonzalez-Herraez, M., and Thévenaz, L., *Opt. Express*, 14(22), 10684, 2006. With permission.)

Following the same approach the gain compensation by a superimposed broadband loss or by two narrowband losses placed on the wings of the gain spectral distribution offers the possibility to double the delaying capability when compared to a single resonance scheme (Schneider et al., 2007).

A recent result shows that fast light is self-generated in the total absence of any pump source. In this implementation the signal is amplified above the critical Brillouin power, often named Brillouin threshold, over which the signal light is converted into a backward Stokes emission resulting from the amplified Brillouin emission. This intense Stokes light is spectrally positioned below the signal frequency by exactly the fiber Brillouin shift ν_B, maintaining through depletion the signal output intensity constant for any signal power over the Brillouin threshold and eventually creating a Brillouin loss resonance exactly positioned at the signal frequency. In a 12 km fiber, advancements up to 12 ns were continuously generated by varying the signal input power between 10 and 28 dBm (Chin et al., 2008). This implementation requires that the signal maintains its average spectral and power characteristics throughout the delaying fibers, but interestingly, shows that the spectrum of the Stokes emission matches the signal spectral distribution and thus the SBS effective loss spectrum self-adapts to the signal spectral distribution.

Slow and fast light delay lines in optical fibers based on SBS shows unprecedented flexibility to realize innovative systems for the generation of optically controlled delays. SBS-based slow and fast light has the unique property to maintain the shape of the spectral resonance through the modulation of the pump or the concatenation of fibers showing different Brillouin spectra. Attractive solutions were proposed earlier to overcome the challenges toward real applications that were incompatible with the narrowband nature of SBS. These challenges were overcome due to the realization of broadband delays up to a bandwidth of 25 GHz and more, the optimization of the gain spectral distribution for effective delaying and slow distortion, the achievement of important delays in meter-long fibers, and eventually the realization of delays with no amplitude change.

SBS-based slow and fast light also shows an attractive characteristic that makes it an excellent candidate for all applications based on analog signals, such as true time delays in microwave photonics: from the signal point of view the SBS amplification is essentially a linear transformation and in most implementations the effective delays are proportional to the control signal power. All these unique features make slow and fast light based on SBS a very open field in which innovative and creative solutions are still to be found. In particular, the impact of the velocity change on nonlinear interactions or for sensing applications is still widely unexplored.

REFERENCES

Agrawal, G. P. *Nonlinear Fiber Optics*, 4th edn. San Diego: Academic Press, 2006.
Boyd, R. W. *Nonlinear Optics*, 2nd edn. New York: Academic, 2003.
Boyd, R. W. and D. J. Gauthier. "Slow" and "Fast" Light, in E. Wolf (Ed.), *Progress in Optics*, Vol. 43, Chap. 6, pp. 497–530. Amsterdam: Elsevier, 2002.
Chin, S., M. Gonzalez-Herraez, and L. Thévenaz. Zero-gain slow & fast light propagation in an optical fiber. *Opt. Express* 14(22) (2006): 10684–10692.
Chin, S., M. Gonzalez-Herraez, and L. Thévenaz. Simple technique to achieve fast light in gain regime using Brillouin scattering. *Opt. Express* 15(17) (2007): 10814–10821.
Chin, S., M. Gonzalez-Herraez, and L. Thévenaz. Self-induced fast light propagation in an optical fiber based on Brillouin scattering. *Opt. Express* 16(16) (2008): 12181–12189.
González Herráez, M., K.-Y. Song, and L. Thévenaz. Optically controlled slow and fast light in optical fibers using stimulated Brillouin scattering. *Appl. Phys. Lett.* 87 (2005): 081113.
González Herráez, M., K.-Y. Song, and L. Thévenaz. Arbitrary-bandwidth Brillouin slow light in optical fibers. *Opt. Express* 14(4) (2006): 1395–1400.
Jáuregui Misas, C., P. Petropoulos, and D. J. Richardson. Slowing of pulses to c/10 with subwatt power levels and low latency using brillouin amplification in a bismuth-oxide optical fiber. *J. Lightwave Technol.* 25(1) (2007): 216–221.

Minardo, A., R. Bernini, and L. Zeni. Low distortion Brillouin slow light in optical fibers using AM modulation. *Opt. Express* 14(13) (2006): 5866–5876.

Niklès, M., L. Thévenaz, and P. Robert. Brillouin gain spectrum characterization in single-mode optical fibers. *IEEE J.Lightwave Technol.* 15(10) (1997): 1842–1851.

Okawachi, Y., et al. Tunable all-optical delays via Brillouin slow light in an optical fiber. *Phys. Rev. Lett.* 94 (2005): 153902.

Pant, R., M. D. Stenner, M. A. Neifeld, and D. J. Gauthier. Optimal pump profile designs for broadband SBS slow-light systems. *Opt. Express* 16 (2008): 2764–2777.

Schneider, T., R. Hemker, K.-U. Lauterbach, and M. Junker. Comparison of delay enhancement mechanisms for SBS-based slow light systems. *Opt. Express* 15 (2007): 9606–9613.

Smith, R. G. Optical power handling capacity of low loss optical fibers as determined by stimulated Raman and Brillouin scattering. *Appl. Opt.* 11(11) (1972): 2489–2494.

Song, K.-Y., and K. Hotate. 25 GHz bandwidth Brillouin slow light in optical fibers. *Opt. Lett.* 32(3) (2007): 217–219.

Song, K.-Y., M. González Herráez, and L. Thévenaz. Observation of pulse delaying and advancement in optical fibers using stimulated Brillouin scattering. *Opt. Express* 13(1) (2005a): 82–88.

Song, K.-Y., M. González Herráez, and L. Thévenaz. Gain-assisted pulse advancement using single and double Brillouin gain peaks in optical fibers. *Opt. Express* 13(24) (2005b): 9758–9765.

Song, K.-Y., M. González Herráez, and L. Thévenaz. Long optically-controlled delays in optical fibers. *Opt. Lett.* 30(14) (2005c): 1782–1784.

Song, K.-Y., K. S. Abedin, M. González Herráez, and L. Thévenaz. Highly efficient Brillouin slow and fast light using As_2Se_3 chalcogenide fiber. *Opt. Express* 14(13) (2006): 5860–5865.

Zadok, A., A. Eyal, and M. Tur. Extended delay of broadband signals in stimulated Brillouin scattering slow light using synthesized pump chirp. *Opt. Express* 14(19) (2006): 8498–8505.

Zhu, Z., A. M. C. Dawes, D. J. Gauthier, L. Zhang, and A. E. Willner. Broadband SBS slow light in an optical fiber. *J.Lightwave Technol.* 25(1) (2007): 201–206.

Part IV

Slow Light and Nonlinear Phenomena

10 Nonlinear Slow-Wave Structures

Andrea Melloni and Francesco Morichetti

CONTENTS

10.1	Fundamentals on SWS	196
10.2	Nonlinear Phase Modulation	200
10.3	SPM and Chromatic Dispersion	202
	10.3.1 Dispersion Regimes	202
	10.3.2 Power Limiting	203
	10.3.3 Soliton Propagation	205
10.4	Cross-Phase Modulation	206
10.5	Nonlinear Spectral Response	208
10.6	Four Wave Mixing	210
10.7	Modulation Instability	215
References		220

The most exciting potentiality offered by slow-wave propagation is the enhancement of the optical field inside the guiding media and hence the enhancement of any nonlinear effect with respect to bulk materials. The increment in the field intensity is linearly related to the velocity reduction as the energy is preserved during the propagation.

Historically, the development of devices for all-optical signal processing is focused on materials with strong nonlinearities, but for most materials the nonlinear figure of merit $n_2/\alpha\tau$ is constant within a factor of 10 [1], α being the absorption coefficient, τ the response time, and n_2 the nonlinear refractive index of the material. As a result, high speed and low absorption are inevitably accompanied by a weak nonlinearity, so that long devices or very high optical powers are required. However, as a slow-wave structure (SWS) enhances the field in a certain frequency ranges, there is a concrete possibility to overcome the weak materials nonlinearity and to arbitrarily enhance the nonlinear interaction between wave and matter or between different waves. This exciting advantage is typically gained as a trade-off between a reduction in the operational bandwidth and the physical dimension of the SWS. Despite this appealing potential, nonlinear propagation in the slow-wave regime has not been really investigated so far and the comprehension of nonlinear effects in SWSs should be considered still at its infancy. Among the few works on the subject, the most comprehensive and complete references are [2–6].

This chapter reports only theoretical results and numerical studies related to the Kerr nonlinearity in coupled resonator structures. The $\chi^{(2)}$ nonlinearity, the Raman and Brillouin processes, other slowing down mechanisms such as electromagnetic induced transparency (EIT) or coherent population oscillations and slow light on the edge of photonic band gaps are not considered. Coupled resonators are considered here in a broad sense, without reference to a specific technology or cavity structure. Ring resonators, Fabry–Pèrot cavities, coupled defects in photonic crystal (PhC) waveguides, Bragg grating defined cavities can be modeled from a circuit point of view in a very similar way, and this

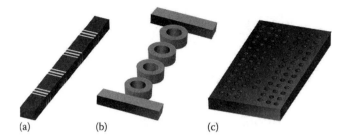

FIGURE 10.1 Three examples of directly coupled optical resonator SWSs: (a) direct-coupled Fabry–Pèrot SWS, (b) direct-coupled microring SWS, and (c) PhC based SWS.

approach is used throughout the chapter. Detailed information on the numerical simulation technique and the circuit approach can be found in Refs. [2,7].

The considered structures are shown in Figure 10.1 and are referred here as SWSs. The propagation is allowed in periodic pass bands with a strong frequency-dependent dispersion characteristic, and this unique spectral behavior has a strong and unique impact on every nonlinear effect. In this chapter, the classical Kerr-based nonlinear effects are revisited in SWSs, pointing out limits, advantages and new phenomena that have been noticed up to now. Self-phase modulation (SPM), cross-phase modulation (XPM), four-wave mixing (FWM), soliton propagation and modulation instability (MI), as well as unusual effects like power limiting and self-pulsation are considered and described but the chapter should be considered a first attempt to approach this field and a starting point for future studies, being far away to be exhaustive.

10.1 FUNDAMENTALS ON SWS

Although most of the linear spectral characteristics of SWSs have been discussed in the previous chapters, a synthetic summary, mainly oriented to set the basis for the nonlinear exploration, is presented. In the following, infinitely long periodic structures are considered and analytical results are derived in the asymptotic limit. The structures shown in Figure 10.1 can all be modeled as a sequence of identical cells. Two examples of physical single cells are shown in Figure 10.2a and b, and these cells together with the equivalent cell are used to investigate almost every kind of SWS. The equivalent cell is made up of a partially reflecting mirror placed between two lines of geometrical length $d/2$, $2d$ being the round-trip of the cavity. The lines, not necessarily uniform or homogeneous, have their own specific optical properties such as effective index n_{eff}, group index n_g, chromatic dispersion β_2, nonlinear index of refraction n_2, attenuation α, and so on. The partial reflector is described by a transmission matrix $\mathbf{T}$ that relates the complex amplitude of the waves at

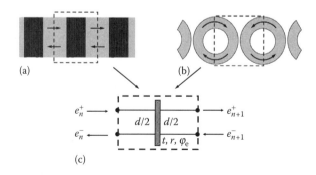

FIGURE 10.2 Two examples of physical cells and their equivalent circuit.

the right and left of the coupling element. **T** is suitable to describe simple ideal coupling elements as well as complex structures such as nonlinear Bragg gratings [7,8], directional couplers, defects in PhC waveguides [9], or multilayered bulk materials [10]. The matrix **T** can be written in the general form

$$\mathbf{T} = \frac{j}{t}\begin{bmatrix} -\exp(j\varphi_e) & r \\ -r & \exp(j\varphi_e) \end{bmatrix}, \tag{10.1}$$

where r and t, being the modulus of the reflectivity and transmissivity (coupling), respectively dependent on wavelength and intensity. The reflector also introduces a phase term $\varphi_e = 2\pi n_0 L_e/\lambda$ taking into account its thickness, n_0 being a convenient reference refractive index, and L_e the equivalent length of the coupling element. The expressions of r, t, and L_e can be derived by equating the elements of the matrix **T** to those of the coupling element, which can be known analytically, and numerically or even experimentally as in the case of a uniform distributed Bragg reflectors or directional couplers. A detailed example of this equivalence can be found in Refs. [7,8] for linear and nonlinear Bragg reflectors.

Both lines of the equivalent circuit are described by a diagonal matrix **P**

$$\mathbf{P} = \begin{bmatrix} e^{j\phi^+} & 0 \\ 0 & e^{-j\phi^-} \end{bmatrix} \tag{10.2}$$

where $\phi^\pm$ are the phase shifts accumulated by the forward and backward fields during the propagation in the line. The accumulated phase shifts in the presence of Kerr nonlinearity are given by

$$\phi^\pm = \frac{\omega}{c}(n_{\text{eff}} + n_2(I^\pm + 2I^\mp))\frac{d}{2} \pm \varphi_e, \tag{10.3}$$

$I^\pm(\omega)$ being the frequency-dependent intensities of the forward and backward waves and reduces to $\phi^\pm = \omega n_{\text{eff}} d/2c \pm \varphi_e$ in the linear case. The phase φ_e, that can be considered just as an additional contribution to the cavity length, is dropped from now on for brevity. The first nonlinear contribution in Equation 10.3 is due to the SPM effect and the second one is due to XPM. The cross term acts when two or more waves propagate in the structure but also in case of a single wave propagating in cavities having the forward and backward paths physically overlapped, as in Fabry–Pèrot based SWSs. In ring resonators the two paths are decoupled and, in single wave regime, XPM vanishes. Parametric terms (FWM) are considered in Section 10.6 whereas Brillouin and Raman processes are not considered in this chapter.

At this point, it is convenient to first solve the linear case and then calculate the intensities $I^\pm$ inside the cavities to retrieve the nonlinear behavior. The analytical results hold for weak nonlinearities while stronger nonlinearities must be treated numerically. The transmission matrix **M** of the elementary cell relating the forward and backward fields $e_{n+1}^\pm$ and $e_n^\pm$ on each side of the cell, is obtained by multiplying the transmission matrices of the elements constituting the cell, that is, $\mathbf{M} = \mathbf{PTP}$. By forcing that the fields on each side of the cell are identical apart from a propagation term $\exp(-\gamma_s d)$, formally

$$e_{n+1}^+ + e_{n+1}^- = (e_n^+ + e_n^-)e^{-\gamma_s d}, \tag{10.4}$$

the characteristic equation

$$\cosh(\gamma_s d) = \frac{M_{11} + M_{22}}{2} = \frac{\sin(kd)}{t}, \tag{10.5}$$

is obtained, with $k = \omega n_{\text{eff}}/c$, from which all the spectral characteristics of an infinitely long SWS can be derived.

Equation 10.5 indicates that passbands and stopbands alternate periodically in frequency, the propagation constant γ_s being real, $\gamma_s = \alpha$, in the stopband and imaginary, $\gamma_s = j\beta$, in the bandpass, where propagation in both directions is permitted. Omitting the derivations, detailed elsewhere in this book and in several papers [2,11], it is found that the periodicity (free spectral range) is FSR $= c/2n_g d$ and the transmission bandwidth, obtained imposing the condition $|\sin(kd)| \leq t$, is

$$B = \frac{2\text{FSR}}{\pi} \sin^{-1}(t) \simeq \frac{2\text{FSR}}{\pi} t, \tag{10.6}$$

the approximation being valid for $t < 0.5$. Within the passband the transmission is unitary and the group velocity, obtained as the derivative of β with respect to ω is

$$v_g = \mp \frac{c}{n_{\text{eff}}} \frac{\sqrt{t^2 - \sin^2(kd)}}{\cos(kd)}. \tag{10.7}$$

The group velocity v_g of the wave propagating inside an SWS is reduced with respect to the phase velocity $v = c/n_{\text{eff}}$ in the same but unloaded structure ($t = 1$) by a factor

$$S = \frac{v}{v_g} = \frac{\cos(kd)}{\sqrt{t^2 - \sin^2(kd)}}, \tag{10.8}$$

called the slowing ratio $S(\omega)$ of the SWS. The factor S is shown in Figure 10.3 for different values of t. At each resonance frequency f_0 the argument $kd = M\pi$ is an integer multiple of π, and $S(f_0)$ is simply equal to $1/t$. By moving toward the band edges v_g ideally drops to zero and S has a singularity. Note that both v_g and the bandwidth B reduce by reducing the coupling coefficient t or by increasing the finesse FSR/B.

We now have all the elements to calculate the intensity of the fields inside the structure and hence we are ready to face the nonlinear effects. From the periodicity condition Equation 10.4 and

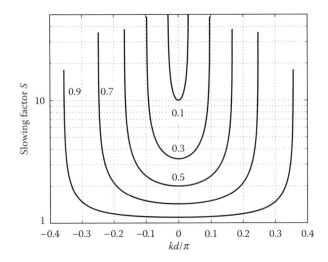

FIGURE 10.3 Slowing factor S in an infinitely long SWS for different values of t. The inverse of the slowing factor is the normalized group velocity.

the definition of the matrix **M**, the ratio between the amplitudes of the backward and forward waves is found,

$$\Gamma = \frac{e_n^-}{e_n^+} = \frac{e^{-\gamma d} - M_{11}}{M_{12}} = \frac{-e^{jkd} + jte^{j\beta d}}{|r|e^{j2kz}e^{-jkd}} \qquad (10.9)$$

where Γ is the complex reflection coefficient inside the SWS. At the resonant frequency the reflection coefficient becomes $\Gamma(z) = (t-1)e^{-j2kz}/|r|$, exhibiting a constant modulus along the structure, a phase decreasing linearly with z and a period equal to $\lambda/2$. Assuming a forward propagating wave $e_n^+(z) = A\exp(-jkz)$, the intensity of the total field (the Bloch mode) is $I(z) = |e_n^+ + e_n^-|^2 = |A|^2|1+\Gamma(z)|^2$, its minimum value being located at $z = M\lambda/2$ and the maximum at $z = M\lambda/2 + \lambda/4$, M being an integer. From these relations, the intensities $I^\pm = |e_n^\pm|^2$ of the forward and backward propagating waves inside each cavity can be related to the intensity I_{in} of the field outside the SWS (or the field in absence of cavities) as

$$I^\pm(\omega) = I_{in}\frac{S(\omega) \pm 1}{2}, \qquad (10.10)$$

the intracavity mean intensity being equal to S times I_{in}, as required by the conservation of the energy flux.

The induced phase shift in the lines can now be calculated by substituting Equation 10.10 into Equation 10.3 and the characteristic equation of the SWS in the nonlinear regime derived as for the linear case:

$$\beta(\omega) = \pm\frac{1}{d}\cos^{-1}\left[\frac{1}{t}\sin\left(\frac{\omega}{c}(n_0 + n_2 c_n S(\omega)I_{in})d\right)\right]. \qquad (10.11)$$

The coefficient c_n depends on the structure and on the nonlinear effect. In ring resonators, for example, the forward and the backward waveguide are physically decoupled and the progressive and the regressive fields do not interact. On the contrary, in a Fabry–Pèrot based SWS the progressive and the regressive waves propagate through the same waveguide. The counter propagating portion of the field effectively behaves as a distinct wave and gives rise to an additional XPM-like contribution. Such nonlinear coupling between the forward and the backward wave has been observed also in a single cavity [12,13] and in Bragg gratings [14]. The distinction between the two structures has to be taken into account when two or more optical waves nonlinearly interact by XPM. It is easy to verify that the coefficient c_n assumes the values reported in Table 10.1 where the term b refers to the relative state of polarization between the two interacting fields [15]. If the interacting fields own the same state of polarization $b = 1$, if they are orthogonal, $b = 1/3$.

These results hold for an infinitely long SWS. In practice, it is necessary to match the impedance of the SWS with initial and final sections with a lower finesse. The matching, or apodization, is mandatory to avoid in-band ripples and reflections as described in Ref. [18] and slightly reduces the total effect. In the central section of the SWS, however, where all the cavities are identical, the asymptotic relations are an excellent approximation and they are used throughout the chapter.

TABLE 10.1
Coefficient c_n for SPM and XPM in SWS

c_n	Ring	Fabry–Pérot
c_S (SPM)	1/2	3/2
c_X (XPM)	$(1+b)/2$	$(3+b)/2$

10.2 NONLINEAR PHASE MODULATION

As in bulk and classical guiding structures, in SWSs also the typical effect of the Kerr nonlinearity is to induce an additional phase shift onto the propagating waves. In SWS, however, the slowdown increases the phase shift for two reasons, the enhancement of both the field intensity and the effective length. The effective phase shift ϕ_{eff} induced in an SWS by an input power P_{in} can be derived as [3,16]

$$\frac{d\phi_{\text{eff}}}{dP_{\text{in}}} = \frac{d\phi_{\text{eff}}}{d\phi}\frac{d\phi}{dP_{\text{m}}}\frac{dP_{\text{m}}}{dP_{\text{in}}}. \tag{10.12}$$

The first term is the enhancement of the induced phase shift due to the propagation slowdown. This derivative is deduced straight from Equation 10.5 and it is equal to the slowing ratio S irrespective of the physical origin inducing the phase shift, also including the nonlinear Kerr effect. The effective length of a SWS is thus S times the physical length of the structure. The second term of Equation 10.12 expresses the dependence of the nonresonant phase shift $d\phi$ on the intracavity mean power P_{m} and it is related to the nonlinear constant $\gamma = \omega n_2/cA_{\text{eff}}$ [15] and the number of cavities. A_{eff} is the effective area of the waveguide and relates the intensity to the power. Finally, the last derivative is the mean power enhancement factor and, as previously discussed, it is equal to the slowing ratio S too.

Another way to find the nonlinear phase sensitivity is to calculate the derivative of the effective phase shift expressed as $\phi_{\text{eff}} = \beta(\omega)Nd$, with $\beta(\omega)$ given by Equation 10.11 and N the number of cavities. The frequency shift of the SWS dispersion characteristic due to the input intensity I_{in} is readily obtained from Equation 10.11 as

$$\Delta\omega = \frac{n_2}{n_{\text{eff}}}c_n S I_{\text{in}}\omega \tag{10.13}$$

and hence the sensitivity (Equation 10.12) can be written as

$$\frac{d\phi_{\text{eff}}}{dP_{\text{in}}} = c_n S^2(\omega + \Delta\omega)\gamma Nd \tag{10.14}$$

where $S(\omega + \Delta\omega)$ is the actual slowing ratio introduced by the SWS redshifted by the nonlinearity.

The implicit expression (Equation 10.13) indicates that the spectral characteristic of the structure shifts toward lower frequencies (or higher wavelengths) changing the slowing factor S while the nonlinear effects take place. The frequency shift $\Delta\omega$ is proportional to S and, if not negligible with respect to the signal and structure relative bandwidth, it can strongly affect the nonlinear effect. The relation (Equation 10.14), with $\Delta\omega$ given by (Equation 10.13), holds for a small perturbation of the SWS spectral response, otherwise the behavior must be investigated through numerical simulations. It is now evident that the slow-wave regime enhances both the SPM and XPM by a factor proportional to S^2 and that the nonlinear effect in SWS is conveniently described by defining an effective nonlinear constant γ_{eff} as

$$\gamma_{\text{eff}} = c_n \gamma S^2. \tag{10.15}$$

In case of strong nonlinear regime and interactions between different waves, a more detailed description is reported in Section 10.4.

Nonlinear effects can be strongly enhanced with respect to classical nonresonant structures, clearly at the expense of a periodicity in the useful bandwidth. As an example with a slowing factor S equal to 10 (rather simple to realize and control in practice) the enhancement ranges between 17 and 23 dB, depending on c_n. Almost every application or device based on nonlinear interactions can take advantage of this efficiency enhancement, which can be exploited as desired, to reduce the input power, to shrink the device dimensions or both, and without strong limitations on the bandwidth. Further, the nonlinear dependence of the SWS spectral characteristic is an enormous potentiality

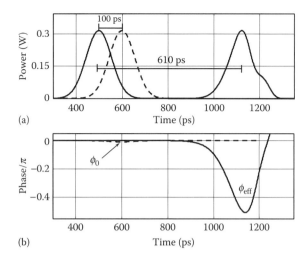

FIGURE 10.4 Comparison of the propagation of a Gaussian pulse in an SWS and a uniform waveguide. (a) Input and output pulse intensities, (b) induced SPMs. Dashed lines refer to the uniform unloaded guiding structure.

because a simultaneous filtering and/or shaping in both time and spectral domain can take place, opening the way to advanced signal processing functions.

A numerical example of SPM in SWS is discussed in the following section. Let us consider a Fabry–Pèrot ($c_n = 3/2$) based SWS with $N = 40$, resonator bandwidth of 20 GHz, and FSR = 200 GHz. The slowing factor is $S = 6.3$ and the group optical length is $n_g Nd = 30$ mm. The effective nonlinear phase shift on a Gaussian pulse with peak power 0.33 W in an SWS structure with $n_{eff} = n_g = 3.3$, $n_2 = 2.2 \cdot 10^{-17}$ m^2/W (AlGaAs) and effective area $A_{eff} = 10$ μm^2 at $\lambda = 1550$ nm, given by

$$\phi_{eff} = \frac{2\pi}{\lambda} \frac{n_2}{A_{eff}} P_{in} dN c_n S^2 = \phi_0 c_n S^2, \qquad (10.16)$$

is $\pi/2$, $c_n S^2 = 60$ times higher than the nonlinear phase shift ϕ_0 induced in a straight uniform waveguide of equal length Nd. Simulation results are reported in Figure 10.4. The pulse width is 120 ps and its spectrum, tuned at the SWS resonant frequency, is well confined to the passband. The pulse envelope, shown in Figure 10.4a exits the SWS after 610 ps, S times later with respect the unloaded uniform waveguide (dashed line). The nonlinear induced SPMs ϕ_0 and ϕ_{eff} are shown in Figure 10.4b where the enhancement is equal to 57, in good agreement with the theoretical value. These small differences with respect to the predicted values are due to the input or output matching sections with a lower finesse and due to the third order chromatic dispersion that induces a small distortion to the pulse reducing the peak power as the pulse propagates on. The spectral shift given by (Equation 10.13) is only 1.3 GHz and its effect is negligible.

In these conditions, the pulse spectrum broadens and distorts as predicted by classical SPM theory [15], with the enhancement factor added. The spectrum of a 1 ns Gaussian pulse is reported in Figure 10.5 for an induced phase modulation equal to 1.5π and 2.5π, obtained with a peak power of 1 W and 1.67 W, respectively. The spectral broadening is evident as well as the characteristic oscillations with deep notches not dropping to zero because of the third-order dispersion and the frequency shift (Equation 10.13). The well-known SPM behaviors in time as well as in frequency domain are obtained also in case of super-Gaussian and chirped pulses. Novel phenomena occur if the pulse spectrum interacts with the band edges of the SWS spectral response where S can reach very high values, higher-order dispersion terms become dominant and positive nonlinear feedbacks dominate. This feature is explored in Section 10.3.

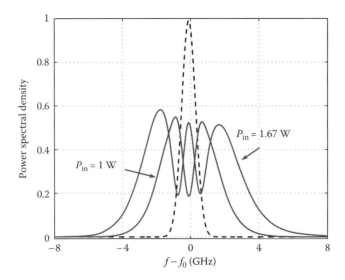

FIGURE 10.5 Pulse spectral broadening for the SWS considered in Figure 10.4 of a 1 ns Gaussian pulse. Peak powers are equal to 1 and 1.67 W. The input pulse spectrum is reported in dashed line.

10.3 SPM AND CHROMATIC DISPERSION

Away from resonances, the second-order group velocity dispersion appears and interactions between dispersion and SPM take place. In this section, well-known phenomena are revisited in the framework of slow-wave propagation, always considering infinite structures and trying to keep the formal description as simple as possible. In general, propagation regimes are affected by the slowdown factor S but, if the pulse spectral width is comparable with the bandwidth of the structure, or simply if the pulse spectrum reaches one of the band edges, other more complex and original effects could take place. After a brief survey of the effects that can occur in the presence of signal-SWS detuning, Section 10.3.2 describes the power limiting mechanism induced by the redshift of the higher frequency edge and Section 10.3.3 investigates soliton propagation, the typical propagation regime in presence of anomalous dispersion and SPM.

10.3.1 Dispersion Regimes

As already mentioned, at the central frequency f_0 the second-order chromatic dispersion vanishes and the total third-order dispersion is $\beta_3 L = N(\pi B)^{-3}$, N being the number of resonators of the structure, B its bandwidth and $L = Nd$ the total length. The second-order dispersion, given by

$$\beta_2 L = \frac{NS^3 \tan(kd)}{\pi^2 B^2 \cos^2(kd)}, \qquad (10.17)$$

is positive at frequencies higher than f_0 and negative below f_0. Figure 10.6 shows the group delay (slowing factor) with the two possible dispersion regimes, normal on the right and anomalous on the left, with some typical effects derived from dispersion and SPM interaction.

In the normal dispersion regime, the phase shift induced by the chromatic dispersion adds to the nonlinear phase shift and the pulse undergoes a rapid broadening, as usual. Further, in SWSs both effects are enhanced by the slowing factor and a severe broadening can be experienced in very short lengths, and severe distortions appear if the pulse spectrum interacts with the structure band edges. The interaction can occur because of both the pulse spectral broadening and the SWS spectral shift (Equation 10.13). There are two main interesting phenomena in the normal dispersion spectral

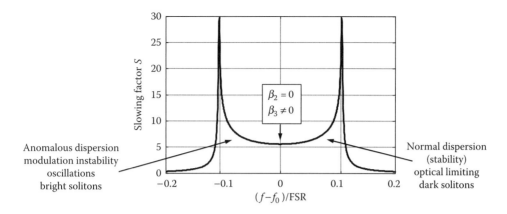

FIGURE 10.6 The interaction between nonlinear effects and chromatic dispersion in the passband of an SWS gives rise to various phenomena.

region, dark solitons and power limiting. Dark solitons, consisting an intensity dip with hyperbolic tangent field distribution in an otherwise uniform continuous-wave (CW) field, propagate almost unaffected if their spectrum is well confined to the SWS bandpass but due to the scarce practical interest and the high sensitivity to the third-order dispersion [17] it is not treated in this chapter. Power limiting, instead, is an interesting phenomenon occurring when the signal interacts with the rightmost band edge and it is discussed in Section 10.3.2.

In the lower frequency range of the passband where dispersion is anomalous, the typical phenomenon that occurs is soliton formation and propagation, as discussed in Section 10.3.3. This dispersion regime can also induce unstable behaviors, like the well-known MI discussed in Section 10.7, bistability, and an interesting self-pulsing phenomenon, discussed in Section 10.5. Also in this case the list of considered effects is not exhaustive, but forms the basis for future studies in the field.

10.3.2 Power Limiting

In Section 10.3.1, it has been pointed out that if the SWS bandwidth is sufficiently large with respect to the frequency shift and to the signal spectral width, the nonlinear enhancement is simply proportional to $S^2(\omega)$. Instead, if the signal spectrum approaches the SWS band edge, it is somewhat filtered by the spectral response of the SWS, and new interesting effects can occur.

In case of a small perturbation around f_0, the frequency shift (Equation 10.13) can be calculated, neglecting the variation of S due to $\Delta\omega$. More precisely, the slowing factor S has to be calculated in $\omega + \Delta\omega$, and the solution to the implicit Equation 10.13 with S given by Equation 10.8 must be found numerically. To better understand the physical behavior in case of strong perturbations both relations are shown in Figure 10.7a where three values of input power P_{in} are considered. The solutions are indicated by marks and are stable. The figure refers to the 40-cavity long SWS considered in Section 10.2, and the CW input signal is tuned at the central frequency. By increasing the input power the spectral characteristic moves to the left and hence S increases (that is the slope of the line decreases) until the two curves become tangent. At this point the slowing factor is S_{lim} and the transmitted power is limited to a maximum value P_{lim} above which the excess input power $P_{in} - P_{lim}$ is reflected back even if the edge of the passband is not reached. From the figure it is evident that a second solution could also exist but it is difficult to reach and is not treated further.

Figure 10.7b shows the time response of the SWS to a slowly increasing input signal. If the input power is below P_{lim} the transmitted signal is simply a delayed replica of the input. By increasing the input power, the output power tends to saturate to P_{lim} and the signal in excess is reflected. The theoretical value (Equation 10.13) is in excellent agreement with the numerical simulation. In the

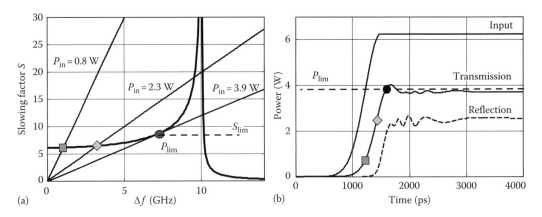

FIGURE 10.7 (a) Graphical representation of Equation 10.13 (the line with slope proportional to P_{in}^{-1}) and $S(\omega)$ given by Equation 10.8. The solutions indicated with marks are the induced frequency shifts. (b) Time response to a slowly varying increasing input signal.

saturated regime, the intensity inside the SWS is $S_{lim}P_{lim}$ and cannot increase further because a balance is established between the nonlinear induced frequency shift and the reflection due to the structure spectral characteristic. The small ripples appearing just before the saturations are due to the finite bandwidth of the used input signal and disappear in the real steady state solution.

The solution of Equation 10.13 also depends on the relative detuning between the input signal and the SWS central frequency f_0. A positive detuning of the signal reduces the saturation level and the SWS acts as a tunable power limiting device. Figure 10.8 shows the SWS time response to an adiabatic step input signal when its carrier frequency is shifted to higher frequencies by Δf. For the highest detuning, $\Delta f = 8$ GHz, very close to the band edge, the saturation P_{lim} decreases by one order of magnitude. Clearly, this behavior refers to the steady state condition, whereas in the pulsed regime the interaction of the pulse spectrum with the SWS band edge induces signal distortions and reflections that need to be calculated numerically.

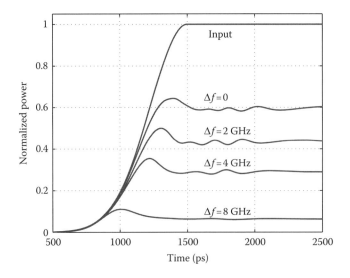

FIGURE 10.8 Time response to a slowly increasing input signal of an SWS as a function of the detuning Δf. The SWS bandwidth is 20 GHz. The output responses have been time overlapped for comparison.

10.3.3 SOLITON PROPAGATION

A rigorous and exhaustive study of the propagation of solitary waves in a band-limited nonlinear guiding structure is a rather difficult task. In SWSs, the problem is even more complicated by the selective spectral characteristic and by the strongly frequency-dependent enhancement of the optical characteristics. Here, few basic considerations are reported and discussed, leaving to future studies deeper investigations. As well-known, the soliton regime occurs when the SPM compensates for the phase modulation induced by the second-order dispersion [15]. The evolution of the field in SWSs is governed, as usual, by the nonlinear Schrödinger equation where the dispersive (Equation 10.17) and nonlinear terms (Equation 10.15) are strongly frequency-dependent through the factor S. It is clear that one must also take into account higher-order dispersive terms and the selective spectral response of the structure. Higher-order dispersion terms are rarely negligible in SWSs and can have a severe impact on the soliton properties that, more correctly, should be called pseudo soliton or solitary wave.

By equating the second-order dispersion length $L_d = T_0^2/|\beta_2|$ to the nonlinear length $L_{NL} = (\gamma_{eff} P_S)^{-1}$, the launch soliton peak power P_S outside the SWS is found,

$$P_S = \frac{|\beta_2|}{c_n S^2 \gamma T_0^2} = \frac{S(f_0) n_0}{c \pi B c_n \gamma T_0^2} \frac{S \tan(kd)}{\cos^2(kd)} \qquad (10.18)$$

where T_0 is the soliton pulsewidth (the soliton envelope being defined as $\text{sech}(t/T_0)$), and both S and β_2 must be calculated at the operating frequency through Equations 10.8 and 10.17, respectively, taking into account the SWS redshift induced by the nonlinearity. The power P_S obviously vanishes at the central frequency f_0 where $kd = M\pi$ and increases with the detuning, due to the stronger dependence of the dispersion term on γ_{eff}.

In Figure 10.9, linear and soliton propagations are compared. The considered SWS having FSR = 100 GHz and bandwidth $B = 20$ GHz, is apodized according to Ref. [18], and its slowing ratio at resonance is $S(f_0) = 3.23$ and its length is $15/n_0$ cm. The input pulse has a soliton shape with $T_0 = 60$ ps (that is a bandwidth of about 10 GHz) and is tuned 5 GHz below resonance ($f_0 - B/4$), where $\beta_2 = -440$ ps^2/mm and $\beta_3 = 26,400$ ps^3/mm. At this frequency, the second (T_0^2/β_2) and the third (T_0^3/β_3) order dispersion lengths are equal and both influence the pulse propagation. In the linear regime, the pulse broadens asymmetrically with a long tail on the rear edge, denoting a combined effect of β_2 and a positive β_3. The SWS is 100 resonators long, more than five times the dispersion length (equal to 19 resonators) and clearly the pulse undergoes a severe distortion even if its spectrum

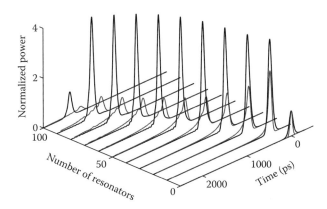

FIGURE 10.9 Evolution of a low-power Gaussian pulse tuned in $f_0 \pm B/4$ (thin line) compared with propagation of a solitary wave when the detuning from resonance is $-B/4$ (anomalous regime). In both cases, pulse power is normalized to the input power. See text for SWS characteristics.

is well confined within the SWS bandpass and the filtering effects are negligible. The pulse evolution in the soliton regime is reported in Figure 10.9 too. With the typical material parameters of AlGaAs waveguides, $n_2 = 2 \cdot 10^{-17}$ m^2/W and $A_{\text{eff}} = 10$ μm^2, the soliton input peak power, calculated using Equation 10.18, is $P_s = 0.96$ W only. It is evident that the pulse is now well preserved even after a long distance and in spite of the unbalanced third-order dispersion. Other familiar characteristics of fundamental solitons such as robustness and reshaping can be readily observed too.

The higher-order dispersion terms and the redshift of the spectral response, however, can limit these properties, especially when the pulse spectrum interacts with the band edges, giving rise to complex and new phenomena still to investigate and explain. As an example, by increasing the input intensity the soliton breaks in a dispersive fast wave and a new slow soliton.

Other higher-order nonlinear effects such as self-steepening and finite time response of the material nonlinearities are neglected in this preliminary investigation but should lead, as in optical fibers, to a soliton decay qualitatively similar to that produced by β_3 [15]. In optical fibers these effects become evident for pulse widths of few tenths of femtoseconds. In SWS, however, the strong third-order dispersion, the intensity enhancement and the cavity life time can play a determinant role with pulses tenths or even hundreds of picoseconds long. According to [15], soliton decay, for example, takes place if $\delta = \beta_3/(6|\beta_2|T_0) > 0.022$, whereas in the case of Figure 10.9 $\delta = 0.17$.

In conclusion, particular care should be taken in designing SWS exploiting the soliton propagation that, under some conditions, is well supported and offers a chance to reduce the detrimental effects of the structure-induced dispersion. As an example, fundamental and higher-order solitons seem to be good candidates to break the limit of the time delay–bandwidth product, considered an intrinsic limit of coupled resonator slow-wave delay lines.

10.4 CROSS-PHASE MODULATION

As well-known, material nonlinearity is the main vehicle for interaction between two optical waves at different frequencies through a phenomenon called XPM. XPM occurs when two waves superimpose, copropagate, or counterpropagate. Typically, an intense wave (pump) modifies the refractive index of the medium and the other wave (signal) experiences a phase shift depending on the intensity of the first one. Analogous to the SPM effect investigated in Section 10.3, XPM can also be treated as in classical uniform nonlinear media such as optical fibers, if the signal spectra involved in the process are well confined to the pass bands of the SWS. Clearly, both waves must be located inside the passband, in the same or in two different bands.

Pump and signal experience their own slowing factors $S(f_p)$ and $S(f_s)$, related to their relative frequency with respect to the passband of the structure. In Section 10.2 the enhancement of the nonlinear effect has been found to be proportional to the square power of S. One term S comes from the intracavity power enhancement, the second one comes from the enhancement of the effective length. In XPM, these two contributions are related to the pump and to the signal, respectively, and the phase shift induced by an intense pump on a signal is given by

$$\phi_{\text{XPM}} = \frac{2\pi}{\lambda} \frac{n_2}{A_{\text{eff}}} c_X P_p S(f_p + \Delta f_p) S(f_s + \Delta f_s) Nd \tag{10.19}$$

where P_p is the pump power, and the coefficient c_X depends on the structure of the resonator and on the relative polarization of the interacting waves according to Table 10.1. The pump is responsible for the frequency downshift of the SWS passband and experiences a slowing factor $S(f_p + \Delta f_p)$, where the self shift Δf_p is calculated with the implicit Equation 10.13 by using the pump power and the coefficient c_S. The meaning of the other term S is that the signal experiences an effective length $S(f_s + \Delta f_s)$ longer than the physical length, where Δf_s is obtained with Equation 10.13 by using the pump power and hence the coefficient c_X.

In the simplest hypothesis of small perturbation, the terms Δf are negligible with respect to B and if the pump and the signal are centered in f_0 the enhancement factor in Equation 10.19 reduces

Nonlinear Slow-Wave Structures

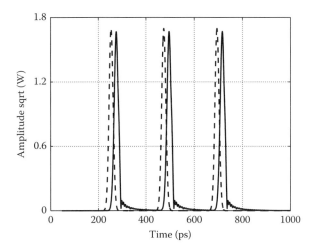

FIGURE 10.10 Amplitude of the pulsed pump sequence in input (dashed) and output (solid) signal.

to $S^2(f_0)$. Let us consider a 10-cavity long Fabry–Pèrot SWS with bandwidth $B = 120$ GHz and FSR $= 3$ THz, that is with $S(f_0) = 16$. The used material properties are $n_2 = 2 \cdot 10^{-17}$ m^2/W, $n_{\text{eff}} = 3.3$, and $A_{\text{eff}} = 10$ µm^2, with a cavity length $d = 15$ µm. A weak CW probe tuned at a resonance frequency f_0 copropagates with the intense pulsed pump signal shown in Figure 10.10. A sequence of three Gaussian pulses with $T_0 = 10$ ps, that is, a bandwidth of 60 GHz, and the peak power $P_{\text{in}} = 2.6$ W is tuned at another resonance $f_0 + M$FSR and propagates through the structure almost unaffected by the third-order dispersion ($\beta_3 L/T_0^3 = 0.18$). The field amplitude (to better appreciate the effect of the dispersion) of the output signal is reported in the figure. The SWS frequency detuning induced by these intense pulses is equal to 9.5 GHz and hence negligible with respect to B. The effective phase modulation induced by the pump on the weak probe is given by Equation 10.19 and is equal to $\pi/2$. In a straight uniform waveguide of the same length Nd, the induced phase shift would be $S^2 = 256$ times smaller, or a pump power 256 times higher would be required to achieve the same phase modulation. The input and output probe intensity and the induced phase temporal behavior are shown in Figure 10.11a and b, respectively. The phase shift is an excellent replica of the pulsed input sequence and the residual ripples on the probe intensity are very small. The slightly smaller value of the induced peak phase shift is due to the SWS apodization. The physical length of the whole device is 150 µm only and it can be used to induce all-optical phase modulation or can be inserted in the arms of an interferometer to convert the phase modulation to intensity modulation.

In the hypothesis of small perturbations, that is, $S(\omega)$ independent on the input power, other well-known effects related to XPM can be exploited by taking advantage of the slowdown enhancement. Simply, one have to keep in mind the origin of the two S factors, especially if the pump and the signal are differently detuned with respect to f_0. As an example, the XPM effect being dependent on the relative state of polarization of the interacting waves, the Kerr-induced birefringence is also enhanced by a factor S^2. More rigorously, one should consider the product $S(f_p)S(f_s)$, and if the induced frequency shifts are not negligible, they must be evaluated by using the proper slowing factor for each polarization. Well-known applications such as Kerr shutters, for example, can really be realized in a much more compact and efficient way.

In principle it is possible to switch a signal in and out of the passbands by means of another intense optical signal. Although intriguing, the possibility of exploiting the spectral selectivity of the SWS with an optical control needs a much deeper investigation. In general, by approaching the band edges, a nonlinear positive feedback can arise, inducing severe distortions of the passband shape, huge high-order dispersive effects, reflections, and so on. Numerical simulations are not easy to carry out as numerical and physical instabilities mix together and the real behavior is difficult to predict

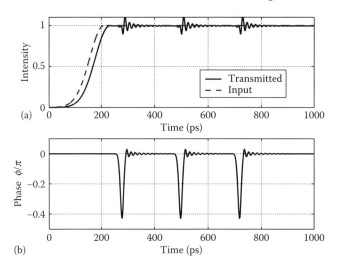

FIGURE 10.11 (a) Normalized intensity of the input (dashed) and output (solid) signal. (b) Nonlinear induced phase shift.

as investigated in Ref. [7] for short SWSs. Moreover, the influence of the disorder on the band edges can be detrimental and unpredictable [19]. In conclusion, near the band edges two or more waves interact according to more complex models, as discussed with some details in Sections 10.6 and 10.7.

10.5 NONLINEAR SPECTRAL RESPONSE

When the light interacts with the lower frequency edge of an SWS, a strong positive feedback is obtained and a considerable part of energy can be transmitted even if its spectrum seems to fall outside the linear passband of the SWS. The light intensity in the neighborhood of the edge is strongly enhanced and the edge downshifts toward lower frequencies increasing the bandwidth and opening the transmission to previously forbidden spectral harmonics. The spectral characteristic is, therefore, strongly distorted because of the frequency dependence of the slowing factor S.

Although it is not formally rigorous to define a transfer function in the nonlinear regime, it is worthwhile to calculate the input–output spectral characteristic as a function of the input power. In other words, one can numerically calculate the transmission of a CW input signal at various frequencies and for various intensities. To this aim, numerical techniques operating in the frequency domain are preferred when the structure is stable and time-domain techniques are preferred when instabilities or bistabilities are found, typically near the low frequency edge. A good survey of the available numerical methods are discussed in Ref. [7].

To better understand how nonlinearities modify the SWS spectral behavior, the three-cavity long structure studied in Ref. [7] is considered. Longer structures behave similarly but high-order instabilities and numerical difficulties hinder a straightforward comprehension of the phenomena. To this aim, we considered as input signal a CW at a given frequency and the complex amplitude transmission at every wavelength is collected after simulation has completed the transient. The linear transfer is reported in Figure 10.12, together with the spectral characteristic obtained with nonlinearities $n_2 I_{in} = 2.4 \cdot 10^{-6}$ and $n_2 I_{in} = 5.4 \cdot 10^{-6}$. Figure 10.12a and b show the intensity transfer function and the group delay. With an intense input signal the spectral characteristic shifts to lower frequencies, the bandwidth narrows, the rightmost edge is smoothened, whereas the low-frequency edge is distorted and sharp transitions with bistable behaviors and oscillations appear. The high-frequency group delay peak is redshifted and enhanced, while on the left the response is unstable. The group delay presents very high peaks, corresponding to strong slow downs and high intracavity

Nonlinear Slow-Wave Structures

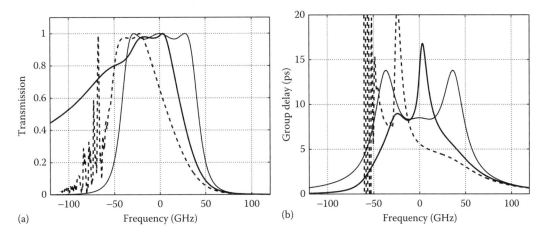

FIGURE 10.12 (a) Transmission and (b) group delay of a three cavity SWS for three values of nonlinearity: $n_2 I_{in} = 0$ (thin line), $n_2 I_{in} = 2.4 \times 10^{-6}$ (thick solid line), $n_2 I_{in} = 5.4 \times 10^{-6}$ (dashed line). At low frequencies the group delay has been truncated because it oscillates and diverges.

intensity enhancements. At these frequencies, the convergence of most of the available numerical techniques is quite slow or even difficult to reach, as well as the physical behavior.

It is worthwhile to insist on the meaning of the curves reported in Figure 10.12a and b. At a given frequency and a given intensity, the values in the graphs correspond to the transmission and the group delay experienced by the CW signal at that frequency. In case of two signals at two different frequencies, each one experiences a rigid translation induced by the other one plus its own perturbation. In case of small perturbation, the mechanism has been discussed in Section 10.4, whereas for strong perturbations near the edges numerical simulations are the only way (not always reliable) to predict the behavior of the structure.

The origin of the spectral oscillations observed in Figure 10.12 can be better understood by looking at the time-domain investigation reported in Figure 10.13. In such conditions a continuous

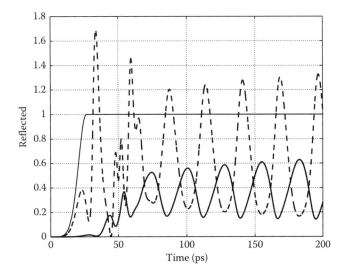

FIGURE 10.13 Time-domain observation of self-pulsing effects in a triple cavity SWS when a signal detuned to -75 GHz is applied. Input (thin solid line), transmitted (solid), reflected (dashed) for $n_2 I_{in} = 5.4 \times 10^{-6}$.

input signal tuned to 75 GHz below the passband center frequency breaks into a stable periodic pulse train. Simulations were performed at input power level $n_2 I_{\text{in}} = 5.4 \times 10^{-6}$ and both transmission and reflection at the SWS are shown. After an initial transient, the periodic pattern at the regime depends only on the frequency detuning of the input signal with respect to the SWS resonant frequency f_0 and on the nonlinearity $n_2 I_{\text{in}}$. Therefore, this phenomenon can be considered a self-pulsing effect in the SWS. In these conditions the group delay loses its physical meaning of transit time of an input signal across the device and this justifies the deep oscillations in the spectrum reported in Figure 10.12. In this example the pulse repetition rate is about 30 ps, the inverse of the frequency difference between the signal carrier frequency and the lower edge. More examples of unstable behaviors and oscillations are related to MI and described in Section 10.7.

10.6 FOUR WAVE MIXING

The benefit of optical resonators in frequency mixing phenomena has been known for several decades in nonlinear optics [20]. Many studies and experiments demonstrated that the efficiency of the nonlinear interaction, which is significantly strengthened by the intracavity resonant field-enhancement, can be improved by several order of magnitude [21–28]. Unfortunately, in single cavity architectures, the nonlinear interaction length is intrinsically related to the resonator spectral response and a severe constraint exists between the efficiency and the bandwidth of the wave mixing process.

More recently, the study of frequency mixing phenomena has been extended to structures including several coupled cavities, such as coupled resonator optical waveguides (CROW) [6,29], coupled defect photonic crystals [30], and SWSs [31]. As discussed in this section, the main advantage of coupled resonator SWSs over single resonators is the possibility to increase the efficiency of the wave mixing process, without narrowing the bandwidth [32]. For brevity's sake, the detailed discussion is restricted here to FWM processes only, occurring in materials exhibiting $\chi^{(3)}$ nonlinearity. More specifically, we consider the case of partially degenerate FWM [15], where the frequencies of two waves combine into one, so that only three distinct waves are involved. Nonetheless, the validity of the presented results can be extended straight to any number of waves, as discussed in this section. The simplified theory and results reported here have been derived under the following limiting assumptions: long structures, where the input and output matching sections represent a small part of the structure, small signals, undepleted pump, negligible nonlinear induced frequency shifts and spectral distortions of the SWS transfer function, negligible SPM and XPM contributions and narrow band signals with respect to the SWS bandwidth B, in order to reduce the dispersive effects [2].

The classical theory of FWM states that, given an intense pump and a weak signal tuned at frequencies f_p and f_s, respectively, an idler wave (hereinafter referred to as converted wave) is generated at a third frequency $f_c = 2f_p - f_s$, as shown in Figure 10.14a. In the simplified hypothesis of slowly varying envelope, undepleted pump, negligible SPM and XPM and lossless media, the spatial evolution along the z-direction of the converted wave P_c satisfies the following equation [15]

$$P_c(z) = \gamma^2 P_p^2 P_s z^2 \text{sin c}^2 \left(\frac{\Delta k z}{2} \right), \tag{10.20}$$

where

P_p is the pump power
P_s is signal power
$\gamma = \omega n_2 / c A_{\text{eff}}$ is the nonlinear propagation constant defined in Section 10.2

The maximum power transfer from the pump to the converted wave is limited by the phase mismatch $\Delta k = 2k_p - k_s - k_c$, k_j (j = s, c, p) being the wave vectors of the three involved waves.

Referring to Figure 10.14b, let us consider the pump and the signal waves each propagating at the central frequency of two different SWS pass bands, named $f_{0,p}$ and $f_{0,s}$, respectively. Thanks to

Nonlinear Slow-Wave Structures

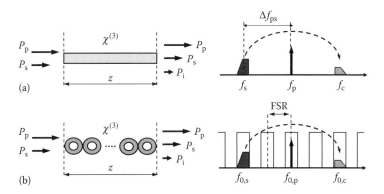

FIGURE 10.14 FWM process in (a) a straight optical waveguide and (b) an SWS. In the SWS case, the frequency of the pump, the signal and the idler waves coincides with the resonant frequencies of the structure.

the periodicity of the SWS frequency response, the converted wave is generated within a third SWS resonance $f_{0,c} = 2f_{0,p} - f_{0,s} = M\text{FSR}$, with M an integer. Inside the SWS, the powers of the three waves, all experiencing slow-wave propagation, are thus enhanced by the S factor as well as the effective interaction length (or equivalently the interaction time). A remarkable point is that, as far as waveguide and material dispersive effects can be neglected, the three waves propagate exactly at the same group velocity $S(f_{0,p}) = S(f_{0,s}) = S(f_{0,c})$ and no walk-off effects are observed.

The efficiency of the FWM process in an SWS can be estimated, without involving any heavy analytical implication, by including in Equation 10.20 the power and the effective length enhancement factors derived in the previous sections. The converted power P_c generated after N resonators of the SWS is simply given by

$$P_{c,\text{SWS}}(N) = \gamma^2 P_p^2 P_s N^2 d^2 \left(\frac{S^2+1}{2}\right)^2 \text{sinc}^2\left(\frac{\Delta\beta}{2} Nd\right), \tag{10.21}$$

where $\Delta\beta = 2\beta_p - \beta_s - \beta_c$ is the mismatch term of the phase inside the SWS. The gain provided by the SWS to the conversion process, defined as the ratio between the converted power with and without the SWS in the absence of phase mismatch ($\Delta\beta = 0$), $G_{\text{SWS}} = P_{c,\text{SWS}}/P_c = (S^2+1)^2/4$, increases proportionally to the fourth power of S. Note that Equation 10.21 gives the converted power outside the SWS, which takes into account the power reduction, by a factor S, experienced by the converted wave when exiting the SWS. Since FWM requires phase matching, the interaction between counter propagating waves carries negligible contribution. Therefore, different from the case of SPM and XPM, Fabry–Pèrot resonator, and ring resonator architectures show an identical conversion gain.

In Figure 10.15, both the conversion gain G_{SWS} and the normalized bandwidth B/FSR are reported versus the slowing factor $S(f_0)$ evaluated at the center of the SWS pass band. More than 26 dB gain is provided by a slowing factor as low as 6.5, achievable by means of a SWS with $B/\text{FSR} = 0.1$. Even higher gains can be reached by increasing the resonator finesse but with a reduction of the useful bandwidth. The main difference with respect to single cavity devices is that in SWSs, the FWM conversion efficiency $\eta_{\text{FWM}} = P_{c,\text{SWS}}/P_s$ scales as N^2 and can be increased by cascading additional cavities, with no bandwidth restrictions. The only drawback is that the chromatic dispersion induced by the SWS also increases with the structure length [2,33] and, when N approaches the SWS dispersion length, suitable compensating architectures are required to avoid pulse distortion [34].

The maximum number of cavities N_{max} of the SWS hosting the FWM process is limited by the intracavity phase mismatch $\Delta\beta$, setting the maximum converted power $P_{c,\text{max}}$. By differentiating Equation 10.5, it is simple to verify that, at the first order, the slowing factor S linearly increases the

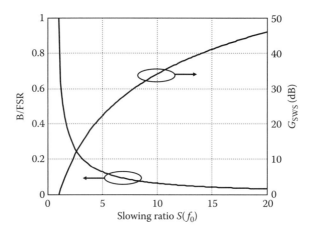

FIGURE 10.15 Normalized bandwidth and conversion gain of the SWS versus the slowing factor $S(f_0)$ at the center of the SWS passing band.

effective phase mismatch $\Delta\beta = S\Delta k$; this result can also be interpreted as a direct consequence of the effective length enhancement by the slow-wave propagation. The distance at which the power transfer starts reversing from the converted wave to the pump wave, conventionally referred to as coherence length, is reached after

$$N_{max} = \frac{\pi}{S|\Delta k|d}, \quad (10.22)$$

resonators. By simple analytical manipulations, Equation 10.22 can be conveniently rewritten in order to point out the role played, in the conversion process, by material parameters and SWS parameters. According to Equation 10.6 and in the hypothesis of highly reflective mirrors ($t \ll 1$), the relation between the cavity length d and the SWS bandwidth may be written as $d = ct/n_g\pi B$. Furthermore, because nonlinear materials are often highly dispersive, we can neglect the contributions by SPM and XPM to the phase mismatch and consider material dispersion only. In these conditions, the second second-order approximation $\Delta k = \hat{\beta}_2(2\pi\Delta f_{ps})^2$ generally holds, $\hat{\beta}_2$ being the second-order dispersion of the used waveguide at the pump frequency and $\Delta f_{ps} = f_p - f_s$. By substituting these relations into Equation 10.22 one obtains

$$N_{max} = \frac{n_g B}{4c|\hat{\beta}_2|\Delta f_{ps}^2}, \quad (10.23)$$

which relates N_{max} to the SWS bandwidth B, the maximum detuning frequency Δf_{ps}, and the material parameters n_g and $\hat{\beta}_2$. Given the material and the frequency detuning, Equations 10.22 and 10.23 state that the maximum length $N_{max}d$ is fixed by S, whereas the maximum number of cavities depends on the SWS bandwidth B only. Note that N_{max} does not take into account the dispersive effect on the pulse shape and the maximum length is limited also by Equation 10.17.

After N_{max} resonators, the converted power is simply

$$P_{c,max} = P_p^2 P_s \left(\frac{\gamma}{\Delta k}\right)^2 \left(\frac{S^2+1}{S}\right)^2, \quad (10.24)$$

where the term in the first bracket depends only on the material parameters, whereas the term in the second bracket is the contribution by the SWS only. Although the SWS gain G_{SWS} scales as S^4, the maximum converted power $P_{c,max}$ is found to be proportional to S^2 only. The explanation is that the enhancement of the effective length is balanced by the enhancement of the phase mismatch and, when

TABLE 10.2
Nonlinear and Dispersive Parameters of Common Optical Nonlinear Materials

Material	A_{eff} (μm²)	γ (m⁻¹W⁻¹)	$\hat{\beta}_2$ (ps²/km)	$n_2/\hat{\beta}_2^a$
SiO₂	80	0.0015	−25	1
TeO₂	10	0.55	45	27.8
Si	10	1.8	1160	3.4
Si-wire[b]	0.1	180	~0	∞
AlGaAs	10	7.8	1240	14
Chalcogenide[c]	10	1.19	500	4

[a] Normalized to silica value. $n_2 = 2.3 \cdot 10^{-20}$ m²/W.
[b] Zero dispersion Si-wire waveguides are reported in Ref. [36].
[c] Values refer to Ref. [35]. Other values can be found in literature.

S increases, the SWS must be shortened (see Equation 10.22) to avoid backward conversion. From the material side, $P_{c,max}$ increases with the square of the ratio $n_2/\hat{\beta}_2$ between the nonlinear and the dispersive coefficients. In Table 10.2, the optical parameters of some optical materials conventionally used in nonlinear applications are reported. Among these materials, the Si-wire waveguide seems to be the best candidate but the two-photon absorbtion can dramatically reduce the conversion efficiency. Also AlGaAs and chalcogenide could be excellent but once the waveguide cross section is engineered to reduce as much as possible the chromatic dispersion. In the table, the highest $P_{c,max}$ can be achieved by means of a TeO₂ SWS, which outperforms AlGaAs and Si structures by 4 times and 67 times, respectively. However, one should take into account that, in order to reach $P_{c,max}$, the required number of cavities N_{max} in a TeO₂ SWS is as large as 25 times that of a semiconductor SWS, with significant implications in the technological realization and the size of the device.

Figure 10.16 summarizes the theoretical performance of an FWM SWS frequency converter made of N_{max} resonators versus the SWS finesse and the frequency spacing $2\Delta f_{ps}$ between the signal

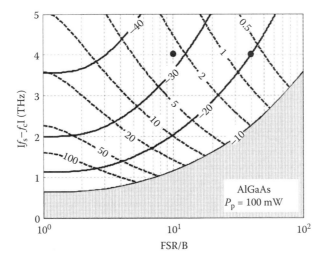

FIGURE 10.16 Performance of a SWS frequency converter based on FWM evaluated after N_{max} resonators: (solid lines) FWM efficiency η_{FWM} [dB], (dashed lines) SWS length in millimeters. The two marks correspond to the example discussed in the text. AlGaAs material parameters are considered.

and the converted wave. In the shadowed area the approximation of undepleted pump and small signals does not hold and results should be calculated numerically. The pump power is fixed at 100 mW, and the AlGaAs waveguide parameters of Table 10.2 are considered (n_{eff} = 3.3). Solid lines show, in dB scale, the FWM efficiency η_{FWM} while the corresponding length of the SWS structure, in mm scale, is reported in dashed lines. Let us consider, for example, frequency conversion over 4 THz. By using an SWS with finesse 10, corresponding to $S(f_0) = 6.5$, the conversion efficiency is about -32dB and the overall length is 2.5 mm. If the slowing ratio S is increased up to 25 (FSR/B = 40), the conversion efficiency increases by about 12 dB and the SWS length reduces to about 0.6 mm. By maintaining $S(f_0) = 6.5$, the same gain is obtained if the waveguide dispersion halves but a structure with double length is required.

Although the maximum number of cavities is limited by the phase mismatch, the converted power can be increased by cascading more than one SWSs interleaved by a dispersive optical device, according to conventional quasi-phase-matched (QPM) schemes [37]. For example, the linear propagation through a medium with second-order dispersion coefficient $\tilde{\beta}_2$ opposite in sign with respect to $\hat{\beta}_2$ can be exploited to compensate for the phase mismatch induced by the SWS. The length of the rephasing device has to be simply $\tilde{L} = Snd\hat{\beta}_2/\tilde{\beta}_2$, independent of the frequency conversion range Δf_{ps}. It means that the device can operate on a WDM channel grid simultaneously.

Figure 10.17 shows in solid lines the performance of a QPM SWS wavelength converter operating over 4 THz (Δf_{ps} = 2 THz) for three different nonlinear materials, namely AlGaAs, Si, and SiO$_2$. For every material, bandwidth B = 20 GHz and FSR = 200 GHz are considered and the number of cavities is equal to N_{max}. A 100 mW pump power and 10 mW signal power are considered. Dashed lines show that, in the absence of rephasing at the end of the first stage, the power transfer reverses from the converted wave to the pump wave after N_{max} resonators. Rephasing elements are required for every 11 resonators for AlGaAs, 12 for Si, and 237 for SiO$_2$. According to FWM theory,

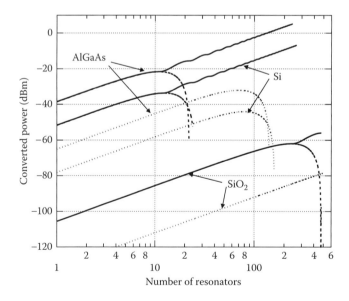

FIGURE 10.17 Solid lines show the performance of a FWM-based SWS wavelength converter (B = 20 GHz, FSR = 200 GHz) operating over 4 THz for three different nonlinear materials AlGaAs, Si, and SiO$_2$. A 100 mW pump power and 10 mW signal power are used. Dashed lines indicate the inversion of the power transfer from the converted wave to the pump wave in the absence of rephasing at the end of the first stage. Rephasing elements are required for every 11 resonators for AlGaAs, 12 for Si, and 237 for SiO$_2$. Dotted lines represent the converted power in the absence of resonators, 26 dB below the SWS converted power.

by doubling the length on the device, and hence the number of the cavities, the output converted power increases by 6 dB. Dotted lines represent the converted power in the absence of resonators, that is by using a conventional straight waveguides, 26 dB below the SWS lines. Note that the phase mismatch enhancement due to slow-wave propagation reduces the coherence length by nearly one order of magnitude ($S = 6.4$) according to Equation 10.22. At the same time, however, the maximum converted power $P_{c,max}$ at the end of the first SWS stage (before rephasing) is more than 10 dB higher than the maximum power converted without SWS. According to Equation 10.24, this is the contribution given by the intracavity power enhancement only.

For brevity's sake, in this section the discussion of wave mixing in SWSs has been restricted to FWM processes only. The nonlinear interaction of a different number of waves in slow-light regime has been also investigated recently and the enhancement factors associated to the group velocity reduction have been evaluated. A remarkable example is provided by second-harmonic generation (SHG) in CROW realized with χ_2 materials. The theory of SHG in CROW is exhaustively discussed in a number of recent contributions, in either CW [6] or pulsed regime [29]. The SHG efficiency is found to be proportional to $v_\omega^{-2} v_{2\omega}^{-1}$, v_ω, and $v_{2\omega}$ being the group velocity of the fundamental wave and the second harmonic wave inside the CROW, respectively. If we assume the same slowing factor S for the group velocity of the two waves propagating along the CROW, the enhancement of the SHG efficiency is found to be proportional to S^3.

More generally, slow-wave propagation enhances the efficiency of frequency mixing processes by S^n, where n is the number of the optical fields involved in the mixing process. For any arbitrary n, a contribution equal to S^2 is always provided by the enhancement of the effective length: in fact, the amplitude of the converted wave, generated by a coherent process, increases linearly in space and its power grows with the square of the SWS effective length. The additional term S^{n-2} is due to the intracavity power enhancement experienced by all the interacting waves. Note that the exponent $n-2$ is simply given by the number of input waves (three in FWM, two in SHG), acting as source fields of the mixing process, reduced by one, in order to take into account the power reduction experienced by the generated field when exiting the SWS.

10.7 MODULATION INSTABILITY

As it is well-known in nonlinear optics, instability effects may arise in any optical systems under the combined effect of nonlinearities and dispersion. It is reasonable to expect that, at least under some conditions, such phenomena can be favored by slow-wave propagation, which are always associated with the enhancement of both nonlinear and dispersive effects. The formal derivation of the equations describing nonlinear propagation in SWS in the presence of instability, is rather involved. Here, a simplified approach is conveniently followed, pointing out the role played by slow-wave propagation with no need for heavy mathematical implications. The idea is to exploit the classical interpretation of MI in terms of an FWM process phase-matched by the nonlinear phase shift. Then, by using the main results derived in Sections 10.2 and 10.6, the classical theory of MI is extended straight to SWSs.

In a lossless nonlinear optical system, the linearized Schrödinger equation

$$j\frac{\partial a}{\partial z} = \frac{1}{2}\beta_2 \frac{\partial^2 a}{\partial T^2} - \gamma P_0(a + a^*), \qquad (10.25)$$

is typically used to describe the propagation of a weak perturbation superposed to a CW. In Equation 10.25, $a(z, T)$ is the complex field amplitude of the perturbation, β_2 is the second-order dispersion coefficient of the guiding structure (waveguide or SWS), γ is the nonlinear constant defined in Section 10.2, and P_0 is the power of the stationary portion of the wave.

After transformation from time domain to frequency domain, Equation 10.25 can be rewritten in the form of the coupled equation system

$$\frac{\partial A(z,\Omega)}{\partial z} = j\left(\frac{1}{2}\beta_2\Omega^2 + \gamma P_0\right)A + j\gamma P_0 A^* = j\kappa_{11}A + j\kappa_{12}A^*, \quad (10.26)$$

$$\frac{\partial A^*(z,-\Omega)}{\partial z} = -j\gamma P_0 A - j\left(\frac{1}{2}\beta_2\Omega^2 + \gamma P_0\right)A^* = -j\kappa_{21}A - j\kappa_{22}A^*, \quad (10.27)$$

where $A(z,\Omega)$ is the perturbation complex amplitude in the Fourier domain and Ω is the modulation frequency of the perturbation. Since in the spectral domain, the conjugate wave $a^*(z,T)$ is mapped to $A^*(z,-\Omega)$, where two waves may exist at symmetrical positions with respect to the zero frequency, that is at the two sides of the stationary wave P_0. This is exactly what happens in a FWM process. In the coupled Equations 10.26 and 10.27, the coupling coefficients responsible for the power transfer via FWM are κ_{ij} ($i \neq j$). The coefficients κ_{ii}, including the contributions of chromatic dispersion and nonlinear phase modulation, are involved in the phase matching condition. To clarify the discussion reported in the following part of this section, it is important to point out that the term γP_0 in κ_{ii} of Equation 10.25 comes from the difference between the XPM experienced by every weak wave, proportional to $2\gamma P_0$, and the SPM of the stationary wave, proportional to γP_0. According to the coupled mode theory (CMT), the general solution of Equations 10.26 and 10.27 is

$$A(z) = A_s \exp(j\beta_s z) + A_a \exp(j\beta_a z), \quad (10.28)$$

with the constant coefficients A_s and A_a depending on the boundary conditions and

$$\beta_{s,a} = \frac{\kappa_{11} + \kappa_{22}}{2} \pm \left[\frac{(\kappa_{11} - \kappa_{22})^2}{4} + \kappa_{12}\kappa_{21}\right]^{\frac{1}{2}} = \pm \frac{1}{2}\Omega\left[\beta_2\left(\beta_2\Omega^2 + 4\gamma P_0\right)\right]^{\frac{1}{2}} \quad (10.29)$$

the eigenvalues of Equations 10.26 and 10.27. MI occurs only if either β_s or β_a become imaginary and this can take place only in anomalous dispersion ($\beta_2 < 0$) regime. The modulation process spreads over a continuous spectral range $\Omega^2 < 4\gamma P_0/\beta_2$ where the MI gain, defined as $G_{MI} = 2\text{Im}(\beta_{s,a})$ is real. The MI gain is maximum at the two frequencies $\Omega_{\max} = \pm(2\gamma P_0/\beta_2)^{1/2}$ satisfying the phase matching condition $\kappa_{11} - \kappa_{22} = 0$.

In order to describe the dynamic nature of MI phenomena in SWSs, we simply need to modify the coupled mode Equations 10.26 and 10.27, by including the effects associated to slow-wave propagation. The first point to be considered concerns the chromatic dispersion. As discussed in Section 10.1, within each pass band, an SWS exhibits anomalous dispersion only in the lower frequency side of each pass band. Therefore, instability phenomena are expected to occur only at frequencies below the resonant frequency. We can anticipate that this is what actually happens in SWSs, but some more remarks are mandatory to justify this statement. In fact, the second-order approximation is not generally sufficient to model the SWS dispersion and, unless extremely narrow bandwidths are considered, higher-order terms must necessarily be included. The extension of the nonlinear Schrödinger equation 10.25 to the third-order dispersion, which is the dominant factor in the neighborhood of the SWS resonant frequency, is used to demonstrate that β_3 does not contribute to the MI [38]. Actually, it can be shown that in the case of MI triggered by a CW, all the odd dispersion terms play no significant role in defining the MI gain [39–41]. On the contrary, higher-order even dispersion terms can significantly affect instability phenomena. For instance, they can lead to MI effects in systems exhibiting a normal dispersion regime $\beta_2 > 0$ and can also cause the splitting of the MI gain bandwidth in several sidebands [39]. By including all the even dispersion coefficients, Equation 10.29 simply becomes

$$\beta_{s,a} = \left[\sum_m \frac{\beta_{2m}}{(2m)}\Omega^{2m}\left(\sum_m \frac{\beta_{2m}}{(2m)!}\Omega^{2m} + 2\gamma P_0\right)\right]^{\frac{1}{2}}. \quad (10.30)$$

The main result contained in Equation 10.30 is that at least one coefficient β_{2m} must be negative in order to have instability effects. For a SWS made of identical directly coupled resonators, it can be proved that, in the normal dispersion regime, all the even coefficients β_{2m} are positive. Therefore instability effects can never arise at frequency higher than resonance frequency and we are allowed to restrict our discussion to the anomalous frequency range comprised between f_0 and $f_0 - B/2$, as schematically shown in Figure 10.18.

Beside the effects of higher-order dispersion, a fundamental role in the phase matching condition is played by the strong enhancement of the nonlinear phase shift given by the slow-wave propagation, discussed in Section 10.2. To model this contribution, the coefficient κ_{11} and κ_{22} of Equations 10.26 and 10.27 must include both the XPM phase shift experienced by the weak waves and the SPM phase shift ϕ_{SPM} of the CW wave. The latter is simply given by $\phi_{SPM} = c_S S_0^2 \gamma P_0$, where S_0 is the slowing ratio at the frequency f_{CW} of the CW wave and the coefficient c_S, reported in Table 10.1, depends on the cavity structure (ring or Fabry–Pèrot). The accurate evaluation of the XPM experienced by the weak waves requires some further considerations. In fact, as sketched in Figure 10.18, the CW wave P_0 and the weak waves $A(\Omega)$ and $A^*(-\Omega)$ propagate at different frequencies and experience different slowing ratios. When $A(+\Omega)$ is concerned, the CW power enhancement S_0 and the effective length enhancement $S_{+\Omega}$ have to be considered. Similarly, the enhancement factors for the XPM phase shift of $A^*(-\Omega)$ are S_0 and $S_{-\Omega}$, respectively. In both cases, the coefficient $c_X = 1$ (ring resonator) or 2 (Fabry–Pèrot) has to be used.

By putting together the dispersion and the nonlinear phase shift terms evaluated inside an SWS, it is possible to write the expression for the coefficients κ_{11} and κ_{22} of Equations 10.26 and 10.27 in the case of MI, as

$$\kappa_{11,22} = \sum_m \frac{\beta_{2m}}{(2m)}\Omega^{2m} + \gamma P_0 \left[c_X S_0 S_{\pm\Omega} - c_S S_0^2\right]. \quad (10.31)$$

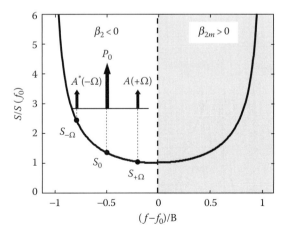

FIGURE 10.18 In the left-side branch ($f > f_0$) of the SWS band, anomalous dispersion ($\beta_2 < 0$) can give rise to MI effects. At frequencies higher than resonance frequency f_0, all the even dispersion coefficients are positive and MI never occurs.

Finally, the coefficients κ_{12} and κ_{21} of Equations 10.26 and 10.27, which determine the power transfer from the CW wave to the weak waves via FWM, can be derived straight away by using the results of Section 10.6. These coefficients are directly related to the FWM gain, that, in the formalism of MI, is defined as $G_{MI} = |A|^2/P_0$. In the hypothesis of high slowing ratio ($S^2 \gg 1$) and considering the proper slowing factor for the two weak waves, it is easy to verify that

$$\kappa_{12,21} = \frac{\gamma P_0}{2} S_0 \sqrt{S_{\pm\Omega} S_{\mp\Omega}}. \tag{10.32}$$

Note that in Equation 10.32 the amplitude of $A(\Omega)$, governed by κ_{12}, depends on the slowing factor of all the three waves. The term S_0 gives the enhancement of the CW power P_0 and $S_{\pm\Omega}^{-1/2}$ is the field enhancement factor of the weak waves. Equation 10.32 takes into account that, when escaping the SWS, the power of the converted waves decreases by $S_{\pm\Omega}$, therefore cancelling the contribution by effective length enhancement.

As a result, the modified coupled equations describing MI in SWSs can be written as

$$\frac{\partial A}{\partial z} = j\left(\sum_m \frac{\beta_{2m}}{(2m)!}\Omega^{2m} + \gamma P_0 \left[c_X S_0 S_{+\Omega} - c_S S_0^2\right]\right) A + j\frac{\gamma P_0}{2} S_0 (S_{+\Omega} S_{-\Omega})^{\frac{1}{2}} A^*, \tag{10.33}$$

$$\frac{\partial A^*}{\partial z} = -j\frac{\gamma P_0}{2} S_0 \sqrt{S_{-\Omega} S_{+\Omega}} A - j\left(\sum_m \frac{\beta_{2m}}{(2m)!}\Omega^{2m} + \gamma P_0 \left[c_X S_0 S_{-\Omega} - c_S S_0^2\right]\right) A^*. \tag{10.34}$$

From Equations 10.33 and 10.34, the MI gain G_{MI} can be evaluated numerically at every frequency. Figure 10.19 shows in solid lines the spectrum of G_{MI} for an SWS made of 40 Fabry–Pèrot resonators with $B = 20$ GHz and FSR $= 200$ GHz. The slowing ratio $S(\omega)$ of the structure is reported in dashed lines. Three values of increasing CW power P_0 are considered, giving $\gamma P_0 = 1$, 2, and 4 m^{-1}, respectively. The carrier frequency of the CW wave is tuned to 1.7 GHz below the resonant frequency f_0 of the SWS. As in the case of conventional MI, the peak gain is found to increase with the incident power P_0. Yet, different from conventional MI, for sufficiently high input power, the MI bandwidth in SWSs splits in two separated side bands, with G_{MI} vanishing in the neighborhood

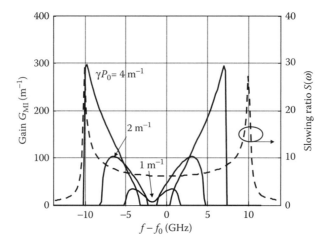

FIGURE 10.19 Gain spectra (solid lines) of MI in a SWS for three power levels γP_0. The SWS, whose slowing ratio is reported in dashed lines, has $B = 20$ GHz, FSR $= 200$ GHz and includes 40 Fabry–Pèrot resonators. The frequency of the CW wave is 2 GHz below the SWS resonance frequency f_0.

of f_0. When the modulation band approaches the leftmost edge of the SWS band, G_{MI} steeply increases under the effect of the slowing ratio. Since propagation is forbidden beyond the SWS edge, for a sufficiently high input power, G_{MI} becomes maximum at the SWS band edge. In these conditions, the optical field evolves with an exponentially growing modulation at a frequency f_{MI} given by the difference between the CW frequency f_{CW} and the SWS edge frequency f_e. Actually, if the input power P_0 is high, the evaluation of f_{MI} is slightly more involved. In fact, the SWS spectrum is redshifted by SPM and the modulation frequency f_{MI} becomes

$$f_{MI} = f_{CW} - f_e + \Delta f_{SPM}, \quad (10.35)$$

with the frequency shift given by Equation 10.13. Once the MI gain is known, the power at the output of the SWS can be simply derived as $P_{out}(\Omega) = P_{in}(\Omega) \exp(G_{MI}(\Omega)Nd)$, where P_{in} is the input power, N and d are the number and the length of the cavities, respectively.

In Figure 10.20a, a time-domain simulation is reported, showing the result of nonlinear propagation in the SWS of Figure 10.19. As input signal (thin solid line), a white Gaussian noise, with power $\sigma^2 = 10^{-4} P_0$, superposed to a stationary wave, with power $P_0 = 0.5$ W ($\gamma P_0 = 4$ m^{-1}), is assumed. The input signal propagates in the anomalous regime of the SWS spectrum, at a frequency $f_{CW} = f_0$-1.7 GHz. At the SWS output (thick solid line), the CW pump wave is partially broken into a periodic pulse train with a modulation period of 110 ps. Figure 10.20b shows in the spectral domain the effects of MI. While the spectrum of the input signal (thin solid line) exhibits a constant noise level, 40 dB below the CW line, two side bands originating from MI are clearly visible in the output spectrum, with peak power only 13 dB below the CW level. Both the simulated output spectrum (thick solid line) and the spectrum predicted by the theoretical model (Equations 10.33 and 10.34) (thick dashed line) are reported and appear in agreement within the entire SWS bandwidth. Simulations confirm the redshift Δf_{SPM} of the output spectrum with respect to the spectrum of the SWS in the linear regime (extending from -10 GHz to 10 GHz). The frequency at which the output power is maximum, $f_{MI} = 9.1$ GHz, is only 0.2 GHz higher than the value given by Equation 10.35, predicting a modulation frequency of $f_{MI} = 8.9$ GHz, with $f_{CW} - f_e = 8.3$ GHz and $\Delta f_{SPM} = 0.6$ GHz.

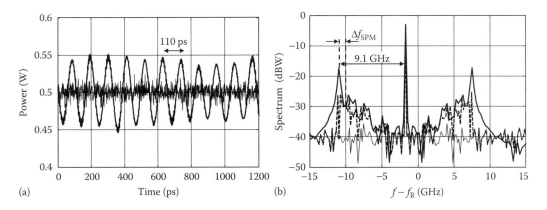

FIGURE 10.20 (a) Time-domain simulation of MI in a SWS. An optical signal composed by a white Gaussian noise ($\sigma = 0.01$) superposed to a CW wave ($P_0 = 0.5$ W) is provided at the input of the SWS of Figure 10.19 ($B = 20$ GHz, FSR $= 200$ GHz, 40 Fabry–Pèrot resonators). The output signal (thick line) exhibits an almost sinusoidal pattern with 110 ps period. (b) Spectral domain. Comparison between simulated (thick solid line) and the theoretical spectra of the output signal. The oscillation frequency f_{MI} is given by the relative distance between the CW carrier and the SWS edge. The thin solid line shows the spectrum of the input signal.

REFERENCES

1. T. Tamir, *Guided Wave Optoelectronics*, 2nd edn., Springer-Verlag, Berlin, 1990.
2. A. Melloni, F. Morichetti, and M. Martinelli, Linear and nonlinear pulse propagation in coupled resonator slow-wave optical structures, *Opt. Quantum Electron.* 35(4/5), 365–379, 2003.
3. J. E. Heebner, R. W. Boyd, and Q.-H. Park, SCISSOR solitons and other novel propagation effects in microresonator modified waveguides, *JOSA B* 19, 722–731, 2002.
4. Y. Chen and S. Blair, Nonlinearity enhancement in finite coupled-resonator slow-light waveguide, *Opt. Express* 12(15), 3353–3366, 2004.
5. M. Soljacic, S. G. Johnson, S. Fan, M. Ibanescu, E. Ippen, and J. D. Joannopoulos, Photonic-crystal slow-light enhancement of nonlinear phase sensitivity, *J. Opt. Soc. Am. B* 19(9), 2052–2059, 2002.
6. Y. Xu, R. K. Lee, and A. Yariv, Propagation and second-harmonic generation of electromagnetic waves in a coupled-resonator optical waveguide, *JOSA B* 17(3), 387–400, 2000.
7. F. Morichetti, A. Melloni, J. Čáp, J. Petráček, P. Bienstman, G. Priem, B. Maes, M. Lauritano, and G. Bellanca, Self-phase modulation in slow-wave structures: A comparative numerical analysis, *Opt. Quantum Electron.* 38, 761–780, 2006.
8. A. Melloni, M. Floridi, F. Morichetti, and M. Martinelli, Equivalent circuit of Bragg gratings and its application to Fabry–Pèrot cavities, *JOSA A*, 20(2), 273–281, 2003.
9. R. Costa, A. Melloni, and M. Martinelli, Bandpass resonant filters in photonic-crystal waveguides, *IEEE Photon. Technol. Lett.* 15(3), 401–403, 2003.
10. M. Ghulinyan, C. J. Oton, G. Bonetti, Z. Gaburro, and L. Pavesi, Free-standing porous silicon single and multiple cavities, *J. Appl. Phys.* 93, 9724–9729, 2003.
11. S. Mookherjea and A. Yariv, Coupled resonator optical waveguides, *IEEE J. Sel. Top. Quantum Electron.* 8, 448–456, 2002.
12. S. Radic, N. George, and G. P. Agrawal, Theory of low-threshold optical switching in nonlinear phase-shifted periodic structures, *J. Opt. Soc. Am. B* 12(4), 671, 1995.
13. A. Melloni, M. Chinello, and M. Martinelli, All-optical switching in phase-shifted fiber Bragg grating, *IEEE Photon. Technol. Lett.* 12(1), 42–44, 2000.
14. C. M. de Sterke and J. E. Sipe, Gap solitons, in E. Wolf (Ed.) *Progress in Optics*, Vol. 33, pp. 203–260, Elsevier, Amsterdam, 1994.
15. G. P. Agrawal, *Nonlinear Fiber Optics*, Academic Press, New York, 1999.
16. J. E. Heebner and R. W. Boyd, Enhanced all-optical switching by use of a nonlinear fiber ring resonator, *Opt. Lett.* 24(12), 847–849, 1999.
17. V. V. Afanasjev, Y. S. Kivshar, and C. R. Menyuk, Effect of third-order dispersion on dark solitons, *Opt. Lett.* 21(24), 1975–1977, 1996.
18. A. Melloni and M. Martinelli, Synthesis of direct-coupled resonators bandpass filters for WDM systems, *J. Lightwave Technol.* 20(2), 296–303, 2002.
19. S. Mookherjea and A. Oh, Effect of disorder on slow light velocity in optical slow-wave structures, *Opt. Lett.* 32(3), 289–291, 2007.
20. A. Ashkin, G. Boyd, and J. Dziedzic, Resonant optical second harmonic generation and mixing, *IEEE J. Quantum Electron.* 2(6), 109–124, 1966.
21. P. Bayvel and I. P. Giles, Frequency generation by four wave mixing in all-fibre single-mode ring resonator, *Electron. Lett.* 25(17), 1178–1180, 1989.
22. J. G. Provost and R. Frey, Cavity enhanced highly nondegenerate four-wave mixing in GaAlAs semiconductor lasers, *Appl. Phys. Lett.* 55(6), 519–521, 1989.
23. R. Lodenkamper, M. M. Fejer, and J. S. Jr. Harris, Surface emitting second harmonic generation in vertical resonator, *Electron. Lett.* 27(20), 1882–1884, 1991.
24. S. Murata, A. Tomita, J. Shimizu, M. Kitamura, and A. Suzuki, Observation of highly nondegenerate four-wave mixing (>1 THz) in an InGaAsP multiple quantum well laser, *Appl. Phys. Lett.* 58, 1458–1460, 1991.
25. S. Jiang and M. Dagenais, Observation of nearly degenerate and cavity enhanced highly nondenerate four-wave mixing in semiconductor lasers, *Appl. Phys. Lett.* 62(22), 2757–2759, 1993.
26. J. A. Hudgings and Y. Lau, Step-tunable all-optical wavelength conversion using cavity enhanced four-wave mixing, *IEEE J. Quantum Electron.* 34(8), 1349–1355, 1998.
27. P. P. Absil, J. H. Hryniewicz, B. E. Little, P. S. Cho, R. A. Wilson, L. G. Joneckis, and P.-T. Ho, Wavelength conversion in GaAs micro-ring resonators, *Opt. Lett.*, 25, 554–556, 2000.

28. M. Fujii, C. Koos, C. Poulton, J. Leuthold, and W. Freude, Nonlinear FDTD analysis and experimental verification of four-wave mixing in InGaAsP–InP racetrack microresonators, *IEEE Photon. Technol. Lett.* 18(2), 361–363, 2006.
29. S. Mookherjea and A. Yariv, Second-harmonic generation with pulses in a coupled-resonator optical waveguide, *Phys. Rev. E* 65, 026607, 2002.
30. S. Blair, Enhanced four-wave mixing via photonic bandgap coupled defect resonances, *Opt. Express* 13(10), 3868–3876, 2005.
31. A. Melloni, F. Morichetti, S. Pietralunga, and M. Martinelli, Slow-wave wavelength converter, *Proceedings of the 11th European Conference on Integrated Optics*, Prague, Vol. 1, pp. 97–100, 2003.
32. A. Melloni, F. Morichetti, and M. Martinelli, Optical slow wave structures, *Opt. Photon. News* 14(11), 44–48, 2003.
33. J. B. Khurgin, Dispersion and loss limitations on the performance of optical delay lines based on coupled resonant structures, *Opt. Lett.* 32(2), 133–135, 2006.
34. J. B. Khurgin, Expanding the bandwidth of slow-light photonic devices based on coupled resonators, *Opt. Lett.* 30(5), 513–515, 2005.
35. V. G. Ta'eed et al., Self-phase modulation-based integrated optical regeneration in chalcogenide waveguides, *IEEE J. Sel. Top. Quantum Electron.* 12(3), 360–370, 2006.
36. A. C. Turner, C. Manolatou, B. S. Schmidt, M. Lipson, M. A. Foster, J. E. Sharping, and A. L. Gaeta, Tailored anomalous group-velocity dispersed in silicon channel waveguides, *Optics Express*, 14(10), 4357–4362, 2006.
37. J. Kim, Ö. Boyraz, J. H. Lim, and M. N. Islam, Gain enhancement in cascaded fiber parametric amplifier with quasi-phase matching: Theory and experiment, *J. Lightwave Technol.* 19(2), 247–251, 2001.
38. M. J. Potasek, Modulation instability in an extended nonlinear Schroedinger equation, *Opt. Lett.* 12(11), 921–923, 1987.
39. W. H. Reeves, D. V. Skryabin, F. Biancalana, J. C. Knight, P. St. J. Russell, F. G. Omenetto, A. Efimov, and A. J. Taylor, Transformation and control of ultra-short pulses in dispersion-engineered photonic crystal fibres, *Nature* 424, 511–515, 2003.
40. M. Nakazawa, K. Suzuki, H. Kubota, and H. A. Haus, High-order solitons and the modulational instability, *Phys. Rev. A* 39, 5768–5776, 1989.
41. A. Demircan, M. Pietrzyk, and U. Bandelow, Effect of higher-order dispersion on modulation instability, soliton propagation and pulse splitting, WIAS Preprint No. 1249, 2007.

11 Slow Light Gap Solitons

*Joe T. Mok, Ian C.M. Littler, Morten Ibsen,
C. Martijn de Sterke, and Benjamin J. Eggleton*

CONTENTS

11.1 Introduction 223
11.2 Background 224
    11.2.1 Linear Properties of Gratings 224
    11.2.2 Nonlinear Properties of Gratings 227
11.3 Experiment 229
11.4 Discussion and Conclusions 233
Acknowledgments 233
References 233

11.1 INTRODUCTION

One of the key issues when dealing with slow light is the inevitable effect of dispersion, the result of which is to broaden a pulse upon propagation. Moreover, the dispersion tends to increase when the group velocity, the speed at which the pulse propagates, decreases. To see this, we note that if we write the group velocity $v_g \equiv c/n_g$, where n_g is the group index, then we find that $n_g = n + \omega \, dn/d\omega$. Since for the materials we are interested in, the refractive index does not vary strongly in magnitude, and slow light is obtained when the derivative $dn/d\omega$ is large and positive. However, as the variations in the magnitude of n itself are limited, the large derivative can only be maintained over a narrow bandwidth $\propto (dn/d\omega)^{-1}$. Therefore, a large delay implies a narrow bandwidth, and hence a limited pulse width. Here we discuss how to avoid this limit by using nonlinear effects in the form of gap solitons. Though this argument is quite simple, it can be made rigorous using the Kramers–Kronig relations. In this way it can be shown that the achievable delay in a linear, two-port device is subject to a limitation of the product of the delay and the bandwidth (Lenz et al., 2001).

Gap solitons are solitons which occur in periodic structures with a Kerr nonlinearity. As in all solitons, the nonlinearity counteracts the effect of dispersion, so that they propagate without broadening. They differ, however, from the more common optical solitons that occur in optical fibers or waveguides (Agrawal, 1995) in that the dispersion originates mainly from the periodic structure, rather from the constituent materials or the waveguide. Because they derive from the grating dispersion, gap solitons have a number of unique properties: (1) they can travel arbitrarily slowly, in principle; (2) since the grating dispersion is much larger than the material dispersion and waveguide dispersion, the intensity required to generate a gap solitons is commensurately higher. The first of these properties indicates why gap solitons are well suited for slow light experiments: not only can they travel arbitrarily slowly, at least in principle, but also because they are solitons that do not undergo dispersive broadening upon propagation. Dispersive broadening is often a key factor in slow light experiments and limits the maximum obtainable delay. In contrast, in our case the limiting factor are losses or the length of the grating, none of which are intrinsic and all of which can be improved by better engineering. Note that a possible exception is the modest loss that is induced while writing the grating writing. We discuss this further in Section 11.4. The second issue listed above indicates one of the drawbacks when using gap solitons for slow light: they require rather high powers (in our experiments the launch power is around 2 kW). However, as discussed in the Section 11.4, there are good prospects for lowering the peak power considerably by using highly nonlinear glasses, rather than silica glass. Gap solitons were experimentally observed first by Eggleton et al. (1996) who reported pulse narrowing and slow propagation in optical fiber Bragg gratings (FBGs). Subsequent studies by these and other researchers extended this work in a number of ways: for example, in further work Eggleton et al. (1997) reported a thorough investigation of gap soliton propagation and also discussed the formation of trains of gap solitons, and higher-order gap solitons. In an independent set of experiments, Taverner et al. (1998a,b) reported multiple gap soliton formation, intensity-dependent switching, and gap soliton-based AND gate operation. An important practical advance was introduced by Millar et al. (1999), who demonstrated gap soliton formation in an AlGaAs waveguide, rather than in a fiber grating. This allows for deeper gratings, though this was not demonstrated in this work, and also allows for integration with other components.

In this chapter we review our recent work on gap soliton propagation in 10 and 30 cm FBGs, with the emphasis on their slow light aspects. We demonstrate substantial delays of several pulse widths, without significant broadening.

11.2 BACKGROUND

We are interested in one-dimensional periodic structures ("gratings") with a period smaller than the wavelength of the light, and which are designed to reflect light between a forward propagating mode and its backward propagating counterpart. This excludes, for example, long period gratings (Kashyap, 1999), gratings with much longer periods, that couple light between different modes propagating in the same direction. Henceforth the term grating shall refer exclusively to the former type.

11.2.1 LINEAR PROPERTIES OF GRATINGS

The linear optical properties of gratings have been well-known for many years. These structures exhibit (one-dimensional) photonic bandgaps, where light cannot propagate and, provided that they have sufficient number of periods, the structures act as mirrors at these frequencies. There are many different ways in which gratings can be realized. That of interest are Bragg gratings, written in the core of an optical fibre or waveguide. Such structures are characterized by a weak modulation of the refractive index (typically 10^{-4} to 10^{-3}), and a large number of periods (in some of our experiments we used gratings with almost one million periods). The fabrication of such gratings is well known (Kashyap, 1999) and is outside the scope of this review.

The linear properties of these structures can be calculated as solutions to the Maxwell equations using a variety of methods, for example, that based on transfer matrices. However, as the index modulation is so small, we can use coupled mode theory, a convenient and much used method which leads to a set of two coupled linear differential equations, one each for the forward and backward modal amplitudes $E_\pm$

$$\begin{aligned} +i\frac{\partial E_+}{\partial z} + \frac{i}{v_g}\frac{\partial E_+}{\partial t} + \kappa E_- &= 0, \\ -i\frac{\partial E_-}{\partial z} + \frac{i}{v_g}\frac{\partial E_-}{\partial t} + \kappa E_+ &= 0. \end{aligned} \quad (11.1)$$

Here z and t are propagation distance and time, respectively, and v_g is the group velocity of the mode in the absence of the grating. Finally, κ couples the forward and backward propagating modes and represents the effect of the grating; it is defined as

Slow Light Gap Solitons

$$\kappa = \frac{\pi}{\lambda}\Delta n, \qquad (11.2)$$

where Δn is the amplitude of the refractive index distribution, defined through

$$n = \bar{n} + \Delta n \cos(2\pi z/\Lambda), \qquad (11.3)$$

Λ is the period of the grating and $\bar{n}$ is the average refractive index. The relationship between the modal amplitudes and the actual electric field is given by

$$E(z,t) = \left(E_+ e^{i\pi z/\Lambda} + E_- e^{-i\pi z/\Lambda}\right) e^{-i\omega_g t} + \text{c.c.}, \qquad (11.4)$$

where c.c. indicates the complex conjugate. Further, ω_g is the Bragg frequency, the frequency where the mode's wave number satisfies the Bragg condition $k = \pi/\Lambda$. Combining this expression for the Bragg condition with $k = 2\pi\bar{n}/\lambda$, we get the Bragg wavelength $\lambda_B = 2\bar{n}\Lambda$.

Some insight into the properties of the solutions to Equation 11.1 can be gained by considering the associated dispersion relation. To do so, we set $E_\pm \sim e^{i(Qz - v_g \delta t)}$ where δ (the detuning) and Q are low frequencies, representing, respectively, the temporal and spatial deviations from the Bragg condition, respectively. We then solve the associated algebraic equations and find that

$$\delta = \pm\sqrt{\kappa^2 + Q^2}, \qquad (11.5)$$

which is shown in Figure 11.1, and where the top (bottom) sign refers to the upper (lower) branch. Note that for the range of detunings $-\kappa < \delta < \kappa$ (corresponding to a frequency range of $\Delta f = v_g \kappa/\pi$) no running wave solutions exist; this spectral band therefore corresponds to the photonic bandgap. For a fiber grating with a typical refractive index modulation amplitude of $\Delta n \approx 5 \times 10^{-4}$, we find from Equation 11.2, with $\lambda \approx 1.06\,\mu\text{m}$, that $\kappa \approx 1.5 \times 10^3\,\text{m}^{-1}$. With a refractive index of around 1.5, this then leads via Equation 11.5 to a width of the photonic bandgap of approximately 0.35 nm.

We note that associated with the solution (Equation 11.5) is a characteristic ratio of the magnitudes of the forward and backward propagating modes. For example, at the high-frequency bandedge ($\delta = \kappa$), this ratio is $+1$, whereas at the lower bandedge ($\delta = -\kappa$), the ratio is -1. More generally, we find that

$$\left|\frac{E_+}{E_-}\right|^2 = \frac{1+v}{1-v}, \qquad (11.6)$$

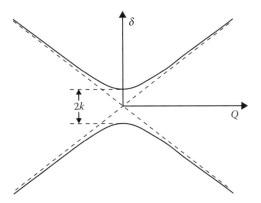

FIGURE 11.1 Schematic of the dispersion relation of a one-dimensional photonic crystal (grating). The photonic bandgap has a width of 2κ in terms of the detuning.

Important for this discussion are also the propagation characteristics of frequencies just outside the photonic bandgap. From Equation 11.5 we find for the group velocity v in the presence of the grating,

$$v = \pm v_g \frac{Q}{\sqrt{\kappa^2 + Q^2}} = \pm v_g \frac{\sqrt{\delta^2 - \kappa^2}}{\delta}, \tag{11.7}$$

the magnitude of which is always smaller than v_g. Consistent with Figure 11.1, we note that at the edges of the bandgaps, where $|\delta| \to \kappa$, the group velocity vanishes. This is not surprising since inside the bandgap no energy is transported. Tuning away from the bandgap, we see that the group velocity smoothly tends toward $\pm v_g$ as the effect of the grating becomes weaker. We thus see that, in principle, any group velocity can be tuned in by choosing the appropriate (frequency) detuning from the Bragg condition, provided it is just outside the bandgap. This is one of two key elements in FBG-based slow light research.

The obvious problem with this approach becomes clear when we consider the dispersion. Note that the group velocity varies between 0 and v_g in a frequency interval of only a few times $v_g\kappa$, that is, with the aforementioned typical numbers, less than a nanometer. This means that the dispersion, the rate at which the group velocity varies with frequency, is high. This is confirmed by calculating the quadratic dispersion (Russell, 1991)

$$\beta_2 \equiv \frac{\partial(1/v)}{\partial \omega} = \mp \frac{1}{v_g^2} \frac{\kappa^2}{(\delta^2 - \kappa^2)^{3/2}}. \tag{11.8}$$

Though diverging at the bandedge (because the inverse group velocity diverges there), it can be ascertained by substituting the aforesaid typical numbers, that even away from the bandedge this is many orders of magnitude larger than the dispersion in conventional fibers. For frequencies above the photonic bandgap $\beta_2 < 0$, which means that the dispersion is anomalous, whereas for frequencies below the bandgap the dispersion is normal. It should be noted that the coupled mode equations do not have explicit dispersion terms, which usually occur as second or higher order derivatives. In fact, the grating dispersion enters via the coupling terms in the equations. Though it is true that the background dispersion in the grating's absence would enter as a higher derivative, such contributions are neglected. Of course, this is justified as this dispersion is dwarfed by that of the grating.

Finally, in conventional fibers quadratic dispersion usually dominates, with the occurence of higher order dispersion being significantly lower. This is not so in fiber gratings, as can be verified by calculating the cubic dispersion by differentiation of Equation 11.8 (Russell, 1991). The linear properties of slow light propagation near the edge of the photonic bandgap of a grating, were confirmed experimentally by Eggleton et al. (1999).

Now we have discussed the dispersive properties of gratings, we briefly turn to the reflection properties. Since the field at frequencies inside the bandgap is evanescent and cannot propagate through the grating (as from Equation 11.5 Q is imaginary), these frequencies are strongly reflected. Indeed, a key application of fiber gratings is all-fiber filters (Kashyap, 1999). However, the reflectivity of a uniform grating, though high within the bandgap, can also be substantial for frequencies close to the gap. The reason for this is easy to see from the previous discussion: since frequencies just outside the bandgap propagate with a reduced group velocity through the grating, they are poorly matched to the field outside the grating, which travels at c/n, leading to a significant reflection. Since in our work we operate in the slow light regime close to the edge of the photonic bandgap, this residual reflectivity is particularly detrimental. For this reason we use apodized gratings, gratings in which the strength of the refractive index modulation gradually increases, rather than uniform gratings. This has the effect of smoothing out the mismatch, thereby reducing the out-of-band reflection considerably. Examples of linear reflection spectra are given in Section 11.3.

11.2.2 NONLINEAR PROPERTIES OF GRATINGS

Having discussed some of the key linear properties of gratings, we now turn to the nonlinear properties. In particular, we consider a grating in which the refractive index is a linear function of intensity (a "Kerr" nonlinearity), that is, $n = n_L + n^{(2)} I$, where $n^{(2)}$ is the nonlinear refractive index, which for silica glass takes the value $n^{(2)} \approx 3 \times 10^{-20}$ m^2/W; since $n^{(2)} > 0$, the refractive index increases with intensity. We have implicitly assumed the response time of the nonlinearity to be negligible. This is a good approximation as, though finite, its femtosecond time scale is certainly much shorter than the pulse lengths of 100's of picoseconds that we consider here. We might expect that one effect of the nonlinearity is that the Bragg wavelength $\lambda_B = 2\bar{n}\Lambda$, and therefore the entire photonic bandgap, depend on intensity. Indeed, we see below that much of the physics can be understood qualitatively using this argument.

Including, then, the Kerr nonlinearity we find that Equation 11.1 are generalized to the nonlinear coupled mode equations (NLCMEs) (Winful and Cooperman, 1982; Christodoulides, 1989; Aceves and Wabnitz, 1989; de Sterke et al., 1994)

$$+i\frac{\partial E_+}{\partial z} + \frac{i}{v_g}\frac{\partial E_+}{\partial t} + \kappa E_- + \Gamma(|E_+|^2 + 2|E_-|^2)E_+ = 0,$$
$$-i\frac{\partial E_-}{\partial z} + \frac{i}{v_g}\frac{\partial E_-}{\partial t} + \kappa E_+ + \Gamma(|E_-|^2 + 2|E_+|^2)E_- = 0,$$
(11.9)

where the nonlinear coefficient $\Gamma = 2kn^{(2)}/Z_0$, where Z_0 is the vacuum impedance, and is chosen such that the forward and backward propagating intensities are related to the amplitudes by $I_\pm = 2n|E_\pm|^2/Z_0$. Equation 11.9 shows that effect of the nonlinearity is to introduce self- and cross-phase modulation terms. The former describe how the intensity-dependent refractive index affects the propagation of a pulse, whereas the latter describe how the change in refractive index of the forward propagating wave affects the backward propagating wave, and vice versa.

Winful et al. (Winful et al., 1979; Winful and Cooperman, 1982) were the first to study the solutions to the NLCMEs. They initially considered the continuous-wave (cw) solutions to the NLCMEs and showed that in general these can be written in terms of Jacobi elliptic functions, both inside and outside the photonic bandgap. Since these functions vary periodically with position, at a sufficiently high intensity where this period matches the grating's length, a grating can perfectly transmit even for frequencies within the bandgap. More generally, Winful et al. found that a nonlinear grating can exhibit bistability between low- and high-transmission states (Winful et al., 1979), though this occurs only at extremely high intensities. However, even at lower intensities the transmission varies with the input power. At an intuitive level this can be understood as follows: since the Bragg wavelength $\lambda_B = 2\bar{n}\Lambda$ increases with intensity to lowest order, the entire photonic band does so as well. Thus, wavelengths inside the photonic bandgap, but close to the short-wavelength edge, may be shifted outside the bandgap with increasing intensity, thereby increasing the transmission. Of course, the same is true for frequency further away from the gap edge—however, this would require much higher intensities. We are therefore exclusively interested in the frequencies close to the high-frequency edge of the bandgap, as the nonlinear effects of interest are most prominent.

In a follow-up paper (Winful and Cooperman, 1982), Winful and Cooperman studied some of the properties of the full time-dependent NLCMEs using a numerical algorithm.[*] They did so by driving the grating with a high-intensity cw field and monitoring the transmission. They found that for some values of the intensity the transmission is constant, while for others the grating exhibited self-pulsations, leading to a time-dependent, periodic transmitted field, which can take the form of a periodic train of pulses. At even higher intensities the pulsations became chaotic.

---

[*] In fact, Winful and Cooperman considered the NLCMEs, but with a finite nonlinear response time. However, this does matter for our purposes here.

Apparently unaware of Winful's work, the study of nonlinear periodic media was next taken up by Chen and Mills (1987). Not assuming the grating to be shallow, they showed that for frequencies in the gap where the grating is transmitting perfectly, the field takes the form of a "soliton-like object," which led them to introduce the term gap soliton, without demonstrating that the object in question, in fact, is a soliton.

This work was continued by Sipe and Winful (1988), and by de Sterke and Sipe (1988) who showed, based on the wave equation for the electric field, that for frequencies close to the upper-frequency edge of the photonic bandgap, the electric field can be approximately described by the nonlinear Schrödinger equation, one of the standard equations in nonlinear science, and which, for example, has soliton solutions and also describes light propagation in uniform optical fibers. The properties of their solution were qualitatively similar to the numerical results of Chen and Mills, confirming that, within this approximation, gap solitons are, indeed, solitons. We return to the nonlinear Schrödinger description of the field below.

The next step in the study of the properties of nonlinear periodic media was based on the NLCMEs, when Christodoulides and Joseph (1989), and Aceves and Wabnitz (1989), independently demonstrated the existence of pulse-like solutions to the full time-dependent equations. In particular, Aceves and Wabnitz showed that Equation 11.9 has a two-parameter family of single-peaked solutions that can travel at any velocity between 0 and c/n. With the existence of these solutions, generalizations of the nonlinear Schrödinger solitons found earlier (Sipe and Winful, 1988; de Sterke and Sipe, 1988), showed that nonlinear periodic media can support slow light pulses, which, because of the nonlinearity, do not suffer from the very considerable grating dispersion and they are thus ideal for slow light experiments.

We thus see that we have two descriptions for wave propagation in nonlinear periodic media: one based on a nonlinear Schrödinger description which is valid for frequencies close to the upper edge of the photonic bandgap, and which has soliton solutions, and the other based on the coupled mode equations (11.9), to which pulse-like solutions have been found. The consistency between these approaches was given by de Sterke and Sipe (1990), who showed that nonlinear Schrödinger equation can be derived from the coupled mode equations, and (de Sterke and Eggleton, 1999) they explicitly showed that the solutions discovered by Aceves and Wabnitz reduce to the nonlinear Schrödinger soliton in the low-intensity limit. The procedure starts by identifying the frequency and the associated wavenumber and ratio between the forward and backward waves (Equation 11.6), and then finding the evolution equation for the slowly varying envelope a of this compound wave. The result is found to be (de Sterke and Sipe, 1990)

$$i\frac{\partial a}{\partial \zeta} - \frac{\beta_2}{2}\frac{\partial^2 a}{\partial \tau^2} + \frac{\Gamma}{2v}\left(3+v^2\right)|a|^2 a = 0. \tag{11.10}$$

Here $\zeta = z$ and $\tau = t - z/(v v_g)$ are the coordinates in a frame that moves with the pulse, and β_2 given by Equation 11.7, represents the grating dispersion. Let us now consider the nonlinear term and in particular its velocity dependence. The factor $(3+v^2)/2$ is associated with the fact that properties of the compound plane wave that is being modulated: at the bandedges where $v = 0$ the forward and backward components have the same amplitude giving rise to a standing wave of the form $\cos(\pi z/\Lambda)$ or $\sin(\pi z/\Lambda)$. Because it is modulated, such a standing wave has a somewhat stronger nonlinear effect than a running wave, as enhanced nonlinear effects at the antinodes, more than compensate for the reduced nonlinear effects near the nodes. In addition, the stronger $1/v$ factor is associated with the enhanced dwell time of the light. We note that if we had included gain or loss in the equations, then we would have found these effects to be similarly enhanced by a factor $1/v$.

In addition to the effects discussed in the previous paragraph, there is an additional enhancement of the nonlinear effect: the net energy flow is proportional to $|E_+|^2 - |E_-|^2 \propto v|a|^2$, and thus, for a given energy flow, the amplitude of the field envelope increases proportionally with $1/v$.

We thus see that slow propagation is associated with a significant enhancement of nonlinear optical effects, one of the prime motivations for research in this area.

Before discussing the results of our experiments, we briefly review the numerical tools that are available. As first pointed out by Winful et al. (1982) NLCMEs (Equation 11.9) can be solved efficiently along their characteristics. A numerical procedure that makes use of the characteristics was later developed by (de Sterke et al., 1991). In contrast, the nonlinear Schrödinger equation was solved using a split-step Fourier method as described, for example, by Agrawal (1995).

11.3 EXPERIMENT

The key elements of the gap soliton-based slow light experiments (Mok et al., 2006a, 2007) are shown in Figure 11.2. The setup includes a laser source, an FBG, and two monitoring devices, namely the oscilloscope and the power meter. The laser source is a Q-switched laser emitting pulses of 0.68 ns at a wavelength of 1.064 μm. The spectral full-width at half maximum of the pulses is 1.9 pm, therefore giving a time bandwidth product of 0.34. The small time–bandwidth product allows for the largest possible fractional delay, defined as the ratio of the delay to the pulse width, for a given pulse bandwidth. The pulses are sent into an FBG through an isolator, which is essential in preventing the residual reflection from the FBG back into the laser. The FBG is mounted on a stretching assembly containing a translation stage. The translation stage controls the strain applied to the FBG, thereby changing the period Λ, hence the Bragg wavelength λ_B, of the grating. Since the laser wavelength is fixed, changing λ_B effectively alters the detuning δ, allowing us to position the laser wavelength to coincide with various regions of the transmission spectrum of the grating. As discussed in Section 11.2.1, to observe slow light, one should position the laser wavelength inside the bandgap and close to the short-wavelength bandedge to lower the required intensity as discussed. On the other hand, to generate a reference pulse that travels through the grating unimpeded, one should position the laser wavelength far from and outside the photonic bandgap, where the effect of the grating is negligible.

The power meter measures the power at both the input and the output to monitor the transmission, which is expected to vary as a function of input power. The oscilloscope measures the time-resolved intensity of the output pulses. Part of the input pulse energy is split off to provide the oscilloscope with a triggering signal that does not depend on the arrival time of the output pulse. Any changes in the temporal position of the measured pulse are therefore a result of the changes in the velocity of the pulse through the FBG. All delay measurements, including the reference pulse against which delayed pulses are compared, are made by measuring the output from the FBG using this configuration.

The FBG is also placed in close thermal contact, via a thermally conducting compound, with an aluminum plate for heat dissipation. This ensures that the nonlinear effects to be observed, such as the increased transmission due to an intensity-dependent refractive index change via $n^{(2)}$, are not

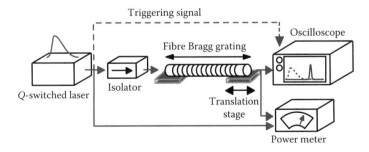

FIGURE 11.2 Schematic diagram of the experimental setup. (From Mok, J.T., de Sterke, C.M., Littler, I.C.M., and Eggleton, B.J., *Nat. Phys.*, 2, 775, 2006a. With permission.)

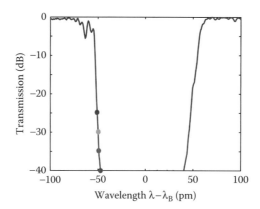

FIGURE 11.3 Transmission spectrum of the 10 cm FBG used in the first experiment. (From Mok, J.T., de Sterke, C.M., Littler, I.C.M., and Eggleton, B.J., *Nat. Phys.*, 2, 775, 2006a. With permission.)

masked by thermal effects, which can also lead to an increased refractive index via the thermo-optic coefficient.

Two experiments were performed, using FBGs of different lengths. In our first experiment (Mok et al., 2006a), the length L of the FBG was 10 cm. The FBG was apodized so that κ varied smoothly between zero and its maximum value of 4.51 cm^{-1} in the first and last 1.5 cm of the FBG. The apodization resulted in a FBG with a transmission spectrum, as shown in Figure 11.3, with suppressed out-of-band reflection and a photonic bandgap that had smooth bandedges and spanned 118 pm.

As discussed, the slow light regime is at detunings close to and just outside the photonic bandgap. However, in the gap soliton experiment discussed here, light was launched at detunings just inside the bandgap, as marked by filled circles in Figure 11.3. Therefore, at low intensities the transmission is very low. In contrast, when the bandgap shifts to a longer wavelength at sufficiently high intensities, the light coincides with the very edge of the bandgap where slow light propagation occurs.

The first experimental observation we made was the increase in transmission of the FBG at sufficiently high intensities, as shown in Figure 11.4a. At all four detunings considered, the transmission increased to around −10 dB at input peak powers beyond ∼1 kW. The increased transmission is consistent with the wavelength shift of the bandgap. The fact that the transmission plateaued was an indication that the bandgap shifted well away from the pulse spectrum. The transmission curves for

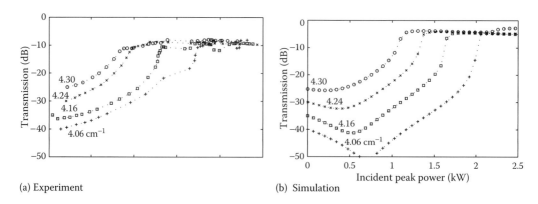

FIGURE 11.4 Transmission vs. input peak power for four detunings: 4.06, 4.16, 4.25, and 4.30 cm^{-1}. (From Mok, J.T., de Sterke, C.M., Littler, I.C.M., and Eggleton, B.J., *Nat. Phys.*, 2, 775, 2006a. With permission.)

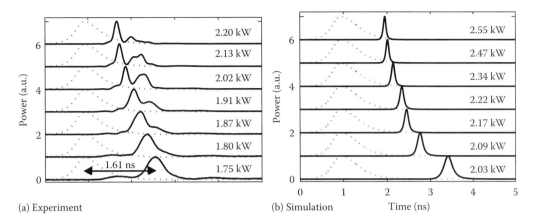

FIGURE 11.5 Delayed pulse (solid) and reference pulse (dotted) at various input peak powers at $\delta = 4.06\,\text{cm}^{-1}$. (From Mok, J.T., de Sterke, C.M., Littler, I.C.M., and Eggleton, B.J., *Nat. Phys.*, 2, 775, 2006a. With permission.)

the four detunings differ from one another in that the threshold peak powers, the peak power at which transmission increases significantly, are different. For detunings closer to the edge of the bandgap, it required less power to shift the bandgap such that the light is outside the bandgap, which explains the lower threshold power for these detunings. These three features, namely the intensity-dependent transmission, the transmission plateau, and the difference in threshold power, are consistent with the results of simulations shown in Figure 11.4b for the four corresponding detunings using the numerical procedure reported by de Sterke et al. (1991).

The most interesting regime is at the input power where the transmission plateau starts to form, where the bandgap is shifted such that the pulse spectrum lies outside but still remains very close to the bandgap. This is where the largest delay is expected. Indeed, as is evident in Figure 11.5a, which shows the output pulse measurement at $\delta = 4.06\,\text{cm}^{-1}$ in solid lines, the largest delay was 1.6 ns at an input peak power of 1.75 kW, the power at which the transmission started to saturate for this detuning. The reference pulse, shown in dotted line in Figure 11.5a, was measured by detuning the bandgap far away from the pulse spectrum such that the light propagated through the grating unimpeded. The delay decreased as the input peak power increased because the bandgap shifted further away from the pulse spectrum, where the effect of the grating became weaker. The absence of pulse broadening indicated that dispersion had been canceled. In fact, the pulses narrowed as predicted by the corresponding simulations shown in Figure 11.5b, which also indicates the same trend in the delay with power observed in the experiment. The input peak powers used in the simulations are different from those measured in the experiment to take into account the difference between the threshold power measured experimentally and predicted with simulations.

In a subsequent experiment (Mok et al., 2007), an FBG of length $L = 30\,\text{cm}$ was used in place of the 10 cm FBG. There are two motivations for using a longer FBG. First, to show that pulses propagating in the system in question are truly solitons, by demonstrating the absence of pulse broadening over a medium that is triple in length. Second, by exploiting the soliton nature of this system, we would like to obtain an improved delay, which is expected to scale linearly with grating length, and still without pulse broadening.

Similar to the shorter FBG, the 30 cm long FBG was apodized so that κ varied smoothly between zero and its maximum value of $8.23\,\text{cm}^{-1}$ in the first and last 2 cm of the FBG. The measured transmission spectrum of this FBG in Figure 11.6 shows a bandgap of width 215 pm, and again, suppressed out-of-band reflection. As a comparison to the shorter FBG, we note that the increase in κ from 4.51 to $8.23\,\text{cm}^{-1}$ would, in principle (Mok et al., 2006b), increase the delay by approximately 30%. Therefore, the improvement of delay should primarily result from the increase in FBG length.

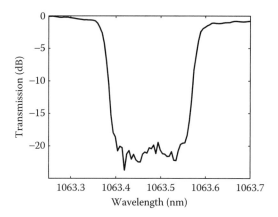

FIGURE 11.6 Transmission spectrum of the 30 cm FBG used in the second experiment. (From Mok, J.T., de Sterke, C.M., Littler, I.C.M., and Eggleton, B.J., *Electron. Lett.* (in press), 2007. With permission.)

Figure 11.7a shows the measured transmission (open circles) as a functional input peak power, compared with the corresponding simulations (crosses). Once again, we observe the increase in transmission with input peak power as well as the plateauing of the transmission at high input peak powers, similar to the case with the 10 cm FBG, indicating the bandgap shift. Figure 11.7b shows the time-resolved measurements of the output. First, we note that the pulse broadening was still absent. Second, the largest delay was 3.2 ns, twice as much as that generated by the previous grating, consistent with the claim that the delay scales with the length of the grating.

It is noted in both experiments that the pulse width changes as the delay is tuned by varying the input power. For some applications, it may be essential to maintain the pulse width even in the absence of pulse broadening. Although not shown above, it is in fact possible to tune the delay without any change in the pulse width. It however requires the simultaneous adjustment of the input power as well as the detuning (Mok et al., 2006b). Finally, we note that in the two experiments the threshold power and hence power required for reaching the slow light regime were about 1–2 kW.

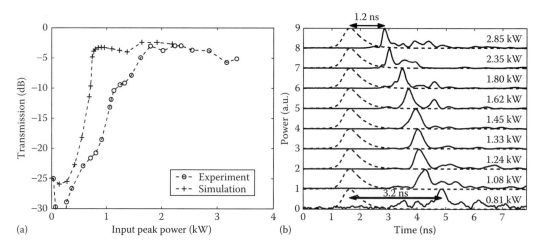

FIGURE 11.7 (a) Transmission versus input peak power for $\delta = 8.02\,\text{cm}^{-1}$; (b) delayed pulse (solid) and reference pulse (dotted) at various input peak powers at $\delta = 8.02\,\text{cm}^{-1}$. (From Mok, J.T., Isben, M., de Sterke, I.C.M., and Eggleton, B.J., *Electron. Lett.* (in press), 2007. With permission.)

This power requirement can be significantly reduced by using highly nonlinear materials such as chalcogenide, possibly lowering the threshold power to a few watts.

11.4 DISCUSSION AND CONCLUSIONS

We have shown that gap solitons can propagate through fiber gratings at low speeds and without broadening—by using nonlinear effects the pulse broadening is suppressed and the limitations of the achievable time–bandwidth product do not apply. In spite of this, when comparing the results in 10 cm grating and 30 cm grating discussed in Section 11.3, it is clear that the delay is not three times larger as the gap solitons travel faster in the longer grating. In fact, as the 30 cm grating is also stronger than the 10 cm grating, we would expect the ratio of the delays to be more than a factor of 3. We are currently investigating this. We initially surmised that the effect of the inevitable random errors in the period and the strength of the fabricated grating lead to a lower limit of the achievable speed in the grating. However, preliminary results seem to indicate that the effect of these errors is inconsistent with observations. Other possibilities are the effect of birefringence, either that in the unexposed fiber or possibly induced by the UV exposure when fabricating the grating.

It is true that the peak powers required to launch gap solitons are rather high. Note from Figures 11.5 and 11.7b that we needed 1–2 kW, which is impractical for some of the applications of slow light discussed elsewhere in this book. As mentioned, we intend to lower the required power by using chalcogenide glass, a highly nonlinear glass that is also strongly photosensitive (Hilton, 1966; Shokooh-Saremi et al., 2006). The strength of the nonlinearity in these glasses is typically 100–1000 times larger than that of silica glass, thus lowering the gap soliton launch power by the same factor. They also possess high refractive indices, typically 2.4–2.7. This means that the light can be strongly confined, leading to a small modal area, further decreasing the required launch power. Experiments of this type, which are currently in progress, are carried out in a planar geometry, rather than in an optical fiber, thus allowing for the possible eventual integration of gap soliton-based slow light devices with other optical functions.

ACKNOWLEDGMENTS

This work was supported by the Australian Research Council under the ARC Centres of Excellence Program. M. Ibsen acknowledges the Royal Society of London for the provision of a University Research Fellowship.

REFERENCES

Aceves, A.B. and S. Wabnitz. 1989. Self induced transparency solitons in nonlinear refractive periodic media. *Phys. Lett. A* 141:37–42.

Agrawal, G.P. 1995. *Nonlinear Fiber Optics*, 2nd edn. San Diego: Academic Press.

Chen, W. and D.L. Mills. 1987. Gap solitons and the nonlinear optical response of superlattices. *Phys. Rev. Lett.* 58:160–163.

Christodoulides, D.N. and R.I. Joseph. 1989. Slow Bragg solitons in nonlinear periodic structures. *Phys. Rev. Lett.* 62:1746–1749.

de Sterke, C.M. and B.J. Eggleton. 1999. Bragg solitons and the nonlinear Schrödinger equation. *Phys. Rev. E* 59:1267–1269.

de Sterke, C.M. and J.E. Sipe. 1988. Envelope-function approach for the electrodynamics of nonlinear periodic structures. *Phys. Rev. A* 38:5149–5165.

de Sterke, C.M. and J.E. Sipe. 1990. Coupled modes and the nonlinear Schrödinger equation. *Phys. Rev. A* 42:550–555.

de Sterke, C.M. and J.E. Sipe. 1994. Gap solitons. In E. Wold (Ed.), *Progress in Optics*, Vol. XXXIII, pp. 203–260. Amsterdam: North Holland.

de Sterke, C.M., K.R. Jackson, and B.D. Robert. 1991. Nonlinear coupled mode equations on a finite interval: A numerical procedure. *J. Opt. Soc. Am. B* 8:403–412.

Eggleton, B.J., R.E. Slusher, C.M. de Sterke, P.A. Krug, and J.E. Sipe. 1996. Bragg grating solitons. *Phys. Rev. Lett.* 76:1627–1630.

Eggleton, B.J., C.M. de Sterke, and R.E. Slusher. 1997. Nonlinear pulse propagation in Bragg gratings. *J. Opt. Soc. Am. B.* 14:2980–2993.

Eggleton B.J., C.M. de Sterke, and R.E. Slusher. 1999. Bragg grating solitons in the nonlinear Schrödinger limit: experiment and theory. *J. Opt. Soc. Am. B.* 16:587–599.

Hilton, A.R. 1966. Nonoxide chalcogenide glass as infrared optical materials. *Appl. Opt.* 5:1877–1882.

Kashyap, R. 1999. *Fiber Bragg Gratings*. San Diego: Academic Press.

Lenz, G., B.J. Eggleton, C.K. Madsen, and R.E. Slusher. 2001. Optical delay lines based on optical filters. *J. Quantum Electron.* 37:525–532.

Millar, P., R.M. De la Rue, T.F. Krauss, J.S. Aitchison, N.G.R. Broderick, and D.J. Richardson. 1999. Nonlinear propagation effects in an AlGaAs Bragg grating filter. *Opt. Lett.* 24:685–687.

Mok, J.T., C.M. de Sterke, I.C.M. Littler, and B.J. Eggleton. 2006a. Dispersionless slow light using gap solitons. *Nat. Phys.* 2:775–780.

Mok, J.T., C.M. de Sterke, and B.J. Eggleton. 2006b. Delay-tunable gap-soliton-based slow-light system. *Opt. Express* 14:11987–11996.

Mok, J.T., M. Ibsen, C.M. de Sterke, and B.J. Eggleton. 2007. Dispersionless slow light with 5-pulse-width delay in fibre Bragg grating. *Electron. Lett.* 43:1418–1419.

Russell, P.S.J. 1991. Bloch wave analysis of dispersion and pulse propagation in pure distributed feedback structures. *J. Mod. Opt.* 38:1599–1619.

Shokooh-Saremi, M., V.G. Taeed, N.J. Baker, I.C.M. Littler, D.J. Moss, B.J. Eggleton, Y. Ruan, and B. Luther-Davies. 2006. High-performance Bragg gratings in chalcogenide rib waveguides written with a modified Sagnac interferometer. *J. Opt. Soc. Am. B* 23:1323–1331.

Sipe, J.E. and H.G. Winful. 1988. Nonlinear Schrödinger solitons in a periodic structure. *Opt. Lett.* 13:132–134.

Taverner, D., N.G.R. Broderick, D.J. Richardson, M. Ibsen, and R.I. Laming. 1998a. All-optical AND gate based on coupled gap-soliton formation in a fiber Bragg grating. *Opt. Lett.* 23:259–261.

Taverner, D., N.G.R. Broderick, D.J. Richardson, R.I. Laming, and M. Ibsen. 1998b. Nonlinear self-switching and multiple gap-soliton formation in a fiber Bragg grating. *Opt. Lett.* 23:328–330.

Winful, H.G. and G.D. Cooperman. 1982. Self-pulsing and chaos in distributed feedback bistable optical devices. *Appl. Phys. Lett.* 40:298–300.

Winful, H.G., J.H. Marburger, and E. Garmire. 1979. Theory of bistability in nonlinear distributed feedback structures. *Appl. Phys. Lett.* 35:379–381.

12 Coherent Control and Nonlinear Wave Mixing in Slow Light Media

Yuri Rostovtsev

CONTENTS

12.1 Introduction ... 235
    12.1.1 Group Velocity: Kinematics ... 236
    12.1.2 Slow Light in a Gas of Three-Level Λ Atoms ... 238
    12.1.3 Propagation in Coherent Media ... 240
        12.1.3.1 Delay Time via EIT ... 243
        12.1.3.2 Interference and Frequency Stabilization via Slow and Fast Light ... 243
12.2 Nonlinear Wave Mixing via Slow Light ... 244
    12.2.1 Forward Brillouin Scattering ... 244
    12.2.2 Coherent Control of Nonlinear Mixing: Coherent Backscattering ... 247
        12.2.2.1 Spectroscopy via Slow Light ... 250
        12.2.2.2 Nonlinear Light Steering ... 251
        12.2.2.3 Implementation of Obtained Results ... 251
References ... 253

12.1 INTRODUCTION

The propagation of light in dispersive media can be described by five different kinds of wave velocities [1–3]:

- Phase velocity, which is the speed at which the zero crossings of the carrier wave move
- Group velocity, at which the peak of a wave packet moves
- Energy velocity, at which the energy is transported by the wave
- Signal velocity, at which the half-maximum wave amplitude moves
- Front velocity, at which the first appearance of a discontinuity moves

Generally, all five velocities can differ from each other, although in a linear passive dispersive media, they coincide and are, usually, less than the light velocity in vacuum. Recently, experimental demonstrations [4–6] in various media show that the group velocity of light can be reduced by 10–100 million compared with its vacuum phase velocity. The physics is based on very steep frequency dispersion in the vicinity of narrow resonance of electromagnetically induced transparency (EIT) [7–11].

Let us note that, recently, much work has been done to demonstrate slow light using various physical mechanisms that are not limited only to EIT. In particular, slow light has been demonstrated by coherent population oscillation, Brillouin scattering, etc. [12]. In this book there are chapters that are devoted to new methods of obtaining slow light.

Ultraslow light related to EIT is particularly interesting because it has a potentially wide variety of applications ranging from lasers without population inversion (LWI) (for the earliest papers on EIT/LWI see [13–16], for reviews on EIT/LWI see [7,9,10,17,18]) to new trends in nonlinear optics [19–24]. The EIT is based on quantum coherence [7–11] that has been shown to result in many counter-intuitive phenomena. The scattering via a gradient force in gases [25], the forward Brillouin scattering in ultra-dispersive resonant media [26–28], controlled coherent multiwave mixing [29,30], EIT and slow light in various media [31–36], Doppler broadening elimination [37], light-induced chirality in nonchiral medium [38], a new class of entanglement amplifier [39] based on correlated spontaneous emission lasers [40,41], and the coherent Raman scattering enhancement via maximal coherence in atoms [31] and biomolecules [42–45] are a few examples that demonstrate the importance of quantum coherence.

Even though, several excellent reviews and books were devoted to slow and fast light [46], here, in this chapter, we present recent results obtained in coherent media that impact nonlinear optics [47]. In particular we focus on modification of the phase-matching condition that can be controlled in coherently driven media. Some nonlinear processes are forbidden because the phase-matching condition cannot be fulfilled under normal condition. However, it turns out that steep dispersion can change the situation drastically and forward Brillouin scattering and backward scattering are allowed in such dispersive media.

12.1.1 Group Velocity: Kinematics

To remind the concept of group velocity, let us consider the propagation of two waves of the same amplitude E_1 and E_2, where $E_i = E_0 \cos(k_i z - \nu_i t)$ with $i = 1, 2$. Superposition of the two waves gives rise to the modulation shown in Figure 12.1

$$E = E_0(\cos(k_1 z - \nu_1 t) + \cos(k_2 z - \nu_2 t))$$
$$= 2E_0 \cos(\Delta k z - \Delta \nu t) \cos(kz - \nu t) \qquad (12.1)$$

where $\Delta k = (k_1 - k_2)/2$, $\Delta \nu = (\nu_1 - \nu_2)/2$, $k = (k_1 + k_2)/2$, $\nu = (\nu_1 + \nu_2)/2$. The two waves create an interference pattern consisting of a rapid oscillation propagating with the so-called phase velocity,

$$\nu_{\text{phase}} = \frac{\nu}{k}, \qquad (12.2)$$

and a slowly varying envelope propagating with the so-called group velocity,

$$\nu_g = \frac{\Delta \nu}{\Delta k}. \qquad (12.3)$$

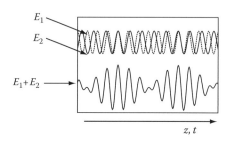

FIGURE 12.1 Interference of two monochromatic waves with different frequencies results in a wave modulated in time and space.

Usually we describe the propagation of wave packets consisting of more than two superimposed harmonics by transforming the ratio $\Delta v/\Delta k$ into dv/dk [3,48].

We now formally consider the propagation of light in a medium possessing both temporal and spatial dispersion of the index of refraction, $n(v,k) = \sqrt{1+\chi(v,k)}$, where $\chi(v,k)$ is the susceptibility of the medium. Differentiating the dispersion equation $kc = vn(v,k)$ gives

$$c = v_g(n + v\partial n/\partial v) + v\partial n/\partial k, \quad (12.4)$$

where we use the definition $v_g \equiv dv/dk$.

Generally, electric susceptibility is a complex quantity. The real part (usually denoted χ') is responsible for the refractive properties of the medium, while the imaginary part (χ'') leads to light absorption. The generalized index of refraction $n = \sqrt{\chi+1}$ is also a complex quantity $n \equiv n' + in''$. As usual, we consider real-valued group velocities and redefine as $v_g \equiv \mathrm{Re}(dv/dk)$ under the condition that imaginary part of dv/dk is negligible. Otherwise group velocity looses its simple kinematic meaning, and strong absorption governs or prevents propagation of the light pulse through the medium. The latter is the reason why the resonant interaction of light with a two-level medium never results in an ultraslow polariton.

According to Equation 12.4 the group velocity of light contains two contributions,

$$v_g \equiv \mathrm{Re}\frac{dv}{dk} = \mathrm{Re}\frac{c - v\frac{\partial n(v,k)}{\partial k}}{n(v,k) + v\frac{\partial n(v,k)}{\partial v}} = \tilde{v}_g - v_s. \quad (12.5)$$

This is the basic equation of this article, as it shows how to find the group velocity from the refractive index of refraction of the medium. The only problem left is to find the actual form of the function $n(v,k)$ for a particular real physical substance, which is not an easy task.

The kinematic meaning of Equation 12.5 becomes clear if one turns to the Maxwell equation for the field amplitude E

$$\left(\frac{\partial^2}{\partial z^2} - \frac{1}{c^2}\frac{\partial^2}{\partial t^2}\right)E = \mu_0 \frac{\partial^2}{\partial t^2}P, \quad (12.6)$$

where $P(z,t)$ is the macroscopic polarization of the medium. The polarization is related to the electromagnetic (EM) field $E(z,t)$ as follows [3]

$$P(z,t) = \epsilon_0 \int \chi(t-t', z-z')E(z',t')dt'dz', \quad (12.7)$$

where $\chi(t-t', z-z')$ is a function depending on the properties of the medium. In general, χ is also a function of the EM field itself. In that case, Equation 12.6 become nonlinear.

The susceptibility $\chi(v,k)$ is the Fourier transformation of $\chi(t-t', z-z')$

$$\chi(t-t', z-z') = \int dv dk\, \chi(v,k) \exp[ik(z-z') - iv(t-t')], \quad (12.8)$$

and the macroscopic polarization (Equation 12.7) can be written in terms of the susceptibility as

$$P(z,t) = \epsilon_0 \int dk dv\, \chi(v,k)E(k,v) \exp(ikz - ivt). \quad (12.9)$$

We consider the propagation of pulses whose duration is long compared to one optical cycle. Then we use the slowly varying amplitude approximation, and the field amplitude E, and the polarization P can be written as $E(z,t) = \mathcal{E}(z,t)\exp(ik_0 z - iv_0 t) + \mathrm{c.c.}$, $P(z,t) = \mathcal{P}(z,t)\exp(ik_0 z - iv_0 t) + \mathrm{c.c.}$, where the functions $\mathcal{E}(z,t)$ and $\mathcal{P}(z,t)$ have spatial and temporal scales much shorter than the inverses of the carrier wavevector k_0 and optical frequency ν_0 ($|\Delta k| \ll k_0$, $|\Delta \nu| \ll \nu_0$). It is convenient to decompose $\chi(\nu, k)$ in a series

$$\chi(\nu, k) = \chi(\nu_0, k_0) + (\nu - \nu_0)\frac{\partial \chi}{\partial \nu} + (k - k_0)\frac{\partial \chi}{\partial k}$$

and obtain

$$P(z, t) = \epsilon_0 \left(\chi(\nu_0, k_0)\mathcal{E} + i\frac{\partial \chi(\nu_0, k_0)}{\partial \nu}\frac{\partial \mathcal{E}}{\partial t} - i\frac{\partial \chi(\nu_0, k_0)}{\partial k}\frac{\partial \mathcal{E}}{\partial z} \right) \exp(ik_0 z - i\nu_0 t).$$

Simplifying the right- and left-hand sides of the Maxwell equation (12.6)

$$\mu_0 \frac{\partial^2}{\partial t^2} P \simeq -\frac{\nu_0^2}{c^2} P,$$

$$\left(\frac{\partial^2}{\partial z^2} - \frac{1}{c^2}\frac{\partial^2}{\partial t^2} \right) E \simeq 2ik \left(\frac{\partial \mathcal{E}}{\partial z} + \frac{1}{c}\frac{\partial \mathcal{E}}{\partial t} \right) \exp(ik_0 z - i\nu_0 t)$$

we obtain for the slowly varying amplitudes

$$\frac{\partial \mathcal{E}}{\partial z} + \frac{1}{c}\frac{\partial \mathcal{E}}{\partial t} = i\frac{\nu_0}{2c}\left(\chi(\nu_0, k_0)\mathcal{E} - i\frac{\partial \chi(\nu_0, k_0)}{\partial \nu}\frac{\partial \mathcal{E}}{\partial t} + i\frac{\partial \chi(\nu_0, k_0)}{\partial k}\frac{\partial \mathcal{E}}{\partial z} \right).$$

Subsequent rearrangement of terms gives

$$\left(c - \frac{\nu_0}{2}\frac{\partial \chi}{\partial k} \right)\frac{\partial \mathcal{E}}{\partial z} + \left(1 + \frac{\nu_0}{2}\frac{\partial \chi}{\partial \nu} \right)\frac{\partial \mathcal{E}}{\partial t} = \frac{i\nu_0}{2}\chi(\nu, k)\mathcal{E},$$

which implies

$$\left(v_g \frac{\partial}{\partial z} + \frac{\partial}{\partial t} \right)\mathcal{E} = \frac{i\nu_0}{c}\chi(\nu, k)\left[1 + \frac{\nu_0}{2c}\frac{\partial \chi}{\partial \nu} \right]^{-1}\mathcal{E}, \qquad (12.10)$$

where v_g is given by Equation 12.5. Under the limit $\chi(\nu, k) \to 0$, the general solution of Equation 12.10 can be presented as two waves of arbitrary shape $f(t \pm z/v_g)$ propagating with group velocity v_g in opposite directions.

12.1.2 Slow Light in a Gas of Three-Level Λ Atoms

A three-level atomic system involving one upper level with allowed transitions to two lower levels is called a Λ system. In the conventional usage, the two lower levels are coupled to the upper levels via two lasers. One of the lasers is considered to be strong and is referred to as the coupling laser. The other is weak and is called the probe laser. The Hamiltonian of such a system shown in Figure 12.2 is given by

$$\hat{H} = \hbar(\Omega_p|b\rangle\langle a| + \Omega_d|c\rangle\langle a|) + \text{h.c.} \qquad (12.11)$$

The EIT of a weak probe pulse relies on an atomic coherence which is induced by the joint action of the probe field and a strong driving field in a three-level system. In Λ system EIT appears due to coherent population trapping (CPT), that was first observed in experiments establishing Zeeman coherence in sodium atoms [49–51]. In these experiments, explained in terms of a three-level Λ-type scheme, a laser field was used to create superpositions of the ground state sublevels. One of these superpositions, referred to as the bright state, can interact with the laser field while the other

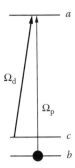

FIGURE 12.2 Energy levels of a three-level Λ system displaying EIT, Ω_p is the Rabi frequency of the probe field, Ω_d is the Rabi frequency of the driving field.

superposition does not and is referred to as the dark state [7]. The entire population in the system is eventually optically pumped into the dark state, and resonant absorption of the EM field almost disappears. The atoms prepared in a coherent superposition of states can produce a steep dispersion that leads to ultraslow group velocity of light. This phenomenon is a manifestation of EIT [7,9,10].

Indeed, we can rewrite the Hamiltonian as

$$H = \hbar\sqrt{\Omega_p^2 + \Omega_d^2}\left(\frac{\Omega_p|b\rangle}{\sqrt{\Omega_p^2 + \Omega_d^2}} + \frac{\Omega_d|c\rangle}{\sqrt{\Omega_p^2 + \Omega_d^2}}\right)\langle a| + \text{h.c.} = \hbar\sqrt{\Omega_p^2 + \Omega_d^2}|B\rangle\langle a| + \text{h.c.}, \quad (12.12)$$

where we introduce a bright state

$$|B\rangle = \frac{\Omega_p|b\rangle}{\sqrt{\Omega_p^2 + \Omega_d^2}} + \frac{\Omega_d|c\rangle}{\sqrt{\Omega_p^2 + \Omega_d^2}}. \quad (12.13)$$

Then, we have three states: excited state $|a\rangle$, bright state $|B\rangle$, and orthogonal to both a so-called dark state,

$$|D\rangle = \frac{\Omega_d|b\rangle}{\sqrt{\Omega_p^2 + \Omega_d^2}} - \frac{\Omega_1|c\rangle}{\sqrt{\Omega_p^2 + \Omega_d^2}}, \quad (12.14)$$

which is indeed decoupled from other states.

Let us find the group velocity in a three-level Λ configuration. To do it, we have to find the susceptibility χ of the medium. Traditionally, the susceptibility is divided into real and imaginary parts: $\chi = \chi' + i\chi''$, and vanishingly small χ' and χ'' at resonance are the signatures of EIT. We consider dilute systems where the susceptibility is small, $|\chi(v,k)| \ll 1$, but $v_g \ll c$, as it is for all the EIT experiments carried out so far. The group velocity in the medium is given by

$$v_g = \frac{c}{1 + (v/2)(d\chi'/dv)},$$

where the derivative is evaluated at the carrier frequency.

We can write density matrix equations for the coherences of the Λ-type atoms moving with velocity v as

$$\dot{\rho}_{ab} = -\Gamma_{ab}\sigma_{ab} + i(\rho_{aa} - \rho_{bb})\Omega_p - i\Omega_d\rho_{cb} \quad (12.15)$$

$$\dot{\rho}_{ca} = -\Gamma_{ca}\sigma_{ab} + i(\rho_{cc} - \rho_{aa})\Omega_d + i\Omega_p\rho_{cb} \quad (12.16)$$

$$\dot{\rho}_{cb} = -\Gamma_{bc}\rho_{cb} + i(\rho_{ca}\Omega_p - \Omega_d\rho_{ab}), \quad (12.17)$$

where $\Gamma_{ab} = \gamma + i(\Delta_p + k_p v)$, $\Gamma_{ca} = \gamma - i(\Delta_d + k_d v)$, $\Gamma_{cb} = \gamma_{bc} + i[\delta + (k_p - k_d)v]$, Δ_d and Δ_p are the detunings of the probe and the drive fields, respectively, and $\delta = \Delta_p - \Delta_s$. We simplify these equations by assuming the case of a strong driving field $|\Omega_d|^2 \gg \gamma\gamma_{bc}$ and a weak probe, so that in the first approximation all atomic population is in state $|b\rangle$. For this case, the density matrix equations for the Λ-system shown in Figure 12.2, are reduced to

$$\dot{\rho}_{ab} = -\Gamma_{ab}\sigma_{ab} - i\Omega_p - i\Omega_d \rho_{cb} \tag{12.18}$$

$$\dot{\rho}_{cb} = -\gamma_{cb}\rho_{cb} - i\Omega_d \rho_{ab}. \tag{12.19}$$

Finally, we obtain

$$\chi = \frac{i\eta\gamma_r \Gamma_{bc}}{\Gamma_{bc}(\gamma + i\Delta_p) + \Omega^2}. \tag{12.20}$$

We may now consider atomic motion. For atoms moving with velocity v, we replace Δ_p in Equation 12.20 with $\Delta_p + k_p v$, assuming $|\Omega_d|^2 \gg (\Delta\omega_D)^2 \gamma_{bc}/\gamma$, where $\Delta\omega_D$ is the Doppler width. Averaging over the velocity distribution gives:

$$\chi(\nu_p) = \int_{-\infty}^{\infty} \frac{i\eta\gamma_r \Gamma_{bc}}{\Gamma_{bc}[\gamma + i(\Delta_p + k_p v)] + \Omega^2} f(v) dv. \tag{12.21}$$

In this expression ν_p is the probe laser frequency, $\eta = (3\lambda^3 N)/(8\pi^2)$ where λ is the probe wavelength and N is the atomic density, $\Gamma_{bc} = \gamma_{bc} + i[\delta + (k_p - k_d)v]$, where γ_r is the radiative decay rate from level a to level b, γ_{bc} is the coherence decay rate of the two lower levels, (governed here by the time-of-flight through the laser beams), γ is the total homogeneous half-width of the drive and probe transitions (including radiative decay and collisions), $\Delta_p = \omega_{ab} - \nu_p$ and $\Delta_d = \omega_{ac} - \nu_d$ are the one-photon detunings of the probe and drive lasers, and $\delta = \Delta_p - \Delta_d$ is the two-photon detuning, Ω is the Rabi frequency of the drive transition, k_p and k_d are wave numbers of the probe and driving fields, respectively. One can obtain a simple analytic expression corresponding to Equation 12.21 by approximating the thermal distribution $f(v)$ by a Lorentzian, $f(v) = (\Delta\omega_D/\pi)/[(\Delta\omega_D)^2 + (kv)^2]$, where $\Delta\omega_D$ is the Doppler half-width of the thermal distribution and v is the projection of the atomic velocity along the laser beams. The result is

$$\chi(\nu_p) = \eta\gamma_r \frac{i\gamma_{bc} - \delta}{(\gamma + \Delta\omega_D + i\Delta_p)(\gamma_{bc} + i\delta) + \Omega^2}, \tag{12.22}$$

where $k = k_p = k_d$. The typical dependence of real and imaginary parts of susceptibility ($\chi(\nu) = \chi'(\nu) + i\chi''(\nu)$) for EIT are shown in Figure 12.3, and correspondent dispersion $k(\nu)$ near atomic resonance is shown in Figure 12.4.

12.1.3 Propagation in Coherent Media

To describe propagation of EM waves in general case, we start with Maxwell equation that is given by

$$\Delta E - \frac{1}{c^2}\frac{\partial^2 E}{\partial t^2} = \frac{4\pi}{c^2}\frac{\partial^2 P}{\partial t^2} \tag{12.23}$$

Presenting the field and the polarization as

$$E = \sum_\nu E_\nu e^{-i\nu t + ik\psi}, \quad P = \sum_\nu P_\nu e^{-i\nu t + ik\psi}, \tag{12.24}$$

Coherent Control and Nonlinear Wave Mixing in Slow Light Media

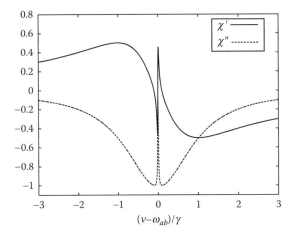

FIGURE 12.3 Real and imaginary parts of susceptibility for a three-level atomic system.

where ψ is the eikonal, the polarization of the medium is related to the field intensity as $P_\nu = \chi_\nu E_\nu$, where the susceptibility χ is $\chi_\nu = \chi'_\nu + i\chi''_\nu$. Neglecting the second-order derivative over coordinates for amplitude E_ν, we obtain the eikonal equation

$$(\nabla \psi)^2 = 1 + 4\pi \chi'. \tag{12.25}$$

The trajectory of the light rays propagating in an inhomogeneous medium can be found by solving a geometrical optics differential equation [52] that is given in vector form by

$$\frac{d}{ds}\left(n\frac{d\vec{R}}{ds}\right) = \nabla n \tag{12.26}$$

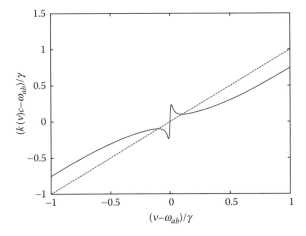

FIGURE 12.4 Dispersion $k(\nu)$ of ultradispersive medium (solid line). For comparison we also show the dispersion of vacuum (dashed line).

where $\vec{R}$ is the point of the ray, n is the index of refraction defined as follows $n^2 \equiv 1 + 4\pi\chi'$. $\vec{R}(x,z) = X(z)\hat{x} + z\hat{z}$, $\hat{x}$ and $\hat{z}$ are the unit vectors along the axis. Then, for the x and z components,

$$\frac{d}{ds}\left(n\frac{dX}{ds}\right) = \frac{\partial n}{\partial x}, \quad \text{and} \quad \frac{d}{ds}\left(n\frac{dz}{ds}\right) = \frac{\partial n}{\partial z}. \tag{12.27}$$

Using $ds = \sqrt{dX(z)^2 + dz^2} = dz\sqrt{1 + X'(z)^2}$, and assuming $n = n_0 + \alpha x$, we obtain

$$\frac{d}{ds}\left(n\frac{dX}{ds}\right) = \alpha, \quad \text{and} \quad \frac{d}{ds}\left(n\frac{dz}{ds}\right) = 0. \tag{12.28}$$

The ordinary differential equation to describe ray trajectory is given by

$$X(z)'' = \frac{\alpha}{n_0 + \alpha X(z)} \simeq \frac{\alpha}{n_0}, \tag{12.29}$$

and its solution is

$$X(z) \simeq X_0 + \frac{\alpha z^2}{2n_0}. \tag{12.30}$$

The equation describing the amplitude of the EM field can be obtained similarly as we obtain Equation 12.25, and it is given by

$$2ik\nabla\psi\nabla E_\nu + ik\nabla^2\psi E_\nu = -\frac{4\pi\nu^2}{c^2}\chi'' E_\nu. \tag{12.31}$$

The solution of the above equation has the following form

$$E_\nu = \frac{E_{0\nu}}{\sqrt{n}}\exp\left(-\int_{s_1}^{s_2}\frac{2\pi\nu\chi''}{nc}ds\right). \tag{12.32}$$

Residual absorption is the following

$$\kappa L = \frac{\gamma_{cb} + (\Omega^2/\Gamma)}{V_g}L \sim 1 \tag{12.33}$$

then the light turning angle is

$$\theta = \frac{\omega - \omega_{ab}}{kV_g}\frac{\Delta V_g}{V_g}\frac{L}{D} = \frac{\omega - \omega_{ab}}{\gamma_{cb}kD}\frac{\Delta V_g}{V_g}\frac{\gamma_{cb}L}{V_g} \tag{12.34}$$

and for realistic parameters $\omega - \omega_{ab} \simeq 10^8$ s^{-1}, $\gamma_{cb} = 10^3$ s^{-1}, $kD \simeq 10^5 0.1 \simeq 10^4$, the estimation gives us

$$\theta \simeq 0.1 - 1, \tag{12.35}$$

which shows a lot of potential for all-optical light steering. Note that the spatial dependence of gradient of the driving field is important. It might lead to some spread of the probe field because the angle may depend on the spatial coordinates. It is easy to check that there is a dependence of driving field that introduces the same angle at all spacial points. Indeed, let us assume dependence of the drive field perpendicular to the direction of propagation is

$$V_g(x) = \frac{V_{g0}}{1 + \beta x} \tag{12.36}$$

where $\beta = \alpha(kV_{g0}/\delta\omega)$, then solving geometrical optics equation gives us the same steering angle for all transverse position x.

Thus, this EIT prism yields large angular dispersion that has been experimentally confirmed [53]. Such ultra high-frequency dispersion could be used for a compact spectrometer with high spectral resolution, similar to compact atomic clocks and magnetometers [54]. The prism has a huge angular dispersion ($d\theta/d\lambda = 10^3$ nm^{-1}) that is of the order of magnitude higher than the typical prisms ($d\theta/d\lambda = 10^{-4}$ nm^{-1}, $R = 10^4$), diffraction gratings ($d\theta/d\lambda = 10^{-3}$ nm^{-1}, $R = 10^6$), or even interferometers ($R = 10^9$) which can spatially resolve spectral widths of a few kHz (corresponding spectral resolution $R = \lambda/\delta\lambda \simeq 10^{12}$).

The ability to control the direction of light propagation by another light beam in transparent medium can be applied to optical imaging and to all-optical light steering [55]. Also, this prism can be used for all-optical controlled delay lines for radar systems. This technique can be easily extended to short pulses by using the approach developed in Ref. [56].

On the other hand, together with application to relatively intense classical fields, the ultradispersive prism can have application to weak fields, such as a single photon source, and control the flow of photons at the level of a single quantum [57,58].

12.1.3.1 Delay Time via EIT

Note that slow light can be used for controllable delay lines for light. Indeed, Equation (12.22) leads to propagation with absorption coefficient $\alpha = (k/2)\chi''(\nu_p)$ and group velocity $v_g = c/(1 + n_g)$. Therefore, one obtains

$$\alpha = \frac{3}{8\pi} N\lambda^2 \frac{\gamma_r \gamma_{bc}}{\gamma_{bc}(\gamma + \Delta\omega_D) + \Omega^2}, \qquad (12.37)$$

$$n_g = \frac{3}{8\pi} N\lambda^2 \frac{\gamma_r \Omega^2 c}{[\gamma_{bc}(\gamma + \Delta\omega_D) + \Omega^2]^2}. \qquad (12.38)$$

After propagation through a dense coherent ensemble of length L, its envelope is delayed compared to free space propagation by $T_g = n_g L/c$ and the intensity of the pulse is attenuated by $\exp(-2\alpha L)$ and, in the limit of $\Omega^2 \gg \gamma_{cb}\Delta_D$, the attenuation is $\exp(-2\gamma_{cb}T_g)$.

12.1.3.2 Interference and Frequency Stabilization via Slow and Fast Light

The medium with steep dispersion can be used for interferometers. A beam of light splitted into two or more arms that are then merged together produces interference pattern depending on relative phases in different optical beams. The phases depend on the dispersion of the medium placed in one of these arms and the length of the arm. Tuning the frequency of light, one can detect frequency shift vs change of the length. And if the length of the arm changes due to gravitational wave, the frequency change can be measured and related to the change of the length.

For simplicity let us consider Mickelson interferometer with the ultradispersive medium, placed inside of one arm. Then, the frequency shift due to the shift of one mirror at the length of δL is given by

$$\delta \nu = kV_g \frac{\delta L}{l} \qquad (12.39)$$

where l is the length of the dispersive medium. The frequency shift is higher for fast light. Also let us note that the medium with ultraslow light has smaller frequency shift (frequency stabilization). The width of resonance is also changed. The white cavity can be built using fast light media. On the other hand, for slow light media, the line can be even narrower, its width determined by the delay time.

Let us note that motion of the dispersive medium produces a drastic modification of the dispersion, changing both the group and phase velocities of the light propagating through the cell [48]

$$V'_g = V_g + v, \quad V'_{ph} = V_{ph} \frac{V_g}{V_g + v} \tag{12.40}$$

where V_g and V'_g are the group velocities of the light with respect to the cell at rest and with respect to the cell moving with velocity v, and similarly V_{ph} and V'_{ph} are the phase velocities of the light. As, it is a linear effect we do not consider it here. In Section 12.2 we consider three- and four-wave mixing (FWM) and their modifications via slow light.

12.2 NONLINEAR WAVE MIXING VIA SLOW LIGHT

In this section, we focus on and show how the ultrasteep dispersion due to EIT can be used for applications to nonlinear optics and, in particular, to control phase-matching relation. First, we consider three-wave mixing. Second, we consider FWM. In both cases, we show that due to ultradispersive media, it is possible to obtain a configuration that is usually forbidden by text book wisdom.

12.2.1 FORWARD BRILLOUIN SCATTERING

We show that phase-matched conditions can be established for a wide range of fiber parameters by taking advantage of the large linear dispersion associated with EIT. This makes it possible to slow the group velocity of a laser pulse down to the speed of sound in solids [5,35], and, therefore, to utilize the ponderomotive nonlinearity of the fiber for new phase modulators, frequency shifters, and sensors on the one hand, and for effective quantum wave mixing, generation of nonclassical states of light, and quantum nondemolition measurements on the other.

Indeed, the phase-matching condition for ordinary media is given by

$$k_1 - k_2 = k_s, \tag{12.41}$$

where k_1, k_2, and k_s are wavevectors of optical fields and an acoustic wave correspondingly. Dispersion of acoustic wave $k_s = V_s \nu_s$. Simultaneously, the condition between optical field frequencies and acoustic frequency have to be satisfied, so

$$\nu_1 - \nu_2 = \nu_s. \tag{12.42}$$

Now, it is clear that for ordinary media (with small group index), both the conditions can be satisfied only for counter-propagating geometry of optical fields. Indeed, for this case $k_2 \simeq -k_1 = -k$

$$k_1 - k_2 \simeq 2k = \frac{\nu_s}{V_s} \tag{12.43}$$

and $\nu_s = 2kV_s$. While for copropagating geometry of optical fields, we have

$$k_1 - k_2 \simeq \frac{\nu_s}{c} \neq \frac{\nu_s}{V_s}, \tag{12.44}$$

and the phase condition cannot be met.

Let us consider two EM waves copropagating in the fiber interacting with an acoustic mode of the fiber. The phonon–photon Hamiltonian interaction is given by

$$H_{int}(t) = \hbar g \frac{\sin \Delta k L}{\Delta k L} \hat{a}_1(t) \hat{a}_2^+(t) \hat{b}^+(t) + \mathrm{adj}, \tag{12.45}$$

Coherent Control and Nonlinear Wave Mixing in Slow Light Media

where

- g is a coupling constant
- L is the length of the fiber
- $\hat{a}_{1,2}$ and $\hat{b}$ are the annihilation operators of the EM fields and acoustic phonons (we assume that all operators are slowly varying functions of time and space)
- $k_{1,2}$, $\nu_{1,2}$ and k_b, ω_b are the wave vectors and frequencies
- $\Delta k = k_1 - k_2 - k_b$

We note that conservation of energy requires $\nu_1 - \nu_2 = \omega_b$. However, in general, we do not have $k_1(\omega_1) - k_2(\omega_2) = k_b(\omega_b)$ because of dispersion effects. As $\omega_b \ll \nu_{1,2}$, we may write

$$k_1 - k_2 = \frac{\nu_1 n(\nu_1)}{c} - \frac{\nu_2 n(\nu_2)}{c} \simeq \frac{\nu_1 - \nu_2}{c} \frac{\partial [\nu n(\nu)]}{\partial \omega}$$

$$= \frac{\nu_1 - \nu_2}{V_g}$$

and $k_b = \omega_b / V_s$, where $V_g = c/[\partial(\nu n(\nu))/\partial \nu]$ and V_s are the EM group and the sound velocities, respectively (see Figure 12.5). In the following we represent the EM field as a sum of large classical expectation and small fluctuation parts $\hat{a}_{1,2} = \langle a_{1,2} \rangle + \delta \hat{a}_{1,2}$ ($\langle a_{1,2} \rangle \gg \delta \hat{a}_{1,2}$, and $\langle a_{1,2} \rangle$ are real values).

It is easy to see from Equation 12.45, that even in the most interesting resonant case $\Delta = \nu_1 - \nu_2 - \omega_b = 0$, we still have phase mismatch $\Delta kL = \omega_b(1/V_g - 1/V_s)L \gg 1$, and the interaction vanishes with increasing L if the light group velocity V_g is different from the phase sound velocity V_s. As noted earlier, the group velocity is relevant here because it characterizes the difference in phase velocities for light waves of different frequencies. It is this phase velocity difference that enters the phase-matching condition. If the condition $\omega_b = \nu_1 - \nu_2$ is satisfied and the condition $V_g = V_s$ is met then phase matching is achieved.

To slow the group velocity of light to the speed of sound we propose to use a medium, consisting of a host doped by Λ-atoms or ions (Figure 12.2) (e.g., Eu^{3+}, Er^{3+}, etc.). This could be a doped silica glass or crystal fiber. The fields are nearly resonant with the corresponding atomic transitions and include a strong driving field with frequency ω_d and Rabi frequency Ω, and two sufficiently weak fields with carrier frequencies ω_1 and ω_2 and Rabi frequencies α_1 and α_2 ($\alpha_{1,2} = \wp_{ab} \langle a_{1,2} \rangle \sqrt{4\pi \omega_{ab}/\hbar V}$, where V is the total volume of the mode in the fiber, and $\wp_{ab}$ is the dipole momentum of the probe transition). The fields interact via the long-existing coherence of the dipole-forbidden transition between the ground state sublevels $|b\rangle$ and $|c\rangle$. Because in doped materials decay of this ground state

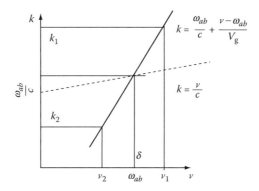

FIGURE 12.5 Dispersion of ultradispersive medium, $k(\nu) = \omega_{ab}/c + (\nu - \omega_{ab})/V_g$, (solid line); and vacuum, $k(\nu) = \nu/c$, (dotted line).

coherence (decay rate is about $\sim 10^3$ s^{-1} or more) usually dominates over decay of spin exchange (decay rate is about ~ 1 s^{-1}), we consider the first type of decay only.

EM wave propagation in the fiber occurs with the absorption coefficients $\beta_{1,2} = (k/2)\chi''(\nu_{1,2})$, and the group velocity $V_g = c/n_g$, $n_g = 1 + \nu/2(\partial \chi'/\partial \nu)$. The pump power should be large enough to sustain EIT in the system ($\Omega^2 \gg \gamma_{bc}\gamma$ and $\Omega \gg \alpha_{1,2}$) and large enough bandwidth to provide EIT for both fields $\alpha_{1,2}$ ($\Omega^2 \gg \omega_b\gamma$). In the above limits we obtain expressions for the group velocity index and the resonant losses in the form [5]

$$\beta_{1,2} \simeq \frac{3}{8\pi} N\lambda^2 \frac{\gamma_r \gamma_{bc}}{|\Omega|^2}, \qquad (12.46)$$

$$n_{g\,1,2} \simeq \frac{3}{8\pi} N\lambda^2 \frac{c\gamma_r}{|\Omega|^2}, \qquad (12.47)$$

where
  N is the density of dopants
  λ is the EM wavelength
  γ_r is the natural linewidth of optical transitions
  γ_{bc} is the decay rate of the ground state coherence
  γ is the total linewidth of the transitions

To meet the phase-matching condition we use Equation 12.3 to write

$$V_s = V_g = \frac{8\pi |\Omega|^2}{3N\lambda^2 \gamma_r}. \qquad (12.48)$$

Noting that Rabi frequency is related to the laser power by

$$|\Omega|^2 = \frac{3\lambda^3 \gamma_r P_\Omega}{8\pi^2 \hbar c \mathcal{A}}, \qquad (12.49)$$

where $P_{\alpha 1}$, $P_{\alpha 2}$, and P_Ω are the powers of the probe and drive fields, respectively, $\mathcal{A}$ is the cross-sectional area of the fiber, from Equations 12.48 and 12.49 we derive

$$\frac{P_\Omega}{\mathcal{A}} = NV_s \frac{\hbar \nu_1}{2}. \qquad (12.50)$$

Once the phase-matching condition has been established (i.e., $V_s = V_g$), interaction between the fields (see Equation 12.45) is sustained through the whole fiber length L.

It should be mentioned that the self-phase modulation is much less than the cross-phase modulation due to the resonant feature of the nonlinearity, while in ordinary fibers they are equal. The total power of the driving field required for establishing the phase-matching condition is quite reasonable $P_\Omega = 38$ mW, and due to the phase-matched operation, the ponderomotive nonlinearity of the fiber holds promise for application in quantum and nonlinear optics.

In summary, we have theoretically demonstrated how ultraslow light can yield phase matching in optical fibers which allows us to achieve strong coupling between high-quality acoustic waves of the fiber and multifrequency EM fields. The method can be implemented in the fiber doped by three-level Λ atoms or ions which possess steep dispersion with low absorption close to the point of two-photon resonance. We predict that the fiber holds promise for an effective wave mixer or amplifier at low temperatures due to the large ponderomotive nonlinearity associated with acoustic oscillations of the fiber.

12.2.2 COHERENT CONTROL OF NONLINEAR MIXING: COHERENT BACKSCATTERING

In this section, we demonstrate strong coherent backward scattering via excitation of quantum coherence between atomic and molecular levels. The developed approach can also be used to control the direction of the signal generated in coherent Raman scattering and other FWM schemes.

Let us consider the FWM in a three-level atomic medium. The pump and Stokes fields $\mathcal{E}_1$ and $\mathcal{E}_2$ (whose Rabi frequencies are defined as $\Omega_1 = \wp_1 \mathcal{E}_1/\hbar$ and $\Omega_2 = \wp_2 \mathcal{E}_2/\hbar$, where $\wp_1$ and $\wp_2$ are the dipole moments of the corresponding transitions) with wave vectors k_1 and k_2 and angular frequencies ν_1 and ν_2 induce a coherence grating in the medium (see Figure 12.6) given by [8]

$$\rho_{cb} \sim -\Omega_1 \Omega_2^* \tag{12.51}$$

Let us stress that the ρ_{cb} coherence grating has an $\exp[i(k_1 - k_2)z]$ spatial dependence. In an ultradispersive medium (see Figure 12.7) where fields propagate with a slow group velocity, the two copropagating fields have wave vectors given by

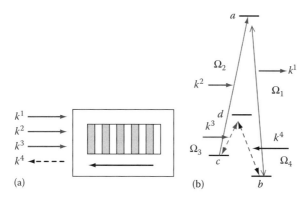

FIGURE 12.6 (a) Copropagating fields 1 and 2 induce coherent grating inside the medium. The field 3 propagating in the same direction is scattered in the opposite direction because the coherence excited by fields 1 and 2 is propagating in the opposite direction (see Figure 12.7). Level scheme, double-Λ (b), for implementation of coherent back scattering.

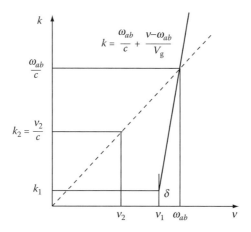

FIGURE 12.7 Dispersion $k(\nu)$ of ultradispersive medium. Choosing $\delta = \nu_1 - \omega_{ab} = -V_g \omega_{cb}/c$, we can have $k_1 - k_2 < 0$ even if $\nu_1 > \nu_2$, thus the third field can be scattered opposite to the direction of propagation of the first two fields.

$$k_1 \simeq k_1(\omega_{ab}) + \frac{\partial k_1}{\partial \nu_1}(\nu_1 - \omega_{ab}) = \omega_{ab}/c + (\nu_1 - \omega_{ab})/V_g, \qquad (12.52)$$

where

V_g is the group velocity of the first wave
ω_{ab} is the frequency of transition between levels a and b
$k_2 = \nu_2/c$

Thus these two fields create a coherence grating in the medium with spatial phase determined by $k_1 - k_2 = \omega_{cb}/c + (\nu_1 - \omega_{ab})/V_g$ which depends strongly on the detuning $\delta = \nu_1 - \omega_{ab}$. By properly choosing the detuning, δ, one can make $k_1 - k_2$ negative.

After the coherence ρ_{bc} is induced in the medium, a probe field $\mathcal{E}_3$, with Rabi frequency $\Omega_3 = \wp_3\mathcal{E}_3/\hbar$ and wave vector k_3, scatters off that coherence to produce the signal field, Ω_4. The signal field depends on the coherence and the input fields as

$$\frac{\partial}{\partial z}\Omega_4 \sim \rho_{cb}\Omega_3 \sim \Omega_1\Omega_2^*\Omega_3 \sim e^{i(k_1 - k_2 + k_3 - k_4)z} \qquad (12.53)$$

That is, the propagation direction of Ω_4 depends on the spatial phase of the ρ_{bc} coherence through the phase-matching condition $k_4 = k_1 - k_2 + k_3$ [47] while its frequency is determined by $\nu_4 = \nu_1 - \nu_2 + \nu_3$.

We here show that for ultradispersive media, one can obtain a strong signal in the backward direction even when all three input fields propagate forward. This is contrary to the usual nondispersive media results, where the phase-matching in the backward direction cannot be achieved for $\mathcal{E}_1$, $\mathcal{E}_2$, and $\mathcal{E}_3$ counter-propagating with respect to $\mathcal{E}_4$ [47]. Indeed, for ordinary nondispersive media, the conditions for frequencies and wave numbers coincide, and $k_4 = k_1 - k_2 + k_3 = n(\nu_1 - \nu_2 + \nu_3)/c = n\nu_4/c$, and thus backward scattering in nondispersive media is forbidden. It is the ultradispersion of coherently driven media that provides opportunity to implement coherent scattering in backward direction.

To demonstrate this result, we write the Hamiltonian interaction of the system as

$$V_I = -\hbar[\Omega_1 e^{-i\omega_{ab}t}|a\rangle\langle b| + \Omega_2 e^{-i\omega_{ac}t}|a\rangle\langle c| + \text{h.c.}] \qquad (12.54)$$
$$-\hbar[\Omega_4 e^{-i\omega_{db}t}|d\rangle\langle b| + \Omega_3 e^{-i\omega_{dc}t}|d\rangle\langle c| + \text{h.c.}] \qquad (12.55)$$

where $\Omega_4 = \wp_4\mathcal{E}_4/\hbar$ is the Rabi frequency of the signal field and ω_{ab}, ω_{ac}, ω_{db}, ω_{dc} are the frequency differences between the corresponding atomic or molecular energy levels (see Figure 12.6b). The time-dependent density matrix equation is given by

$$\frac{\partial \rho}{\partial \tau} = -\frac{i}{\hbar}[V_I, \rho] - \frac{1}{2}(\Gamma\rho + \rho\Gamma), \qquad (12.56)$$

where Γ is the relaxation matrix. A self-consistent system also includes the field propagation equations

$$\frac{\partial \Omega_1}{\partial z} + ik_1\Omega_1 = -i\eta_1\rho_{ab}, \quad \frac{\partial \Omega_2}{\partial z} + ik_2\Omega_2 = -i\eta_2\rho_{ac}, \qquad (12.57)$$

$$\frac{\partial \Omega_4}{\partial z} + ik_4\Omega_4 = -i\eta_3\rho_{db}, \quad \frac{\partial \Omega_3}{\partial z} + ik_3\Omega_3 = -i\eta_4\rho_{dc}, \qquad (12.58)$$

where

$\eta_j = \nu_j N\wp_j/(2\epsilon_0 c)$ are the coupling constants ($j = 1, 2, 3, 4$)
N is the particle density of the medium
ϵ_0 is the permittivity in vacuum

The equations of motion for the density matrix elements of the polarization ρ_{ab} and the coherence ρ_{cb} are given by

$$\dot\rho_{ab} = -\Gamma_{ab}\rho_{ab} + i\Omega_1(\rho_{aa} - \rho_{bb}) - i\rho_{cb}\Omega_2^*, \tag{12.59}$$

$$\dot\rho_{cb} = -\Gamma_{cb}\rho_{cb} + i\rho_{ca}\Omega_1 - i\rho_{ab}\Omega_2. \tag{12.60}$$

where

$\Gamma_{ab} = \gamma_{ab} + i(\omega_{ab} - \nu_1);\ \Gamma_{ca} = \gamma_{ca} - i(\omega_{ac} - \nu_2);\ \Gamma_{cb} = \gamma_{cb} + i(\omega_{cb} - \nu_1 + \nu_2)$
ω_{cb} is the frequency of $c - b$ transition
$\gamma_{\alpha\beta}$ are the relaxation rates at the corresponding transitions

In the steady state regime, and assuming that $|\Omega_2| \gg |\Omega_1|$, almost the entire population remains in the ground level $|b\rangle$, $\rho_{bb} \simeq 1$. Let us consider the fields as plane waves: $\Omega_1(z,t) = \tilde\Omega_1(z,t)\exp(ik_1 z)$, $\Omega_2(z,t) = \tilde\Omega_2(z,t)\exp(ik_2 z)$, where $\tilde\Omega_1(z,t)$ and $\tilde\Omega_2(z,t)$ are the slowly varying in envelopes of the fields Ω_1 and Ω_2 in space, while $k_1 = \nu_1[1 + \frac{\chi_{ab}(\nu_1)}{2}]/c$ and $k_2 = \nu_2[1 + \frac{\chi_{ac}(\nu_2)}{2}]/c$. The susceptibilities are $\chi_{ab} = \eta_1\rho_{ab}/\Omega_1 = 2c(\nu_1 - \omega_{ab})/(\nu_1 V_g)$ and $\chi_{ac} = \eta_2\rho_{ac}/\Omega_2 \simeq 0$. By solving the self-consistent system of Maxwell's Equations. 12.57 and 12.58 and the density matrix Equations 12.59 and 12.60, we obtain Equation 12.52 for the wave vectors, where $V_g \simeq c|\Omega_2|^2/(\nu_1\eta_1)$ is the group velocity of the optical field Ω_1. Thus, the spatial dependence of ρ_{cb} is determined by

$$\Delta k = k_1 - k_2 = \frac{\nu_1 - \nu_2}{c} + \frac{\nu_1 - \omega_{ab}}{V_g}. \tag{12.61}$$

The signal field Ω_4 is generated by the polarization ρ_{db} of the transition it couples (Equation 12.58). The equation of motion for this polarization element reads

$$\dot\rho_{db} = -\Gamma_{db}\rho_{db} + i\Omega_4(\rho_{dd} - \rho_{bb}) - i\rho_{cb}\Omega_3, \tag{12.62}$$

where $\Gamma_{db} = \gamma_{db} + i(\omega_{db} - \nu_4)$, and ν_4 is the frequency of generated field. In the steady state regime and for $|\Omega_4| \ll |\Omega_3|$, the field Ω_4 at the output of the cell is given by

$$\Omega_4 = -i\frac{\eta_4 e^{ik_4 L}}{\Gamma_{db}} \int_0^L dz\, e^{i(k_3 - k_4)z} \rho_{cb} \tilde\Omega_3 \tag{12.63}$$

where L is the length of the cell. Note here that Equation 12.63 is valid if the field $|\Omega_3|$ does not change coherence ρ_{cb} via power broadening which is true if $|\Omega_3|^2 \ll |\Omega_1|^2 + |\Omega_2|^2$.

Hence, after integrating Equation 12.63, we obtain for the scattered field Ω_4

$$|\Omega_4|^2 = \left[\frac{\sin(\delta kL)}{\delta kL}\right]^2 \frac{\eta_4^2 L^2 |\Omega_1|^2 |\Omega_3|^2}{|\Gamma_{db}|^2 |\Omega_2|^2} \tag{12.64}$$

where $\delta k = k_3 + \Delta k - k_4$, and the expression in the brackets describes the phase matching and determines the direction in which the signal field is generated (a small mismatch δk of the order of $1/L$ is allowed).

The most interesting effect following from Equation 12.64 is the coherent backscattering. Indeed, even when all three input fields propagate forward, one may observe a backscattered signal field by satisfying the condition

$$k_3 + \Delta k = k_3 + \frac{\omega_{cb}}{c} + \frac{\nu_1 - \omega_{ab}}{V_g} = -|k_4|, \tag{12.65}$$

and for appropriate detuning, $v_1 - \omega_{ab} < 0$, this equality can be met. That is, in order to obtain phase matching in the backward direction, we have to satisfy

$$-\frac{v_1 - \omega_{ab}}{V_g} = k_3 + \frac{\omega_{cb}}{c} + |k_4| \simeq 2|k_4|. \tag{12.66}$$

Hence, in order to demonstrate the effect, the detuning δ should meet the condition $\delta = -2|k_4|V_g$. It is useful to rewrite the condition in terms of susceptibility for the probe field, indeed,

$$k_1 = \frac{v_1}{c}n_1 \simeq \frac{v_1}{c}\left(1 + c\frac{v_1 - \omega_{ab}}{v_1 V_g}\right) = \frac{v_1}{c}\left(1 - c\frac{2|k_4|}{v_1}\right), \tag{12.67}$$

then, $\chi_{ab} = 2(n_1 - 1) = -4\frac{\lambda_{ab}}{\lambda_{db}}$, for gases $\chi_{ab} \ll 1$, so $\lambda_{ab} \ll \lambda_{db}$, that is, the effect can be implemented for scattering of IR fields. Then, for the Doppler-broadened EIT media as shown in Refs. [59,60], we can write

$$\chi_{ab}(\delta) \simeq \frac{3\lambda_{ab}^3 N}{8\pi^2}\left(\frac{\gamma_r \delta}{|\Omega_2|^2} + i\frac{\gamma_r \Delta_D \delta^2}{|\Omega_2|^4}\right), \tag{12.68}$$

where
$\Delta_D = k_1 u_D$ is the Doppler width
u_D is the rms velocity
γ_r is the radiative decay rate

Thus, for detuning smaller than the EIT width $|\delta| \leq |\Omega_2|^2/\sqrt{\gamma_r \Delta_D}$, absorption can be neglected, and

$$\frac{3\lambda_{ab}^2 N \gamma_r \delta}{16\pi |\Omega|^2} = -2|k_4|, \tag{12.69}$$

then, the atomic or molecular density is given by

$$N = \frac{32\pi |k_4|}{3\lambda_{ab}^2}\frac{|\Omega_2|^2}{\gamma_r |\delta|} \simeq \frac{32\pi |k_4|}{3\lambda_{ab}^2}\sqrt{\frac{\Delta_D}{\gamma_r}}. \tag{12.70}$$

12.2.2.1 Spectroscopy via Slow Light

Next, let us consider the FWM in the case where all fields propagate in the same direction. The tuning of the probe frequency changes the signal frequency, $v_4 = v_1 - v_2 + v_3$, and eventually breaks the phase-matching condition aforementioned. The spectral width of the signal is determined by the delay time, and, surprisingly, it is not related to the relaxation; besides this spectral narrowing of the FWM in a dense medium can be seen in a collinear configuration of fields, and it comes from the phase matching.

$$\left|\frac{\Omega_4}{\Omega_3}\right|^2 = \left(\frac{\sin \delta v L/V_g}{\delta v L/V_g}\right)^2 \left|\frac{\eta L}{\Gamma_{bd}}\right|^2 \tag{12.71}$$

The spectral resolution is defined by the delay time, as shown in Ref. [61],

$$\delta v \simeq 2\pi \tau^{-1}. \tag{12.72}$$

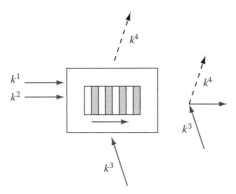

FIGURE 12.8 Scheme for observation of reflection of noncolinear generation of new field.

12.2.2.2 Nonlinear Light Steering

Next, let us consider a noncollinear propagation as shown in Figure 12.8. This configuration provides the opportunity to develop a spectroscopic technique based on the spatial resolution of the signal generated at different angles corresponding to different phase-matched directions for various two-photon detunings.

The spectral resolution is obtained by the condition that the difference between the angles should be larger than the one given by diffraction, namely, one should meet

$$\delta\phi = \frac{\delta\nu}{V_g k_3} > \frac{\lambda_3}{L}, \quad (12.73)$$

which leads to the same condition as seen earlier, hence, the spectral resolution is given by

$$\delta\nu > k_3 \lambda_3 \frac{V_g}{L} = \frac{2\pi}{\tau_{\text{delay}}}. \quad (12.74)$$

Foe example, to perform such an experiment in Rb vapor, where one needs to observe

$$\phi \simeq \frac{2\delta\nu}{kV_g} \simeq 10^{-2} \quad (12.75)$$

which is a very realistic condition for experimental verification. Our scheme might be used as a novel spatial filter as it provides the way to specifically resolve coherent antistokes Raman scattering (CARS) signals from two molecules with close vibrational energy spacings even if they have very fast relaxation.

12.2.2.3 Implementation of Obtained Results

There are several schemes to demonstrate the effect. For example, the double-Lambda scheme can be implemented in molecular rotational levels (see Figure 12.6a). Moreover, the effect can be implemented in the ladder-Λ using molecular vibrational levels (see Figure 12.9a). The phase-matching condition should be slightly modified for the scheme as $k_4 = k_1 - k_2 - k_3$. Also, the phenomenon can be demonstrated in a V–Λ scheme that can be realized in atomic levels (see Figure 12.9b, for Rb atoms, $b = 5S_{1/2}$, $c = 7D_{3/2,5/2}$, $a = 5P_{1/2,3/2}$, $d = 8P_{1/2,3/2}$), and phase-matching condition has a form $k_4 = k_1 + k_2 - k_3$. Let us note that the requirement for detuning in all cases is $\delta/V_g = -2|k_4|$, and the Equation 12.70 to estimate molecular or atomic density is still valid.

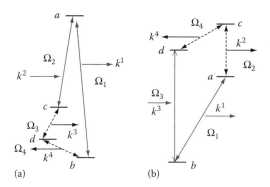

FIGURE 12.9 Implementation: molecular systems, (a) vibrational levels; copropagating fields 1 and 2 induce coherence between vibrational levels. The field 3 propagating in the same direction is scattered in the opposite direction. (b) Atomic Rb scheme for implementation of coherent back scattering.

Examples of systems to this effect could be seen in molecules of NO (a resonant transition at 236 nm, $A^2\Sigma^+ - X^2\Pi$), NO$_2$ (a resonant transition at wavelength 337 nm), and atomic Rb vapor (EIT and CPT have been recently demonstrated for molecules, see [62]). The required molecular density of NO and NO$_2$ molecules is $N \simeq 1.2 \times 10^{13}$ cm^{-3} if one can use transition between rotational levels $\simeq 10$ cm^{-1}. Using vibrational IR transitions for NO (vibration frequency of 1900 cm^{-1}) at 5.26 μm and for NO$_2$ (vibrational frequency of 750 cm^{-1}) at 13.3 μm, the densities are $N = 8 \times 10^{15}$ cm^{-3} and $N = 1.4 \times 10^{15}$ cm^{-3}, correspondingly. For atomic Rb vapor, wavelengths are $\lambda_1 = 780$ nm, $\lambda_2 = 565$ nm, $\lambda_3 = 335$ nm, $\lambda_4 = 23.4$ μm, and the atomic density is $N = 1.4 \times 10^{13}$ cm^{-3}.

The intensity needed for EIT is determined by the condition $|\Omega|^2 \gg \gamma_{bc}\Delta_D$ [59,60,63] which corresponds to a laser intensity of the order of 1 mW/cm^2 for atoms and of the order of 10 W/cm^2 for molecules, as typical dipole moments are two orders of magnitude smaller for molecules. These conditions are realistic and well-suited for an experimental implementation. We note that, for the schemes shown in Figures 12.1b and 12.3a, the coherence preparation fields, Ω_1 and Ω_2, are in a Λ-configuration and have almost the same frequency ($\nu_1 \simeq \nu_2$) hence there is no Doppler broadening on the two-photon transition [7,8]. Meanwhile, for the scheme shown in Figure 12.3b, the field frequencies are different, and the Doppler broadening at the two-photon transition leads to the depletion of the signal field Ω_1 [43]. Then, Equation 12.64 can be rewritten as

$$|\Omega_4|^2 = \frac{\left(1 - e^{-\kappa L}\right)^2 + 4e^{-\kappa L}\sin^2(\delta k L/2)}{\delta k^2 + \kappa^2} \frac{\eta_4^2 |\Omega_1|^2 |\Omega_3|^2}{|\Gamma_{db}|^2 |\Omega_2|^2},$$

where $\kappa = 3\lambda^2 N \gamma_r (\gamma_{cb} + |\Delta k|u_D)/(8\pi |\Omega_2|^2)$ is the absorption coefficient of Ω_1. One can see that the signal generation occurs at the effective length determined by the absorption of the signal field $L_{\text{eff}} = \kappa^{-1}$ instead of L. To avoid additional Doppler broadening, the experiments could be performed in cold gases.

Several applications of the effect can be envisioned, like in nonlinear CARS microscopy [64], while the controlling of coherent backscattering could provide a new tool for creating an image. A variation in the molecular density would modify the intensity of the signal in both the forward and the backward directions. Additionally, Equation 12.64 also allows one to control the direction of the generated signal field and thus provides an all-optical control when scanning an optical field over an object.

In summary, we theoretically develop approach to control phase-matching condition for nonlinear wave mixing in coherently prepared media and predict strong coherent scattering in the backward direction while using only forward propagating fields [29,30]. This is achieved by exciting atomic or molecular coherence by properly detuned fields, in such a way that the resulting coherence has a spatial phase corresponding to a backward, counter-propagating wave. Applications of the technique to coherent scattering and remote sensing are discussed. The method holds promise for observation induced scattering in a backward direction with application to CARS microscopy.

REFERENCES

1. L. Brillouin, *Wave Propagation and Group Velocity* (New York, Academic Press), 1960.
2. R. Y. Chiao, *Quantum Opt.* 6, 359 (1994).
3. J. D. Jackson, *Classical Electrodynamics*, 2nd edn. (John Wiley and Sons, New York, NY), 1975.
4. L. V. Hau, S. E. Harris, Z. Dutton, and C. H. Behroozi, *Nature* 397, 594 (1999).
5. M. M. Kash, V. A. Sautenkov, A. S. Zibrov, L. Hollberg, G. R. Welch, M. D. Lukin, Y. Rostovtsev, E. S. Fry, and M. O. Scully, *Phys. Rev. Lett.* 82, 5229 (1999).
6. D. Budker, D. Kimball, S. Rochester, and V. Yashchuk, *Phys. Rev. Lett.* 83, 1767 (1999).
7. E. Arimondo, in E. Wolf (Ed.), *Progress in Optics*, vol. XXXV, p. 257 (Elsevier Science, Amsterdam), 1996.
8. M. O. Scully and M. S. Zubairy, *Quantum Optics* (Cambridge University Press, Cambridge, England), 1997.
9. S. E. Harris, *Phys. Today*, 50, 36 (1997).
10. J. P. Marangos, *J. Mod. Opt.* 45, 471 (1998).
11. M. Fleischhauer, A. Imamoglu, and J. P. Marangos, *Rev. Mod. Phys.* 77, 633 (2005).
12. M. S. Bigelow, N. N. Lepeshkin, and R. W. Boyd, *Science* 301, 200 (2003); J. B. Khurgin, *JOSA B* 22, 1062 (2005). G. M. Gehring, A. Schweinsberg, C. Barsi, N. Kostinski, R. W. Boyd, *Science* 312, 895 (2006); Z. Zhu et al., *J. Lightw. Technol.* 25, 201 (2007); B. Zhang et al., *Opt. Exp.* 15, 1878 (2007); F. Xia et al., *Nat. Photon.* 1, 65 (2007); Q. Xu et al., *Nat. Phys.* 3, 406 (2007); S. Fan et al., *Opt. Photon. News* 18, 41 (2007).
13. A. Javan, *Phys. Rev.* 107, 1579 (1957).
14. M. S. Feld and A. Javan, *Phys. Rev.* 177, 540 (1969).
15. I. M. Beterov and V. P. Chebotaev, *Pis'ma Zh. Eksp. Teor. Fiz.* 9, 216 (1969) [*Sov. Phys. JETP Lett.* 9, 127 (1969)].
16. T. Hänch and P. Toschek, *Z. Phys.* 236, 213 (1970).
17. O. Kocharovskaya, *Phys. Rep.* 219, 175 (1992).
18. M. O. Scully, *Phys. Rep.* 219, 191 (1992).
19. S. E. Harris, J. E. Field, and A. Imamoglu, *Phys. Rev. Lett.* 64, 1107 (1990).
20. K. Hakuta, L. Marmet, and B. P. Stoicheff, *Phys. Rev. Lett.* 66, 596 (1991).
21. P. Hemmer, D. P. Katz, J. Donoghue, M. Cronin-Golomb, M. Shahriar, and P. Kumar, *Opt. Lett.* 20, 982 (1995).
22. M. Jain, H. Xia, G. Y. Yin, A. J. Merriam, and S. E. Harris, *Phys. Rev. Lett.* 77, 4326 (1996).
23. A. S. Zibrov, M. D. Lukin, and M. O. Scully, *Phys. Rev, Lett.* 83, 4049 (1999).
24. A. V. Sokolov, D. R. Walker, D. D. Yavuz, G. Y. Yin, and S. E. Harris, *Phys. Rev. Lett.* 85, 562 (2000).
25. S. E. Harris, *Phys. Rev. Lett.* 85, 4032–4035 (2000).
26. A. B. Matsko, Y. Rostovtsev, and M. O. Scully, *Phys. Rev. Lett.* 84, 5752 (2000).
27. A. B. Matsko, Y. Rostovtsev, M. Fleischhauer, and M. O. Scully, *Phys. Rev. Lett.* 86, 2006 (2001).
28. Y. Rostovtsev, A. B. Matsko, R. M. Shelby, and M. O. Scully, *Opt. Spectrosc.* 91, 490 (2001).
29. Y. Rostovtsev, Z. E. Sariyanni, and M. O. Scully, *Phys. Rev. Lett.* 97, 113001 (2006).
30. Y. Rostovtsev, Z. E. Sariyanni, and M. O. Scully, *Proc. SPIE* 6482, 64820T (2007).
31. S. E. Harris, G. Y. Yin, M. Jain, and A. J. Merriam, *Philos. Trans. R. Soc., London, Ser. A* 355 (1733), 2291 (1997).
32. A. J. Merriam, S. J. Sharpe, M. Shverdin, D. Manuszak, G. Y. Yin, and S. E. Harris, *Phys. Rev. Lett.* 84, 5308 (2000).
33. H. Wang, D. Goorskey, and M. Xiao, *Phys. Rev. Lett.* 87, 073601 (2001).
34. R. Coussement, Y. Rostovtsev, J. Odeurs, G. Neyens, H. Muramatsu, S. Gheysen, R. Callens, K. Vyvey, G. Kozyreff, P. Mandel, R. Shakhmuratov, and O. Kocharovskaya, *Phys. Rev. Lett.* 89, 107601 (2002).

35. A. B. Matsko, O. Kocharovskaya, Y. Rostovtsev, G. R. Welch, A. S. Zibrov, M. O. Scully, in B. Bederson and H. Walther (Eds.), *The Advances in Atomic, Molecular, and Optical Physics*, vol. 46, 191 (Academic Press, New York), 2001.
36. A. V. Turukhin, V. S. Sudarshanam, M. S. Shahriar, et al., *Phys. Rev. Lett.* 88, 023602 (2002).
37. C. Y. Ye, A. S. Zibrov, Yu. V. Rostovtsev, and M. O. Scully, *Phys. Rev. A* 65, 043805 (2002).
38. V. A. Sautenkov, Y. V. Rostovtsev, H. Chen, P. Hsu, G. S. Agarwal, and M. O. Scully, *Phys. Rev. Lett.* 94, 233601 (2005).
39. H. Xiong, M. O. Scully, and M. S. Zubairy, *Phys. Rev. Lett.* 94, 023601 (2005).
40. M. O. Scully, *Phys. Rev. Lett.* 55, 2802 (1985); M. O. Scully, M. S. Zubairy, *Phys. Rev. A* 35, 752 (1987).
41. W. Schleich, M. O. Scully, and H.-G. von Garssen, *Phys. Rev. A* 37, 3010 (1988); W. Schleich and M. O. Scully, *Phys. Rev. A* 37, 1261 (1988).
42. M. O. Scully, G. W. Kattawar, P. R. Lucht, T. Opatrny, H. Pilloff, A. Rebane, A. V. Sokolov, and M. S. Zubairy, *Proc. Natl. Acad. Sci. USA* 9, 10994 (2002).
43. Z. E. Sariyanni and Y. Rostovtsev, *J. Mod. Opt.* 51, 2637 (2004).
44. G. Beadie, Z. E. Sariyanni, Y. Rostovtsev, et al., *Opt. Commun.* 244, 423 (2005).
45. D. Pestov et al., *Science* 316, 265 (2007).
46. R. W. Boyd and D. J. Gauthier, *Prog. Opt.* 43, 497 (2002).
47. R. W. Boyd, *Nonlinear Optics* (Boston, Academic Press), 1992.
48. L. D. Landau and E. M. Lifshitz, *Mechanics* (Pergamon, Oxford), 1976.
49. E. Arimondo and G. Orriols, *Nuovo Cimento Lett.* 17, 333 (1976).
50. H. R. Gray, R. M. Whitley, and C. R. Stroud, *Opt. Lett.* 3, 218 (1978).
51. H. I. Yoo and J. H. Eberly, *Phys. Rep.* 118, 239 (1985).
52. Max Born and Emil Wolf, *Principles of Optics : Electromagnetic Theory of Propagation, Interference and Diffraction of Light*, (Cambridge, UK ; New York ; Cambridge University Press), 1997.
53. V. A. Sautenkov, H. Li, Y. V. Rostovtsev, and M. O. Scully, Ultra-dispersive adaptive prism, quant-ph/0701229.
54. S. Knappe, P. D. D. Schwindt, V. Gerginov, V. Shah, L. Liew, J. Moreland, H. G. Robinson, L. Hollberg, and J. Kitching, *J. Opt. A: Pure Appl. Opt.* 8, S318 (2006).
55. Q. Sun, Y. V. Rostovtsev, and M. S. Zubairy, *Proc. SPIE* 6130, 61300S (2006).
56. Q. Sun, Y. Rostovtsev, J. Dowling, M.O. Scully, and M.S. Zubairy, *Phys. Rev. A* 72, 031802 (2005).
57. S. E. Harris and Y. Yamamoto, *Phys. Rev. Lett.* 81, 3611 (1998).
58. V. Balic, D. A. Braje, P. Kolchin, G. Y. Yin, and S. E. Harris, *Phys. Rev. Lett.* 94, 183601 (2005).
59. A. B. Matsko, D. V. Strekalov, and L. Maleki, *Opt. Express* 13, 2210 (2005).
60. H. Lee, Y. Rostovtsev, C.J. Bednar, and A. Javan, *Appl. Phys. B* 76, 33 (2003).
61. M. D. Lukin, M. Fleischhauer, A. S. Zibrov, H. G. Robinson, V. L. Velichansky, L. Hollberg, and M. O. Scully, *Phys. Rev. Lett.* 79, 2959 (1997).
62. J. Qi, F. C. Spano, T. Kirova, A. Lazoudis, J. Magnes, L. Li, L. M. Narducci, R. W. Field, and A. M. Lyyra, *Phys. Rev. Lett.* 88, 173003 (2002); J. Qi and A. M. Lyyra, *Phys. Rev. A* 73, 043810 (2006).
63. Y. Rostovtsev, I. Protsenko, H. Lee, and A. Javan, *J. Mod. Opt.* 49, 2501 (2002).
64. J. X. Cheng, A. Volmer, and X.S. Xie, *JOSA B* 19, 1363 (2002).

Part V

Dynamic Structures for Storing Light

13 Stopping and Storing Light in Semiconductor Quantum Wells and Optical Microresonators

Nai H. Kwong, John E. Sipe, Rolf Binder, Zhenshan Yang, and Arthur L. Smirl

CONTENTS

13.1 Introduction 257
13.2 Linear Optical Response 258
    13.2.1 Quantum Well Structure 258
    13.2.2 SCISSOR Structure 262
13.3 Band Structures 265
13.4 Stopping and Storing Light in BSQWs 269
13.5 Conclusion 273
Acknowledgments 274
References 274

13.1 INTRODUCTION

There is a wide variety of proposals for implementations of slow-light effects, and many of them are described in the various chapters of this book. The physical implementations vary from system to system, and so do the mathematical formalisms used to describe the specific physical systems. However, it is also possible that one has two completely different physical slow-light systems, yet the mathematical framework describing the slowing and stopping of light is very similar. In this chapter, we present an example falling into this category. The two physically different slow-light systems are (1) semiconductor multiple quantum wells (QWs) (more specifically: Bragg-spaced multiple QWs, or BSQWs [1–16]) and (2) optical resonators (more specifically: side-coupled integrated spaced sequence of resonators with two waveguide channels, or two-channel SCISSORs [17,18]). In Figure 13.1, BSQWs are schematically shown as a sequence of thin semiconductor QWs, spaced by a distance a with a dielectric of refractive index n_b between the wells. SCISSORs are sequences of ring-shaped dielectric optical microresonators coupled to two waveguides (Figure 13.2). Clearly, these are very different physical systems, and so one may wonder in which sense slow-light concepts (i.e., the concepts for delaying, stopping, storing, and releasing light) can be similar or almost identical in these systems. The answer to this question lies in the fact that both systems can be viewed as physical realizations of one-dimensional resonant photonic bandgap structures (RPBGs). RPBGs differ from conventional (nonresonant) photonic bandgap structures in that each unit cell in the photonic lattice exhibits an optical resonance with a resonance frequency that is identical or at

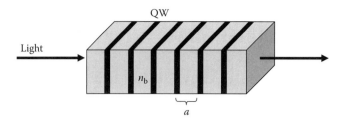

FIGURE 13.1 Schematic plot of a Bragg-spaced multiple quantum well structure. The spacing a between the thin QWs corresponds to a Bragg frequency that is close to the exciton frequency in the quantum wells. The material between the wells is a dielectric with refractive index n_b.

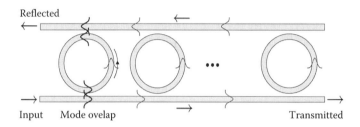

FIGURE 13.2 Schematic plot of a SCISSOR structure.

least close to the Bragg frequency characterizing the optical lattice. In QWs, the optical resonance is taken to be the heavy-hole exciton resonance, while in SCISSORs we take one of the optical resonances of the microrings.

In the following we will first review the linear optical properties of BSQWs (Section 13.2.1) and SCISSORs (Section 13.2.2). In Section 13.3 we discuss the bandstructure of the two systems and illustrate their similarities and differences in terms of their bandstructures. In Section 13.4 we show for the case of the BSQW structures how external manipulation of the bandstructure can be utilized to stop and trap, and also to release an optical light pulse. We present our conclusions in Section 13.5.

13.2 LINEAR OPTICAL RESPONSE

13.2.1 QUANTUM WELL STRUCTURE

The aforesaid QW structure is a sequence of thin QWs spaced at intervals in a dielectric medium (Figure 13.1). While the linear optical response of such a structure has been considered in detail in the literature, we present it here in a form that will both highlight the physics and facilitate a comparison with the two-channel SCISSOR structure. We specialize to a set of QWs with $\hat{\mathbf{z}}$ as the growth axis, surrounded by a dielectric medium characterized by a real, nondispersive refractive index n_b; we take our electric field to be propagating in the $\pm\hat{\mathbf{z}}$ directions, and hence polarized in the xy plane. We will often consider stationary fields,

$$E(\mathbf{r},t) = E(z,\omega)e^{-i\omega t} + \text{c.c.},$$

for which the polarization $P(\mathbf{r},t)$ takes a similar form, with

$$P(z,\omega) = \left[\chi(\omega)\sum_{n=1}^{N}\delta(z-z_n)\right]E(z,\omega),$$

$$= \sum_{n=1}^{N} P^{(n)}(\omega)\delta(z-z_n), \tag{13.1}$$

where the z_n identify the location of the nth QWs. The quantity in the square brackets in the first line is the susceptibility of the QWs, and

$$P^{(n)}(\omega) = \chi(\omega) E(z_n, \omega), \tag{13.2}$$

is the dipole moment per unit area associated with the nth QW. The use of Dirac delta functions to describe the polarization and its response to the electric field is a good approximation as long as the width of the QW is much less than the wavelength of light, which describes the variation of the electric field $E(z, \omega)$. This is valid for the systems we consider.

Standard linear response calculations for the polarizability $\chi(\omega)$ of the QW lead, within the rotating wave approximation, to an expression

$$\chi(\omega) = \frac{e^2 \langle r \rangle^2 \left| \tilde{\phi}(0) \right|^2}{\hbar \omega_X - \hbar \omega - i\hbar \gamma_{\text{nrad}}}, \tag{13.3}$$

where $\hbar \omega_X$ is the exciton energy, $e \langle r \rangle$ is the material's atomic dipole moment, $\tilde{\phi}(0)$ is the (2D) configuration space exciton wavefunction inside the QW at zero electron–hole separation. This last quantity is given by

$$\tilde{\phi}(0) = \frac{2\sqrt{2}}{a_o \sqrt{\pi}},$$

where a_o is the exciton Bohr radius. In this section we take parameters typical for $In_{0.025}Ga_{0.975}$ As QWs in GaAs, putting $\hbar \omega_X = 1.497$ eV, $\langle r \rangle = 0.354$ nm, and $a_o = 15$ nm.

The term γ_{nrad} describes a damping rate due to nonradiative effects. The inclusion of radiative damping will be described below; nonradiative damping in this system is largely due to polarization dephasing. We adopt a typical value of $\hbar \gamma_{\text{nrad}} = 0.1$ meV, which is appropriate for low-temperature experiments.

We may write our expression for the polarizability as

$$\chi(\omega) = \frac{e^2 f_A}{2m\omega_X} \frac{1}{\omega_X - \omega - i\gamma_{\text{nrad}}} \tag{13.4}$$

where we have introduced the free electron mass m, and put

$$f_A \equiv \frac{2m\omega_X \langle r \rangle^2 \left| \tilde{\phi}(0) \right|^2}{\hbar}.$$

While we need not introduce the free electron mass and the quantity f_A, the latter does have a simple physical significance. Note that within the rotating wave approximation, the second form Equation 13.4 of the susceptibility would, if f_A replaced by unity, describe the response of a classical charge of mass m bound to a force center by a potential associated with a resonance frequency ω_X, and damping characterized by γ_{nr}. The term f_A has units of inverse area, and can thus be identified as an oscillator strength per unit area for the excitonic transition. For the typical parameters adopted above we find $f_A = (4.23 \text{ nm})^{-2}$.

Radiative damping of the exciton transition is described through the electric field $E(z, \omega)$ that appears in Equation 13.1. In the one-dimensional propagation geometry we consider, the Maxwell equations reduce to a second-order differential equation for $E(z, \omega)$,

$$\frac{d^2 E(z, \omega)}{dz^2} + \frac{\omega^2 n_b^2}{c^2} E(z, \omega) = -\frac{4\pi \omega^2}{c^2} P(z, \omega), \tag{13.5}$$

with the solution, subject to outward going radiation conditions at infinity, given by

$$E(z, \omega) = E_{\text{inc}}(z, \omega) + \frac{2\pi i k}{n_b^2} \int e^{ik|z-z'|} P(z', \omega) dz', \tag{13.6}$$

where $k = \omega n_b/c$ and $E_{\text{inc}}(z, \omega)$ is a homogeneous solution to Equation 13.5 that can be identified as the incident field. At the position of the nth QW we have

$$E(z_n, \omega) = E'(z_n, \omega) + \frac{2\pi i k}{n_b^2} P^{(n)}(\omega), \tag{13.7}$$

with

$$E'(z_n, \omega) = E_{\text{inc}}(z_n, \omega) + \frac{2\pi i k}{n_b^2} \sum_{n' \neq n} e^{ik|z_n - z_{n'}|} P^{(n')}(\omega), \tag{13.8}$$

where we have used Equations 13.1 and 13.2 in the integral Equation 13.6. Here $E'(z_n, \omega)$ contains the contribution to the field at the n QW due to the incident field and any other QWs, while the second term on the right hand side of Equation 13.7 contains the field at the position of the nth QW due to that QW itself. It is this term that is responsible for the radiative damping, as can be explicitly seen by using the decomposition (Equation 13.7) of $E(z_n, \omega)$ to derive an expression for the response of $P^{(n)}(\omega)$ not to the whole electric field but rather to the field incident on it (including contributions from other QWs),

$$P^{(n)}(\omega) = \overline{\chi}(\omega) E'(z_n, \omega), \tag{13.9}$$

where, using the second form (Equation 13.4) of the polarizability, we find

$$\overline{\chi}(\omega) = \frac{e^2 f_A}{2m\omega_X} \frac{1}{\omega_X - \omega - i\gamma},$$

where

$$\gamma = \gamma_{\text{nrad}} + \gamma_{\text{rad}},$$

with

$$\gamma_{\text{rad}} = \frac{\pi e^2 f_A}{n_b m c}$$

describing the radiative contribution to the damping of the exciton resonance; consistent with the rotating wave approximation, in deriving Equation 13.9 we have put $k \approx \omega_X n_b/c$ in Equation 13.7. For the parameters introduced above and $n_b = 3.61$ (for GaAs) we find $\hbar \gamma_{\text{rad}} = 26.9$ μeV, and so the ratio of the radiative to nonradiative decay rate is given by

$$\frac{\gamma_{\text{rad}}}{\gamma_{\text{nrad}}} = 0.269.$$

Thus even at low temperatures the nonradiative damping will dominate over radiative damping for this system. We can introduce a quality factor Q for the QW oscillator in the usual way, which in terms of the parameters introduced is given by

$$Q = \frac{\omega_X}{2\gamma}. \tag{13.10}$$

For the values of the parameters given above, the quality factor for the QW is $Q = 5.9 \times 10^3$.

Stopping and Storing Light in Semiconductor Quantum Wells and Optical Microresonators 261

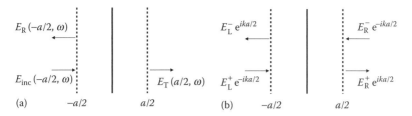

FIGURE 13.3 Sketch illustrating the transfer matrix.

Finally, it will be useful to have expressions for the transmission and reflection coefficients of a single QW. We consider such a well at $z = 0$ subject to an incident field $E_{\text{inc}}(z, \omega)$ coming from $z = -\infty$,

$$E_{\text{inc}}(z, \omega) = E_o e^{ikz}. \tag{13.11}$$

Using the response formula (Equation 13.9) in our expression (Equation 13.6) for the polarization, the total electric field for $z > 0$ is given by

$$E(z, \omega) = E_o e^{ikz} + \frac{2\pi i k}{n_b^2} \overline{\chi}(\omega) E_o e^{ikz} \equiv E_T(z, \omega) \quad \text{for} \quad z > 0,$$

$$= E_o e^{ikz} + \frac{2\pi i k}{n_b^2} \overline{\chi}(\omega) E_o e^{-ikz} \equiv E_o e^{ikz} + E_R(z, \omega) \quad \text{for} \quad z < 0,$$

where we have introduced the transmitted field $E_T(z, \omega)$ and the reflected field $E_R(z, \omega)$. Referencing our incident and reflected fields to $z = -a/2$ and our transmitted field to $z = a/2$ (see Figure 13.3a), we introduce Fresnel reflection and transmission coefficients, $r(\omega)$ and $t(\omega)$, respectively, given by

$$r(\omega) \equiv \frac{E_R(-\frac{a}{2}, \omega)}{E_{\text{inc}}(-\frac{a}{2}, \omega)},$$

$$t(\omega) \equiv \frac{E_T(\frac{a}{2}, \omega)}{E_{\text{inc}}(-\frac{a}{2}, \omega)}.$$

We find

$$r(\omega) = \frac{i\gamma_{\text{rad}}}{\omega_X - \omega - i\gamma} e^{ika},$$

$$t(\omega) = \frac{\omega_X - \omega - i\gamma_{\text{nrad}}}{\omega_X - \omega - i\gamma} e^{ika}. \tag{13.12}$$

Note that these expressions satisfy

$$|r(\omega)|^2 + |t(\omega)|^2 \leq 1,$$

the equality—and thus energy conservation in the electromagnetic field—resulting only if $\gamma_{\text{nrad}} = 0$, as would be expected.

Of course, an incident field more general than Equation 13.11 can be considered. If an incident field from $z = -\infty$ and one from $z = \infty$ is considered, then there will be reflected and transmitted fields in both directions. Hence, the total field near $z = -a/2$ will consist of two components, one which will vary as $E_L^-(\omega) e^{-ikz}$ and the other will vary as $E_L^+(\omega) e^{ikz}$; similarly, the total field near $z = a/2$ will consist of a component varying as $E_R^-(\omega) e^{-ikz}$ and another component varying as

$E_R^+(\omega)e^{ikz}$ (see Figure 13.3b). Working out the total fields for $z > 0$ and $z < 0$ for incident fields from both $z = \pm\infty$, one can identify the 2×2 transfer matrix $M(\omega)$ that relates them; we find

$$\begin{bmatrix} E_R^+ e^{ika/2} \\ E_R^- e^{-ika/2} \end{bmatrix} = M(\omega) \begin{bmatrix} E_L^+ e^{-ika/2} \\ E_L^- e^{ika/2} \end{bmatrix}, \quad (13.13)$$

where

$$M(\omega) = \frac{1}{t(\omega)} \begin{bmatrix} t^2(\omega) - r^2(\omega) & r(\omega) \\ -r(\omega) & 1 \end{bmatrix}, \quad (13.14)$$

and the Fresnel coefficients $r(\omega)$ and $t(\omega)$ are given by Equation 13.12. Note that Equation 13.13 is equivalent to

$$\begin{bmatrix} E_R^+ e^{ika/2} \\ E_L^- e^{ika/2} \end{bmatrix} = \begin{bmatrix} t(\omega) & r(\omega) \\ r(\omega) & t(\omega) \end{bmatrix} \begin{bmatrix} E_L^+ e^{-ika/2} \\ E_R^- e^{-ika/2} \end{bmatrix}. \quad (13.15)$$

This equation simply says that each outgoing wave [left hand side of Equation 13.15] is given in terms of the incoming wave from the other side (multiplied by the transmission coefficient) plus the reflected wave on the same side (multiplied by the reflection coefficient).

13.2.2 SCISSOR STRUCTURE

We now turn to a system that is at least superficially quite different from the QW structure considered above. The structure is shown in Figure 13.2, and consists of two channel waveguides that are side-coupled, via evanescent fields, to a set of circular microresonators. Following the terminology in the literature, we call these two-channel SCISSOR structures. For simplicity, we consider the situation in which the strength of the coupling between the upper channel and the microresonator is the same as the coupling between the lower channel and the microresonator. Light traveling in the forward (backward) direction in the lower (upper) channel can couple, via the microresonators, to light traveling in the backward (forward) direction in the upper (lower) channel. In general, the coupling between the channel waveguides and the microresonators will be small. However, if the frequency of light is close to an integer multiple of the fundamental resonant frequency of the microresonator (ω_r), then the effective coupling between the two channels can become quite large, as we will see below. Hence in contrast to the QWs, the resonance here is of a geometrical, rather than a material, nature.

Like the QWs, the SCISSOR structure has been well-studied in the literature with respect to both its linear and nonlinear responses, but we sketch the usual description of its linear response (see Figure 13.4) to make contact with the QW structure. We use l and u to denote, respectively, the appropriate channel waveguide mode amplitude in the lower and upper channels, and q to denote the corresponding mode amplitude in the resonator. The amplitudes just preceding and just following the coupling regions are denoted by l_- and l_+ (u_- and u_+) in the lower (upper) channel; the amplitudes just preceding and just following the coupling regions in the resonator are denoted by q_{l-} and q_{l+} (q_{u-} and q_{u+}) near the lower (upper) channel. In the coupling regions we define complex self (σ) and cross (κ) coupling coefficients, and assume

$$\begin{bmatrix} q_{l+} \\ l_+ \end{bmatrix} = \begin{bmatrix} \sigma & i\kappa \\ i\kappa & \sigma \end{bmatrix} \begin{bmatrix} q_{l-} \\ l_- \end{bmatrix},$$

$$\begin{bmatrix} q_{u-} \\ u_- \end{bmatrix} = \begin{bmatrix} \sigma & i\kappa \\ i\kappa & \sigma \end{bmatrix} \begin{bmatrix} q_{u+} \\ u_+ \end{bmatrix}. \quad (13.16)$$

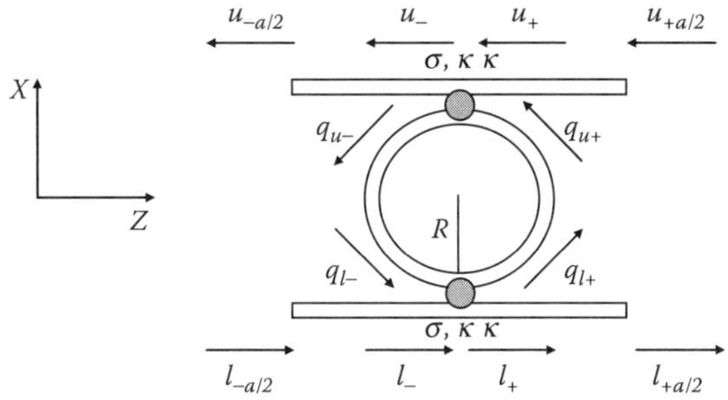

FIGURE 13.4 Sketch of a two-channel SCISSOR unit.

With the assumption that no energy is lost in the coupling regions, but only transferred between the waveguides as indicated, the coefficients satisfy

$$|\sigma|^2 + |\kappa|^2 = 1,$$

$$\sigma \kappa^* = \sigma^* \kappa.$$

If σ and κ are real, then the second of these conditions is automatically satisfied. For the rest of this chapter, we assume that σ and κ are indeed real; this simplies our expressions without losing any of the essential physics, and corresponds to the assumption that the coupling occurs only at a single point.

Away from the coupling points, the mode amplitudes propagate with effective refractive indices n_{eff} appropriate for the channels or the resonator. Scattering losses that inevitably arise due to fabrication difficulties, and in principle bending losses that arise in the resonator, can be taken into account phenomenologically by allowing n_{eff} to have an imaginary part, and we write $n_{\text{eff}} = n_b + i n_I$, with both n_b and n_I being real. In general the channels and rings can of course have different effective indices, but for simplicity we use the same n_{eff} to refer to them all; in any expression, it will be clear from the context, which actual index should be used.

As shown in Figure 13.4, we consider a length a of channel waveguides, with the coupling occurring near the center. If at $z = -a/2$ and $z = a/2$ we, respectively, denote the mode amplitudes in the lower (upper) channel by $l_{-a/2}$ and $l_{a/2}$ ($u_{-a/2}$ and $u_{a/2}$), then propagation in regions away from the coupling points leads to the relations

$$l_-/l_{-a/2} = l_{a/2}/l_+ = u_-/u_- = u_+/u_{a/2} = e^{ik_{\text{eff}} a/2}, \quad (13.17)$$

where $k_{\text{eff}} = \omega n_{\text{eff}}/c$, and propagation within the ring leads to the relations

$$q_{u+}/q_{l+} = q_{l-}/q_{u-} = e^{ik_{\text{eff}} \pi R}, \quad (13.18)$$

where R is the radius of the ring. Combining the coupling (Equation 13.16) and propagation (Equations 13.17 and 13.18) equations, we can write a transfer matrix for the structure shown in Figure 13.4 that is analogous to that in Equation 13.13 for the QWs,

$$\begin{bmatrix} l_{a/2} \\ u_{a/2} \end{bmatrix} = M(\omega) \begin{bmatrix} l_{-a/2} \\ u_{-a/2} \end{bmatrix}, \quad (13.19)$$

where as in the QWs the components of $M(\omega)$ can easily be written in terms of reflection and transmission coefficients,

$$M(\omega) = \frac{1}{t(\omega)} \begin{bmatrix} t^2(\omega) - r^2(\omega) & r(\omega) \\ -r(\omega) & 1 \end{bmatrix}, \quad (13.20)$$

exactly as in Equation 13.14, but here

$$r(\omega) = \frac{(1-\sigma^2)e^{i\phi(\omega)/2}}{\sigma^2 e^{i\phi(\omega)} - 1} e^{ik_{\text{eff}}a},$$

$$t(\omega) = \frac{\sigma(e^{i\phi(\omega)} - 1)}{\sigma^2 e^{i\phi(\omega)} - 1} e^{ik_{\text{eff}}a},$$

(13.21)

where we have introduced a (complex) phase $\phi(\omega) \equiv 2\pi R n_{\text{eff}} \omega/c$ associated with propagation around the ring.

An important difference between the QW and SCISSOR structures is that in the latter the forward and backward going amplitudes are physically separated, in two distinct channels. In the foregoing analysis, we have considered a forward propagating mode in the lower channel and a backward propagating mode in the upper channel. Within our approximations, there is completely identical and uncoupled dynamics in which a forward propagating mode in the upper channel is coupled to a backward propagating mode in the lower channel; we will not explicitly consider it here. As well, the reflection and transmission coefficients (Equation 13.21) have more structure than Equation 13.12 for the QWs. Whereas there is only one exciton resonance in the QW model, in Equation 13.21 we see resonance structure, in the usual case where σ is close to unity, whenever $\phi(\omega)$ is close to $2\pi M$, where M is an integer. Nonetheless, we will see that near such a resonance the structure of the Fresnel coefficients (Equation 13.21) for the SCISSOR structure is indeed close to that of the QWs.

To do this, we introduce a fundamental resonance frequency

$$\omega_r \equiv \frac{c}{Rn_b},$$

that defines the first resonance of the ring structure in the absense of any loss. For our sample calculations we take a ring circumference of 26 µm and an effective index $n_b = 3.0$; this yields $\hbar\omega_r = 15.9$ meV, corresponding to a vacuum wavelength of $\lambda_r = 78$ µm. We consider a frequency ω close to the Mth multiple of this $\omega \approx M\omega_r \equiv \omega_M$; for $M = 52$ we have a vacuum resonance wavelength corresponding to 1.5 µm. In the neighborhood of such a frequency we have $\phi(\omega) = 2\pi M + \delta\phi$, where $\delta\phi = \delta\phi(\omega)$ is a small quantity. In the numerator of the first of Equation 13.21 we put $\exp(i\phi(\omega)/2) = \exp(i\pi M + i\delta\phi/2) \approx (-1)^M$, while in the denominator we take $\exp(i\phi(\omega)) = \exp(i\delta\phi) \approx 1 + i\delta\phi$. Together with similar approximations for the second of Equation 13.21, together with the assumption that $n_1/n_R \ll 1$, we find the approximate expressions

$$r(\omega) = \frac{i(-1)^M \gamma_{\text{rad}}}{\omega_M - \omega - i\gamma} e^{ika} e^{-\eta a}$$

$$t(\omega) = \frac{\omega_M - \omega - i\gamma_{\text{nrad}}}{\omega_M - \omega - i\gamma} e^{ika} e^{-\eta a}$$

(13.22)

(compare with Equation 13.12), where $k \equiv \omega n_b/c$, $\eta \equiv n_1 \omega_M/c$, and in evaluating the latter we have taken $\omega \approx \omega_M$. As for the QWs we have

$$\gamma = \gamma_{\text{rad}} + \gamma_{\text{nrad}},$$

where

$$\gamma_{\text{rad}} = \frac{\omega_r(1-\sigma^2)}{2\pi\sigma^2}$$

describes the radiative coupling of the ring resonance at ω_M to the channels, and

$$\gamma_{\text{nrad}} = \frac{\omega_M n_1}{n_b}$$

describes nonradiative loss, due to scattering of the light out of the resonator. There is also a nonresonant loss term, described here by the factor $\exp(-\eta a)$, due to loss from propagation in the channels. No analogous term arises in the QWs because of the assumption that dielectric index is real. A final difference between the resonant expressions (Equation 13.22) for the SCISSOR structure and the QWs is the appearance of the factor $(-1)^M$ in the reflection coefficient of former, associated with the symmetry of the ring resonance under consideration.

For a typical self-coupling constant of $\sigma = 0.96$ we find $\hbar\gamma_{rad} = 215$ μeV, an order of magnitude larger than for a single QW. The ratio of the radiative to nonradiative coupling is given by

$$\frac{\gamma_{rad}}{\gamma_{nrad}} = \frac{n_b}{n_1}\frac{(1-\sigma^2)}{2\pi M\sigma^2} = \frac{20 n_b(1-\sigma^2)}{M\sigma^2 \ln 10}\frac{1}{\alpha\lambda},$$

where λ is the vacuum wavelength associated with the resonant frequency ω_M, and α is the loss coefficient (dB/length) associated with propagation in the ring. For the parameters identified above ($n_b = 3.0$, $\sigma = 0.96$, $\lambda = 1.5$ μm, $M = 52$), even if we assume a loss coefficient of 1 dB/mm we find $\gamma_{rad}/\gamma_{nrad} = 28.4$. Hence, unlike the QW system, radiative coupling can strongly dominate for the SCISSOR structure. The quality factor for this resonator is given by

$$Q = \frac{\omega_M}{2\gamma} \quad (13.23)$$

(cf. Equation 13.10). Assuming the loss coefficient is low enough that the nonradiative coupling is much less than the radiative coupling, as it is for the values of the parameters adopted and a loss coefficient of 1 dB/mm, we find a quality factor $Q = 1.9 \times 10^3$. Essentially due to radiative coupling, this is a lower factor than for the QW, where the nonradiative coupling makes the largest contribution, in part because the radiative coupling for the SCISSOR structure is typically larger than the total coupling for the QW, and in part because the resonance energy for the SCISSOR structure is lower than that for the QW.

13.3 BAND STRUCTURES

We are now in a position to calculate the band structures that result if the unit structure identified for each of our systems is considered to be repeated interminably (see Figure 13.5). When the QW structure is so repeated, with $z_n = na$, we refer to the many-well structure as a BSQW structure, while when the SCISSOR structure is so repeated we refer to the many-ring structure as a SCISSOR-Bragg structure. In both instances we have the same form for the transfer matrix across a single unit; for the BSQW structure the Equations are 13.13 and 13.14, while for the SCISSOR-Bragg structure there

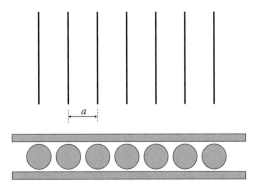

FIGURE 13.5 Sketch illustrating the similarity of a BSQW (top) with a SCISSOR (bottom).

are Equations 13.19 and 13.20. Using Bloch's theorem we can then seek solutions characterized by a crystal wavenumber K, demanding that the fields at z are related to those at $z + a$ by a factor $\exp(iKa)$. For the BSQW structure this condition is

$$\begin{bmatrix} E_R^+ e^{iKa/2} \\ E_R^- e^{-iKa/2} \end{bmatrix} = e^{iKa} \begin{bmatrix} E_L^+ e^{-iKa/2} \\ E_L^- e^{iKa/2} \end{bmatrix},$$

while for the SCISSOR-Bragg structure it is

$$\begin{bmatrix} l_{a/2} \\ u_{a/2} \end{bmatrix} = e^{iKa} \begin{bmatrix} l_{-a/2} \\ u_{-a/2} \end{bmatrix}.$$

In both cases, combination with Equation 13.13 or Equation 13.19 leads to the condition

$$\det \left(M(\omega) - U e^{iKa} \right) = 0,$$

for a nontrivial solution, where U is the 2×2 unit matrix. Writing the transfer matrix $M(\omega)$ in terms of the Fresnel coefficients, this yields the general form

$$\cos Ka = \frac{1 + t^2(\omega) - r^2(\omega)}{2t(\omega)},$$

which implicitly identifies the dispersion relation $\omega(K)$. For the BSQW structure, using Equation 13.12 for the Fresnel coefficients, after a bit of algebra this reduces to

$$\cos Ka = \cos ka + \frac{\gamma_{\text{rad}}}{\omega - \omega_X + i\gamma_{\text{nrad}}} \sin ka. \tag{13.24}$$

For the general SCISSOR-Bragg structure, using Equation 13.21 for the Fresnel coefficients, the corresponding dispersion relation is

$$\cos Ka = \frac{1 + \sigma^2}{2\sigma} \cos k_{\text{eff}} a + \frac{1 - \sigma^2}{2\sigma} \cot \left(\frac{\phi(\omega)}{2} \right) \sin k_{\text{eff}} a, \tag{13.25}$$

while near the Mth resonance where we can use the approximate expressions (Equation 13.22) for the Fresnel coefficients, we find

$$\cos Ka = (\cos ka \cosh \eta a - i \sin ka \sinh \eta a)$$
$$+ \frac{\gamma_{\text{rad}}}{\omega - \omega_M + i\gamma_{\text{nrad}}} (\sin ka \cosh \eta a + i \cos ka \sinh \eta a). \tag{13.26}$$

In the presence of nonradiative damping there is usually no solution with a real K corresponding to a real ω; in the following we consider the band structures that result when such nonradiative damping is neglected. Here $\gamma_{\text{nrad}} = 0$ for the BSQW structure, while $\gamma_{\text{nrad}} = 0$ and $\eta = 0$, with $n_{\text{eff}} \to n_b$ for the SCISSOR-Bragg structure. For each real ω there is either a solution of the dispersion relation with a real K, or there is no solution, identifying a band gap.

We begin with the generic SCISSOR-Bragg structure dispersion relation, (Equation 13.25) with $k_{\text{eff}} = \omega n_b/c$ and $\phi(\omega) = 2\pi R n_b \omega/c$ in the lossless case. A portion of the band structure is shown in Figure 13.6. Two types of gaps appear. There are Bragg gaps associated with the usual condition for constructive interference of backscattered light from the different resonators spaced by a. The fundamental Bragg frequency $\omega_B = c\pi/(n_b a)$ is the product of the phase velocity c/n_b and the wave number π/a associated with the edge of the first Brillouin zone, and the higher order

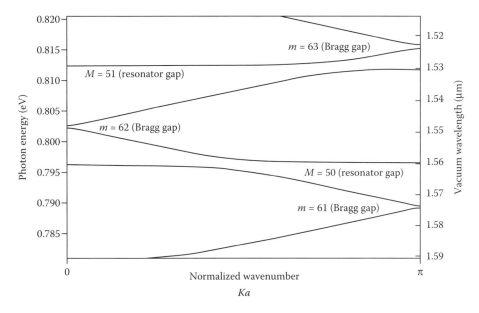

FIGURE 13.6 Photonic bandstructure of the SCISSOR. Only the spectral region around 0.8 eV (or 1.5 µm) is shown.

Bragg-frequencies are given by an integer multiple of $m\omega_B$, where m is a integer. At the band edges the group velocity vanishes, but the group velocity dispersion is typically large. There are also resonator gaps, associated with an integer multiple of the fundamental resonance frequency ω_r, with their centers near $\omega_M = M\omega_r$, where M is an integer. Here the backscattered light is strong as the microring resonance being satisfied, and as the reflectivity from even one resonator reaches unity at ω_M in the absense of loss (see Equation 13.22), it is not surprising that these gaps can appear even though there is no constructive interference from the backscattering of the separate resonators. At the edge of a resonator gap the group velocity vanishes, as of course it does at the edge of a Bragg gap, but the group velocity dispersion is also small near the edge of a resonator gap because the strength of the resonance tends to pull the dispersion relation toward a flat band near the band edges.

In the generic case the resonator gaps and Bragg gaps occur at quite separate frequencies. We can then derive approximate relations for the band structure close to the resonance frequency ω_M by using Equation 13.26 with $\eta = 0$ and $\gamma_{nrad} = 0$. Comparing this with the dispersion relation (Equation 13.24) for the BSQW structure in the absence of loss ($\gamma_{nrad} = 0$), it is clear that the dispersion relation for the two systems will be identical. In the lossless limit, we write both Equations 13.26 and 13.24 as

$$\cos Ka = \cos ka + \frac{\gamma_{rad}}{\omega - \omega_o} \sin Ka, \qquad (13.27)$$

where $k = \omega n_b/c$ and ω_o is either ω_X or ω_M, as we consider either the BSQW structure or the SCISSOR-Bragg structure. As an illustration, we show in Figure 13.7 a graphical solution of this equation, which readily demonstrates the existence of bands and forbidden regions (gaps) (compare, for example, [19]). Of course, in the BSQW structure there will be only a single analog of the resonator gap, which we might call an exciton gap, in the neighborhood of ω_X, while in the SCISSOR-Bragg structure there will be a host of them at a different ω_M.

Nonetheless, in either case for the resonance frequency of interest ω_o lying above $(m-1)\omega_B$ but below $m\omega_B$, we put

$$\Delta \equiv \frac{\pi}{\omega_B}(m\omega_B - \omega_o), \qquad (13.28)$$

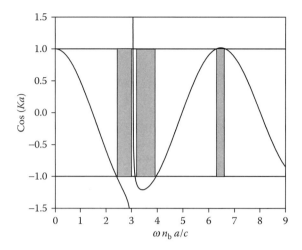

FIGURE 13.7 Graph of Equation 13.27. The solid line shows the right hand side of Equation 13.27 vs frequency with $\gamma_{rad} = 0.1\omega_o$. The Bragg frequency is given by $\omega_B n_b a/c = \pi$, and the resonance frequency is chosen to be given by $\omega_o n_b a/c = \pi - 0.1$. Those parts of the solid line bounded between the two horizontal lines with values -1 and 1 yield the polariton bands. The shaded areas highlight the bandgaps. (From Kwong, N.H., Yang, Z.S., Nguyen, D.T., Binder, R.L., and Smirl, A.L., *Proceedings of SPIE—The International Society for Optical Engineering*, vol. 6130, pp. 6130A1–6130A11, The International Society for Optical Engineering, Bellingham, WA, 2006. With permission.)

and then $ka = \pi m + \delta - \Delta$, where

$$\delta = \frac{\pi}{\omega_B}(\omega - \omega_o),$$

and our dispersion relation (Equation 13.27) can be written as a relation between δ and K,

$$(-1)^m \cos Ka = \cos(\delta - \Delta) + \frac{\mu}{\delta}\sin(\delta - \Delta), \tag{13.29}$$

where we have put

$$\mu \equiv \frac{\gamma_{rad}\pi}{\omega_B}.$$

For the portions of the bands where ω is very close to ω_o with respect to the natural scale of the fundamental Bragg frequency ω_B we have $|\delta| \ll 1$ and we can put $\cos(\delta - \Delta) \approx \cos\Delta + \delta\sin\Delta$ and $\sin(\delta - \Delta) \approx \delta\cos\Delta - \sin\Delta$. Using these in Equation 13.29 we recover a quadratic equation for δ; its solution yields

$$\omega = \omega_o + \frac{\omega_B}{\pi}\left(g(K) \pm \sqrt{g^2(K) + \mu}\right), \tag{13.30}$$

where

$$g(K) = \frac{(-1)^m \cos Ka - (1+\mu)\cos\Delta}{2\sin\Delta}.$$

These two solutions are sketched in Figure 13.8 along with the exact dispersion relation for the aforementioned SCISSOR-Bragg example in this regime (where $\delta \ll 1$) the exact SCISSOR-Bragg dispersion relation (Equation 13.25) and the result from the simpler (Equation 13.27) are indistinguishable. Hence there would be a corresponding agreement between Equation 13.30 and the exact dispersion relation for the BSQW structure. For frequencies ω far from ω_o the approximation

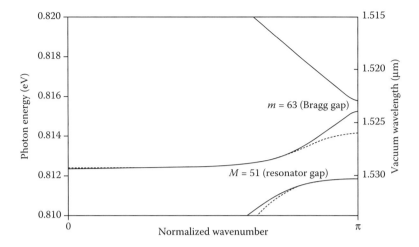

FIGURE 13.8 Same as Figure 13.6, but on an expanded scale. Also shown as dotted line are analytical approximations to the bandstructure (see text).

(Equation 13.30) obviously fails, but it gives a good description of both bands near their respective band edges.

A special case arises when one of the resonator frequencies ω_M in the SCISSOR-Bragg structure, or the exciton frequency ω_X in the BSQW structure, is near one of the Bragg frequencies. Calling that resonance frequency ω_o, as aforesaid, we use the notation Equation 13.28, but allow Δ to be either positive or negative, and assume both $|\delta|$ and $|\Delta|$ to be much less than unity in Equation 13.29. Then we can put $\cos(\delta - \Delta) \approx 1$ and $\sin(\delta - \Delta) \approx \delta - \Delta$ in Equation 13.29, and we find

$$\omega = \omega_o + \frac{(m\omega_B - \omega_o)\gamma_{rad}\pi/\omega_B}{1 + \gamma_{rad}\pi/\omega_B - (-1)^m \cos Ka}. \tag{13.31}$$

The single band described by this dispersion relation is the band associated with the resonance frequency ω_o; the other bands in the neighborhood, associated with the Bragg gap, are distorted as well. In Figures 13.9 and 13.10, we illustrate this for the case of a BSQW (compare also, for example, Refs. [1,3,7,8,12,20]). Here, $m = 1$, and we assume the exciton frequency ω_X to be very close to ω_B. This results in an almost flat band extending from ω_B to a frequency just above ω_X. In our previous slow-light studies [13,14,21–23], we have named this the intermediate band (IB), because it appears to be intermediate between two almost light-like polariton branches, as can be seen from Figure 13.9. In the present discussion, dephasing effects on the bandstructure are not included. However, those effects can be important, and in Ref. [13] we have shown that they lead to a strongly frequency-dependent polariton decay in the IB.

The illustration in Figures 13.9 and 13.10 is characteristic for $(m\omega_B - \omega_o) > 0$ and if m is odd; if instead $(m\omega_B - \omega_o) < 0$, it is clear from Equation 13.31 that the deviation of the band from ω_o simply reverses the sign, while if m is even rather than odd, the structure is shifted by π/a in reciprocal space, since $(-1)^{m+1} \cos Ka = (-1)^m \cos(Ka - \pi)$.

13.4 STOPPING AND STORING LIGHT IN BSQWs

In this section, we give a brief review of slow light effects in RPBGs, and we use BSQW structures as the example. The photonic bandstructure shown in Figures 13.9 and 13.10 suggests that the IB, being almost flat, corresponds to a small group velocity $v_g = d\omega/dK$, and hence may be well-suited for studies of slow light. This is especially true if the dispersion of the IB can be externally (ideally

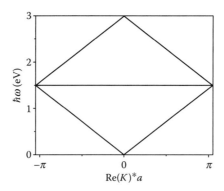

FIGURE 13.9 Photonic band structure of the BSQW with a single exciton resonance at $\hbar\omega_X = 1.496$ eV, a Bragg resonance at $\hbar\omega_B = 1.497$ eV and $\gamma_{rad} = 8.5 \times 10^{-6}\omega_X$. Note that, in a real semiconductor, the large frequency range above the electronic bandgap energy, which is about 10 meV above the exciton resonance, is filled with the strongly absorbing electron–hole pair continuum; hence the photonic bands shown in the figure are not an appropriate model in this frequency region. (From Kwong, N.H., Yang, Z.S., Nguyen, D.T., Binder, R., and Smirl, A.L., *Proceedings of SPIE—The International Society for Optical Engineering*, vol. 6130, pp. 6130A1–6130A11, The International Society for Optical Engineering, Bellingham, WA, 2006. With permission.)

optically) controlled. In this case, the group velocity of a pulse that has entered the structure can be further slowed down by making the band flatter. In the case of a BSQW structure, this can be achieved by shifting the exciton frequency toward the Bragg frequency. Optically induced shifts of the exciton resonance (so-called AC Stark shifts) are well established [9,24–30], but other means of shifting the exciton resonance could also be useful (e.g., voltage-induced shifts, known as quantum-confined Stark shifts). In the following, we are less interested in the specific means used for shifting the resonance. We simply assume this to be possible, and study consequences of this possibility for the slowing and possible trapping of light pulse inside the BSQW. The basic idea is simple. We assume the optical pulse spectrum to be somewhere inside the spectral region of the IB. Once the optical pulse has entered the structure, we shift the exciton frequency, thereby making the IB completely flat and reducing the group velocity to zero. The pulse is now trapped in the structure. After the desired delay time, the exciton frequency is shifted back to its original value. This restores the original IB, and the pulse propagates through (and finally exits) the sample.

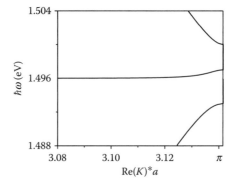

FIGURE 13.10 Photonic bandstructure of the BSQW (same as Figure 13.9, but on an expanded scale). (From Kwong, N.H., Yang, Z.S., Nguyen, D.T., Binder, R., and Smirl, A.L., *Proceedings of SPIE—The International Society for Optical Engineering*, vol. 6130, pp. 6130A1–6130A11, The International Society for Optical Engineering, Bellingham, WA, 2006. With permission.)

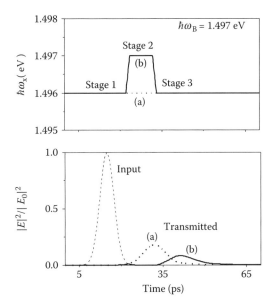

FIGURE 13.11 (*Top*) The time dependence of the exciton resonance ω_X. (*Bottom*) The transmitted pulse intensity (normalized to the peak input pulse intensity $|E_0|^2$), for 2000 quantum wells. Two cases are shown. The dotted curves, labeled (a), in both panels are for the case where the light is not stopped and (b) labels the case (solid curves) where the light is stopped for 8.2 ps inside the BSQW structure. The input field at the front end of the BSQW structure vs time (dashed) is also shown. (From Yang, Z.S., Kwong, N.H., Binder, R., and Smirl, A.L., *J. Opt. Soc. B*, 22, 2144, 2005. With permission.)

We note that related but conceptually more complicated schemes for optical delay in RPBGs have been proposed in atom-based systems [31] and coupled resonator optical waveguide structures [32]. In addition, schemes based on nonlinear pulse propagation have been suggested in BSQW structures [33]. The schemes used in Refs. [31,32] are more complicated than ours because they involve more than one optical resonance. In Ref. [31], a resonance that is split by an electromagnetically induced transparency (EIT) quantum coherence is spatially periodically modulated, while in Ref. [32] a classical analog to the EIT resonance was constructed from two resonators.

The basic operational principle of the BSQW structure is shown in Figure 13.11 where the number of quantum wells $N = 2000$. The top shows the imposed time-dependence of the exciton frequency and the bottom shows the pulses as functions of time. In stage 1, the IB is open and the pulse can propagate into the BSQW. In stage 2, the IB is closed and the pulse is trapped inside the BSQW. Finally, in stage 3 the exciton frequency is shifted back to the original value and the IB opens again. Now, the pulse propagates again and exits the structure. The actual delay can be inferred from a comparison with a pulse transmission in which the IB always remains open (this case is labeled (a) in the figure).

We note that the numerical simulation shown in Figure 13.11 includes the transition from the pulse being in air initially and then entering the BSQW. A technical problem with RPBs in general is that they exhibit strong Fabry–Perot fringes. This is especially true in the case of BSQW structures in the spectral region of the IB. These fringes can lead to strong reflection of the incoming pulse, which is of course undesirable. For this reason, we have developed an effective antireflection coating scheme that works especially well for the IB of a BSQW. However, the scheme can in principle be applied to any one-dimensional RPBG structure. Details of the AR coating are given in Ref. [23]. The results shown in Figure 13.11 are obtained with such antireflection coating.

From a practical point of view, one should note that it is still very challenging to fabricate BSQW structures with large numbers of quantum wells. Typical numbers of existing BSQW structures are around 200, which is significantly smaller than the number used in the simulation discussed above. A major problem with using short BSQW structures (e.g., $N = 200$) is the fact that the light pulse may be longer than the structure, and thus cannot be trapped in its entirety. Only the portion of the pulse that is inside the structure at the time of the trapping, can actually be delayed. If the pulse could be made very short in duration and thus very short in space, the length of the structure would not pose a problem. However, the pulse spectrum is limited to be within the IB bandwith, which in turn means that it has a minimum duration. In Ref. [14], we used this argument to define a minimal length of the BSQW structure, which is

$$L_{min} = \frac{9(\ln 2)\sqrt{2}a}{\sqrt{3\pi(\gamma_{rad}/\omega_X)}}. \tag{13.32}$$

For typical parameter values appropriate for GaAs, L_{min} is on the order of $1000a$, which corresponds to $N = 1000$.

Another limitation of the scheme is related to possible absorption of the pulse while it is traveling through the structure and also while it is stopped inside the structure. The absorption is governed by γ_{nrad}. In the case of quantum well structures, this depends strongly on the temperature and also on the quality of the quantum wells. Since, in BSQW structures, the requirement of small absorption translates into the requirement of low temperature, it is possible that, in the long term, SCISSOR structures could be better candidates for slow-light applications than BSQW structures. This is at least true if in the future the growth technology of SCISSORs permits high quality structures to be grown with negligible radiation losses.

We conclude our review of trapping light in BSQW structure with a brief discussion of temporal and spectral distortion of the delayed pulse. Since the IB exhibits strong dispersion, a pulse traversing the BSQW structure will inevitably be deformed. This happens for a pulse that propagates without being trapped, and it also occurs for one that has been trapped, because the propagation distance through the material is the same in both cases. In order to address this problem, we proposed in Ref. [14] a trapping scheme that has a built-in compensation of dispersion-induced pulse distortions. The idea was to release the pulse in the direction from where it entered the structure, rather than having the released pulse propagate out in the direction it was originally traveling when it entered the structure. The direction reversal can be achieved by shifting the exciton frequency, which is initially below the Bragg frequency, above the Bragg frequency when the pulse should be released. In this case, the dispersion of the IB changes shape as shown in Figure 13.12. Compared to the original

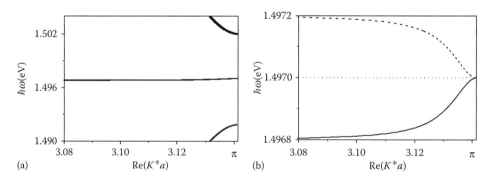

FIGURE 13.12 (a) The photonic bandstructure of the BSQW structure when $\omega_X < \omega_B$ ($\hbar\omega_X = 1.4968$ eV, $\hbar\omega_B = 1.497$ eV and $\gamma_{rad} = 1.8 \times 10^{-5}\omega_X$). The middle curve represents the IB. (b) The IB shown in (a) on an expanded scale (solid line), the IB when $\omega_X = \omega_B$ (dotted line), and the IB when $\omega_X - \omega_B$ has the same magnitude but opposite sign as in (a) (dashed line). (From Yang, Z.S., Kwong, N.H., Binder, R., and Smirl, A.L., *Opt. Lett.*, 30, 2790, 2005. With permission.)

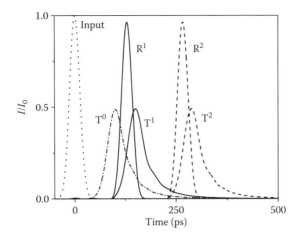

FIGURE 13.13 The time dependence of the output pulse intensity $I(t) \sim |E(t)|^2$ normalized to the peak input pulse intensity I_0 in reflection (R^1, R^2) and transmission (T^0, T^1, T^2) geometry, for 1500 QWs. The stopping time is 0, 50, and 190 ps for the dash-dotted, solid and dashed curves, respectively. The input pulse is the dotted curve. (From Yang, Z.S., Kwong, N.H., Binder, R., and Smirl, A.L., *Opt. Lett.*, 30, 2790, 2005. With permission.)

dispersion, not only the group velocity changes sign, but all higher derivatives of the dispersion relation. Therefore, the dispersion-induced deformation that occurs while the pulse is entering the structure is exactly compensated when the pulse exits the structure. In Figure 13.13 we can clearly see that the pulse exiting in the direction from where it came (we call it "reflection geometry") has the same temporal shape as the incoming pulse. This is in contrast to a pulse that would be released in the forward direction (which we call "transmission geometry").

13.5 CONCLUSION

In conclusion, we have shown that BSQW structures and two-channel SCISSOR structures are both representatives of one-dimensional resonant photonic bandgap structures. In these structures, the light velocity depends on the dispersion of the photonic bandstructures. Especially useful for slow-light application is the case in which we have an almost flat band, which can be achieved when the resonance frequency of each unit cell (i.e., in the case of the exciton frequency in the quantum well or the optical resonance of the microrings in the case of SCISSOR structures) is very close to the Bragg frequency. If, in addition, the resonance frequency of each unit cell can be shifted by external means, a simple way of trapping light is to let the light propagate inside the structure and, while it is in the structure, close the band (i.e., making it completely flat) by shifting the resonance frequency to coincide with the Bragg frequency. This way, the pulse is trapped in the structure. It can then be released at a later time, simply by shifting the resonance frequency back to its original value. We have demonstrated this through numerical simulations for the case of the BSQW structure and shown that this idea is a reasonable and conceptually very simply way of stopping, trapping and releasing light pulses. We have also discussed practical limitations of the scheme, which are related to fabrication issues (such as the limitations of the achievable length of BSQW structures with present-day growth technologies). Once the fabrication technology has advanced enough to produce high-quality and sufficiently long BSQW structures and high-quality low-radiation-loss SCISSOR structures, both systems might become viable choices as slow-light device components.

ACKNOWLEDGMENTS

We thank Philip Chak, John Prineas, Claudia Ell, and Andreas Knorr for helpful discussions, and DARPA, ONR, and JSOP for financial support.

REFERENCES

1. E. L. Ivchenko, A. I. Nesvizhskii, and S. Jorda, Bragg reflection of light from quantum wells, *Phys. Solid State* 36, 1156–1161 (1994).
2. E. L. Ivchenko, A. I. Nesvizhskii, and S. Jorda, Resonant Bragg reflection from quantum well structures, *Superlattices Microstruct.* 16, 17–20 (1994).
3. L. C. Andreani, Polaritons in multiple quantum wells, *Phys. Stat. Sol.* (B) 188, 29–42 (1995).
4. T. Stroucken, A. Knorr, P. Thomas, and S. W. Koch, Coherent dynamics of radiatively coupled quantum-well excitons, *Phys. Rev. B* 53, 2026–2033 (1996).
5. M. Hübner, J. Kuhl, T. Stroucken, A. Knorr, S. W. Koch, R. Hey, and K. Ploog, Collective effects of excitons in multiple quantum well Bragg and anti-Bragg structures, *Phys. Rev. Lett.* 76, 4199–4202 (1996).
6. J. P. Prineas, C. Ell, E. Lee, G. Khitrova, H. M. Gibbs, and S. W. Koch, Exciton polariton eigenmodes in light-coupled InGaAs/GaAs semiconductor multiple-quantum-well periodic structures, *Phys. Rev. B* 61, 13863–13872 (2000).
7. L. I. Deych and A. A. Lisyansky, Polariton dispersion law in periodic-Bragg and near-Bragg multiple quantum well structures, *Phys. Rev. B* 62, 4242–4244 (2000).
8. T. Ikawa and K. Cho, Fate of superradiant mode in a resonant Bragg reflector, *Phys. Rev. B* 66, 085338–(13) (2002).
9. J. Prineas, J. Zhou, J. Kuhl, H. Gibbs, G. Khitrova, S. Koch, and A. Knorr, Ultrafast ac Stark effect switching of the active photonic band gap from Bragg-periodic semiconductor quantum wells, *Appl. Phys. Lett.* 81, 4332–4334 (2002).
10. N. C. Nielsen, J. Kuhl, M. Schaarschmidt, J. Förstner, A. Knorr, S. W. Koch, G. Khitrova, H. M. Gibbs, and H. Giessen, Linear and nonlinear pulse propagation in multiple-quantum-well photonic crystal, *Phys. Rev. B* 70, 075306 (2004).
11. M. Schaarschmidt, J. Förstner, A. Knorr, J. Prineas, N. Nielsen, J. Kuhl, G. Khitrova, H. Gibbs, H. Giessen, and S. Koch, Adiabatically driven electron dynamics in a resonant photonic band gap: Optical switching of a Bragg periodic semiconductor, *Phys. Rev. B* 70, 233302–(4) (2004).
12. M. Artoni, G. LaRocca, and F. Bassani, Resonantly absorbing one-dimensional photonic crystals, *Phys. Rev. E* 72, 046604–(4) (2005).
13. Z. S. Yang, N. H. Kwong, R. Binder, and A. L. Smirl, Stopping, storing and releasing light in quantum well Bragg structures, *J. Opt. Soc. B* 22, 2144–2156 (2005).
14. Z. S. Yang, N. H. Kwong, R. Binder, and A. L. Smirl, Distortionless light pulse delay in quantum-well Bragg structures, *Opt. Lett.* 30, 2790–2792 (2005).
15. W. Johnston, M. Yildirim, J. Prineas, A. Smirl, H. M. Gibbs, and G. Khitrova, All-optical spin-dependent polarization switching in Bragg spaced quantum wells structures, *Appl. Phys. Lett.* 87, 101113 (2005).
16. W. J. Johnston, J. P. Prineas, A. L. Smirl, H. M. Gibbs, and G. Khitrova, Spin-dependent ultrafast optical nonlinearities in Bragg-spaced quantum well structures band gap, Frontiers in Optics/Laser Science XXII, p. FTuO2 (2006).
17. S. Pereira, J. E. Sipe, J. E. Heebner, and R. W. Boyd, Gap solitons in a two-channel microresonator structure, *Opt. Lett.* 27, 536–538 (2002).
18. P. Chak, J. E. Sipe, and S. Pereira, Lorentzian model for nonlinear switching in a microresonator structure, *Opt. Commun.* 213, 163–171 (2002).
19. N. Ashcroft and N. Mermin, *Solid State Physics* (Sounders College Publishing, New York, 1976).
20. A. E. Kozhekin, G. Kurizki, and B. Malomed, Standing and moving gap solitons in resonantly absorbing gratings, *Phys. Rev. Lett.* 81, 3647–3650 (1998).
21. N. H. Kwong, Z. S. Yang, D. T. Nguyen, R. Binder, and A. L. Smirl, Light pulse delay in semiconductor quantum well Bragg structures, in *Proceedings of SPIE—The International Society for Optical Engineering*, vol. 6130, pp. 6130A1–6130A11 (The International Society for Optical Engineering, Bellingham, WA, 2006).
22. R. Binder, Z. S. Yang, N. H. Kwong, D. T. Nguyen, and A. L. Smirl, Light pulse delay in semiconductor quantum well Bragg structures, *Phys. Stat. Sol.* (b) 243, 2379–2383 (2006).

23. Z. S. Yang, J. Sipe, N. H. Kwong, R. Binder, and A. L. Smirl, Antireflection coating for quantum well Bragg structures, *J. Opt. Soc. B* 24, 2013–2022 (2007).
24. A. Mysyrowicz, D. Hulin, A. Antonetti, A. Migus, W. T. Masselink, and H. Morkoc, Dressed excitons in a multiple-quantum-well structure: Evidence for an optical Stark effect with femtosecond response time, *Phys. Rev. Lett.* 56, 2748–2751 (1986).
25. N. Peyghambarian, S. W. Koch, M. Lindberg, B. Fluegel, and M. Joffre, Dynamic Stark effect of exciton and continuum states in CdS, *Phys. Rev. Lett.* 62, 1185–1188 (1989).
26. M. Combescot and R. Combescot, Optical Stark effect of the exciton: Biexcitonic origin of the shift, *Phys. Rev. B* 40, 3788–3801 (1989).
27. D. S. Chemla, W. H. Know, D. A. B. Miller, S. Schmitt-Rink, J. B. Stark, and R. Zimmermann, The excitonic optical Stark effect in semiconductor quantum wells probed with femtosecond optical pulses, *J. Luminescence* 44, 233–246 (1989).
28. R. Binder, S. W. Koch, M. Lindberg, W. Schäfer, and F. Jahnke, Transient many-body effects in the semiconductor optical Stark effect: A numerical study, *Phys. Rev. B* 43, 6520–6529 (1991).
29. S. W. Koch, M. Kira, and T. Meier, Correlation effects in the excitonic optical properties of semiconductors, *J. Optics B* 3, R29–R45 (2001).
30. P. Brick, C. Ell, S. Chatterjee, G. Khitrova, H. M. Gibbs, T. Meier, C. Sieh, and S. W. Koch, Influence of light holes on the heavy-hole excitonic optical Stark effect, *Phys. Rev. B* 64, 075323–(5) (2001).
31. A. Andre and M. Lukin, Manipulating light pulses via dynamically controlled photonic band gap, *Phys. Rev. Lett.* 89, 143602–(4) (2002).
32. M. F. Yanik and S. Fan, Stopping light in a waveguide with an all-optical analog of electromagnetically induced transparency, *Phys. Rev. Lett.* 93, 233903–(4) (2004).
33. W. N. Xiao, J. Y. Zhou, and J. P. Prineas, Storage of ultrashort optical pulses in a resonantly absorbing Bragg reflector, *Opt. Express* 11, 3277–3283 (2003).

14 Stopping Light via Dynamic Tuning of Coupled Resonators

Shanhui Fan and Michelle L. Povinelli

CONTENTS

14.1 Introduction 277
14.2 Theory 278
    14.2.1 Tuning the Spectrum of Light 278
    14.2.2 General Conditions for Stopping Light 279
    14.2.3 Tunable Fano Resonance 280
    14.2.4 From Tunable Bandwidth Filter to Light-Stopping System 280
    14.2.5 Numerical Demonstration in a Photonic Crystal 281
    14.2.6 Dynamic Tuning Suppresses Dispersion 283
    14.2.7 Stopping Light via Loss Tuning 284
14.3 Experimental Progress 285
    14.3.1 General Requirements for Microresonators 285
    14.3.2 Experiments with Microring Resonators 285
    14.3.3 Experiments with Photonic Crystals 286
    14.3.4 Prospects for Loss Tuning 286
14.4 Outlook and Concluding Remarks 286
Acknowledgments 287
References 287

14.1 INTRODUCTION

In this chapter, we describe how coupled resonator systems can be used to stop light—that is, to controllably trap and release light pulses in localized, standing wave modes. The inspiration for this work lies in previous research on stopped light in atomic gases using electromagnetically induced transparency [1], in which light is captured in dark states of the atomic system via adiabatic tuning [2–4]. However, such atomic systems are severely constrained to operate only at particular wavelengths corresponding to available atomic resonances and have very limited bandwidth. The coupled resonator systems we study are amenable to fabrication in on-chip devices such as photonic crystals [5–9] or microring resonators [10]. As such, the operating wavelength and other operating parameters can be engineered to meet flexible specifications, such as for optical communications applications.

    The idea of using dynamic tuning in a coupled resonator system is to modulate the properties of the resonators (e.g., the resonator frequencies) while a light pulse is in the system. In so doing, the spectrum of the pulse can be molded almost arbitrarily, leading to highly nontrivial information processing capabilities. In earlier work [11], we have shown that dynamic tuning can be used for time reversal of pulses. Here we focus on approaches to stopping light [12–18]. A wide variety of work has been done on slow-light structures employing coupled resonators [19–29]. However, in all such systems, the maximum achievable time delay scales inversely with the operating bandwidth [22]. As we will see in the following section, the use of dynamic tuning overcomes this constraint by manipulation of the photon spectrum in time. Here, we start with an overview of our theoretical work on light stopping in dynamically tuned coupled resonator systems. We then discuss quite recent experimental results that have demonstrated the plausibility of adiabatic tuning in on-chip systems and review the growing body of work inspired by dynamic optical modulation ideas.

14.2 THEORY

14.2.1 Tuning the Spectrum of Light

Here we provide a simple example to show how the spectrum of electromagnetic wave can be modified by a dynamic photonic structure. Consider a linearly polarized electromagnetic wave in one dimension, the wave equation for the electric field is

$$\frac{\partial^2 E}{\partial x^2} - (\varepsilon_0 + \varepsilon(t)) \mu_0 \frac{\partial^2 E}{\partial t^2} = 0 \tag{14.1}$$

Here, $\varepsilon(t)$ represents the modulation and ε_0 is the background dielectric constant. We assume that both ε_0 and $\varepsilon(t)$ are independent of position. Hence different wavevector components do not mix in the modulation process. For a specific wavevector component at k_0, with electric field described by $E(t) = f(t)e^{i(\omega_0 t - k_0 x)}$, where $\omega_0 = k_0/\sqrt{\mu_0 \varepsilon_0}$, we have

$$-k_0^2 f - [\varepsilon_0 + \varepsilon(t)] \mu_0 \left[\frac{\partial^2 f}{\partial t^2} + 2i\omega_0 \frac{\partial f}{\partial t} - \omega_0^2 f \right] = 0 \tag{14.2}$$

By using a slowly varying envelope approximation, that is, ignoring the $\partial^2 f/\partial t^2$ term, and by further assuming that the index modulations are weak, that is, $\varepsilon(t) \ll \varepsilon_o$, Equation 14.2 can be simplified as

$$i\frac{\partial f}{\partial t} = \frac{\varepsilon(t)\omega_0}{2[\varepsilon(t) + \varepsilon_0]} f \approx \frac{\varepsilon(t)\omega_0}{2\varepsilon_0} f \tag{14.3}$$

which has an exact analytic solution:

$$f(t) = f(t_0) \exp\left[-i\omega_0 \int_{t_0}^{t} \frac{\varepsilon(t')}{2\varepsilon_0} dt' \right] \tag{14.4}$$

where t_0 is the starting time of the modulation. Thus the instantaneous frequency of the electric field for this wavevector component is

$$\omega(t) = \omega_0 \left(1 - \frac{\varepsilon(t)}{2\varepsilon_0} \right) \tag{14.5}$$

We note that the frequency change is proportional to the magnitude of the refractive index shift alone. Thus, the process defined here differs in a fundamental way from traditional nonlinear optical processes. For example, in a conventional sum frequency conversion process, in order to convert the frequency of light from ω_1 to ω_2, modulations at a frequency $\omega_2 - \omega_1$ need to be provided. In contrast, in the process described here, regardless of how slow the modulation is, as long as light is in the system, the frequency shift can always be accomplished. We will demonstrate some very spectacular consequences of such a frequency shift in the dynamic photonic crystal, in its application for stopping a light pulse all-optically.

The existence of the frequency shift in dynamic photonic crystal structures [30] and in laser resonators [31,32] was also pointed out in a number of previous works. In practical optoelectronic or nonlinear optical devices, the accomplishable refractive index shift is generally quite small. Thus, in most practical situations the effect of dynamics is prominent only in structures in which the spectral feature is sensitive to small refractive index modulations. This motivates our design on Fano interference schemes, described below, which are employed to enhance the sensitivity of photonic structures to small index modulations.

14.2.2 General Conditions for Stopping Light

By stopping light, we aim to reduce the group velocity of a light pulse to zero, while completely preserving all the coherent information encoded in the pulse. Such ability holds the key to the ultimate control of light, and has profound implications for optical communications and quantum information processing.

There has been extensive work attempting to control the speed of light using optical resonances in static photonic crystal structures. Group velocities as low as $10^{-2}c$ have been experimentally observed at waveguide band edges [33,34] or with coupled resonator optical waveguides (CROWs) [35–38]. Nevertheless, such structures are fundamentally limited by the delay—bandwidth product constraint—the group delay from an optical resonance is inversely proportional to the bandwidth within which the delay occurs. Therefore, for a given optical pulse with a certain temporal duration and corresponding frequency bandwidth, the minimum group velocity achievable is limited. In a CROW waveguide structure, for example, the minimum group velocity that can be accomplished for pulses at 10 Gbit/s rate with a wavelength of 1.55 µm is no smaller than $10^{-2}c$. For this reason, static photonic structures cannot be used to stop light.

To stop light, it is necessary to use a dynamic system. The general condition for stopping light [12] is illustrated in Figure 14.1. Imagine a dynamic photonic crystal system, with an initial band structure possessing a sufficiently wide bandwidth. Such a state is used to accommodate an incident pulse, for which each frequency component occupies a unique wavevector component. After the pulse has entered the system, one can then stop the pulse by flattening the dispersion relation of the crystal adiabatically, while preserving the translational invariance. In doing so, the spectrum of the pulse is compressed, and its group velocity is reduced. In the meantime, since the translational symmetry is still preserved, the wavevector components of the pulse remain unchanged, and thus one actually preserves the dimensionality of the phase space. This is crucial in preserving all the coherent information encoded in the original pulse during the dynamic process.

The idea of dynamically changing a photonic crystal system while the light is inside, has an analogy in quantum mechanics. In quantum mechanics, perturbing the physical system changes

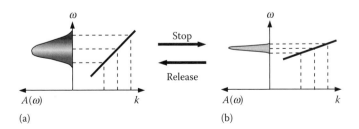

FIGURE 14.1 The general conditions for stopping a light pulse. (a) The large bandwidth state that is used to accommodate an incident light pulse. (b) The narrow bandwidth state that is used to hold the light pulse. An adiabatic transition between these two states stops a light pulse inside the system.

the Hamiltonian as a function of time. At any given instant of time, it is possible to solve for the instantaneous eigenstates of the system. From the adiabatic theorem, it is guaranteed that if the Hamiltonian changes slowly enough, a system originally in the nth eigenstate of the unperturbed Hamiltonian will be found in the nth instantaneous eigenstate of the perturbed Hamiltonian. As a result, adiabatic perturbation of the Hamiltonian can change the expectation value of the energy [39]. In the photonic crystal system, light initially in a particular band of the dispersion relation will remain in that band as the band is compressed, provided the time scale for compression is long compared to the inverse of the frequency separation between the bands [15]. Adiabatic perturbation of the band structure can thus change the frequency of light within the system, allowing the stopping of light pulses.

14.2.3 TUNABLE FANO RESONANCE

To create a dynamic photonic crystal, one needs to adjust its properties as a function of time. This can be accomplished by modulating the refractive index, either with electro-optic or nonlinear optic means. However, the amount of refractive index tuning that can be accomplished with standard optoelectronics technology is generally quite small, with a fractional change typically on the order of $\delta n/n \approx 10^{-4}$. Therefore, we employ Fano interference schemes in which a small refractive index modulation leads to a very large change of the bandwidth of the system. The essence of a Fano interference scheme is the presence of multipath interference, where at least one of the paths includes a resonant tunneling process [40]. Such interference can be used to greatly enhance the sensitivity of resonant devices to small refractive index modulation [15,41,42].

Here we consider a waveguide side-coupled to two cavities [43]. The cavities have resonant frequencies $\omega_{A,B} \equiv \omega_0 \pm (\delta\omega/2)$, respectively. (This system represents an all-optical analogue of atomic systems exhibiting electromagnetically induced transparency, EIT [1]. Each optical resonance here is analogous to the polarization between the energy levels in the EIT system [27].) For simplicity, we assume that the cavities couple to the waveguide with equal rate of γ, and we ignore the direct coupling between the side cavities. Consider a mode in the waveguide passing through the cavities. The transmission and reflection coefficients for a single side cavity can be derived using Green's function method [44] and used to calculate the two-cavity transmission spectrum via the transfer matrix method [43].

The transmission spectra of one- and two-cavity structures are plotted in Figure 14.2. In the case of one-cavity structure, the transmission features a dip in the vicinity of the resonant frequency, with the width of the dip controlled by the strength of waveguide–cavity coupling (Figure 14.2a). With two cavities, when the condition

$$2\beta(\omega_0) L = 2n\pi \tag{14.6}$$

is satisfied, the transmission spectrum features a peak centered at ω_0. The width of the peak is highly sensitive to the frequency spacing between the resonances $\delta\omega$. When the cavities are lossless, the center peak can be tuned from a wide peak when $\delta\omega$ is large (Figure 14.2b), to a peak that is arbitrarily narrow with $\delta\omega \to 0$ (Figure 14.2c). The two-cavity structure, appropriately designed, therefore behaves as a tunable bandwidth filter, (as well as a tunable delay element with delay proportional to the inverse peak width [27]), in which the bandwidth can, in principle, be adjusted by any order of magnitude with very small refractive index modulation.

14.2.4 FROM TUNABLE BANDWIDTH FILTER TO LIGHT-STOPPING SYSTEM

By cascading the tunable bandwidth filter structure as described in Section 14.2.3, one can construct a structure that is capable of stopping light (Figure 14.3a). In such a light-stopping structure, the photonic band becomes highly sensitive to small refractive index modulation.

The photonic bands for the structure in Figure 14.3a can be calculated using a transmission matrix method [44]. The band diagrams are shown in Figure 14.3, in which the waveguide and

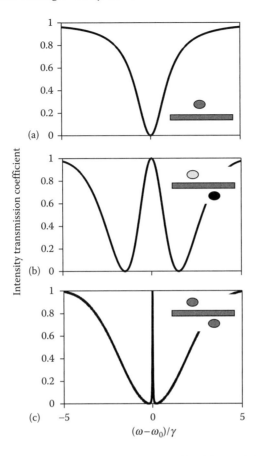

FIGURE 14.2 (a) Transmission spectrum through a waveguide side-coupled to a single mode cavity. (b and c), transmission spectra through a waveguide side-coupled to two cavities. The parameters for the cavities are: $\omega_0 = 2\pi c/L_1$, $\gamma = 0.05\omega_0$. And the waveguide satisfies a dispersion relation $\beta(\omega) = \omega/c$, where c is the speed of light in the waveguide, and L_1 is the distance between the cavities. In (b), $\omega_{a,b} = \omega_0 \pm 1.5\gamma$. In (c), $\omega_{a,b} = \omega_0 \pm 0.2\gamma$.

cavity parameters are the same as those used to generate the transmission spectrum in Figure 14.2. In the vicinity of the resonances, the system supports three photonic bands, with two gaps occurring around ω_A and ω_B. The width of the middle band depends strongly on the resonant frequencies ω_A, ω_B. By modulating the frequency spacing between the cavities, one goes from a system with a large bandwidth (Figure 14.3b), to a system with a very narrow bandwidth (Figure 14.3c). In fact, it can be analytically proved that the system can support a band that is completely flat in the entire first Brillouin zone [13], allowing a light pulse to be frozen inside the structure with the group velocity reduced to zero. Moreover, the gaps surrounding the middle band have sizes on the order of the cavity–waveguide coupling rate γ, and are approximately independent of the slope of the middle band. Thus, by increasing the waveguide–cavity coupling rate, this gap can be made large, which is important for preserving the coherent information during the dynamic bandwidth compression process [12].

14.2.5 NUMERICAL DEMONSTRATION IN A PHOTONIC CRYSTAL

The system presented in Section 14.2.4 can be implemented in a photonic crystal of a square lattice of dielectric rods ($n = 3.5$) with a radius of $0.2a$ (a is the lattice constant), embedded in

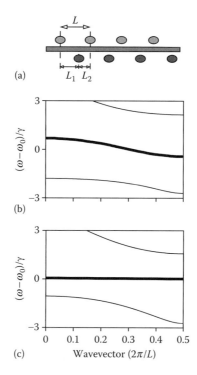

FIGURE 14.3 (a) Schematic of a coupled-cavity structure used to stop light. (b and c), band structures for the system shown in (a), as the frequency separation between the cavities are varied, using the same waveguide and cavity parameters as in Figure 14.2b and c, with the additional parameter $L_2 = 0.7L_1$. The thicker lines highlight the middle band that will be used to stop a light pulse.

air ($n = 1$) [13] (Figure 14.4). The photonic crystal possesses a band gap for TM modes with electric field parallel to the rod axis. Removing one row of rods along the pulse propagation direction generates a single-mode waveguide. Decreasing the radius of a rod to $0.1a$ and the dielectric constant to $n = 2.24$ provides a single mode cavity with resonance frequency at $\omega_c = 0.357 \cdot (2\pi c/a)$. The nearest neighbor cavities are separated by a distance of $l_1 = 2a$ along the propagation direction, and the unit cell periodicity is $l = 8a$. The waveguide–cavity coupling occurs through barrier of one rod, with a coupling rate of $\gamma = \omega_c/235.8$. The resonant frequencies of the cavities are tuned by refractive index modulation of the cavity rods.

We simulate the entire process of stopping light for $N = 100$ pairs of cavities with the finite-difference-time-domain (FDTD) method, which solves Maxwell's equations without approximation [45]. The dynamic process for stopping light is shown in Figure 14.4. We generate a Gaussian pulse in the waveguide (the process is independent of the pulse shape). The excitation reaches its peak at $t = 0.8t_{\text{pass}}$, where t_{pass} is the traversal time of the pulse through the static structure. During the pulse generation, the cavities have a large frequency separation. The field is concentrated in both the waveguide and the cavities (Figure 14.4b, $t = 1.0t_{\text{pass}}$), and the pulse propagates at a relatively high speed of $v_g = 0.082c$. After the pulse is generated, we gradually reduce the frequency separation Δ to zero. During this process, the speed of light is drastically reduced to zero. As the bandwidth of the pulse is reduced, the field concentrates in the cavities (Figure 14.4b, $t = 5.2t_{\text{pass}}$). When zero group velocity is reached, the photon pulse can be kept in the system as a stationary waveform for any duration. In this simulation, we store the pulse for a time delay of $5.0t_{\text{pass}}$, and then release the pulse by repeating the same index modulation in reverse (Figure 14.4b, $t = 6.3t_{\text{pass}}$). The pulse intensity as a function of time at the right end of the waveguide is plotted in Figure 14.3a, and it shows the same

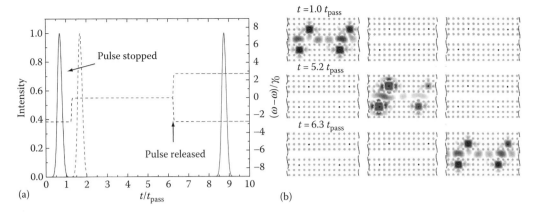

FIGURE 14.4 Light-stopping process in a photonic crystal simulated using FDTD methods. The crystal consists of a waveguide side-coupled to 100 cavity pairs. Fragments of the photonic crystal are shown in part (b). The three fragments correspond to unit cells 12–13, 55–56, 97–98. The dots indicate the positions of the dielectric rods. The black dots represent the cavities. (a) The dashed green and black lines represent the variation of ω_A and ω_B as a function of time, respectively. The blue solid line is the intensity of the incident pulse as recorded at the beginning of the waveguide. The red dashed and solid lines represent the intensity at the end of the waveguide, in the absence and the presence of modulation, respectively. t_{pass} is the passage time of the pulse in the absence of modulation. (b) Snapshots of the electric field distributions in the photonic crystal at the indicated times. Red and blue represent large positive and negative electric fields, respectively. The same color scale is used for all the panels.

temporal shape as both the pulses that propagate through the unmodulated system, and the initial pulse recorded at the left end of the waveguide.

14.2.6 DYNAMIC TUNING SUPPRESSES DISPERSION

The dynamic tuning scheme largely eliminates the dispersive effects associated with static delay lines. The time-varying dispersion relation $\omega(k,t)$ can be expanded around a center wave vector k_c as

$$\omega(k,t) \approx \omega(k_c,t) + \omega^{(1)}_{k_c}(t)(k-k_c) + \frac{\omega^{(2)}_{k_c}(t)}{2}(k-k_c)^2,$$

where $\omega^{(n)}_{k_c}(t) \equiv d^n\omega(k,t)/dk^n|_{k=k_c}$. It can be shown [17] that the output width of the pulse in time (Δt_{out}) after a total delay time τ is given by

$$\Delta t_{out}^2 = \Delta t_{in}^2 + \left[\frac{\int_0^\tau \omega^{(2)}_{k_c}(t')dt'}{v_g^2(0)\Delta t_{in}} \right]^2,$$

where we have assumed that $v_g(\tau) = v_g(0)$. For a static system, this reduces to the result

$$\Delta t_{out}^2 = \Delta t_{in}^2 + \left[\frac{\omega^{(2)}_{k_c}(0)\tau}{v_g^2(0)\Delta t_{in}} \right]^2,$$

and the pulse spreads with increasing delay. For the dynamic system, however, $\omega^{(2)}_{k_c}(t)$ (and all higher-order derivatives) are identically zero in the flat band state. Assuming that the bandwidth compression and decompression each occupy a time T,

$$\Delta t_{\text{out}}^2 = \Delta t_{\text{in}}^2 + \left[\frac{2 \int_0^T \omega_{kc}^{(2)}(t')dt'}{v_g^2(0) \Delta t_{\text{in}}} \right]^2.$$

The pulse spreading is independent of the delay time τ, as it occurs only during spectrum compression and decompression. The delay can thus be increased arbitrarily without any additional increase in dispersion.

14.2.7 Stopping Light via Loss Tuning

An alternative way to stop light is to tune the resonator loss [18], rather than the resonator frequencies. An example system in which loss tuning can stop a light pulse is shown in Figure 14.5. It consists of a chain of resonators with resonance frequencies ω_0 and coupling constant κ. Initially (Figure 14.5a), both the A and B resonators have low values of loss ($\gamma_A, \gamma_B \ll \kappa$, where γ is the inverse decay time of the resonator). Light can propagate freely by coupling from one resonator to the next. The dispersion relation has a relatively wide bandwidth and nonzero group velocity. If the A resonators are tuned to a high loss value ($\gamma_A \gg \kappa$), as indicated by Figure 14.5b, light can no longer propagate down the chain. The dispersion relation is a flat band with zero group velocity. Light contained primarily in the B resonators will oscillate in a low-loss standing wave mode.

In the dynamic tuning process, the system is tuned from state (a) to state (b) to trap the pulse. After the desired delay time, the system is tuned back to state (a) to release the pulse. In this case, two output pulses result: the delayed pulse travels in the forward direction, and a second, time-reversed [11] pulse travels in the backward direction. Here, we have assumed that the input pulse is centered at the frequency where the initial band folds (corresponding to $k = \pi/L$). To minimize pulse loss, the tuning between states should be performed as rapidly as possible. In addition, the loss of the A resonators should be much higher than the loss of the B resonators in the stopped state.

Loss tuning may increase the operating bandwidth of the light-stopping system as compared to frequency tuning schemes. During the light-stopping process, the system must be switched between a flat band state and a state with bandwidth large enough to accommodate the initial pulse. In the frequency tuning scheme, the maximum bandwidth of the initial pulse is limited by the tuning range of the refractive index. In the loss tuning scheme, the initial bandwidth can be chosen as large as desired by increasing the coupling constant between the resonators. Imposing a large difference in loss between the A and B resonators will nevertheless result in a flat band state.

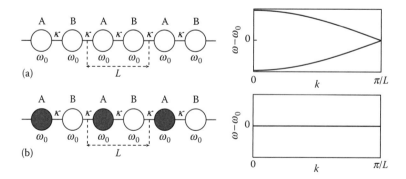

FIGURE 14.5 Light stopping via loss tuning. (a) Initial state of the system. All resonators (both resonators labeled "A" and "B") have identical resonance frequency and low initial loss. The characteristic band structure is shown on the right. (b) Every other resonator (resonators labeled "A") are tuned to a high loss value. The characteristic band structure is shown on the right and is a flat band.

14.3 EXPERIMENTAL PROGRESS

14.3.1 GENERAL REQUIREMENTS FOR MICRORESONATORS

In the aforementioned numerical example, we have demonstrated the use of photonic crystal microresonators for slowing and stopping light. However, the phenomena we describe are quite general and apply to arbitrary coupled resonator systems. To be useful for stopping light, the particular resonator implementation should satisfy several criteria.

First, the intrinsic quality factor of the resonator should be as high as possible, since it limits the delay time. Light stopped for time longer than the cavity lifetime will substantially decay. However, the optical loss might be counteracted with the use of gain media in the cavities or external amplification.

Second, small resonator size is generally desirable, as shorter length devices tend to consume less power. Moreover, for fixed device length, decreasing the size of the resonator increases the storage capacity [25].

Third, the resonator should be highly tunable on the time scale of operation of the device. The resonance frequency can be tuned by locally changing the refractive index of the material near the resonator. To date, experiments have focused on silicon materials systems, including microring resonators and photonic crystal slabs, because of the paramount importance of silicon in information technology. Methods for tuning the refractive index of silicon are reviewed in Ref. [46]. The thermo-optic effect is restricted to modulation frequencies below 1 MHz, the linear electro-optic effect is absent in pure crystalline silicon, and the electroabsorption (Franz–Keldysh) and second-order electro-optic (Kerr) effects have very low efficiency. The most effective method for fast modulation of the refractive index of silicon is the free-carrier plasma-dispersion effect. The free carrier concentration can be varied by generating carriers with an optical pump at a wavelength above the silicon bandgap or by injecting carriers with a diode or MOSFET structure [46]. For a small refractive index shift of $\delta n/n = 10^{-4}$ achievable in practical optoelectronic devices, and assuming a carrier frequency of approximately 200 THz, as used in optical communications, the achievable bandwidths are on the order of 20 GHz, which are comparable to the bandwidth of a single wavelength channel in high-speed optical systems.

14.3.2 EXPERIMENTS WITH MICRORING RESONATORS

Experiments with silicon microring resonators have demonstrated the use of a tunable Fano resonance in a double-resonator system [47] to controllably trap and release light pulses [48].

Initially, the frequencies of the two microring resonators are slightly detuned, as in Figure 14.2b. In this state, input light couples into a supermode of the two resonators. The frequencies of the two resonators are then tuned into resonance with one another, as in Figure 14.2c. In this state, the supermode is isolated from the input and output waveguides, and light is stored in and between the two resonators. After a given storage time, the resonator frequencies are again detuned to release the light.

The resonators are tuned using the free-carrier dispersion effect in silicon [49] to blueshift the resonant wavelength. In this experiment, an optical pump pulse at 415 nm was used to excite free carriers in the microrings. Electro-optic tuning of the ring resonances via built-in p–i–n junctions [50] should allow electrically controlled storage, with an expected bandwidth of over 10 GHz.

In the experiment, the storage time was limited to < 100 ps by the intrinsic Q of the microresonators ($Q = 143,000$). However, the recent demonstration of $Q \sim 4.8 \times 10^6$ in a silicon ring resonator [51] suggests that storage times of several nanoseconds may be possible.

One drawback of using a double-resonator system for pulse delay is that the pulse shape and spectrum are not preserved in the process. To retain the information encoded in the shape of the original pulse, a cascaded multiresonator system is necessary. Nevertheless, this experiment represents first the major step toward the realization of the theoretical ideas for stopping the aforesaid light.

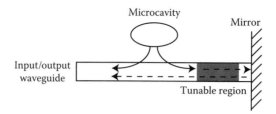

FIGURE 14.6 Schematic of system used for experiments on dynamic light trapping in photonic crystals.

14.3.3 Experiments with Photonic Crystals

Photonic crystal microcavities may represent the ultimate limit of miniaturization for resonator modes. Such microcavities have been demonstrated with Q up to 2×10^6 and modal volumes as small as a cubic wavelength [52].

A recent experiment has demonstrated the fundamental requirement for dynamic trapping and delay: the ability to tune between a supermode that is strongly coupled to an input waveguide, and one that is decoupled, or isolated [53]. The geometry used is shown schematically in Figure 14.6. A single cavity is side coupled to a waveguide that is terminated by a mirror. The coupling between the input waveguide and the supermode of the resonator–waveguide–mirror complex is determined by the reflection phase from the mirror. When the wave emitted from the cavity in the backward direction interacts constructively with the wave emitted from the cavity in the forward direction and reflected backwards by the mirror, light can easily couple from the supermode to the input waveguide. Conversely, when the waves add destructively, the coupling is reduced. We note that this structure is in fact conceptually very similar to the structure shown in Figure 14.2b. The mirror, in essence, creates a mirror image of the first resonator.

In the experiment, a pump pulse was used to dynamically tune the refractive index of the waveguide between the nanocavity and the mirror, adjusting the reflection phase. Pump–probe measurements of the power emitted from the cavity to free space show that the coupling properties of the supermode could be tuned on the picosecond timescale.

14.3.4 Prospects for Loss Tuning

Experimental demonstration of stoppage and storage via tuning of resonator loss is an intriguing area for future research. Free carrier injection in silicon modifies both the real and imaginary parts of the refractive index. Achieving a high material loss requires a high injected-carrier density, which may be achievable in heavily doped structures [49]. Achieving a low loss requires most of the carriers to be removed, which could potentially be achieved using a reverse bias structure [54]. The modulation time will be limited by the carrier lifetime, which can be reduced to <1 ps using ion implantation [55]. Another semiconductor materials system with highly tunable loss is InGaAs/InP quantum wells. Experiments have demonstrated that the material loss can be tuned from $\alpha = 1500\,\mathrm{cm}^{-1}$ to $\sim 0\,\mathrm{cm}^{-1}$ at room temperature operation [56]. It is likely that a higher contrast ratio of the material loss can be accomplished at a lower operating temperature. The intrinsic response time is in the femtosecond range [57].

14.4 OUTLOOK AND CONCLUDING REMARKS

Beyond the work described above, the idea of using dynamic tuning of the refractive index for stopping, storing, and time-reversing pulses has sparked a wide range of research. For example, alternate dynamic tuning schemes that do not require translational invariance have recently been investigated [58]. Moreover, the generality of the physics governing coupled resonators has suggested the possibility of light stopping and time reversal in quite diverse physical systems. In semiconductor multiple quantum well structures, tuning of the excitonic resonance via the AC Stark effect can potentially flatten the photonic band structure to stop light pulses in a similar fashion as described [59,60]. In superconducting qubit systems, tuning of the qubit transition frequency can theoretically stop pulses at the single photon level [61]. Such an ability to manipulate single photons is of increasing interest for quantum information processing and quantum computing.

The concept of using dynamic index tuning for frequency conversion is also actively explored. Ideally, one could use a coupled resonator system to change the center frequency of a pulse while leaving its shape unchanged, a feat achieved via a uniform shift of the band structure [62]. While experiments are not yet feasible, a similar effect can be observed in single cavity systems. For a single cavity, changing the resonance frequency of the cavity mode on a time scale faster than the cavity decay time results in frequency conversion [63]. The frequency shift is linearly proportional to the index shift. The phenomenon has been demonstrated experimentally in both silicon microring resonators [64] and photonic crystal microcavities [65].

In summary, dynamic tuning of coupled resonator systems opens the possibility for coherent optical pulse stopping and storage. More generally, dynamic processes in coupled resonator systems allow one to mold the spectrum of a photon pulse almost at will, while preserving coherent information in the optical domain. In future, the use of dynamic photonic structures, as we envision here, may provide a unifying platform for diverse optical information processing tasks.

ACKNOWLEDGMENTS

The work is supported in part by NSF, DARPA, and the Lucile and Packard Foundation. The authors acknowledge the important contributions of Prof. Mehmet Fatih Yanik as well as Sunil Sandhu to the works presented here.

REFERENCES

1. S. E. Harris, Electromagnetically induced transparency, *Phys. Today*, 50, 36–42 (1997).
2. M. D. Lukin, S. F. Yelin, and M. Fleischhauer, Entanglement of atomic ensembles by trapping correlated photon states, *Phys. Rev. Lett.*, 84, 4232–4235 (2000).
3. C. Liu, Z. Dutton, C. H. Behroozi, and L. V. Hau, Observation of coherent optical information storage in an atomic medium using halted light pulses, *Nature*, 409, 490–493 (2001).
4. D. F. Phillips, A. Fleischhauer, A. Mair, R. L. Walsworth, and M. D. Lukin, Storage of light in an atomic vapor, *Phys. Rev. Lett.*, 86, 783–786 (2001).
5. J. D. Joannopoulos, R. D. Meade, and J. N. Winn, *Photonic Crystals: Molding the Flow of Light*, Princeton University Press, Princeton, 1995.
6. J. D. Joannopoulos, P. R. Villeneuve, and S. Fan, Photonic crystals: Putting a new twist on light, *Nature*, 386, 143–147 (1997).
7. C. Soukoulis (Ed.), *Photonic Crystals and Light Localization in the 21st Century*, NATO ASI Series, Kluwer Academic Publisher, the Netherlands, 2001.
8. S. G. Johnson and J. D. Joannopoulos, *Photonic Crystals: The Road from Theory to Practice*, Kluwer Academic Publisher, Boston, MA, 2002.
9. K. Inoue and K. Ohtaka, *Photonic Crystals*, Springer-Verlag, Berlin, 2004.
10. H. A. Haus, M. A. Popovic, M. R. Watts, C. Manolatou, B. E. Little, and S. T. Chu, Optical resonators and filters, in K. Vahala (Ed.), *Optical Microcavities*, World Scientific, Singapore, 2004.
11. M. F. Yanik and S. Fan, Time-reversal of light with linear optics and modulators, *Phys. Rev. Lett.*, 93, art. no. 173903 (2004).
12. M. F. Yanik and S. Fan, Stopping light all-optically, *Phys. Rev. Lett.*, 92, art. no. 083901 (2004).
13. R. L. Liboff, *Quantum Mechanics*, Addison-Wesley Publishing Company, Reading, MA, 1992.
14. M. F. Yanik, W. Suh, Z. Wang, and S. Fan, Stopping light in a waveguide with an all-optical analogue of electromagnetically induced transparency, *Phys. Rev. Lett.*, 93, art. no. 233903 (2004).
15. M. F. Yanik and S. Fan, Stopping and storing light coherently, *Phys. Rev. A*, 71, art. no. 013803 (2005).
16. M. F. Yanik and S. Fan, Dynamic photonic structures: stopping, storage, and time-reversal of light, *Stud. Appl. Math.*, 115, 233–254 (2005).

17. S. Sandhu, M. L. Povinelli, M. F. Yanik, and S. Fan, Dynamically-tuned coupled resonator delay lines can be nearly dispersion free, *Opt. Lett.*, 31, 1985–1987 (2006).
18. S. Sandhu, M. L. Povinelli, and S. Fan, Stopping and time-reversing a light pulse using dynamic loss-tuning of coupled-resonator delay lines, *Opt. Lett.*, 32, 3333–3335 (2007).
19. N. Stefanou and A. Modinos, Impurity bands in photonic insulators, *Phys. Rev. B*, 57, 12127–12133 (1998).
20. A. Yariv, Y. Xu, R. K. Lee, and A. Scherer, Coupled-resonator optical waveguide: A proposal and analysis, *Opt. Lett.*, 24, 711–713 (1999).
21. M. Bayindir, B. Temelkuran, and E. Ozbay, Tight-binding description of the coupled defect modes in three-dimensional photonic crystals, *Phys. Rev. Lett.*, 84, 2140–2143 (2000).
22. G. Lenz, B. J. Eggleton, C. K. Madsen, and R. E. Slusher, Optical delay lines based on optical filters, *IEEE J. Quantum Electron.*, 37, 525–532 (2001).
23. J. E. Heebner, R. W. Boyd, and Q.-H. Park, Slow light, induced dispersion, enhanced nonlinearity, and optical solitons in a resonator-array waveguide, *Phys. Rev. E*, 65, art. no. 036619 (2002).
24. J. E. Heebner, R. W. Boyd, and Q.-H. Park, SCISSOR solitons and other novel propagation effects in microresonator-modified waveguides, *J. Opt. Soc. Am. B*, 19, 722–731 (2002).
25. Z. Wang and S. Fan, Compact all-pass filters in photonic crystals as the building block for high capacity optical delay lines, *Phys. Rev. E*, 68, art. no. 066616 (2003).
26. D. D. Smith, H. Chang, K. A. Fuller, A. T. Rosenberger, and R. W. Boyd, Coupled resonator-induced transparency, *Phys. Rev. A*, 69, art. no. 063804 (2004).
27. L. Maleki, A. B. Matsko, A. A. Savchenkov, and V. S. Ilchenko, Tunable delay line with interacting whispering-gallery-mode resonators, *Opt. Lett.*, 29, 626–628 (2004).
28. J. K. S. Poon, J. Scheuer, Y. Xu, and A. Yariv, Designing coupled-resonator optical waveguide delay lines, *J. Opt. Soc. Am. B*, 21, 1665–1673 (2004).
29. J. B. Khurgin, Expanding the bandwidth of slow-light photonic devices based on coupled resonators, *Opt. Lett.*, 30, 513–515 (2005).
30. E. J. Reed, M. Soljacic, and J. D. Joannopoulos, Color of shock waves in photonic crystals, *Phys. Rev. Lett.*, 91, art. no. 133901 (2003).
31. A. Yariv, Internal modulation in multimode laser oscillators, *J. Appl. Phys.*, 36, 388–391 (1965).
32. A. E. Siegmann, *Lasers*, University Science Books, Sausalito, 1986.
33. M. Notomi, K. Yamada, A. Shinya, J. Takahashi, C. Takahashi, and I. Yokoyama, Extremely large group-velocity dispersion of line-defect waveguides in photonic crystal slabs, *Phys. Rev. Lett.*, 87, art. no. 253902 (2001).
34. Y. A. Vlasov, M. O'Boyle, H. F. Harmann, and S. J. McNab, Active control of slow light on a chip with photonic crystal waveguides, *Nature*, 438, 65–69, (2005).
35. H. Altug and J. Vuckovic, Experimental demonstration of the slow group velocity of light in two-dimensional coupled photonic crystal microcavity arrays, *Appl. Phys. Lett.*, 86, art. no. 111102 (2005).
36. F. Xia, L. Sekaric, M. O'Boyle, and Y. Vlasov, Coupled resonator optical waveguides based on silicon-on-insulator photonic wires, *Appl. Phys. Lett.*, 89, art. no. 041122 (2006).
37. S. C. Huang, M. Kato, E. Kuramochi, C.-P. Lee, and M. Notomi, Time-domain and spectral-domain investigation of inflection-point slow-light modes in photonic crystal coupled waveguides, *Opt. Express*, 15, 3543–3549 (2007).
38. D. O'Brien, M. D. Settle, T. Karle, A. Michaeli, M. Salib, and T. F. Krauss, Coupled photonic crystal heterostructure cavities, *Opt. Express*, 15, 1228–1233 (2007).
39. R. L. Liboff, *Introductory Quantum Mechanics*, 2nd edn., Addison-Wesley Publishing Company, Reading, MA, 1992.
40. U. Fano, Effects of configuration interaction on intensities and phase shifts, *Phys. Rev.*, 124, 1866–1878 (1961).
41. S. Fan, Sharp asymmetric lineshapes in side-coupled waveguide-cavity systems, *Appl. Phys. Lett.*, 80, 910–912 (2002).
42. S. Fan, W. Suh, and J. D. Joannopoulos, Temporal coupled mode theory for Fano resonances in optical resonators, *J. Opt. Soc. Am. A*, 20, 569–573 (2003).
43. W. Suh, Z. Wang, and S. Fan, Temporal coupled-mode theory and the presence of non-orthogonal modes in lossless multi-mode cavities, *IEEE J. Quantum Electron.*, 40, 1511–1518 (2004).
44. S. Fan, P. R. Villeneuve, J. D. Joannopoulos, C. Manalatou, M. J. Khan, and H. A. Haus, Theoretical investigation of channel drop tunneling processes, *Phys. Rev. B*, 59, 15882–15892 (1999).

45. A. Taflove and S. C. Hagness, *Computational Electrodynamics: The Finite-Difference Time-Domain Method*, Artech House, Norwood, 2005.
46. M. Lipson, Guiding, modulating, and emitting light on silicon—challenges and opportunities, *J. Lightwave Technol.*, 23, 4222–4238 (2005).
47. Q. Xu, S. Sandhu, M. L. Povinelli, J. Shakya, S. Fan, and M. Lipson, Experimental realization of an on-chip all-optical analogue to electromagnetically induced transparency, *Phys. Rev. Lett.*, 96, art. no. 123901 (2006).
48. Q. Xu, P. Dong, and M. Lipson, Breaking the delay-bandwidth limit in a photonic structure, *Nat. Phys.*, 3, 406–410 (2007).
49. R. A. Soref and B. R. Bennett, Electrooptical effects in silicon, *IEEE J. Quantum Electron.*, 23, 123–129 (1987).
50. Q. Xu, B. Schmidt, S. Pradhan, and M. Lipson, Micrometre-scale silicon electro-optic modulator, *Nature*, 435, 325–237 (2005).
51. M. Borselli, High-Q microresonators as lasing elements for silicon photonics. PhD thesis, California Institute of Technology, Pasadena (2006).
52. S. Noda, M. Fujita, and T. Asano, Spontaneous-emission control by photonic crystals and nanocavities, *Nat. Photon.*, 1, 449–458 (2007).
53. Y. Tanaka, J. Upham, T. Nagashima, T. Sugiya, T. Asano, and S. Noda, Dynamic control of the Q factor in a photonic crystal nanocavity, *Nat. Mater.*, 6, 862–865 (2007).
54. H. Rong, R. Jones, A. Liu, O. Cohen, D. Hak, A. Fang, and M. Paniccia, A continuous-wave Raman silicon laser, *Nature*, 433, 725–728 (2005).
55. A. Chin, K. Y. Lee, B. C. Lin, and S. Horng, Picosecond response of photoexcited carriers in ion-implanted Si, *Appl. Phys. Lett.*, 69, 653–655 (1996).
56. I. Bar-Joseph, C. Klingshirn, D. A. B. Miller, D. S. Chemla, U. Koren, and B. I. Miller, Quantum-confined Stark effect in InGaAs/InP quantum wells grown by organometallic vapor phase epitaxy, *Appl. Phys. Lett.* 50, 1010–1012 (1987).
57. W. H. Knox, D. S. Chemela, D. A. B. Miller, J. B. Stark, and S. Schmitt-Rink, Femtosecond ac Stark effect in semiconductor quantum wells: Extreme low- and high-intensity limits, *Phys. Rev. Lett.* 62, 1189–1192 (1989).
58. S. Longhi, Stopping and time reversal of light in dynamic photonic structures via Bloch oscillations, *Phys. Rev. B*, 75, art. no. 026606 (2007).
59. Z. S. Yang, N. H. Kwong, R. Binder, and A. L. Smirl, Distortionless light pulse delay in quantum-well Bragg structures, *Opt. Lett.*, 30, 2790–2792 (2005).
60. Z. S. Yang, N. H. Kwong, R. Binder, and A. L. Smirl, Stopping, storing, and releasing light in quantum-well Bragg structures, *J. Opt. Soc. Am. B*, 22, 2144–2156 (2005).
61. J. T. Shen, M. L. Povinelli, S. Sandhu, and S. Fan, Stopping single photons in one-dimensional quantum electrodynamics systems, *Phys. Rev. B*, 75, art. no. 035320 (2007).
62. Z. Gaburro, M. Ghulinyan, F. Riboli, L. Pavesi, A. Recati, and I. Carusotto, Photon energy lifter, *Opt. Express*, 14, 7270–7278 (2006).
63. M. Notomi and S. Mitsugi, Wavelength conversion via dynamic refractive index tuning of a cavity, *Phys. Rev. A*, 73, art. no. 051803(R) (2006).
64. S. F. Preble, Q. F. Xu, and M. Lipson, Changing the colour of light in a silicon resonator, *Nat. Photon.*, 1, 293–296 (2007).
65. M. W. McCutcheon, A. G. Pattantyus-Abraham, G. W. Rieger, and J. F. Young, Emission spectrum of electromagnetic energy stored in a dynamically perturbed microcavity, *Opt. Express*, 15, 11472–11480 (2007).

Part VI

Applications

15 Bandwidth Limitation in Slow Light Schemes

Jacob B. Khurgin

CONTENTS

15.1 Introduction 293
15.2 Atomic Resonances 294
15.3 Photonic Resonances 298
15.4 Double-Resonant Atomic SL Structures 298
15.5 Double-Resonant Photonic SL Structures—Cascaded Gratings ... 301
15.6 Tunable Double-Resonant Atomic SL Structures—Electromagnetic Transparency (EIT) 305
15.7 Coupled Photonic Resonator Structures 309
15.8 Dispersion Limitation of Nonlinear Photonic SL Devices 315
15.9 Conclusions 318
References 318

15.1 INTRODUCTION

Recent years have seen a great increase in interest in the regime of light propagation with reduced group velocity, commonly referred to as slow light (SL). Starting with the pioneering work performed in the late 1990s [1] a great effort had been directed at both understanding the basic physics of SL phenomena and applying it to various practical tasks—optical communications, signal processing, photonic switching, and many others.

A diverse set of SL schemes have been investigated analytically, modeled, and demonstrated experimentally as amply evidenced by this book. Diverse as they are, all SL techniques essentially rely upon the existence of a resonance which causes a decrease in the group velocity $v_g = d\omega/dk$ which can be described by a slow down factor

$$S = c/nv_g = \omega dk/kd\omega = d(\ln k)/d(\ln \omega) \quad (15.1)$$

The resonances can be intrinsic resonances in atoms and molecules, or can be external to the material, imposed by a resonant photonic structure, such as coupled cavities, gratings, etc. In the vicinity of the resonant frequency ω_{12} the dispersion curve $k(\omega)$ (Figure 15.1a) experiences strong modification, prompted by resonant coupling between the light and medium (in case of atomic resonance) [2–5], between forward and backward propagating light waves (in case of gratings or coupled resonator structures [CRS]) [6–8], or between the light and lattice vibrations (Brillouin and Raman amplifiers) [9–11]. Both the index of absorption (Figure 15.1b) and the refractive index (Figure 15.1c) change in the vicinity of resonance. The slow down factor near the resonant frequency ω_{12} of the transition between two atomic levels 1 and 2 experiences changes. For some frequencies, $S > 1$ and this is the SL case which is a subject of this book, but in addition to the SL, also the cases of

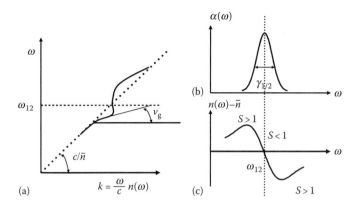

FIGURE 15.1 (a) Group velocity in the vicinity of a resonance, (b) spectrum of absorption coefficient, and (c) spectrum of refractive index.

$0 < S < 1$ (fast light) and $S < 0$ (negative group velocity) can occur. Since the changes occur only in the vicinity of the resonance, one can expect the slow down effect to be bandwidth limited. Therefore, in the course of all the SL work one question had invariably been raised: what is the practical limitation imposed on delay and bandwidth of the SL schemes? This issue had been studied, from different angles, for the atomic resonance schemes in Refs. [12,13], for photonic resonance schemes in Ref. [14], for both atomic and photonic schemes [15] and for the case of optical amplifiers [16]. Most recently, Miller [17] had investigated some of the limitations in the more general case. All the authors have identified two interrelated bandwidth limiting factors—the dispersion of loss (traced to the imaginary part of the dielectric constant ε'') and the dispersion of the group velocity itself (high-order dispersion of the real part of dielectric constant ε'). Nevertheless, the authors have arrived at different conclusions regarding the bandwidth and delay limitations, which in our view, is related to the fact that different figures of merit have been used. It is the goal of this work to introduce the single figure of merit for all linear and nonlinear SL schemes based on a simple logic; since all the SL schemes are bandwidth limited, for each scheme, be that a delay line capable of storing N bits or a nonlinear switch or an electro optic modulator, one can define the maximum or cutoff bit rate B_{max} (or cutoff bandwidth for analog applications) at which using SL schemes ceases to bring any improvement to the performance. With B_{max} defined, we can develop simple expressions for the performance of SL schemes at bit rates less than cutoff as a function of B/B_{max}.

15.2 ATOMIC RESONANCES

In the vicinity of an atomic resonance one can write the dielectric constant in a familiar Lorentzian form

$$\varepsilon(\omega) = \bar{\varepsilon} + \frac{\Omega_p^2}{\omega_{12}^2 - \omega^2 - j\omega\gamma_{12}} \quad (15.2)$$

where $\bar{\varepsilon} = \bar{n}^2$ is the nonresonant part or background part of the dielectric constant, and γ_{12} is the dephasing rate of the polarization. The expression in the numerator Ω_p is the plasma frequency which can be found as

$$\Omega_p^2 = \frac{N_a e^2}{\varepsilon_0 m_0} f_{12} \quad (15.3)$$

where
  e is an electron charge
  ε_0 is the dielectric permittivity of vacuum
  m_0 is a free electron mass

Bandwidth Limitation in Slow Light Schemes

$$f_{12} = \frac{2}{3} m_0 \hbar^{-1} \omega_{12} |r_{12}|^2 \tag{15.4}$$

is the oscillator strength of the resonant transition, and r_{12} is the matrix element of the transition. From the real and imaginary parts of Equation 15.2 one immediately obtains the expressions for the absorption coefficient shown in Figure 15.1b

$$\alpha(\omega) = \frac{2\omega}{c} \text{Im}(\varepsilon^{1/2}) \approx \frac{1}{4\bar{n}c} \frac{\Omega_p^2 \gamma_{12}}{(\omega_{12} - \omega)^2 + \gamma_{12}^2/4} \tag{15.5}$$

and the refractive index (Figure 15.1c)

$$n(\omega) = \text{Re}(\varepsilon^{1/2}) \approx \bar{n} + \frac{1}{4\bar{n}\omega} \frac{\Omega_p^2 (\omega_{12} - \omega)}{(\omega_{12} - \omega)^2 + \gamma_{12}^2/4} = \bar{n} + \frac{c}{\omega} \alpha(\omega) \frac{\omega_{12} - \omega}{\gamma_{12}} \tag{15.6}$$

the last relation being a particular form of more general Kramers–Kronig relation

$$n(\omega) = \frac{c}{\pi} \int_0^\infty \frac{\alpha(\omega_1)}{\omega_1^2 - \omega^2} d\omega_1 \approx \bar{n} + \frac{c}{2\pi\omega} \int_{\text{near resonance}} \frac{\alpha(\omega_1)}{\omega_1 - \omega} d\omega_1 \tag{15.7}$$

Differentiating Equation 15.6 one can then obtain expressions for the group velocity

$$v_g^{-1} = \frac{\partial(nk)}{\partial \omega} = \bar{n}c^{-1} + c\omega \frac{\partial n}{\partial \omega} = \bar{n}c^{-1} + \frac{\Omega_p^2 c^{-1}}{4\bar{n}} \frac{(\omega_{12} - \omega)^2 - \gamma_{12}^2/4}{\left[(\omega_{12} - \omega)^2 + \gamma_{12}^2/4\right]^2} \tag{15.8}$$

and the slow down factor

$$S(\omega) = 1 + \frac{\Omega_p^2}{4\bar{\varepsilon}} \frac{(\omega_{12} - \omega)^2 - \gamma_{12}^2/4}{\left[(\omega_{12} - \omega)^2 + \gamma_{12}^2/4\right]^2} = 1 + \frac{c\alpha(\omega)}{\bar{n}} \frac{(\omega_{12} - \omega)^2 - \gamma_{12}^2/4}{\gamma_{12}} \tag{15.9}$$

where the last expression shows the relation between the absorption coefficient and the slow down factor.

We start with the influence of the loss which is twofold. First of all, the insertion loss weakens the signal, and if one attempts to compensate the loss with optical amplifiers, the amplifier noise decreases the signal-to-noise ratio (SNR). Second, the dispersion of the loss changes the spectrum of the signal, typically causing spectral narrowing. Let us assume for the sake of argument that the input is a binary on–off-keyed (OOK) signal with a bit rate B and that each "1" bit is a Gaussian pulse with full width half maximum (FWHM) equal to one half of the bit interval $T = B^{-1}$ (Figure 15.2). In the frequency domain the spectrum of the pulse is centered at frequency ω_0 and has FWHM $\Delta\omega_{\text{sig},0} \sim 8\ln(2)B$. Now, if the total insertion loss, $\exp[\alpha(\omega)L]$, where L is the length and varies in such a way that the spectral components at the edges of spectrum, that is, at $\omega_0 \pm \Delta\omega_{\text{sig},0}/2$, experience insertion loss that is twice as large as the loss at ω_0, one can say that the spectrum had been narrowed by a factor of 2 and, correspondingly, the bit pulse in the time domain had expanded beyond the bit interval, which we can take as a indication that intersymbol interference (ISI) becomes too large for error-free detection, as shown in Figure 15.2. In case of off-resonance Lorentzian absorption (Equation 15.5) one can write the condition under which the ISI is within allowable limits as

$$\alpha(\omega_0) L \times 8B < |\omega_0 - \omega_{12}| \tag{15.10}$$

Important as it is, dispersion of loss is not the ultimate factor limiting the performance of SL devices because it can be reduced in both atomic and SL schemes. In the photonic SL schemes considered

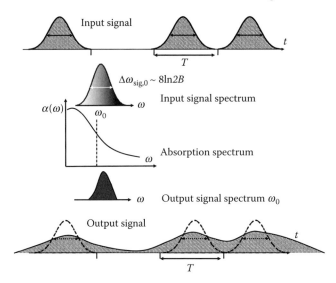

FIGURE 15.2 Explanation of dispersion of loss $\alpha(\omega)$ having a deleterious effect on the signal.

in Section 15.3, the loss is not inherent, but caused by imperfections in fabrication. Therefore, as fabrication techniques improve one can expect that loss dispersion will become less of a factor. In atomic schemes the loss is inherent and is associated with the dephasing of the optical polarization. But, while one cannot eliminate the absorption loss entirely, one can greatly reduce it by using the ingenious electromagnetic transparency scheme first suggested by Harris [1–3]. This is discussed in Section 15.2.

The critical limiting factor in SL schemes usually turns out to be the dispersion of group velocity (GVD) which always occurs in the vicinity of the resonance, and, as we shall see, can be somewhat reduced but never eliminated. To understand the role of GVD we shall consider a Taylor series expansion of the dispersion curve $k(\omega)$ around the central frequency of the signal ω_0

$$k(\omega) = k(\omega) = k(\omega_0) + \left.\frac{\partial k}{\partial \omega}\right|_{\omega_0}(\omega - \omega_0) + \frac{1}{2}\left.\frac{\partial^2 k}{\partial \omega^2}\right|_{\omega_0}(\omega - \omega_0)^2 + \frac{1}{6}\left.\frac{\partial^3 k}{\partial \omega^3}\right|_{\omega_0}(\omega - \omega_0)^3 + \cdots$$

$$= k(\omega_0) + v_g^{-1}(\omega_0)(\omega - \omega_0) + \frac{1}{2}\beta_2(\omega - \omega_0)^2 + \frac{1}{6}\beta_3(\omega - \omega_0)^3 + \cdots \quad (15.11)$$

where we have introduced the higher order dispersion terms $\beta_n(\omega) = \partial k/\partial \omega$. We can now find the group velocity

$$v_g^{-1}(\omega) = v_g^{-1}(\omega_0) + \beta_2(\omega - \omega_0) + \frac{1}{2}\beta_3(\omega - \omega_0)^2 + \cdots \quad (15.12)$$

and the slow down factor

$$S(\omega) = S(\omega_0) + \beta_2 c \bar{n}^{-1}(\omega - \omega_0) + \frac{1}{2}\beta_3 c \bar{n}^{-1}(\omega - \omega_0)^2 + \cdots \quad (15.13)$$

where $n = k(\omega_0)/\omega_0$ is the refractive index and c is the speed of light in a vacuum. One can estimate the limitations imposed by GVD by noticing that the time delay of the signal can be written as

$$T_d(\omega) = v_g^{-1}(\omega)L = v_g^{-1}(\omega_0)L + \beta_2(\omega - \omega_0)L + \frac{1}{2}\beta_3(\omega - \omega_0)^2 L + \cdots \quad (15.14)$$

Bandwidth Limitation in Slow Light Schemes

and then introducing a criterion for the OOK Gaussian signal: the difference between the delay times of spectral components at the edges of the signal bandwidth, that is, at $\omega_0 \pm \Delta\omega_{sig,0}/2$, should be less than one-half of the bit interval, that is,

$$\Delta T_d(L) \approx \beta_2 \Delta\omega_{sig} L < T/2 \tag{15.15}$$

which leads to the condition relating the maximum allowable length and bit rate

$$|\beta_2| B^2 L < \frac{1}{16 \ln 2} \tag{15.16}$$

This condition shows the limitations by the second-order GVD term β_2. In case of off-resonance Lorentzian absorption (Equation 15.5) it becomes

$$32\alpha(\omega_0) L B^2 / \gamma_{12} < |\omega_0 - \omega_{12}| \tag{15.17}$$

Comparing Equation 15.17 with Equation 15.10 shows that, as long as bit rate is high (in excess of Gbps) and the atomic transition is narrow γ_{12} (say a few MHz), the GVD limitations are far more severe than those due to the loss dispersion.

If this lowest term GVD is eliminated, as is indeed the case in many SL schemes, then one uses the third-order GVD term and estimates the maximum bandwidth and bit rate combination at which the difference between the delay times of spectral components at ω_0 and $\omega_0 \pm \Delta\omega_{sig,0}/2$ does not exceed $T/2$

$$|\beta_3| B^3 L < 1/16 (\ln 2)^2 \tag{15.18}$$

Before leaving this simple case of a single atomic resonance let us look once again at the dispersion curve $k(\omega)$ in the absence of loss (Figure 15.3a). One can see that in the vicinity of resonance there

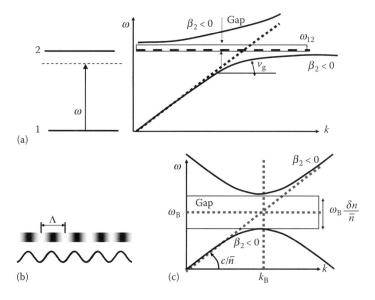

FIGURE 15.3 (a) Dispersion and group velocity near strong lossless atomic resonance, (b) Bragg grating and its index profile, and (c) dispersion of Bragg grating.

exists a forbidden gap where the dielectric constant is negative (Restrahlen region). The width of the gap according to Equation 15.2 is

$$\omega_{\text{gap}} \approx \frac{\Omega_p^2}{2\bar{n}^2 \omega_{21}} \tag{15.19}$$

i.e., it is proportional to the oscillator strength of the transition. Close to the gap, the light experiences significant slow down. The GVD β_2 is also strong near that gap, but it has different signs below and above the gap which can be used to cancel it. It is also important to emphasize the common principle of all atomic SL schemes—the light is slowed down because the energy is constantly transferred between the propagating electromagnetic wave and the atomic polarization that of course does not propagate. The smaller the group velocity, the more time the energy spends in the form of stationary atomic excitations. When the group velocity approaches zero one can hardly talk about an electromagnetic wave any more since the energy has been nearly completely transferred to atomic excitations.

15.3 PHOTONIC RESONANCES

The situation could not have been more different in the case of photonic resonances, where the energy is transferred between the forward and backward waves. The most common photonic resonance is the Bragg grating (Figure 15.3b)—a structure in which the refractive index is periodically modulated with period Λ,

$$n = \bar{n} + \delta n \cos\left(\frac{2\pi}{\Lambda} z\right) \tag{15.20}$$

As a result, a photonic bandgap opens in the vicinity of Bragg frequency [18]

$$\omega_B = \frac{\pi c}{\Lambda \bar{n}} \tag{15.21}$$

and the dispersion law becomes modified as

$$\frac{k - k_B}{k_B} = \sqrt{\left(\frac{\omega - \omega_B}{\omega_B}\right)^2 - \left(\frac{\delta n}{2\bar{n}}\right)^2} \tag{15.22}$$

This dispersion law is plotted in Figure 15.3c. One can see the similarities between it and the dispersion of a single atomic resonance (Figure 15.3a). Close to the gap the group velocity becomes reduced with the slow down factor being

$$S = \frac{|\omega - \omega_B/\omega_B|}{\sqrt{(\omega - \omega_B/\omega_B)^2 - (\delta n/2\bar{n})^2}} \tag{15.23}$$

Furthermore, the width of forbidden gap is $\Delta\omega_{\text{gap}} = \omega_B \delta n/\bar{n}$, and hence the index contrast $\delta n/\bar{n}$ can be called the strength of the grating. This grating strength plays a role equivalent to that played by the oscillator strength of the atomic resonance. But the physics is quite different—the slow down effect in a photonic structure is the result of the transfer of energy between the forward and backward propagating waves—no energy is transferred to the medium—hence the strength of the electric field in photonic SL structures gets greatly enhanced with important implications for nonlinear optics.

15.4 DOUBLE-RESONANT ATOMIC SL STRUCTURES

Since according to Figure 15.3 the lowest order GVD β_2 is positive below resonance and negative above resonance, if one can combine two resonances as in Figure 15.4a, only the third-order GVD

Bandwidth Limitation in Slow Light Schemes

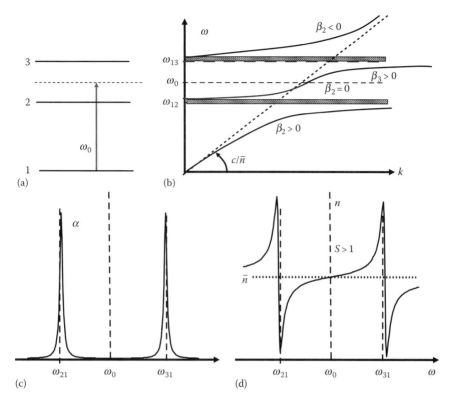

FIGURE 15.4 (a) Double atomic resonance and its (b) dispersion and group velocity near it. (c) Absorption spectrum and (d) refractive index spectrum.

β_3 (Figure 15.4b) will be a factor for signals centered at frequency ω_0 in the middle between two transitions. Closely spaced narrow resonances do occur in metal vapors, such as, in Rb^{85} [19] where two D^2 resonances near 780 nm separated by $\nu_{32} = 3$ GHz have been used in the most successful present day SL experiments in atomic medium. As one can see from dispersion of the refractive index in Figure 15.4c and the absorption in Figure 15.4d, both residual absorption and β_3 will limit the bit rate of the signals propagating through the double-resonant SL atomic medium. One can approximate the Taylor expansion of the residual absorption near ω_0 by

$$\alpha(\omega) \approx \alpha(\omega_0) + \frac{1}{2}\alpha_2(\omega - \omega_0)^2 \tag{15.24}$$

where

$$\alpha(\omega_0) = \frac{2}{\bar{n}c}\frac{\Omega_p^2}{\omega_{32}^2}\gamma_{21} \tag{15.25}$$

and

$$\alpha_2 = \frac{48}{\bar{n}c}\frac{\Omega_p^2}{\omega_{32}^4}\gamma_{21} \tag{15.26}$$

One can then see that the spectrum of the OOK Gaussian signal passing through the double-resonant SL medium will remain Gaussian but the FWHM will decrease from $\Delta\omega_{sig,0} = 8\ln(2)B$ to

$$\Delta\omega_{sig,L}^{-2} = \Delta\omega_{sig,0}^{-2} + \alpha_2 L/8\ln 2 \tag{15.27}$$

Since a decrease of spectral FWHM by a factor of $2^{1/2}$ will cause pulse broadening by the same factor, we can obtain the condition for the maximum length and bandwidth combination at which the ISI can still be deemed acceptable as

$$\alpha_2 B^2 L = \frac{48}{\bar{n}c} \frac{\Omega_p^2}{\omega_{32}^4} \gamma_{21} B^2 L < 1/8 \ln 2 \tag{15.28}$$

The group velocity in the vicinity of ω_0 can also be expressed using Taylor expansion as

$$v_g^{-1}(\omega) \approx v_g(\omega_0) + \frac{1}{2}\beta_3(\omega - \omega_0)^2 \tag{15.29}$$

where

$$v_g^{-1}(\omega_0) = c^{-1}\bar{n} + \frac{2}{c\bar{n}} \frac{\Omega_p^2}{\omega_{32}^2} \approx \frac{2}{c\bar{n}} \frac{\Omega_p^2}{\omega_{32}^2} \tag{15.30}$$

and

$$\beta_3 = \frac{48}{\bar{n}c} \frac{\Omega_p^2}{\omega_{32}^4} = 24 \frac{v_g^{-1}}{\omega_{32}^2} \tag{15.31}$$

From the higher order GVD point of view, ISI is acceptable as long as condition (Equation 15.18) is satisfied. Upon substitution of β_3 from Equation 15.31, Equation 15.18 becomes

$$|\beta_3| B^3 L = \frac{48}{\bar{n}c} \frac{\Omega_p^2 B^3 L}{\omega_{32}^4} < 1/16 (\ln 2)^2 \tag{15.32}$$

Quick comparison of Equations 15.28 and 15.32 reveals that for $B \gg \gamma_{21}$ GVD will be the dominant limitation, while for $B \ll \gamma_{21}$ the main limitation will be related to the loss dispersion. In the Rb vapor [19] $\gamma_{21} = 2\pi \times 6$ MHz. Therefore for the bit rates in excess of 100 MHz GVD, in that work was indeed the main limitation.

Now, the actual number of bits stored in the delay line can be found as

$$N_{st} = v_g^{-1}LB = \frac{2}{c\bar{n}} \frac{\Omega_p^2}{\omega_{32}^2} LB \tag{15.33}$$

To gain insight into the bit rate limitations imposed by GVD all we need to do is to eliminate L from Equations 15.32 to 15.33 to obtain the maximum number of OOK bits that one can store at rate B,

$$N_{st}^{(max)} = \frac{1}{6} \left(\frac{\omega_{32}}{8 \ln 2B}\right)^2 = \left(\frac{B_{max}^{(1)}}{B}\right)^2 \tag{15.34}$$

where we have introduced a maximum bit rate at which at least one bit of information can be stored. This bit rate depends only on the transparency bandwidth of the double resonance $\nu_{32} = \omega_{32}/2\pi$ as

$$B_{max}^{(1)} \approx 0.42\nu_{32} \tag{15.35}$$

indicating that even to store just one bit of information the bandwidth must be significantly narrower than the transparency bandwidth. Alternatively, we introduce the maximum bit rate at which one can store N_{st} bits as

$$B_{max}^{(N_{st})} = B_{max}^{(1)} N_{st}^{-1/2} = 0.42 N_{st}^{-1/2} \tag{15.36}$$

This is an important result—as the number of bits one wants to store increases, the allowable bit rate decreases.

To find out what length is required to store N_{st} bits, we substitute N_{st} from Equation 15.34 back into Equation 15.33 to obtain

$$L(B) = \frac{8\sqrt{3}}{\ln 2}\left(\frac{B^{(1)}}{B}\right)^3 \frac{\omega_{32}}{\Omega_p^2}c\bar{n} = 16\ln 2\sqrt{3}\,(N_{st})^{3/2}\frac{L_{abs}}{F} \qquad (15.37)$$

where we have introduced the absorption length $L_{abs} = 1/\alpha(\omega_0)$ and the finesse of the scheme $F = \omega_{32}/\gamma_{21}$. Thus, finesse is the ratio of the transparency bandwidth to the FWHM of the absorption. A very narrow resonance is thus desirable and in the Rb vapor, finesse does approach $F \sim 500$ making it possible to store a few bits of information per absorption length. Note that neither Equation 15.34 nor Equation 15.37 depend explicitly on the plasma frequency Ω_p^2. Therefore, increasing the concentration of the active atoms will allow one to reduce the length of the delay line, but will affect neither the maximum bandwidth nor insertion loss. One can rewrite Equation 15.37 as

$$N_{st} \approx \left(\frac{FL}{30L_{abs}}\right)^{2/3} \qquad (15.38)$$

This $L^{2/3}$ dependence of storage capacity is easy to understand—when the length increases one is forced to reduce the bit rate as $L^{1/3}$ according to Equation 15.32. Thus the overall storage capacity grows only sublinearly with L. According to Equation 15.38 one can store 20 bits of information in Rb vapor with insertion loss of 20 dB which is close to the results obtained by Howell in Ref. [19].

15.5 DOUBLE-RESONANT PHOTONIC SL STRUCTURES—CASCADED GRATINGS

Since any atomic SL scheme requires operation near a particular narrow linewidth absorption resonance, finding such a resonance near a particular wavelength is not an easy task, and, in fact, only a very few absorption lines have been employed in practice, Rb vapors being a workhorse. Finding two closely spaced narrow lines are even more difficult, and even if such two lines can be found, the splitting between them ν_{32} is fixed—hence the SL delay will be optimized for one particular combination of storage capacity and bit rate.

In contrast, the photonic double resonant can be easily implemented by simply combining two Bragg gratings with slightly different periods Λ_1 and Λ_2 as shown in Figure 15.5a. Such a combination was first suggested for dispersion compensation [20,21] and then considered for application in electro optic modulators [22]. As long as one deals with linear devices, such as delay lines, one can simply cascade two Bragg gratings sequentially and the resulting dispersion curve will be simply the mean of the individual dispersion curves (Figure 15.5c,d). For the nonlinear and electro optic devices, one can alternate the short segments of Bragg gratings with periods Λ_1 and Λ_2. The dispersion curve of Figure 15.5d is remarkably similar to the dispersion curve of the atomic double resonance in Figure 15.5b. Two gratings engender two photonic bandgaps, centered at $\omega_{B,i} = \pi c/\Lambda_i \bar{n}$ of almost equal widths $\Delta\omega_{gap,i} = \omega_{B,i}(\delta n/\bar{n}) \approx \omega_0(\delta n/\bar{n})$ with a narrow passband $\Delta\omega$ between them. By choosing the periods Λ_1 and Λ_2 for a given index modulation δn one can design $\Delta\omega$ to be arbitrarily narrow or wide. This fact gives the designer true flexibility. The slow down factor dispersion can be found using Equation 15.23 as

$$S(\omega_0) = [S_1(\omega_0) + S_2(\omega)]/2 = \frac{1 + \Delta\omega/\Delta\omega_{gap}}{\sqrt{(2 + \Delta\omega/\Delta\omega_{gap})\,\Delta\omega/\Delta\omega_{gap}}} \approx \left[1 + \frac{\Delta\omega_{gap}}{2\Delta\omega}\right]^{1/2} \qquad (15.39)$$

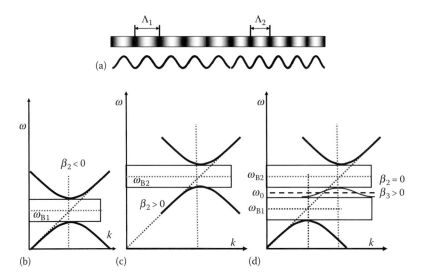

FIGURE 15.5 (a) Cascaded Bragg grating, (b) and (c) dispersion curves of two Bragg gratings that are cascaded, and (d) dispersion curve of cascaded grating.

where the last equality holds well for $\Delta\omega < \Delta\omega_{\text{gap}}$. The second-order GVD β_2 is cancelled as in any double-resonant scheme, while the third-order dispersion is

$$\beta_3(\omega_0) = \frac{12\left(1 + \Delta\omega/\Delta\omega_{\text{gap}}\right)\bar{n}}{c\sqrt{\left(2 + \Delta\omega/\Delta\omega_{\text{gap}}\right)\Delta\omega/\Delta\omega_{\text{gap}}}\left(2 + \Delta\omega/\Delta\omega_{\text{gap}}\right)^2 \Delta\omega^2} \approx \frac{3\bar{n}}{c\Delta\omega^2}\left[\frac{\Delta\omega_{\text{gap}}}{2\Delta\omega}\right]^{1/2} \quad (15.40)$$

Notice that the relation between the group velocity and the third-order GVD

$$\beta_3 = 3\frac{v_g^{-1}}{\Delta\omega^2} \quad (15.41)$$

is strikingly similar to Equation 15.31—in both cases, β_3 is proportional to the slow down factor and inversely proportional to the square of the passband.

In order to determine the GVD-imposed limitations we can now write the equation in a manner similar to Equations 15.32 and 15.33:

$$|\beta_3|B^3 L = 3\frac{v_g^{-1} L B^3}{\Delta\omega^2} < 1/16\,(\ln 2)^2 \quad (15.42)$$

and

$$N_{\text{st}} = v_g^{-1} L B \quad (15.43)$$

and by eliminating $v_g^{-1} L$ arrive at the relation between the passband, bit rate, and the storage capacity

$$N_{\text{st}} = \frac{1}{3}\left(\frac{\Delta\omega}{4B \ln 2}\right)^2 \quad (15.44)$$

from which one can obtain the expression for the maximum bit rate at which one can store N_{st} bits as

$$B_{\text{max}}^{(N_{\text{st}})} = 0.75\Delta\nu N_{\text{st}}^{-1/2} \quad (15.45)$$

Bandwidth Limitation in Slow Light Schemes

which looks remarkably similar to Equation 15.36. There is one substantial difference, however, and it has to do with the ability to change the width of passband $\Delta \nu = \Delta \omega / 2\pi$ in the photonic scheme. At first look, it appears that the bit rate is not really limited as one can expand the passband to fit the expanding bandwidth. This, however, would be an erroneous conclusion because expansion of the passband will decrease the slow down effect according to Equation 15.39, and at a certain value of bit rate that we call cutoff bit rate $B_{cut}^{(N_{st})}$, the slow down factor S will not be substantially larger than unity. This will occur when the passband width $\Delta \nu$ will become comparable to the gap width $\Delta \nu_{gap}$. To quantify this statement one can obtain by rewriting Equation 15.44 the expression for the passbandwidth required to pass the N_{st} bits through the delay line without GVD-caused ISI

$$\Delta \nu(N_{st}, B) = 1.3 B N_{st}^{1/2} \quad (15.46)$$

When one substitutes Equation 15.46 into Equation 15.39, a bit rate dependence of slow down factor is obtained

$$S(B, N_{st}) \approx \left[1 + \frac{\Delta \nu_{gap}}{2.6 B N_{st}^{1/2}}\right]^{1/2} = \left[1 + \frac{B_{cut}^{(N_{st})}}{B}\right]^{1/2} \quad (15.47)$$

where we have defined our cutoff bit rate

$$B_{cut}^{(N_{st})} = \frac{\Delta \nu_{gap}}{2 N_{st}^{1/2}} = \nu_0 \frac{\delta n}{2.6 N_{st}^{1/2} \bar{n}} \quad (15.48)$$

as a rate at which the slow down factor is reduced to $2^{1/2}$ and the SL structure does not make the delay line significantly shorter than using simple unstructured material, for instance a fiber. This can be seen from Figure 15.6 in which we plot the length of the SL fiber with cascaded gratings required to store N_{st} bits as a function of bit rate B

$$L(B, N_{st}) = \frac{N_{st} c \bar{n}^{-1} B^{-1}}{S(B, N_{st})} = c \bar{n}^{-1} N_{st} \left[B^2 + B B_{cut}^{(N_{st})}\right]^{-1/2} \quad (15.49)$$

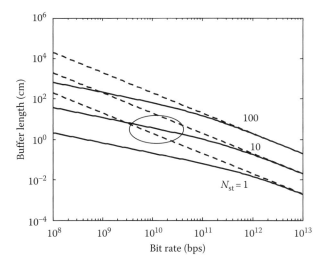

FIGURE 15.6 Solid lines: length of SL optical buffer based on cascaded Bragg grating required to store N_{st} bits as a function of bit rate for three different storage capacities. Dashed lines: length of an unstructured fiber also capable of storing the same number of bits.

Using the example of a strong fiber grating with an index modulation of $\delta n/\bar{n} = 0.01$, that is, with a cutoff bit rate for 1 bit $B_{\text{cut}}^{(1)} \sim 900\,\text{GHz}$. The length of the buffer decreases with the bit rate, mostly due to simply the shorter delay time required to store the same number of bits with a shorter bit interval, but as bit rate increases, the length of the simple, unstructured, fiber shown by the dashed line decreases even faster, and past the cutoff bit rate, SL fiber loses all its benefits. At low bit rates SL fiber offers significant benefits, but the length required becomes unrealistic. From a practical point of view, one can see that the only sweet spot region occurs around 10 Gbps where one can store 10 bits in a 3 cm long cascaded grating which is significantly shorter than 20 cm of unstructured fiber required for the same goal. Clearly, one must consider photonic structures with larger index contrast to obtain better delay lines.

To understand it better, one can make a comparison between the slow down factors. For the double-resonant atomic structure (Equation 15.30)

$$S_{\text{atom}} = 1 + \frac{2}{\bar{n}^2}\frac{\Omega_p^2}{\omega_{32}^2} = 1 + \left(\frac{v_0}{\Delta v_{\text{pass}}}F_{\text{atom}}\right)^2 \tag{15.50}$$

where $\Delta v_{\text{pass}} = v_{32}$ and the strength of atomic SL scheme has been introduced as

$$F_{\text{atom}} = 2^{1/2}\Omega_p/\omega_0\bar{n} = \frac{e}{\omega_0}\left(\frac{2N_a}{\varepsilon_0\bar{\varepsilon}m_0}\right)^{1/2} f_{12}^{1/2} \tag{15.51}$$

This strength depends only on the density of atoms and the transition dipole moment. For the cascaded grating, one obtains from Equation 15.39

$$S_{\text{casc}} = \left[1 + \frac{v_0}{\Delta v_{\text{pass}}}F_{\text{casc}}\right]^{1/2} \tag{15.52}$$

where the strength of photonic SL scheme has been introduced as

$$F_{\text{casc}} = \delta n/2\bar{n} \tag{15.53}$$

As we have already mentioned, the index contrast in the photonic SL schemes plays essentially the same role as the oscillator strength in atomic transitions. In fact, one can estimate F_{atom} in the Rb scheme used in Ref. [19] to be on the order of 10^{-4} which is much less than F_{casc} even in weak gratings. But different dependencies on the passbandwidth—$S_{\text{casc}} \sim (\Delta v/v_0)^{-1/2}$ and $S_{\text{atom}} \sim (\Delta v/v_0)^{-2}$ indicate drastically different behavior of atomic and photonic schemes as a function of bandwidth. The atomic schemes can achieve enormous slow down factors, but only with very narrow bandwidth signals, while photonic schemes offer a relatively modest decrease in group velocity, but over a much wider bandwidth.

There are two important conclusions that can be drawn from the analyses of cascaded Bragg gratings—the first example of SL structure with tunable (at least by design) passband. The first conclusion is the existence of the cutoff bandwidth $B_{\text{cut}}^{(N_{\text{st}})}$. This comes from the fact that as we increase the bit rate B we must increase the required passband width according to Equation 15.46 in order to mitigate the GVD-caused ISI which will reduce the slow down factor and cause us to increase the length of the delay line to accommodate N_{st} bits. Eventually, as the bit rate approaches cutoff $B_{\text{cut}}^{(N_{\text{st}})}$ the length of delay line will not be much shorter than that of the unstructured fiber. At this point, there would be no advantage in using the SL delay line.

The second conclusion is that the length of the SL delay line increases superlinearly with the storage capacity. As long as the bit rate is substantially lower than cutoff (i.e., as long as delay line is useful) expression 15.49 can be approximated as

$$L(B, N_{\text{st}}) = \frac{cN_{\text{st}}^{5/4}}{\bar{n}\,(F_{\text{casc}}Bv_0)^{1/2}} = \frac{\lambda}{\bar{n}}\left(N_{\text{st}}^{5/4}\frac{v_0}{B}\right)^{1/2} F_{\text{casc}}^{-1/2} \tag{15.54}$$

Bandwidth Limitation in Slow Light Schemes

This is of course easy to understand—an increase in storage capacity at first causes one to increase the length proportionally; but with longer length the GVD becomes too large and to avoid this, one is forced to increase the passband. That in turn reduces the slow down factor and thus the length needs to be further augmented. That in its turn causes an additional increase in GVD, so the passband needs to be further expanded and the whole process must go on iteratively until the length approaches its value given by Equation 15.54.

15.6 TUNABLE DOUBLE-RESONANT ATOMIC SL STRUCTURES—ELECTROMAGNETIC TRANSPARENCY (EIT)

As we have already mentioned, the fixed double atomic resonant scheme cannot be adapted to variable bandwidth because the width of passband cannot be changed. To change the passband width one can consider an alternative of spectral hole burning in the inhomogeneously broadened transition [23,24]. As shown in Figure 15.7, a strong pump pulse creates a situation where the absorption in the frequency range $\Delta\omega$ becomes depleted. The profile of the absorption spectrum shown in Figure 15.7b looks remarkably like the double resonant profile of Figure 15.5. With the refractive index profile shown in Figure 15.7c, one can see that a strong reduction of group velocity can be expected near the center of the spectral hole.

By changing the spectrum of the pump, for instance, using intensity or frequency modulation, one can change $\Delta\omega$ [25] to achieve the maximum delay without distortion for a given bit rate. In Ref. [25], delays of 2 bit intervals were achieved for a moderate bandwidth of 100 MHz but only in a 40 cm long Rb vapor delay line. Since the background absorption in the hole burning is always high, it is the dispersion of loss, α_2 term in Equation 15.24, which causes the signal distortion and is in fact a limitation in this scheme. The scheme also suffers from the large energy dissipation as the pump gets absorbed.

To avoid large background absorption and to achieve wide passband tunability, one uses an entirely different SL scheme based on EIT, first considered by Harris [1–3]. Without trying to explain all the intricacies of EIT we first consider the main rationale of using it. Since finding two closely spaced atomic resonances is not trivial, one should consider the means for their creation artificially.

It is well known that if one considers a harmonic wave (Figure 15.8a), its spectrum (Figure 15.8b) contains only one frequency component ω_0. But if the wave is amplitude modulated with some frequency Ω (Figure 15.8c), there will appear two side bands at frequencies $\omega_0 \pm \Omega$ in its spectrum

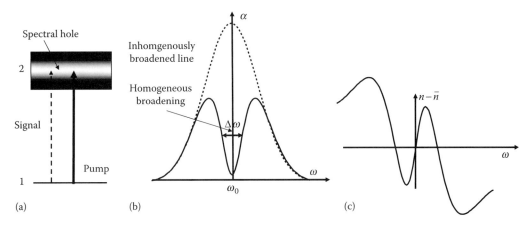

FIGURE 15.7 (a) SL scheme based on spectral hole burning, (b) absorption spectrum, and (c) refractive index spectrum.

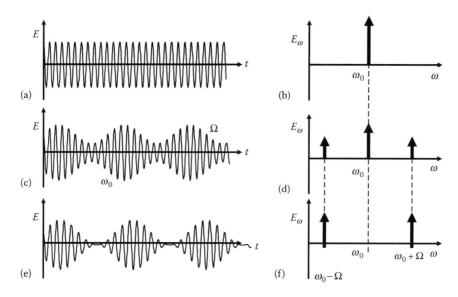

FIGURE 15.8 (a) Harmonic wave at carrier frequency ω_0, (b) spectrum of harmonic wave, (c) harmonic wave modulated with frequency Ω, (d) its spectrum with two sidebands, (e) fully modulated harmonic wave, and (f) spectrum with complete suppression of carrier frequency.

(Figure 15.8d). When the modulation depth reaches 100% (Figure 15.8e) the carrier frequency ω_0 gets completely suppressed and the spectrum shows just two sidebands separated by 2Ω (Figure 15.8f).

Now, if one thinks of the atom as an oscillator absorbing at some resonant frequency ω_0 and strongly modulates the atom absorption with some external frequency Ω, one should expect the absorption spectrum to behave in a fashion similar to the spectrum of amplitude-modulated wave, that is, it should show two absorption lines separated by 2Ω. The material should become transparent at the resonant frequency ω_0—hence the term EIT.

To accomplish the EIT transmission modulation there exist numerous schemes, but we shall consider only one—the most widely used three-level Λ scheme [1] shown in Figure 15.9 in which the ground-to-excited state transition ω_{12} is resonant with the frequency of the optical signal carrier ω_0 and has a dephasing rate of γ_{21}. In the absence of pump, the absorption spectrum (dashed line in Figure 15.9) is a normal Lorentzian line. There also exists a strong transition coupling the excited level 2 with level 3, which is critical, but the transition between levels 1 and 3 is forbidden. When a strong resonant pump at frequency ω_{12} is turned on, the mixing of states 2 and 3 causes modulation

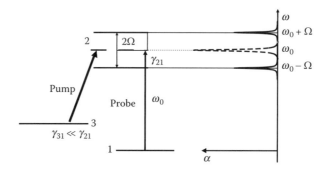

FIGURE 15.9 Principle of electromagnetic transparency in a "Λ" scheme.

of the absorption of signal. As expected, the Lorentzian peak in the absorption spectrum splits into two smaller peaks at frequencies $\omega_0 \pm \Omega$, where the Rabi frequency

$$\Omega = er_{32}\hbar^{-1}\left(2\eta_0 I_{pump}/\bar{n}\right)^{1/2} \quad (15.55)$$

depends on the pump intensity I_{pump} and η_0 is the vacuum impedance. Apart from achieving full tunability in the double-resonant scheme, the EIT is important because the residual absorption rate at the resonant frequency

$$\alpha(\omega_0) = \frac{1}{\bar{n}c}\frac{\Omega_p^2}{8\Omega^2}\gamma_{31} \quad (15.56)$$

is proportional to the dephasing rate of the intra-atomic excitation 31 which is not coupled to the outside world. Thus, typically $\gamma_{31} \ll \gamma_{21}$ and the residual absorption is much weaker in the EIT than in the case of two independent resonances. It indicates that EIT is a coherent effect and the reduction of absorption occurs because of the destructive interference of the absorption by two sidebands. But, from the point of view of practical applications in delay lines, a different perspective of EIT may be more appropriate. One can consider the following sequence of events (Figure 15.10)—as the signal photon propagates in the EIT medium it transfers its energy to the excitation of atomic transition between levels 1 and 2. Due to the presence of strong pump wave coupling between levels 2 and 3, the excitation is almost instantly transferred to the long-lived excitation between the levels 1 and 3 and then the process occurs in reverse until the energy is transferred back into the photon. Then the process repeats itself. Overall, for most of the time the energy gets stored in the form of 1–3 excitations and thus it propagates with a very slow group velocity. Furthermore, the actual absorption event occurs only when the excitation 1–3 loses coherence and the energy cannot get back to the photon. Naturally, it is the dephasing rate of this excitation, that is, $\gamma_{31} \ll \gamma_{21}$ which determines the residual absorption loss in Equation 15.56. We once again stress here that since the energy gets stored in the form of atomic excitation, one cannot expect enhancement in the strength of the optical field.

Aside from the intricacies associated with dephasing rate, the EIT scheme whose dispersion is shown in Figure 15.11 behaves exactly as a simple double-resonant scheme considered before, except that its passband is tunable. The group velocity then can be written similar to Equation 15.30

$$\Delta \nu_{pass} = 2\Omega \quad (15.57)$$

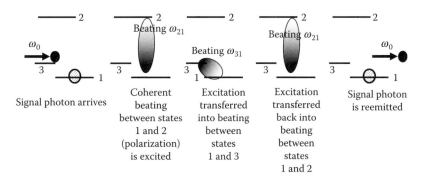

FIGURE 15.10 Explanation of energy storage and slow light propagation in the EIT scheme.

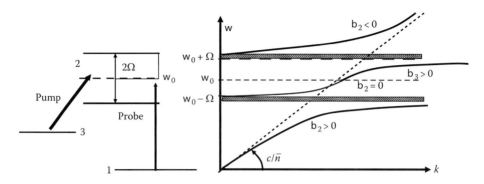

FIGURE 15.11 Dispersion in the EIT SL scheme.

$$v_g^{-1}(\omega_0) = \frac{\bar{n}}{c}\left(1 + \frac{1}{\varepsilon}\frac{\Omega_p^2}{4\Omega^2}\right) = \frac{\bar{n}}{c}\left[1 + \left(\frac{v_0}{\Delta v_{\text{pass}}}F_{\text{EIT}}\right)^2\right] \quad (15.58)$$

with $F_{\text{EIT}} = F_{\text{atom}}/\sqrt{2} = \Omega_p/\omega_0\bar{n}$ since in EIT the oscillation strength of the $1 \to 2$ transition is split between two sidebands. The third-order GVD in EIT scheme becomes (Equation 15.31)

$$\beta_3 = \frac{24}{\bar{n}c}\frac{\Omega_p^2}{(2\Omega)^4} = \frac{6\bar{n}}{\pi^2\Delta v_{\text{pass}}^2 c}\left(\frac{v_0}{\Delta v_{\text{pass}}}F_{\text{EIT}}\right)^2 = \frac{6v_g^{-1}}{\pi^2\Delta v_{\text{pass}}^2} \quad (15.59)$$

Once again using Equation 15.33 to insert $v_g^{-1}L = N_{\text{st}}B$ into acceptable ISI condition (Equation 15.18) we obtain

$$|\beta_3|B^3L = \frac{6v_g^{-1}LB^3}{\pi^2\Delta v_{\text{pass}}^2} = \frac{6N_{\text{st}}B^2}{\pi^2\Delta v_{\text{pass}}^2} < 1/16(\ln 2)^2 \quad (15.60)$$

and the relation for the passband required to store N_{st} bits at bit rate B is

$$\Delta v_{\text{pass}}(N_{\text{st}}, B) \approx 3.5BN_{\text{st}}^{1/2} \quad (15.61)$$

Substituting this relation into Equation 15.58 we obtain

$$S_{\text{EIT}}(B, N_{\text{st}}) = 1 + \left(\frac{v_0}{3.5BN_{\text{st}}^{1/2}}F_{\text{EIT}}\right)^2 = 1 + \left(\frac{B_{\text{cut}}^{(N_{\text{st}})}}{B}\right)^2 \quad (15.62)$$

where we have introduced the cutoff bit rate

$$B_{\text{cut}}^{(N_{\text{st}})} = \frac{v_0}{3.5N_{\text{st}}^{1/2}}F_{\text{EIT}} = \frac{1}{3.5N_{\text{st}}^{1/2}}\frac{\Omega_p}{2\pi\bar{n}} \quad (15.63)$$

as the bit rate at which the slow down factor is reduced to 2, which is actually slightly (by a factor of $\sqrt{2}$) different from the definition of cutoff bit rate in the case of cascaded grating. We can also determine the length of the EIT medium required to store N_{st} bits at a bit rate B as

$$L(B, N_{\text{st}}) = \frac{N_{\text{st}}c\bar{n}^{-1}B^{-1}}{1 + \left(B_{\text{cut}}^{(N_{\text{st}})}/B\right)^2} = \frac{c\bar{n}^{-1}N_{\text{st}}}{B + \left[B_{\text{cut}}^{(N_{\text{st}})}\right]^2/B} \quad (15.64)$$

TABLE 15.1
Characteristic Parameters of the EIT Slow Light Media

Atomic SL Medium	Concentration N_a (cm^{-3})	Oscillator Strength f_{21}	Plasma Frequency $\Omega_p/2\pi$ (GHz)	Strength of Slow Down F_{EIT}	Cutoff Bit Rate for 1 Bit Storage Capacity $B_{cut}^{(1)}$ (Gbps)
Rb^{87}	0.5×10^{15}	0.1	100	2.5×10^{-4}	22
Pb^{205}	7×10^{15}	0.2	440	1.1×10^{-3}	100
Pr : Y$_2$SiO$_5$	7×10^{19}	3×10^{-7}	40	10^{-4}	9
QD	1×10^{16}	3	1000	3×10^{-3}	220

This dependence is drastically different from the one for the required length of cascaded grating (Equation 15.49). For small bit rates $B \ll B_{cut}^{(N_{st})}$ the required length actually increases linearly with the bit rate despite the fact that the bit interval (and thus the required delay time) becomes shorter. Beyond cutoff the EIT medium does not slow down the light by much and thus it behaves like a normal medium with L decreasing with the bit rate.

To estimate the performance, we consider several EIT schemes from the literature, shown in Table 15.1. The schemes relying upon Rb [26] and Pb [3] vapors have been among the first suggested and experimentally realized. Their main disadvantage is the density of metal vapor. The third scheme is all solid-state relying upon narrow transition lines in the rare earth ion of Pr placed in an yttrium silicate matrix [24]. This scheme has the advantage of high density of active atoms, but the oscillator strength of transition is low. In fact, the product of oscillator strength and concentration cannot be too high for the linewidth of the transition γ_{12} to remain narrow—hence the plasma frequency does not vary by more than an order of magnitude in these schemes. The hypothetical scheme relying on strong excitonic transition lines in semiconductor quantum dots (QD) [27] can in principle provide a large plasma frequency, but this scheme is very difficult to realize in practice due to inhomogeneous broadening in QD. Nevertheless, we shall include the QDs in consideration just to see what can be hypothetically achieved if one perfects their growth.

The results are plotted in Figure 15.12a–c for different storage capacities. As one can see from Figure 15.12a more or less all schemes are capable of storing one bit at a typical telecommunication bit rate of 10 Gbps although in case of all-solid-state Pr : Y$_2$SiO$_5$ the improvement relative to simple free space (shown by dashed line) would be insignificant. Figure 15.12b shows that only Pb vapor or QDs can accommodate 10 bits at a few Gbps in a length of less than 10 cm. When it comes to 100 bits storage capacity (Figure 15.12c) none of the schemes can operate even at 1 Gbps bit rate with realistically compact length. And for 40 Gbps, none of the schemes offer any improvement over free space or fiber.

This disappointing result follows directly from not just the superlinear but also the quadratic relation between the length of the EIT delay buffer and its capacity for the bit rates below cutoff

$$L(B, N_{st}) \approx 20 \frac{\lambda}{\bar{n}} N_{st}^2 \frac{B}{v_0} F_{EIT}^{-2} \qquad (15.65)$$

which is much stronger than the 5/4 power dependence for the cascaded grating (Equation 15.54).

15.7 COUPLED PHOTONIC RESONATOR STRUCTURES

As we have seen, SL schemes based on EIT possess many desirable attributes including tunability and the possibility of achieving extremely large delays, but their best performance occurs at low bit rates. This fact is determined by a $1/B^2$ bit rate dependence of the slow down factor. As we have seen

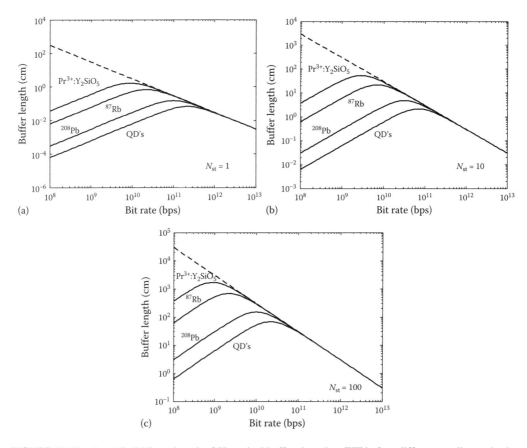

FIGURE 15.12 (a–c) Solid lines: length of SL optical buffers based on EIT in four different media required to store N_{st} bits as a function of bit rate for three different storage capacities. Dashed lines: length of an unstructured dielectric also capable of storing the same number of bits.

with an example of cascaded Bragg gratings, the bit rate dependence is not very prominent there, and their performance does not deteriorate as rapidly at high bit rates. But cascaded Bragg gratings also have a number of disadvantages, the first of which is a relatively small index contrast available, and the second is difficulty in fabricating two gratings with a prescribed value of the frequency offset. Furthermore, the cascaded geometry is applicable only to the linear devices, which do not incorporate any nonlinear or electro optic component. For this reason it is preferable to use alternating short segments of gratings with different periods. But a periodic sequence of short Bragg grating segments can be considered a new grating with periodically modulated properties—or Moiré grating (Figure 15.13a). In a Moiré grating, the segments are not independent but interact coherently—hence its properties are somewhat different from the cascaded grating as was shown in Ref. [28], with the main distinction being the fact that the dispersion curve of Moiré grating with period d is also periodic in wave vector space with a period $2\pi/d$. The ability of a Moiré Grating to slow down the light was first predicted in Ref. [28] and demonstrated in Ref. [29]. It was also noted that the Moiré grating is only one example of periodically structured photonic media in which slow light can be observed. In the periodically structured media the light energy density is distributed periodically and the periodically spaced regions of high intensity can be thought of as the resonators coupled to each other. Thus, we shall refer to them as CRS [30]. Aside from using Moiré gratings, CRS can be fabricated by coupling Fabry–Perot resonators (Figure 15.13b), ring resonators (Figure 15.13c) [31], or so-called defect modes in the photonic crystal (Figure 15.13d) [32,33].

Bandwidth Limitation in Slow Light Schemes

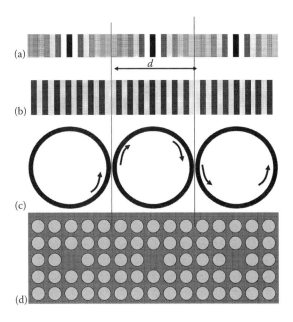

FIGURE 15.13 Photonic SL structures based on coupled resonators (CRS); (a) Moiré grating, (b) coupled Fabry–Perot resonators, (c) coupled ring resonators, and (d) coupled defect modes in photonic crystal.

These and other CRS implementations are discussed at length in other chapters of this book, so we shall quickly discuss the basics and then go directly to estimating the dispersion impact.

A periodic chain of coupled resonators is characterized by three parameters: period d, the time of one way pass through each resonator τ, and the coupling (or transmission) coefficient κ. The dispersion relation in this chain can be written as

$$\sin \omega\tau = \kappa \sin kd \tag{15.66}$$

The dispersion curves are shown in Figure 15.14 and consist of the series of passbands around resonant frequencies $\nu_m = m/2\tau$ separated by wide gaps. The width of the passband is

$$\Delta\nu_{\text{pass}} = (\pi\tau)^{-1} \sin^{-1}(\kappa) \tag{15.67}$$

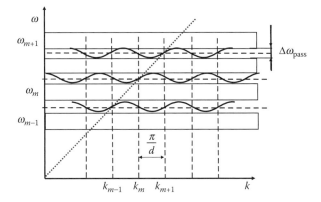

FIGURE 15.14 Dispersion in a typical CRS.

These three parameters are not independent of each other. First of all d and τ are obviously related to each other. This relation can be obtained from taking a limit of Equation 15.66 at $\kappa = 1$

$$\frac{d}{\tau} = \frac{\omega}{k} = \frac{c}{\bar{n}} \tag{15.68}$$

which simply indicates that with 100% coupling the light simply propagates through the medium without reflections. Also related are the size of resonator, that is, d and the coupling coefficient κ—to achieve small κ, one needs to confine the light tightly within the resonators which requires large spacing between them. If the index contrast $\delta n/\bar{n}$ is large, a high degree of confinement can be achieved within a relatively small resonator; otherwise the light will leak from one resonator to another. This issue has been addressed in detail in Ref. [15], but here we simply assume that one uses the smallest resonator size that can be fabricated using a technology with a given index contrast.

Using Taylor expansion of the dispersion relation (Equation 15.67) one obtains the group velocity

$$v_g^{-1} = \frac{\tau}{d\kappa} = \frac{\bar{n}}{c}\kappa^{-1} \tag{15.69}$$

And the third-order GVD

$$\beta_3 = \frac{1}{d}\left(\frac{\tau}{\kappa}\right)^3 (1 - \kappa^2) = v_g^{-1}\left(\frac{\tau}{\kappa}\right)^2 (1 - \kappa^2) \tag{15.70}$$

Once again we invoke the acceptable ISI condition (Equation 15.18) to obtain

$$16 (\ln 2)^2 |\beta_3| B^3 L = 16 (\ln 2)^2 v_g^{-1} \left(\frac{\tau}{\kappa}\right)^2 (1 - \kappa^2) B^3 L$$
$$= 16 (\ln 2)^2 N_{st} B^2 \left(\frac{\tau}{\kappa}\right)^2 (1 - \kappa^2) = \left(\frac{B}{B_{cut}^{(N_{st})}}\right)^2 \frac{1 - \kappa^2}{\kappa^2} < 1 \tag{15.71}$$

where we have introduced the cutoff bit rate as

$$B_{cut}^{(N_{st})} = (4 \ln 2\tau)^{-1} N_{st}^{-1/2} \tag{15.72}$$

Therefore the smallest possible coupling coefficient that allows one to store N_{st} bits at bit rate B without excessive ISI is

$$\kappa(N_{st}, B) = \left[1 + \left[B_{cut}^{(N_{st})}/B\right]^2\right]^{-1/2} \tag{15.73}$$

and the slow down factor is

$$S_{CRS} = \kappa^{-1} = \left[1 + \left[B_{cut}^{(N_{st})}/B\right]^2\right]^{1/2} \tag{15.74}$$

which differs from the slow down factor in cascaded grating (Equation 15.47) only by the fact that the bit rate in Equation 15.74 is squared. Now the length required to store N_{st} bits can be found as

$$L(B, N_{st}) = \frac{N_{st} c \bar{n}^{-1} B^{-1}}{\left[1 + \left[B_{cut}^{(N_{st})}/B\right]^2\right]^{1/2}} = \frac{c\bar{n}^{-1} N_{st}/B_{cut}^{(N_{st})}}{\left[1 + \left[B/B_{cut}^{(N_{st})}\right]^2\right]^{1/2}} = \frac{4 \ln(2) d N_{st}^{3/2}}{\left[1 + \left[B/B_{cut}^{(N_{st})}\right]^2\right]^{1/2}} \tag{15.75}$$

The most interesting feature of Equation 15.75 is that as long as the bit rate is below cutoff, the length of the CRS delay line does not depend on bit rate. This happens because as the bit rate increases,

TABLE 15.2
Characteristic Parameters of the CRS Slow Light Media

CRS Type	One-Way Pass Time τ (fs)	Strength of Slow Down F_{EIT}	Cutoff Bit Rate for 1 Bit Storage Capacity $B_{cut}^{(1)}$ (Gbps)
Coupled rings waveguide	100	0.017	4×10^3
Photonic crystal with defects	25	0.07	16×10^3

the required delay time $T_d = N_{st} B^{-1}$ decreases, but at the same time the passband has to be widened to accommodate large bit rate, and thus the group velocity increases. The required length, being a product $T_d v_g$ of two, stays constant.

In fact, sufficiently far from cutoff, one can find the required number of coupled resonators as

$$N_{res}(N_{st}) = \frac{L(N_{st})}{d} \sim 3 N_{st}^{3/2} \qquad (15.76)$$

This is a simple and instructive relation. It tells us that no matter what the bit rate is (as long as it is well below cutoff), the number of resonators required to store a given number of bits does not depend on the type of resonator or bit rate. While it is obvious that one cannot store more than one bit of information per resonator, it is a fact that their number goes up superlinearly with the storage capacity. The increase with a power of 3/2 is not as strong as the square law increase of the length of the EIT buffer (Equation 15.65)—and this can be considered to be an important advantage of resonant structures over the photonic ones.

A more detailed analysis in Ref. [15] shows that if one can choose the optimum period of the CRS, the required length actually goes down slightly with increased bit rate, but here we neglect this small effect and consider that, independent of bit rate, one uses the same CRS period d.

Let us now consider two different CRS, both based in Si/SiO$_2$ technology—one using coupled ring resonators waveguide, with a radius of 6 µm [34] with a one-way pass time of $\tau \sim 100$ fs, and the other one that would use defect modes in photonic crystals [33] with defects separated by 3 µm ($\tau \sim 25$ fs). The parameters are shown in Table 15.2 and the required buffer lengths are plotted versus bit rate for two different storage capacities in Figure 15.15. As one can see, it is entirely realistic from the dispersion point of view to store between 10 and 100 bits of information at tens of Gbps in CRS with large index contrast. The required buffer length for 100 bit storage in photonic crystal is less than 1 cm while with ring resonators, it is about 2 cm. A more relevant measure is the area of the buffer since one can fold the optical path in a meandering buffer. Then, to store 100 bits using ring resonators one would require only about 0.3 mm^2 and even less than that with a photonic crystal. Although this density is orders of magnitude less than the one attainable in electronic storage, it is much better than anything attainable with an atomic medium at these bit rates. Unfortunately, once the storage capacity increases to 1000 bits (Figure 15.15c) the required length of the buffer becomes too high. Even if one could fold the optical path, it is difficult to imagine that one would be able to fabricate the 100,000 resonators required by Equation 15.76 without incurring a prohibitively high insertion loss. Even if one compensates the insertion loss with optical amplifiers, the accumulated noise will deplete the SNR. In addition, the power consumption of such a scheme would be prohibitively high [35].

Overall, though, using CRS to store high bit rate signals is obviously advantageous to the EIT and other atomic schemes. To elucidate this point we can use Equation 15.67 to express the slow down factor via the passband width as

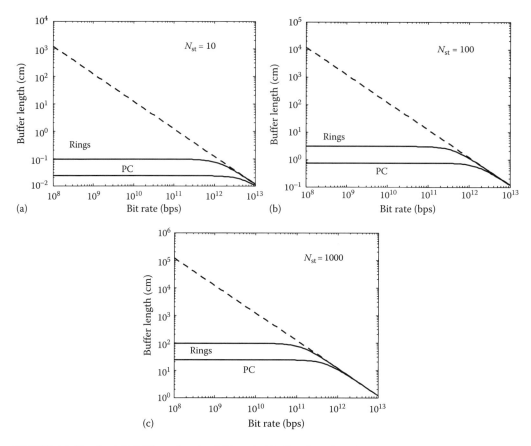

FIGURE 15.15 (a–c) Solid lines: length of SL optical buffers based on ring resonators and photonic crystals required to store N_{st} bits as a function of bit rate for three different storage capacities. Dashed lines: length of an unstructured dielectric also capable of storing the same number of bits.

$$S_{CRS} = \kappa^{-1} = \frac{1}{\sin(\pi \Delta \nu_{pass} \tau)} \approx \frac{\nu_0}{\Delta \nu_{pass}} F_{CRS} \qquad (15.77)$$

where we have introduced the strength of CRS scheme as

$$F_{CRS} = \frac{1}{\pi \nu_0 \tau} = \frac{\lambda}{\pi \bar{n} d} \sim \frac{\delta n}{\bar{n}} \qquad (15.78)$$

where the last proportionality relation indicates that the ability to make small resonators hinges upon the availability of two materials with different refractive indices. As one can see from Table 15.2, the values of F_{CRS} calculated at $\lambda = 1.55\,\mu\text{m}$ are very high compared to the strengths of cascaded grating (Equation 15.52) and atomic (Equation 15.58) schemes. In addition to having larger slow down strength, the slow down factor in CRS (Equation 15.77) is only inversely proportional to the passband, while in atomic schemes it is always inversely proportional to the square of $\Delta \nu_{pass}$. Obviously, CRS schemes are superior for wide bandwidth signals, while the EIT and other atomic schemes are the best for delaying relatively narrow bandwidth signals, where spectacularly low group velocities have been demonstrated.

15.8 DISPERSION LIMITATION OF NONLINEAR PHOTONIC SL DEVICES

Earlier on, we mentioned that the energy density in all the SL structures scales up with a slow down factor due to energy conservation. In the atomic structures, the energy is transferred to the atomic excitation, while in the SL photonic devices, the energy is kept in the electromagnetic form, and the square of the electric field is enhanced by a factor of S_{CRS}. In addition to this, SL propagation enhances the interaction between light and matter also by a factor of S. Therefore, if one incorporates slow light propagation into a nonlinear device, one can expect to enhance its performance by S^m where m depends on the order of the nonlinear optical effect. Chapter 9 of this book describes in greater detail many nonlinear devices using the slow light effects [36–38]. In this chapter we consider just one most important nonlinear effect—nonlinear index modulation, also referred to as optical Kerr effect. In the nonlinear Kerr medium the refractive index depends on the power density I of light as

$$n(I) = \bar{n} + n_2 I \tag{15.79}$$

where n_2 is a nonlinear refractive index. The refractive change may occur due to the absorption of light near resonance. For instance, if one considers a semiconductor medium, absorption generates electrons near the bottom of the conduction band and holes near the top of the valence band. As a result, the band-to-band absorption is saturated while free carrier absorption increases. Then, according to the Kramers–Kronig relation (Equation 15.7), the refractive index changes. The change can be rather large due to the accumulation of carriers, but it is also slow because it takes a long time for the carriers to recombine. This type of near resonant nonlinearity is usually referred to as a slow nonlinearity as it occurs on the scale of hundreds of picoseconds and longer. But there also exists an ultrafast component of nonlinear index change associated with virtual excitation of carriers.

The speed of the ultrafast virtual component is proportional to the detuning from resonance and can be faster than one hundred femtoseconds. It is this practically instant type of nonlinearity that is of interest in the application of ultrafast all-optical devices. Typical values of the ultrafast n_2 are in the range from 3×10^{-16} cm^2/W in SiO$_2$ to 10^{-13} cm^2/W for III–V semiconductors [39].

The simplest example of a nonlinear switch involves a weak signal and a strong optical pump (Figure 15.16a). The index change caused by the pump causes a change of phase for the signal

$$\Delta \Phi_s = \Delta k L = \frac{2\pi}{\lambda_0} n_2 I_p L \tag{15.80}$$

If the nonlinear device is placed into one arm of an interferometer, say a Mach–Zehnder interferometer (MZI), as shown in Figure 15.16a, then, one can achieve all-optical switching when the optically induced phase shift becomes $\Delta \Phi_s = \pi$. If one considers different switching schemes, say relying on directional couplers or gratings, full switching would still occur only when a phase shift on the order of π is induced. One can then define switching input power density I_0 and switching length related as

$$I_0 L_0 = \frac{\lambda_0}{2 n_2} \tag{15.81}$$

Since the nonlinear index is small and it is important to miniaturize the switching device, intensity inside the device typically approaches the optical damage threshold. For semiconductors, this intensity (with picosecond pulse durations) is on the order of $I_0 \sim 10^9$ W/cm^2. This power density can be achieved by focusing a 5 W peak power mode-locked laser into a waveguide with about 1 μm^2 mode size which results in a switching length L_0 at 1550 nm of about 0.75 cm. It is this requirement of a few watt peak power that prevents ultrafast optical switching from becoming practical. Let us now see whether SL can significantly affect the situation. Consider a CRS structure placed into one of the MZI branches (Figure 15.16b).

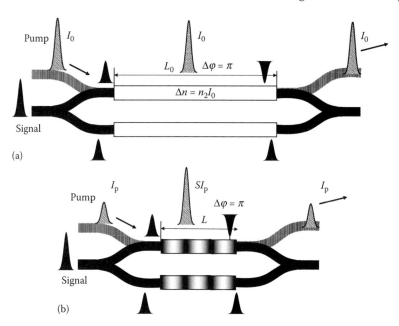

FIGURE 15.16 (a) MZI ultrafast all-optical switch with nonlinear structure. (b) All-optical switch with an SL nonlinear structure.

Now, in the CRS structure the intensity (and thus the square of electric field) inside is enhanced by a factor of S. The nonlinear change of index

$$\Delta n = n_2 I_p S \tag{15.82}$$

changes the pass time as

$$\Delta \tau = \frac{\Delta n}{\bar{n}} \tau \tag{15.83}$$

and then, by differentiating dispersion relation (Equation 15.66) one obtains the change in the wave vector of the signal

$$\Delta k = \kappa^{-1} \frac{\omega}{d} \Delta \tau = S \frac{\omega \tau}{d\bar{n}} \Delta n = S \frac{\omega}{c} \frac{c\tau}{d\bar{n}} \Delta n = S \frac{2\pi}{\lambda_0} \Delta n = S^2 \frac{2\pi}{\lambda_0} n_2 I_p \tag{15.84}$$

and a new switching condition

$$S^2 I_p L = \frac{\lambda_0}{2 n_2} \tag{15.85}$$

Therefore by slowing down both pump and signal one can reduce the switching power–length product by as much as S^2. To quantify the effect of SL we can normalize the switching intensity and length to I_0 and L_0 as $p_s = I_p/I_0$ and $l_s = L/L_0$—then the switching condition (Equation 15.85) becomes

$$p_s l_s = S^{-2} = \kappa^2 \tag{15.86}$$

Since the power density of the pump inside the SL structure I_p/κ should not exceed the optical damage threshold I_0, using Equation 15.86 we obtain the limiting condition

$$p_s \leq l_s \tag{15.87}$$

Bandwidth Limitation in Slow Light Schemes

Substituting Equation 15.86 into the GVD-imposed limitation on bit rate (Equation 15.71) we obtain [40]

$$16 (\ln 2)^2 \frac{\bar{n}}{c} \tau^2 L_0 (1 - i_s l_s) B^3 \frac{l_s}{(p_s l_s)^{3/2}} = (1 - p_s l_s) \frac{B^3}{B_{\text{cut,nl}}^3 p_s^{3/2} l_s^{1/2}} < 1 \qquad (15.88)$$

where we have introduced a nonlinear cutoff bit rate

$$B_{\text{cut,nl}} = (4 \ln 2)^{-2/3} \tau^{-2/3} \tau_{T0}^{-1/3} \approx 0.5 \tau^{-1} \left(\frac{L_0}{d} \right)^{-1/3} \qquad (15.89)$$

where the transit time through L_0 is

$$\tau_{T0} = \frac{\bar{n}}{c} L_0 \qquad (15.90)$$

Defined in these terms $B_{\text{cut,nl}}$ is the maximum bit rate at which the SL CRS brings benefits, that is, a reduction in either the length of the device, or in the operating power. The origin of the cutoff bit rate is GVD. As one increases the bit rate, GVD becomes too large and one has to increase the coupling coefficient κ and thus reduce the slow down factor. Then, to maintain switching condition (Equation 15.85), one is forced to increase power thus reducing SL benefits. Eventually the slow down factor will approach unity as the bit rate approaches $B_{\text{cut,nl}}$ indicating that the SL benefits are gone. One can transform Equation 15.89 into

$$B_{\text{cut,nl}} \approx 1.25 \nu_0 F_{\text{CRS}}^{2/3} \left(\frac{\Delta n_{\text{max,nl}}}{\bar{n}} \right)^{1/3} \qquad (15.91)$$

where $\Delta n_{\text{max,nl}} = n_2 I_0$ is the maximum nonlinear index change attainable in a given material.

In order to have a large $B_{\text{cut,nl}}$ one needs to have both good nonlinear material, that is, large $\Delta n_{\text{max,nl}}$, and a strong CRS. Notice that cutoff bit rate in Equation 15.89 has a rather weak power of 1/3 transit time dependence, in stark contrast to the much stronger transit time limitations of a single resonant cavity device, as, for instance, a single ring resonator. This bandwidth advantage of CRS follows directly from the fact that, unlike a single cavity, CRS is essentially a traveling wave device.

For a semiconductor photonic crystal with $\tau = 25$ fs, one gets $B_{\text{cut,nl}} = 1.5$ TBps, while in the coupled ring resonator structures with $\tau = 100$ fs, $B_{\text{cut,nl}} \approx 600$ GBps. This is an impressive bandwidth, but is it enough to make ultrafast photonic switching a reality?

In Figure 15.17 we have plotted the bit rate dependence of the input switching power density for a GaAs PC with a mode size of 1 µm² for different lengths of the device. The results are respectable but not really overwhelming. At 100 Gbps one can obtain optical switching in a 300 µm long device driven by about 200 mW but as the bit rate starts approaching cutoff, even at 300 Gbps, one still requires a 1 mm long switch driven by 1 W—which, though an improvement over the unstructured device, is probably not good enough to make a practical device with a switching energy of 3 pJ.

For relatively low bit rates below 10 Gbps the improvement is quite significant—the device length can be only 10 µm and the switching power would be 10 mW, which can make the device more practical. The switching energy, however, is not reduced by much—it is still 1 pJ per bit or at least three orders of magnitude larger than in electronics.

Thus, one can conclude this section by simply stating that SL photonic structures are definitely capable of improving the performance of nonlinear devices, but these improvements disappear with increased bandwidth. The question of applicability of SL to each particular task needs to be considered on an individual basis.

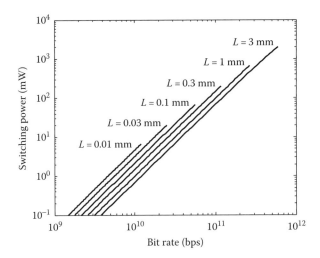

FIGURE 15.17 Switching power in the GaAs ultrafast switch as a function of bit rate for six different device lengths.

15.9 CONCLUSIONS

In this chapter we have considered, in most general terms, bandwidth limitations on the performance of slow light structures imposed by the dispersion of group velocity and the dispersion of loss. Both linear and nonlinear devices were considered. Our main conclusion is that, for any task, whether it is storing N bits of information, performing photonic switching, or other, one can always define a cutoff bit rate (bandwidth for analog devices) beyond which all the ostensive benefits of SL cease to exist. For the case of delay line, this does not mean that one cannot store more than a certain number of bits at a given bit rate. What it means is that beyond the cutoff rate the slow down factor would become so close to unity that the delay line would not be any shorter than, say, a fiber coil. By comparing SL atomic and photonic schemes, we found their performance characteristics complimentary. Atomic SL performs best for relatively narrow bandwidth signals with small capacity but they achieve very large slow down factors. Photonic SL structures on the other hand are best for storing moderate number of bits of wide bandwidth signals with moderate slow down factors, and for nonlinear manipulation.

REFERENCES

1. K.-J. Boller, A. Imamoglu, and S. E. Harris, Observation of EIT, *Phys. Rev. Lett.*, 66, 2593–2596 (1991).
2. S. E. Harris, J. E. Field, and A. Kasapi, Dispersive properties of EIT, *Phys. Rev. A*. 46, R39–R32 (1992).
3. A. Kasapi, M. Jain, G. Y. Jin, and S. E. Harris, EIT: Propagation dynamics, *Phys. Rev. Lett.*, 74, 2447–2450 (1995).
4. L. V. Hau, S. E. Harris, Z. Dutton, and C. H. Behroozi, Light speed reduction to 17 metres per second in an ultracold atomic gas, *Nature*, 397, 594–596 (1999).
5. D. F. Phillips, A. Fleischhauer, A. Mair, R. L. Walsworth, and M. D. Lukin, Storage of light in atomic vapor, *Phys Rev Lett*, 86, 783–786 (2001).
6. Y. Tao, Y. Sugimoto, S. Lan, N. Ikeda, Y. Tanaka, and Y. K. Asakawa, Transmission properties of coupled-cavity waveguides based on two-dimensional photonic crystals with a triangular lattice of air holes, *J. Opt. Soc. Am. B*, 20, 1992–1998 (2003).
7. S. Nishikawa, S. Lan, N. Ikeda, Y. Sugimoto, H. Ishikawa, and K. Asakawa, Optical characterization of photonic crystal delay lines based on one-dimensional coupled defects, *Opt. Lett.*, 27, 2079–2081 (2002).
8. Y. Sugimoto, S. Lan, S. Nishikawa, N. Ikeda, H. Ishikawa, and K. Asakawa, Design and fabrication of impurity band-based photonic crystal waveguides for optical delay lines, *Appl. Phys. Lett.*, 81, 1948–1950 (2002).

9. Y. Okawachi, M. S. Bigelow, J. E. Sharping, Z. Zhu, A. Schweinsberg, D. J. Gauthier, R. W. Boyd, and A. L. Gaeta, Tunable all-optical delays via Brillouin slow light in an optical fiber, *Phys. Rev. Lett.*, 94, 153902 (2005).
10. J. E. Sharping, Y. Okawachi, and A. L. Gaeta, Wide bandwidth slow light using a Raman fiber amplifier, *Opt. Express*, 13, 692 (2005).
11. F. Öhman, K. Yvind, and J. Mørk, Voltage-controlled slow light in an integrated semiconductor structure with net gain, *Opt. Express*, 14(21), 9955 (2006).
12. R. W. Boyd, D. J. Gauthier, A. L. Gaeta, and A. E. Willner, Maximum time delay achievable on propagation through a slow-light medium, *Phys. Rev. A*, 71, 023801 (2005).
13. A. B. Matsko, D. V. Strekalov, and L. Maleki, On the dynamic range of optical delay lines based on coherent atomic media, *Opt. Express*, 13, 2210–2223 (2005).
14. R. S. Tucker, P.-C. Ku, and C. J. Chang-Hasnain, Slow-light optical buffers: Capabilities and fundamental limitations, *J. Lightwave Technol.*, 23, 4046–4065 (2005).
15. J. B. Khurgin, Optical buffers based on slow light in EIT media and coupled resonator structures—comparative analysis, *J. Opt. Soc. Am. B*, 22, 1062–1074 (2005).
16. J. B. Khurgin, Performance limits of delay lines based on optical amplifiers, *Opt. Lett.* 31(7), 948–950 (2006).
17. D. A. B. Miller, Fundamental limits of optical components, *JOSA B*, 24(10), A1–A18 (2007).
18. T. Erdogan, Fiber grating spectra, *J. Lightwave Technol.*, 15(8), 1277–1294 (1997).
19. R. M. Camacho, M. V. Pack, and J. C. Howell, Low-distortion slow light using two absorption resonances, *Phys. Rev. A*, 73, 063812 (2006).
20. N. M. Litchinitser, B. J. Eggleton, and G. P. Agrawal, Dispersion of cascaded fiber gratings in WDM lightwave systems, *J. Lightwave Technol.*, 16, 1523–1529 (1999).
21. S. Wang, S. Erlig, H. Fetterman, H. R. Yablonovitch, E. Grubsky, V. Starodubov, and D. S. Feinberg, Group velocity dispersion cancellation and additive group delays by cascaded fiber Bragg gratings in transmission, *J. Microwave Guided Wave Lett.*, 8, 327–329 (1998).
22. J. B. Khurgin, J. U. Kang, and Y. J. Ding, Ultrabroad-bandwidth electro-optic modulator based on a cascaded Bragg grating, *Opt. Lett.*, 25, 70–72 (2000).
23. M. S. Bigelow, N. N. Lepeshkin, and R. W. Boyd, Observation of ultraslow light propagation in a ruby crystal at room temperature, *Phys. Rev. Lett.*, 88, 023602 (2002).
24. A. V. Turukhin, V. S. Sudarshanam, M. S. Shahriar, and P. R. Hemmer, Observation of ultraslow and stored light pulses in a solid, *Phys. Rev. Lett.*, 88, 023602 (2002).
25. R. M. Camacho, M. V. Pack, and J. C. Howell, Slow light with large fractional delays by spectral hole-burning in rubidium vapor, *Phys. Rev. A*, 74, 033801 (2006).
26. M. D. Lukin, M. Fleichhauer, A. S. Zibrov, and M. O. Scully, Spectroscopy in dense coherent media: Line narrowing and interference effects, *Phys. Rev. Lett.*, 79, 2959–2962 (1997).
27. P. C. Ku, C. J. Chang-Hasnain, and S. L. Chuang, Variable semiconductor all-optical buffers, *Electron. Lett.*, 38, 1581–1583 (2002).
28. J. B. Khurgin, Light slowing down in Moire fiber gratings and its implications for nonlinear optics, *Phys. Rev. A*, 62, 3821–3824 (2000).
29. S. Longhi, D. Janner, G. Galzerano, G. Della Valle, D. Gatti, and P. Laporta. Optical buffering in phase-shifted fibre gratings, *Electron. Lett.*, 41, 1075–1077 (2005).
30. A. Yariv, Y. Xu, R. K Lee, and A. Scherer, Coupled-resonator optical waveguide: A proposal and analysis, *Opt. Lett.*, 24, 711–713 (1999).
31. C. K. Madsen and G. Lenz, Optical all-pass filters for phase response design with applications for dispersion compensation, *IEEE Photon Technol. Lett.*, 10, 994–996 (1998).
32. A. Melloni, F. Morichetti, and M. Martnelli, Linear and nonlinear pulse propagation in coupled resonator slow-wave optical structures, *Opt. Quantum Electron.*, 35, 365–378 (2003).
33. Z. Wang, and S. Fan, Compact all-pass filters in photonic crystals as the building block for high-capacity optical delay lines, *Phys. Rev. E*, 68, 066616-23 (2003).
34. F. Xia, L. Sekaric, and Y. Vlasov, Ultracompact optical buffers on a silicon chip, *Nat. Photon.*, 1, 65–71 (2006).
35. J. B. Khurgin, Dispersion and loss limitations on the performance of optical delay lines based on coupled resonant structures, *Opt. Lett.*, 32, 163–165 (2007).
36. M. Scalora, J. P Dowling, C. M. Bowden, and M. J. Bloemer, Optical limiting and switching of ultrashort pulses in nonlinear photonic band gap materials, *Phys. Rev. Lett.*, 73, 1368–1371 (1994).

37. A. Hache and M. Bourgeois, Ultrafast all-optical switching in a silicon-based photonic crystal, *Appl. Phys. Lett.*, 77, 4089–4091 (2000).
38. M. Soljacic, S. G. Johnson, S. Fan, M. Inanescu, E. Ippen, and J. D. Joannopulos, Photonic-crystal slow-light enhancement of nonlinear phase sensitivity, *J. Opt. Soc. Am. B*, 19, 2052 (2002).
39. D. A. Nikogosyan, *Properties of Optical and Laser Related Materials. A Handbbok*. Wiley, NY, 1997.
40. J. B. Khurgin, Performance of nonlinear photonic crystal devices at high bit rates, *Opt. Lett.*, 30, 643–645 (2005).

16 Reconfigurable Signal Processing Using Slow-Light-Based Tunable Optical Delay Lines

Alan E. Willner, Bo Zhang, and Lin Zhang

CONTENTS

16.1 Introduction ... 321
16.2 Slow-Light-Based Tunable Delay Lines ... 323
    16.2.1 Overview of Slow-Light Techniques ... 323
    16.2.2 Applications of Slow-Light-Based Tunable Delay Lines ... 324
    16.2.3 Slow-Light-Induced Data Distortion and Its Mitigation ... 325
        16.2.3.1 Data-Pattern-Dependent Distortion ... 326
        16.2.3.2 Figures of Merit ... 328
        16.2.3.3 Distortion Mitigation ... 329
16.3 Phase-Preserving Slow Light ... 331
    16.3.1 Delaying DPSK Signals ... 331
    16.3.2 DPSK Data-Pattern Dependence and Its Mitigation ... 332
    16.3.3 Spectrally Efficient Slow Light ... 334
16.4 Signal Processing Applications ... 335
    16.4.1 Variable-Bit-Rate OTDM Multiplexer ... 336
    16.4.2 Multichannel Synchronizer ... 338
    16.4.3 Simultaneous Multiple Functions ... 340
    16.4.4 Other Applications ... 340
16.5 Summary ... 342
Acknowledgments ... 342
References ... 342

16.1 INTRODUCTION

As a potential enabling technology, slow light has captured much research interest over the past few years for achieving a continuously tunable optical delay line [1]. In principle, slow light is generated by tailoring an enhanced group-index resonance within a given nonlinear medium. This effective refractive index change is experienced by the data stream passing through and thus induces a controllable group delay onto the signal. The main feature of a slow-light-based optical delay line is the capability for fine-grain temporal manipulation of optical pulses. This fine tunable delay element is envisioned to be useful for various high-bandwidth signal processing functions and applications.

In general, signal processing is considered as an efficient and powerful enabler for a host of communication functions as well as a system performance enhancer [2,3]. The hope is that performing signal processing purely in the optical domain might reduce any optical–electronic conversion inefficiencies and take advantage of the ultrahigh bandwidth inherent in optics [4,5]. One of the most basic building blocks to achieve efficient and reconfigurable signal processing is a continuously tunable optical delay line, and yet this element has historically been difficult to realize [6]. Applications of such a delay line could include: (1) accurate synchronization for bit-level interleaving, (de)multiplexing, and switching [7], (2) tapped delay lines for signal equalization, optical filtering, and dispersion compensation [8,9], and (3) data packet synchronization, switching, time-slot interchanging, and buffering in dynamic network environments [10].

Tunable delay lines have typically been accomplished by varying a free-space or optical waveguide propagation path, choosing from a combination of predesigned optical path lengths [7]. This technique produces only a finite set of discrete time delays, and tends to be bulky and lossy with increased number of stages. Moreover, the delay resolution is usually limited to subnanoseconds, which is determined by the shortest optical path [11]. From a systems perspective, a partial "wish list" for a tunable optical delay line should include: (1) continuous tunability, (2) wide bandwidth, (3) amplitude-, frequency-, and phase-preserving, (4) large tuning range, and (5) fast reconfiguration speed. Such delays provide flexible time domain data grooming in the optical domain.

Slow light is now viewed as a strong candidate for achieving such tunable optical delay lines. However, in order to achieve the maximum possible delay on the data signals with minimal system power penalty, one needs to combat and compensate for any slow-light-induced signal degrading effects [12–15]. The data fidelity is considered one of the main issues to be tackled before any practical signal processing modules can be designed.

As an initial step in identifying potential target areas for slow-light delay-based applications, a good understanding of the properties unique to the optical domain is required. Some of the desirable features are listed as follows.

1. *High speed capability*: A judicious choice of the slow-light medium with different bandwidth limitations to support high-speed data streams
2. *Preservation of optical properties*: The ability to maintain the phase information so as to support various data modulation formats
3. *Multiple channel operation*: The ability to delay multiple wavelength channels independently in a single slow-light device
4. *Input bit-rate variability*: Accommodation of different input bit rates by dynamically adjusting the delay elements
5. *Simultaneous multiple functions*: Simultaneously perform more than one processing tasks in a single medium

These properties in conjunction with the attractive tunable delay feature are the key to building a reconfigurable signal processing platform to extract and process multidimensional information at ultrahigh speed.

This chapter is structured as follows. In Section 16.2, we present a brief overview of slow-light techniques, with a focus on various signal processing applications using such a tunable delay line. Slow-light-induced data degradation is carefully discussed and data-pattern dependence is identified as one of the main reasons for signal distortion. Section 16.3 shows phase-preserving slow light. Experimental results of slow-light delay on 10 Gbps differential phase-shift-keying (DPSK) signals with reduced DPSK data-pattern dependence are shown. Spectrally efficient slow light by utilizing advanced multilevel phase-modulated formats (e.g., differential quadrature phase-shift-keying [DQPSK]) is also demonstrated. Section 16.4 presents recent advances in slow-light-based novel signal processing applications. Unique features such as multichannel operation, variable-bit-rate operation and simultaneous functions are highlighted and detailed examples are provided. We finally conclude in Section 16.5 with a brief discussion about future research directions.

16.2 SLOW-LIGHT-BASED TUNABLE DELAY LINES

16.2.1 Overview of Slow-Light Techniques

Slow-light technology is aimed at reducing and controlling the group velocity of optical pulses. The key feature of the slow-light technique centers on the ability to introduce a relatively large change in the refractive index seen by the light as it passes through a medium. This causes different wave components within an optical pulse to travel with different speeds and therefore affects the group velocity of the pulse envelop. By making the dispersion of the material sufficiently strong, the group velocity can be reduced significantly lesser than the speed of light in vacuum. This opens up great opportunities to manipulate the speed of information being transmitted. Typically, the resonance is achieved in the amplitude response of the slow-light media by either a sharp absorption or gain peak [11,16] or a transparent window in a lossy background that has an effect similar to a gain spectrum [17]. In most of the cases, this understanding may provide a clue to search for a slow-light phenomenon in a potential medium.

To date, various physical mechanisms in a host of media have been reported in literature to have observed slow light. These include electromagnetically induced transparency (EIT) [17,18], coherent population oscillations (CPOs) [19,20], stimulated scattering effects (stimulated Brillouin scattering, SBS [21–25] and stimulated Raman scattering, SRS [26–28]), optical parametric amplifier (OPA) [29–31], coupled resonator optical waveguide (CROW) [32–34], photonic crystal structures [35–37], and various other schemes [38–40].

Given the specific physics involved, the properties of various slow-light media could be quite different. For example, slow-light bandwidth may vary from ~hz to ~THz depending on the slow-light material and the nonlinear processes [19,27]. One can also categorize different schemes based on the amplitude and phase responses of the slow-light resonance. For instance, as shown in Figure 16.1, in the SBS effect, both the gain and delay spectra feature narrowband single peak profiles. The maximum delay occurs at the gain peak wavelength where dispersion is zero in theory. In contrast, in nonlinear parametric processes, if the pump is placed on the red side of the zero-dispersion wavelength, the gain profile features double-sided peaks while the delay tends to be linear over the frequency. This indicates that the delay might not necessarily occur at the gain peak and the dispersion should be considered throughout the gain bandwidth. More detailed comparative analyses for a variety of slow-light techniques can be found in Refs. [41–44].

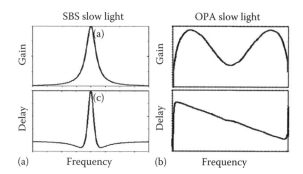

FIGURE 16.1 Gain and delay profiles of (a) SBS and (b) OPA-based slow light, indicating that various slow-light schemes could be categorized by their characteristic amplitude and phase responses.

We note that, as a typical slow-light resonance, Lorentzian-shaped gain profiles are widely observed [16,17,21–26,28,45]. The generality fundamentally originates from the Kramers–Kronig relation which governs the imaginary and real parts of the refractive index. As a good representative of Lorentzian resonances, the SBS-based slow light has attracted fair amount of research attention in recent years. Moreover, SBS-based slow light also features the following advantages: (1) wide wavelength tunability, (2) low control power requirement, (3) room-temperature operation, and (4) seamless integration with fiber-optic systems. However, the major issue for the conventional SBS slow light is the limited tens of MHz bandwidth. Therefore, the bit rates are commonly restricted to the order of tens of Mbps due to this narrow intrinsic Brillouin gain spectral-width.

Recent breakthroughs of broadening the SBS gain bandwidth from tens of MHz [21,22] to tens of GHz [23–25] have enabled the transmission of multi-Gbps data streams. In general, broadband SBS is achieved by frequency modulating a coherent pump laser so as to broaden the pump spectral-width and consequently the SBS gain/delay bandwidth. Techniques involving pump broadening include: (1) direct modulation of the pump laser using either pseudorandom bit sequence (PRBS) modulation [23] or Gaussian noise modulation [24], (2) external phase modulation [46], and (3) a simple and incoherent spectrally sliced amplified spontaneous emission (ASE) pumping [47]. These advances made it possible to transmit multi-Gbps data signals through the SBS slow-light medium [14,46,48,49] which makes the SBS mechanism promising for practical systems.

Other promising wideband slow-light techniques which have also been shown to be capable of transmitting Gbps and beyond optical signals include: (1) SRS in fiber and on silicon chips [27,28], (2) OPA process in fiber [15,31], and (3) nonlinear processes in semiconductor optical amplifiers (SOAs) [50,51]. These techniques together with the SBS effect hold great promise for future multi-Gbps and beyond optical switching and signal processing.

16.2.2 Applications of Slow-Light-Based Tunable Delay Lines

It is believed that the efficiency and throughput of future reconfigurable optical networks can be significantly enhanced by the availability of a tunable, wideband delay line [10]. Accurate, widely tunable optical delays are thus a critical requirement for future optically switched networks to enable synchronization, header recognition, buffering, optical time multiplexing, and equalization. Promising slow-light-based tunable delays are believed to have direct applications in the following signal processing areas, as shown in Figure 16.2.

1. *Optical synchronization and multiplexing* [52–56]: Synchronizing multiple misaligned input streams is one of the basic functions and thus a prerequisite for almost any subsequent processing. Optical time division multiplexing (OTDM) is one example which requires precise allocation of each lower rate signals into specific time slots. This may also see significant use in synchronous optical packet-switched networks where header recognition, buffering, and time switching take place. Advanced modules relying on bit or packet-level

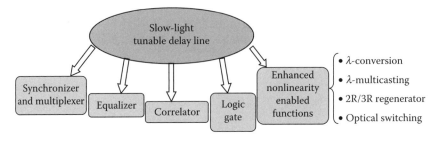

FIGURE 16.2 Applications of slow-light-tunable delay lines in the field of optical signal processing.

synchronizers can also be constructed, such as serial-to-parallel (TDM to wavelength division multiplexing, WDM), parallel-to-serial (WDM to TDM) converters and time-slot interchangers.

2. *Optical equalization* [8,57,58]: Equalizers can be used to mitigate the impairments of intersymbol interference and fiber dispersive effects. Typically, interferometric structures incorporating tapped optical delay lines are employed for the design of optical equalizer. Slow-light-based tunable delay lines are potential candidates in that the delays can be accurately and flexibly adjusted, enabling bandwidth-tunable operation and thus supporting variable input bit-rate signals. Delay-line interferometer (DLI)-based DPSK demodulator can also be considered as one type of filter-like equalizers.

3. *Optical correlation* [59,60]: Optical correlation is viewed as an indispensable function for pattern/header matching and thresholding. The fundamental "delay and stack" function imposes the requirement of fast tuning speed as well as high tuning resolution. Slow-light-based delay features these merits and can thus be considered as a potential candidate.

4. *Optical logic gates* [61,62]: Performing logic operations purely in the optical domain is desired for future optical networks which have the aggressive goal of >100 Gbps processing with the potential of format and bit-rate transparency. Slow-light-based tunable delays are expected to find value in XOR-type parity checks, differential phase encoders, and more complicated looping adder-based checksum processing.

5. *Enhanced nonlinear interaction* [63,64]: By reducing the group velocity of light, enhanced nonlinearities can be achieved in certain photonic devices by dramatically increasing the induced phase shifts caused by small changes in the index of refraction. Foreseeable key applications by utilizing this enhanced nonlinear phase shift may include: wavelength conversion, wavelength multicasting, 2R/3R regeneration, and optical switching. Key parameters such as nonlinear coefficient, bandwidth, effective interaction length, polarization sensitivity, and tuning speed are directly related to the slow-light medium. Expected advantages include high extinction ratio, negligible frequency chirp, data format transparency, and scalability to multiple channels.

However, for slow light to be useful in practical systems, critical parameters related to tunable delay lines should be carefully examined. Each application imposes certain metrics that must be met by the slow-light-based devices in order to justify a specific use in real optical systems. Some typical delay metrics are listed as follows.

1. *Delay bandwidth*: The optical bandwidth over which a certain delay can be achieved.
2. *Maximum delay*: The maximum achievable delay value.
3. *Delay range*: The tuning range (from minimum to maximum) that the delay can be achieved.
4. *Delay resolution*: The minimum incremental delay tuning step.
5. *Delay accuracy*: The precision percentage of the actual delay to that of the desired delay value.
6. *Delay reconfiguration time*: The amount of time it takes to switch a delay from one state to another steady state.
7. *Fractional (normalized) delay*: The absolute delay value divided by the pulse width (bit time). This is important to the delay/storage capacity.
8. *Loss over delay*: The amount of loss incurred per unit delay for a given slow-light mechanism. Lesser loss per unit delay is desired.

16.2.3 SLOW-LIGHT-INDUCED DATA DISTORTION AND ITS MITIGATION

From a signal processing and telecommunication perspective, one would like to transmit as high bit-rate signals as possible by making full use of the slow-light bandwidth and, in the mean time, achieve as large delays as possible. This would require a perfect yet unrealistic slow-light medium which should feature a constant amplitude response and a linear phase response. However, almost all types of slow-light media exhibit nonlinear dependencies of the group index over frequency. This induces not only dispersive effects accompanying the achieved delay, but some "filtering" effects from the amplitude response as well. The signal delay and signal quality trade-off originate from the delay-bandwidth product and have long been considered as the fundamental limitation in slow-light systems. This fundamental trade-off has been extensively studied in systems described by Lorentzian-shaped [13,41,48,65] and non-Lorentzian-shaped resonances [15,41,42,44].

We emphasize that most of the studies have considered only a single or a few pulses propagating through the slow-light media. However, telecommunication systems transmit true data streams with a variety of information-bearing patterns. Different kinds of slow-light media might lead to various patterning effects [12–15] and require careful analysis based on the specific amplitude and phase responses. Furthermore, designing and optimizing the slow-light element to reduce pattern-dependent distortion while simultaneously maximizing the induced delay becomes an important aspect of slow-light research.

16.2.3.1 Data-Pattern-Dependent Distortion

We start off modeling the slow-light element with a typical Lorentzian-shaped imaginary part of the refractive index, while the real part is determined by the Kramers–Kronig relation. The slow-light bandwidth can be flexibly tuned from 5 to 50 GHz so as to accommodate 10-Gbps nonreturn-to-zero (NRZ) on–off-keying (OOK) signals. We note that the dispersion is zero when both the delay and the gain reach their maximum value, and the signal carrier is centered at the gain peak wavelength to experience the highest delay.

To highlight the impact of slow-light-induced pattern-dependent distortions, we characterize the output data patterns by looking into both the amplitude and the phase response, which are reflected on the data signals as pattern-dependent gain and pattern-dependent delay.

With the assistance of Figure 16.3, we can see that the peak power of the optical pulse after slow light clearly depends on the data patterns. For example, consecutive "1" bits ("0110" pattern) have a much higher "1"-level power than that of a single "1" bit (010 pattern). This gives rise to "1" level splitting in the eye diagram and thus severely distorts the signal. This type of distortion is caused by pattern-dependent gain and can be explained as follows: When the signal carrier is located exactly at the gain peak and the bit rate is comparable with the bandwidth of the slow-light element, we note that the lower and higher frequency components in the signal spectrum, compared to the carrier, are in the low-gain region. The frequency components of consecutive "1"s are closer to the carrier and thus see more gain than a single "1," whose frequency components spread out over the low-gain region. As a result, the consecutive "1"s obtain higher peak power and causes the "1" level splitting at the output.

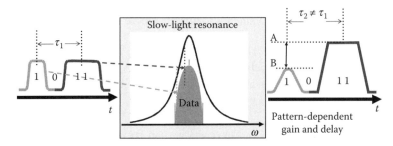

FIGURE 16.3 Origin of data-pattern-dependent distortion. Pattern-dependent gain and delay are shown as the two main degrading effects.

Reconfigurable Signal Processing Using Slow-Light-Based Tunable Optical Delay Lines

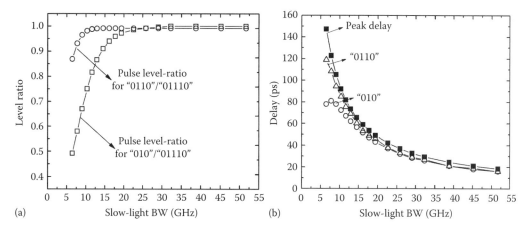

FIGURE 16.4 Quantification of slow-light-induced pattern-dependent gain and delay. (a) Level ratio versus slow-light bandwidth for two typical combinations of data patterns and (b) Delay versus slow-light bandwidth for two typical patterns.

Lorentzian-shaped resonances not only feature a narrowband amplitude response, which causes the aforementioned pattern-dependent gain, but also a narrowband delay response as well (Figure 16.1). As a result, the frequency components of the consecutive "1"s are closer to the delay peak and thus experience larger delay compared to the single "1." This pattern-dependent delay effect is illustrated also in Figure 16.3.

We quantify both of these patterning effects in Figure 16.4. Level ratio (defined as B/A, see Figure 16.3) is used as the key parameter to access the pattern-dependent gain. We consider here the level ratios among three typical patterns by using two combinations, "010" over "01110," and "0110" over "01110." As shown in Figure 16.4a, the level ratios of both these two combinations go down quickly as the slow-light bandwidth decreases. For a typical situation where the slow-light bandwidth is the same as the signal bit rate, more than a 30% level ratio difference can exist in the case of "010" over "01110," indicating the "010" pattern is the limiting factor to the data fidelity.

Figure 16.4b shows the pulse delay versus the slow-light bandwidth for different data patterns. The calculated peak delay (the maximum delay in the frequency-dependent delay profile, Figure 16.1) serves as an upper bound for different types of data patterns. Two typical patterns "010" and "0110" are shown and the delay difference grows with the reduced slow-light bandwidth. In some extreme cases, the pattern-dependent delay might cause pulse collision between specific patterns (e.g., for a data stream of "011010," the "0110" pattern might catch up with the "010" pattern) and thus could induce a high bit-error-rate (BER).

After the identification and quantification of the two degrading effects from the amplitude and phase responses, it is of importance to determine which of the two responses plays a dominating role for the data degradation. In our slow-light model, we isolate either amplitude or phase response by artificially eliminating one response at a time. Figure 16.5 shows signal Q-factor as a function of slow-light bandwidth for the cases of amplitude-only response, phase-only response, and the combined response. The amplitude-only response curve is almost identical to the combined effects curve, while the phase-only response curve shows an overall better performance. This indicates that the narrowband amplitude response and thus the pattern-dependent gain is the dominating effect for data degradation. We emphasize that the second-order dispersion at the gain peak is zero for Lorentzian-shaped resonances, which means that the pulse broadening induced by dispersive effects is relatively small.

As a comparison, non-Lorentzian-shaped slow light might exhibit different types of pattern-dependent distortion. As shown in Figure 16.6, for OPA-based slow-light schemes, the amplitude

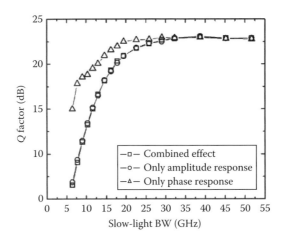

FIGURE 16.5 Isolating the effect of amplitude response and phase response on the signal quality (Q-factor). Amplitude response is shown as the dominating effect for data distortion.

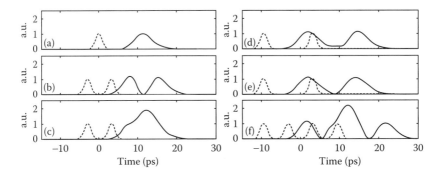

FIGURE 16.6 Examples of various data-pattern dependence induced by OPA-based slow-light scheme. (a) a single pulse, (b) two consecutive "1"s, (c) two consecutive "1"s with opposite phase, (d) two "1"s separated by a "0" (1 0 1), (e) two "1"s with opposite phase, separated by a "0" (1 0 −1) and (f) typical patterns for pairwise alternate phase-carrier suppressed return-to-zero (PAP-CSRZ) (1 1 − 1 − 1). (From Liu, F., Su, Y., and Voss, P.L., *Proc. OFC*, Anaheim, CA, paper OWB5, 2007. With permission.)

response does not significantly contribute to the patterning effect. Rather, the phase response-induced dispersive effect plays a dominant role. Furthermore, various phase patterns associated with different formats are also signal degrading factors. As a result, the data-pattern-dependent distortion reveals itself as not only pulse broadening and merging, but "1"-level fluctuation and "0"-level rising as well [15].

16.2.3.2 Figures of Merit

There has been much interest in developing a set of universal figures of merit (FOMs) to compare the trade-off among various slow-light schemes. Therefore, one may have to carefully consider slow-light-induced data degradation in order to keep a balance between pulse delay and signal quality. The question that becomes critical to answer is what operating conditions of a slow-light element are optimal.

Here, we define two FOMs in order to find an optimized slow-light bandwidth. To make it more general, the slow-light bandwidth is normalized by signal bit rate, and the pulse delay is normalized

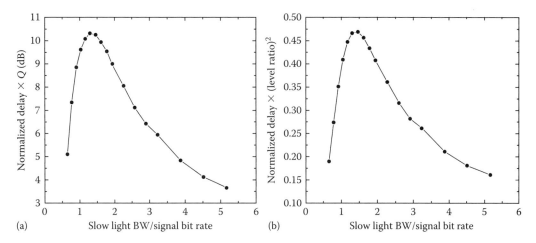

FIGURE 16.7 (a) FOM I versus normalized slow-light bandwidth. (b) FOM II versus normalized slow-light bandwidth. Both FOMs show that the optimized slow-light bandwidth equals 1.4 times the signal bandwidth.

by the bit time, which is named the normalized delay. The pulse delay is measured for a single "1" pulse. Compared to the fractional delay (defined as pulse delay divided by pulse width) that is used often in literature, the normalized delay defined here is a more practical consideration, from the point of view of communication systems.

The first FOM (FOM I) is defined as the product of the normalized delay and the signal Q-factor. In Figure 16.7a, we show FOM I as a function of the normalized slow-light bandwidth. An optimized value is found for NRZ signal when FOM I is maximized, that is, the normalized slow-light bandwidth is equal to 1.4. For narrow spectra slow-light devices, FOM I drops down very rapidly due to the decreased signal Q-factor. FOM I is also small in the wider bandwidth region due to the reduced delay.

The level ratio (B/A, as shown in Figure 16.3) can also be used to measure the distortion of the data signal after a slow-light element. We define FOM II as the product of the normalized delay and the (level ratio)^2, which has similar trends, as shown in Figure 16.7b, and predicts a very similar optimized slow-light bandwidth as that given by FOM I. This confirms that the vertical eye closure caused by "1"-level splitting in the output data streams is the dominant factor for signal degradation. Using FOM II, the system impact of the narrowband slow-light elements can be estimated by calculating a much simpler parameter (level ratio), so that device researchers do not need to simulate the whole system and measure the complicated signal Q-factor to optimize the slow-light design.

We note that the definition of FOMs should be adjusted based on the specific slow-light scheme involved [43]. This indicates that FOMs are dependent on the specific amplitude and phase responses of the slow-light mechanism. For example, delay over eye-opening penalty (DOE) might be more appropriate to be used in evaluating and optimizing OPA-based slow-light schemes [15].

16.2.3.3 Distortion Mitigation

We introduce in this section various methods of mitigating the signal distortion by either optimizing the system operating conditions or designing novel slow-light systems.

We first show here one way for reducing the data-pattern dependence for NRZ-OOK signals in Lorentzian slow-light elements [13,65]. We understand that when the signal spectrum is centered at the resonance peak, the data suffers the most distortion, as the pattern-dependent gain seen by "010" and "0110" has the largest difference. However, if the data channel is managed to be detuned from the resonance peak by certain value, frequency components of both single "1" and consecutive "1"s might experience a much similar gain, as illustratively depicted in Figure 16.8. This configuration

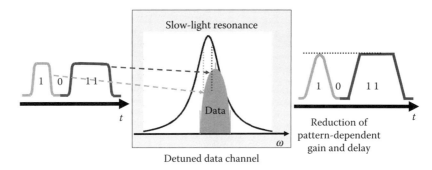

FIGURE 16.8 Concept of data-pattern dependence reduction by detuning the channel away from the slow-light resonance peak.

potentially results in the equalization of the peak power at the output of the slow-light element and thus the reduction of pattern-dependent distortion. We note that detuning the slow-light resonance with respect to the channel will have the similar effect in terms of distortion reduction.

Figure 16.9 quantifies the improvement of the 10 Gbps NRZ-OOK signal quality by detuning the channel away from a 14 GHz bandwidth fixed slow-light element. An approximate 2 dB signal Q-factor enhancement, compared to the no detuning case, is achieved for both positive and negative detuning. The optimal detuning value is found to be around 40% of the signal bandwidth and the improvement also features symmetry, as expected.

Approaches for combating and mitigating signal distortions have also been proposed by judiciously tailoring the shape of the gain or absorption resonances. Interesting techniques involving the design of two or multiple gain resonances [49,66–69] have been shown to effectively extend the delay while still maintaining good signal quality. For example, Stenner et al. [66] have demonstrated that distortion management using a gain doublet can provide approximately a factor of 2 increase in slow-light pulse delay and a factor of 5 improvement in the delay at the optimum bandwidth, as compared with the optimum single-line case.

Realizing that the delay-bandwidth trade-off is inherent to linear, resonant systems [9], some of the research efforts have been made to move beyond linear systems to nonlinear systems. A recent

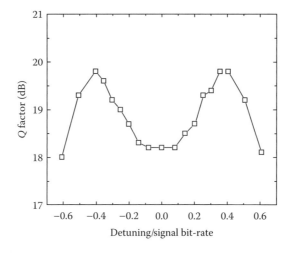

FIGURE 16.9 Simulation results of data-pattern dependence reduction by detuning the channel away from the gain peak. 2-dB Q-factor improvement is achieved by either red or blue detuning 40% of the signal bandwidth.

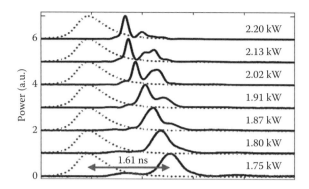

FIGURE 16.10 Experimental results of slow light in fiber Bragg gratings using gap soliton. (From Mok, J., Sterke, C., Littler, I., and Eggleton, B., *Nat. Phys.*, 2, 775, 2006. With permission.)

demonstration [40] exploited the nonlinear behavior of Bragg gratings in fibers, where the pulse can travel slowly but still remain undistorted over long propagation lengths, by the formation of a gap soliton. Using this method, as shown in Figure 16.10, Mok et al., have obtained a slow-light delay that is two-and-a-half pulse widths by adjusting the intensity of the input pulse.

16.3 PHASE-PRESERVING SLOW LIGHT

We emphasize that almost all previously published slow-light system results were for intensity-modulated data signals. However, phase-encoded modulation formats, such as DPSK and DQPSK have not been explored before in a slow-light element. DPSK and DQPSK are becoming ever-more important in the optical communications community due to their potential for increased receiver sensitivity, tolerance to various fiber impairments, and enhanced spectral efficiency [70]. It is expected that future optical signal processors should be capable of dealing with different types of modulation formats [71,72]. For the slow-light-tunable delay lines to be truly versatile in heterogeneous communication systems, it is highly desirable to understand how the information-bearing phase patterns could be preserved and how much fractional delay on the phase patterns the signal could experience. Furthermore, it is of great importance to identify the data distortions coming from the combined effects of slow light nonlinearities and advanced modulation formats, as well as to design techniques to alleviate these degradations. A laudable goal would thus be to examine critical system limitations on multi-Gbps phase-encoded data signals as they traverse a tunable slow-light element.

16.3.1 DELAYING DPSK SIGNALS

Figure 16.11 shows both the simulation and experimental results of slowing down a 10-Gbps DPSK signal. The slow-light element is analytically modeled as discussed in Section 16.2.3. The phase patterns of the DPSK signal are shown in Figure 16.11a, both before and after an 8-GHz bandwidth slow-light element. We can see that the phase patterns can be delayed by as much as 46 ps and the differential π phase relation is maintained quite well for the delayed copy.

We further carry out DPSK slow-light experiment which is based on broadband SBS [23,24] mechanism in a piece of highly nonlinear fiber (HNLF). Detailed experimental setup and key parameters can be found in Ref. [14]. We note that the BER measurements are taken on both the constructive duobinary (DB) and the destructive alternate mark inversion (AMI) ports after the one-bit delay interferometer (DI).

Figure 16.11b shows the measured delay of a 10.7 Gbps NRZ-DPSK signal with 0 dBm power under an 8 GHz SBS gain bandwidth. The measured delay scales fairly linearly with the increased pump power, demonstrating the ability to continuously control the delay of the DPSK phase pattern.

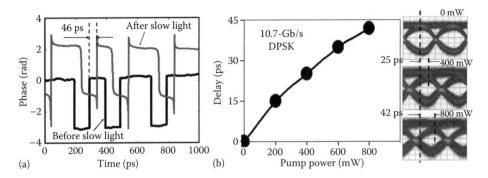

FIGURE 16.11 (a) Simulation result of phase patterns of a 10-Gbps DPSK signal before and after 8 GHz BW slow-light element. Phase is preserved and delayed by 46 ps and (b) experimental results of DPSK slow light: continuous delay up to 42 ps for a 10.7 Gbps DPSK signal is achieved.

The detected balanced DPSK eyes are shown for three different pump powers, with a maximum delay of 42 ps at a pump power of 800 mW. The achieved 42 ps delay of a 10.7 Gbps NRZ-DPSK signal corresponds to a fractional delay of 45%.

Very recently, OPAs in fiber have also been shown as a potential slow-light delay element to slow-down phase-encoded optical signals, with the potential advantage of wider operating bandwidths. As a result, Liu et al. have shown via simulation that 160-Gbps DPSK signals can be delayed using OPA-based slow light [15].

16.3.2 DPSK Data-Pattern Dependence and Its Mitigation

Similar to OOK formats, DPSK also suffers data-pattern dependence after traversing band-limited slow-light elements. This can be confirmed in Figure 16.11b that balance-detected signals exhibit vertical eye closure with increased slow-light delays. In order to assess the signal fidelity, we analyze both the constructive and destructive ports of the DI after demodulation independently. Figure 16.12 shows the 10.7 Gbps NRZ-DPSK intensity patterns before and after passing through the slow-light

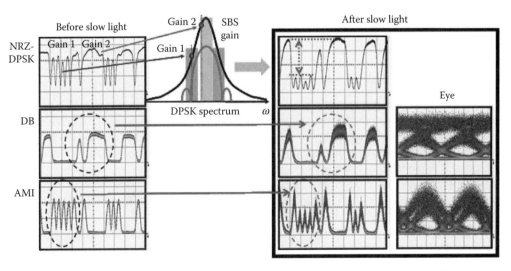

FIGURE 16.12 Slow-light-induced DPSK data-pattern dependence: 10.7-Gbps NRZ-DPSK through an 8 GHz slow-light element. Bit-patterns for NRZ-DPSK, DB, and AMI are shown before and after slow light.

element, with positions recorded right before demodulation (NRZ-DPSK) and right after demodulation (DB and AMI), respectively. We note here that although the DPSK data-pattern dependence also originates from the narrowband amplitude response, it deserves a careful treatment due to the added effects from DPSK format generation and detection.

As shown in Figure 16.12, NRZ-DPSK features residual intensity modulation which occurs during phase transitions while using a Mach–Zehnder modulator as the typical generation method. We can categorize these intensity dips as isolated "1"s (between two consecutive dips) and consecutive "1"s (between two long separated dips). Isolated "1"s occupy higher frequency components compared to consecutive "1"s, and will therefore experience much less gain after passing through a narrowband slow-light resonance. This effect can be clearly seen in the distorted NRZ-DPSK intensity patterns after slow light (Figure 16.12). This pattern-dependent gain experienced by NRZ-DPSK translates into two different types of data-pattern dependence on the two demodulated ports. For the DB signal, the peak power is much higher for long "1"s, compared to that of a single "1." This can be explained by the fact that single "1"s are only demodulated from two consecutive "1"s in NRZ-DPSK which has a much slower rise time due to slow-light third-order dispersion [73]. This leads to an insufficient constructive interference for the generation of single "1"s. The AMI signal exhibits strong pattern dependence within a group of "1" pulses. Compared with the "1"s in the middle, the leading and the trailing "1"s always have much higher peak powers in that they both experience unequal-power constructive interference from the edge pulses in a group of isolated dips in delayed NRZ-DPSK pattern. Both DB and AMI eye diagrams exhibit vertical data-pattern dependence. Furthermore, the AMI port also features nonnegligible pulse walk-off, which can be attributed to the slower rising and falling times of the two edge pulses compared with fast-transitioned middle pulses, in a group of "1" pulses. Detailed analysis and comparison reveals that RZ-DPSK suffers less patterning effect owing to the absence of pattern-dependent intensity variations [14].

BER measurements and eye diagrams on the demodulated DB port of the 10.7 Gbps DPSK signal under different delay conditions are shown in Figure 16.13a. We emphasize that we could still achieve error-free operation for the demodulated DB signal at a delay of 42 ps with a power penalty of 9.5 dB. The trade-off between delay and signal quality can be clearly seen where the previously analyzed DPSK-data-pattern dependence is one of the main signal degrading effects, as confirmed by the vertically closed eyes. The performance of the demodulated AMI signal is worse than that of the DB port owing to the reduced dispersion tolerance and the severe pulse walk-off.

Realizing that the data-pattern dependence comes mainly from the pattern-dependent slow-light gain, we red-detune the peak of the SBS gain profile by 0.016 nm from the channel center, which

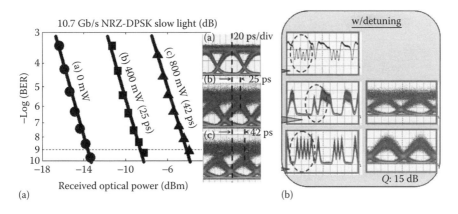

FIGURE 16.13 (a) BER measurement of DB port from 10.7-Gbps DPSK signals after SBS-based slow-light element. DPSK data-pattern dependence is the main reason for signal degradation. (b) Reduction of DPSK data-pattern dependence by detuning the SBS gain peak: 3-dB Q-factor improvement on the AMI port demodulated from 10.7-Gbps DPSK signals is achieved.

results in gain equalization and thus pattern dependence reduction between isolated "1"s and consecutive "1"s within NRZ-DPSK "intensity dips." As shown in Figure 16.13b, bit-patterns and eye diagrams, after detuning for both demodulated DB and AMI signals, are recorded for comparison with those of Figure 16.12. The optimum 3-dB Q-factor (determined from BER measurement) improvement (from 12 to 15 dB) for the AMI eyes confirms the effectiveness of this detuning method. The detuning not only resolves vertical data-pattern dependence, but also reshapes the rising and falling times of the edge pulses in a group of "1" pulses such that pulse walk-off is also alleviated.

16.3.3 Spectrally Efficient Slow Light

In recent years, there is a strong trend toward higher-order (i.e., multilevel) modulation formats since they are more spectrally efficient, and thus, more robust to dispersion effects. A highly popular format is DQPSK, especially the RZ version of it (i.e., RZ-DQPSK) due to the absence of pattern-dependent intensity variation [70]. DQPSK formats utilize four phase levels to encode data and thus have a narrower spectrum than DPSK. This spectral efficiency feature becomes extremely valuable since most of the slow-light media are limited in bandwidth.

Simulation results in Figure 16.14 show that the slow-light delay on binary-phase patterns can be generalized to multilevel phase-encoded signals. Figure 16.14a shows the results of four-level phase patterns of a 20 Gbps (10 Gbaud) DQPSK signal, both before and after a fixed 10 GHz bandwidth slow-light element. The phase patterns can be delayed by as much as 35 ps while the four-level differential phase relationships are well maintained. This could be explained by the fact that when compared with phase transitions, the four deterministic relative phase levels occupy low-frequency components within the slow-light resonance bandwidth and are thus immune to slow-light narrowband filtering-induced distortions. Based on this understanding, slow light can be generalized to delaying arbitrary-level phase-encoded formats. Figure 16.14b extends 10 Gbaud RZ-DQPSK to RZ-D8PSK signals, in which the bit rate is tripled to 30 Gbps compared with 10 Gbps binary DPSK signals. Under the same 10-GHz limited slow-light bandwidth, since the spectral-width for DQPSK and D8PSK are almost identical to that of DPSK, the maximum achievable delay for both DQPSK and D8PSK formats is almost the same ($\Delta T_1 = \Delta T_2$), regardless of the increased bit rates. We conclude that, in theory, a bandwidth-limited B-GHz slow-light element can provide almost identical maximum achievable delays for $\log 2(M)*$B-Gbps M-ary ($M = 2, 4, 8, \ldots$) DPSK signals. We also note here that due to the reduced phase noise tolerance and thus the reduced receiver sensitivity, distortion-constraint delay might be compromised for higher-order DPSK signals in practical slow-light systems.

The above findings motivate us to transmit multilevel, spectrally efficient DQPSK signals through a bandwidth-limited slow-light element. The experimental setup resembles that of DPSK

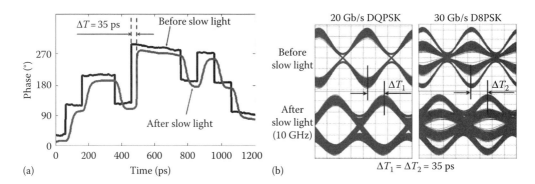

FIGURE 16.14 Simulation results of (a) four-level phase patterns of 10-Gbaud DQPSK signals before and after 10 GHz BW slow-light element. (b) Demodulated eye diagrams of both 10-Gbaud DQPSK and D8PSK signals after slow light.

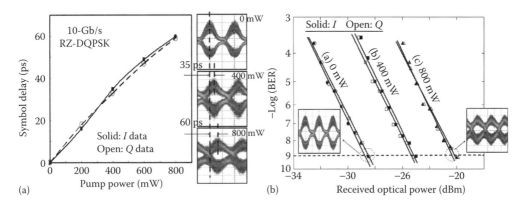

FIGURE 16.15 Experimental results on DQPSK slow light. (a) Symbol delay versus pump power for 10-Gbps RZ-DQPSK signals. (b) BER measurement for various delay values.

except for the part of signal generation and detection [74]. 10-Gbps RZ-DQPSK is generated by independently driving two parallel Mach–Zehnder modulators in an interferometric structure with a relative $\pi/2$ phase shift between arms. A subsequent pulse carver modulator is biased at quadrature for 50% duty-cycle RZ generation. The RZ-DQPSK signal, with polarization control for SBS effect maximization, counterpropagates with respect to the high power erbium-doped fiber amplifier (EDFA)-controlled broadband pump inside the HNLF. The amplified and delayed signals are sent to a preamplified balanced DQPSK receiver with a one-symbol DLI aligned for either quadrature (I or Q) demodulation. The error detector is postprogrammed for BER measurement on both I and Q data channels.

Figure 16.15a shows the experimental results of slow-light symbol delay on 10 Gbps (5 Gbaud) RZ-DQPSK signals. Demodulated signals after balance detection for both I and Q channels show very similar delays of around 35 and 60 ps at pump powers of 400 and 800 mW, respectively. We can see in Figure 16.15a that the symbol delays on both channels scale fairly linearly with increased pump power. The maximum achievable symbol delay of 60 ps corresponds to 60% of the fractional pulse width delay for either I or Q channel by itself, considering 10 Gbps 50% duty-cycle RZ-DQPSK pulses. Three typical balance-detected eye diagrams before and after slow-light delays are also shown.

In order to access the signal quality after the slow-light element, Figure 16.15b shows BER measurements on both I and Q data channels for 10 Gbps RZ-DQPSK signals. Both I and Q channels show very similar power penalty (~ 0.2 dB difference) under the same pump power case (i.e., same delay value). At 400 mW pump power, the system power penalty is around 3.5 dB at a BER of 10^{-9}. Another 3 dB penalty is added at 800 mW, which corresponds to the maximum delay condition. A clear trade-off exists between delay and signal quality. Signal power penalties are mainly attributed to the slow-light-induced data-pattern dependence [12–14] and some pulse broadening.

Based on both simulation and experimental results, we can conclude that slow light is applicable not only to binary amplitude- and phase-modulated signals, but also to quadrature or even multilevel modulated optical signals. Introducing advanced modulation formats to the slow-light community will enable interesting applications which require spectral efficiency. Optimization among the symbol delay, signal quality, receiver sensitivity, and spectral efficiency is considered crucial for any practical applications in the field of optical signal processing.

16.4 SIGNAL PROCESSING APPLICATIONS

Although various techniques have been shown for achieving slow-light delay on a single pulse or data streams, there have been very few reports of utilizing such a tunable delay for implementing true signal processing functions.

Before digging deep into the applications using slow-light tunable delay, let us take a step back by revisiting some of the unique features desirable for nonlinear optical signal processing in general. This could serve as an inspiration for the designing of slow-light-based novel signal processing modules.

1. *Ultrahigh speed*: Optical devices and materials provide nonlinearities with bandwidths ranging from several GHz to tens of THz [5]. Techniques to increase the modulation bandwidth have also been developed [75]. As a result, optical processing has been demonstrated at bit rates $\geq$ 320 Gbps [76].
2. *Multichannel operation*: Some nonlinear processes (such as wave-mixing, cross-phase modulation) can operate on multiple wavelengths simultaneously [77]. Significant cost reduction may be envisioned if a single optical device can replace a large number of electronic modules operating on each channel independently. This unique optical domain property has been used for multicasting [78] and equalization [8].
3. *Preservation of optical domain properties*: Optical properties, such as phase and polarization, add an additional dimension that carries information in the optical domain. Optical processing can be performed while preserving these properties, enabling unique techniques like phase conjugation [79] to mitigate fiber impairments. Moreover, various signal degrading effects can be isolated and monitored [80] if the desired optical properties are maintained.
4. *Bit rate and format transparency*: The hope is that future optical processing techniques do not employ bit-rate specific components, so that they will be immune to changes in input signal bit rates. Moreover, since optical properties (e.g., phase) are expected to be preserved, certain functions might be implemented in a format-transparent way [77].
5. *Simultaneous nonlinear processes*: Several nonlinear optical devices exhibit simultaneous nonlinear processes which can be exploited to perform more than one processing task on the input signals. This capability has been used to achieve diverse functions like clock-recovery and demultiplexing in a single device [81]. This feature could potentially enable functional integrations which exploits the multitude of functionalities available through a single device [82].
6. *Functional reconfigurability*: Advanced processing modules realized by a particular optical device are expected to be dynamically reconfigurable simply by tweaking their operating knobs. For example, reconfigurable optical logic gates based on a single SOA whose function can be changed from an XOR to an AND simply by detuning the filter have been demonstrated [83].

The following sections are dedicated to the description of various slow-light-based novel signal processing modules, with a special focus on the abovementioned desirable features.

16.4.1 VARIABLE-BIT-RATE OTDM MULTIPLEXER

Conventional fiber-based OTDM multiplexer (MUX) incorporates a fixed set of fiber lengths which is only suitable for a discrete set of given bit misalignments of the incoming data streams [7,11]. As shown in Figure 16.16a, when two input streams from two different locations pass through a fiber-based fixed OTDM MUX, it is highly likely that they will be misaligned and cause bit-overlapping at the output. By utilizing the continuously controllable delay feature of slow-light-based OTDM MUX, one could possibly manipulate the relative time misalignment by tweaking the slow-light control knob (e.g., pump power) and align the two streams nicely.

Another motivation of designing a slow-light-based OTDM MUX is its dynamic adaptivity to the incoming data bit rates so that it could enable a variable-bit-rate OTDM system, whereas conventional fiber-based fixed OTDM could not provide this capability. Shown in Figure 16.16b is

Reconfigurable Signal Processing Using Slow-Light-Based Tunable Optical Delay Lines

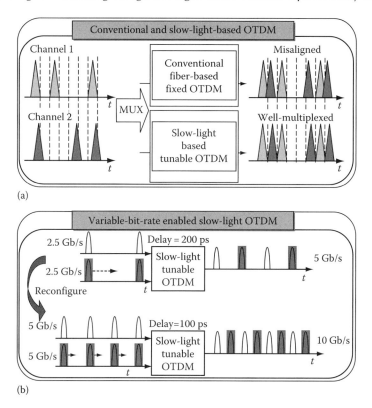

FIGURE 16.16 Concept of the advantages of slow-light-based OTDM compared with conventional fiber-based OTDM. (a) Slow-light OTDM multiplexer offers continuous tunability which can dynamically adjust the offsets among input channels. (b) Slow-light OTDM multiplexer can dynamically reconfigure its tunable delay according to different input data rates.

one typical example of two sets of input bit rates data streams. By simple control of the slow-light knob for producing either 200 or 100 ps delay, we could effectively multiplex either two 2.5 or two 5 Gbps data streams, respectively. One point to note is that the incoming bit rates can in theory be reconfigured to any value, because of the continuously tunable slow-light delay. This nice feature virtually puts no limitations on the delay resolution.

The detailed experimental setup and parameters can be referred to in Ref. [55]. Here we show the results of this tunable OTDM multiplexer. When the SBS pump is put off, the upper and lower arms are initially offset by 75 ps away from the well-multiplexed condition. The choice of the misalignment is only for the proof of concept and can be dependent on various parameters, such as the slow-light bandwidth, SBS pump power and initial pulse width. Figure 16.17 quantifies the power penalty reduction as a function of the fractional delay, which is defined as the absolute slow-light delay divided by the pulse width of 33% RZ pulses. We show that as the fractional delay increases, the relative power penalty can be reduced gradually, resulting in a maximum power penalty reduction of 9 dB when the lower channel is slowed down by 75 ps. The eye diagrams corresponding to three pump power levels are shown. The main reason for the improvement is that the beating caused by the bit-overlapping at the same wavelength reduces dramatically after efficient slow-light-based multiplexing. Slight beating still remains in the nondelayed reference bit slots since dispersive slow-light medium broadens the delayed bits whose tails penetrate into the neighboring slot. Therefore, a crucial requirement for efficient 2:1 OTDM is that the original incoming RZ pulses should ideally occupy less than half the bit slot so that beating region could be minimized after multiplexing. On the other hand, we do not want to have too narrow-width pulses which waste the slow-light bandwidth.

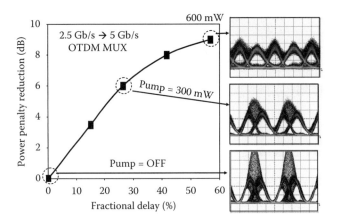

FIGURE 16.17 Penalty reduction versus fractional delay and the corresponding eye diagrams. Up to 9-dB power penalty reduction at 5-Gbps is achieved by employing slow-light-based dynamically tunable OTDM multiplexer.

Variable-bit-rate OTDM is also experimentally demonstrated using SBS-based slow light. Three different input bit rates of 2.5, 2.67, and 5-Gbps 33% RZ PRBS data are subsequently passed through the slow-light element. Broadband SBS bandwidth is fixed at 5 GHz for different bit rates comparison. We show in Figure 16.18 three eye diagrams after OTDM multiplexing of 2.5 to 5-Gbps, 2.67 to 5.34-Gbps, and 5 to 10-Gbps, respectively. We can see that the MUX performance gets worse with increased bit rate, which is mainly due to the restricted slow-light bandwidth. This limitation can be alleviated if a much wider slow-light bandwidth medium, such as OPAs [29–31], is used. Future variable-bit-rate N:1 OTDM multiplexing could be realized by either cascading multiple 2:1 OTDM MUX or enabling multiple channel operation in a single slow-light element [54].

16.4.2 Multichannel Synchronizer

Future slow-light applications might need independent delay controls over more than one data channel. This is particularly true for the most obvious implementations of synchronization and OTDM, or even equalization or regeneration on multiple data channels [84,85], possibly on different modulation formats [14]. In these applications, it is imperative to have independent fine-grained control over the delays of each channel. For N channels, this can certainly be achieved by brute-force approach of incorporating $N-1$ slow-light elements in parallel and then coupling all channels together at the output. However, a laudable goal would be to achieve individually controllable delays on multiple channels within a single slow-light element. The following design represents an important first step toward multichannel-enabled slow-light-based signal processing.

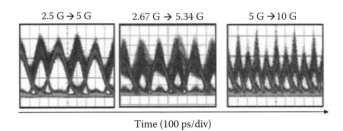

FIGURE 16.18 Variable-bit-rate OTDM: Efficient multiplexing of three data streams at three different input bit rates.

Reconfigurable Signal Processing Using Slow-Light-Based Tunable Optical Delay Lines

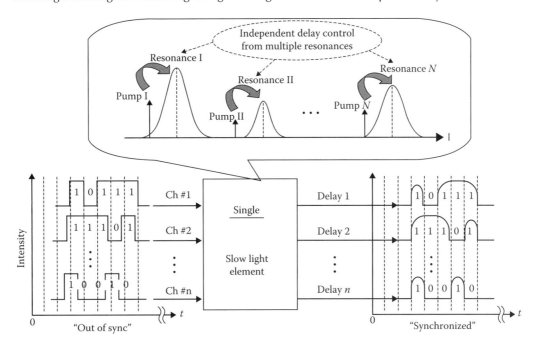

FIGURE 16.19 Conceptual diagram of independent delay control and synchronization on multiple data channels within a single slow-light element. The key enabler is the generation of multiple slow-light resonances from multiple pumps inside a single piece of slow-light fiber medium (inset). Independent fine-grained delay controls are thus achieved from their corresponding resonances.

Generation of multiple, individually tunable slow-light resonances within a single medium is considered the key challenge for independent delay control of multiple data channels. Inspired by the fact that SBS slow-light gain resonance is stimulated by its unique pump which is ~10-GHz blueshifted, we thus utilize many such pumps spaced sufficiently apart in a single piece of HNLF to achieve multiple resonances. Figure 16.19 conceptually illustrates this idea with one application of those independently controllable slow-light delays for multichannel data synchronization within a single slow-light element.

We show here the experimental demonstration of synchronizing three 2.5-Gbps NRZ-OOK data channels in a single piece of HNLF, which serves as our slow-light medium. Detailed experimental setup can be found in Ref. [54]. Two (1546.8 and 1554.7 nm) of the three channels have their own controllable pumps (pump 1 and pump 2) and the middle channel (1550.9 nm) serves as a reference without any pump. Broadband SBS pumps are adjusted to be ~3.5 GHz. One 7 GHz filter chooses the intended data channel for synchronization. The delayed signals are detected by a 2.5-GHz receiver and BER measurements are taken to evaluate the signal qualities.

Figure 19.20a shows the bit-patterns before and after multichannel slow-light synchronizer. When both pumps are off, channel #3 (1554.7 nm) and channel #1 (1546.8 nm) are offset from reference channel #2 (1550.9 nm) by 80 and 112 ps, respectively. By controlling their individual pump powers, both channels #3 and #1 can be delayed independently. With 150 mW for pump #3 and 250 mW for pump #1, all three channels are perfectly synchronized. Note that synchronization ranges (original offset) could be increased by further optimizing the slow-light bandwidth and pump power.

BER measurements are then taken to evaluate system performance of multichannel synchronizer. Both delayed and synchronized channels (#1 and #3) are shown to be error-free in Figure 16.20b, with <3.5 dB power penalty at a delay of up to 112 ps (#1). Data-pattern dependence due to limited SBS gain bandwidth is believed to be the main factor of power penalties. This can be confirmed from the recorded eye diagrams (after sync) which are vertically closing.

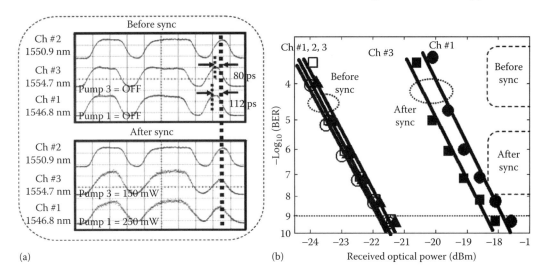

FIGURE 16.20 (a) Bit-patterns before and after synchronization confirm independent delay control on individual channels. (b) BER measurement before and after the multichannel synchronizer.

For such multichannel operations, crosstalk effects among multiple data channels might impose potential issues onto the system performance and thus must be carefully considered [54]. This requires not only optimization inside single channel, but also global optimization (such as respective pump power and resonance bandwidth) in a multichannel slow-light environment.

16.4.3 Simultaneous Multiple Functions

Slow-light phenomenon uniquely depends on the material and the mechanism. By judiciously choosing the slow-light medium, nonlinear processes and intrinsic functions inherent to the material can be explored. This, together with the basic job of delaying optical pulses, forms the idea of designing multiple functions via simultaneous nonlinear processes.

In Ref. [86], Yi et al. have demonstrated simultaneous demodulation and slow-light delay of DPSK signals at flexible bit rates using SBS-based amplifying and optical filtering effect in optical fiber. Both 10 and 2.5 Gbps DPSK signals have been demodulated and delayed with good performances. Results obtained for demodulation of 10 Gbps DPSK signals with up to 50 ps delay are highlighted in Figure 16.21. With an 18 dBm of pump power that results in 6.5 GHz gain bandwidth, the best demodulated performance is obtained.

By exploring a different slow-light medium, Hu et al. have demonstrated a slow-light delay line with simultaneous pulse reshaping function based on an OPA with clock-modulated pump. The idea is based on the fact that the OPA is able to reshape the signal pulses and minimize the distortion of the delayed pulses since the clock-modulated pump determines the quality of the delayed pulse. They have shown the propagation of 10 Gbps optical RZ packets and PRBS data through OPA medium. 25 ps slow-light delay and simultaneous pulse reshaping from 50 to 20 ps, with less than 2 dB power penalty, have been highlighted in Ref. [87].

16.4.4 Other Applications

In addition to the above demonstrations, various other interesting applications have also been shown using slow-light effects, which might find value in broadly defined optical signal processing areas.

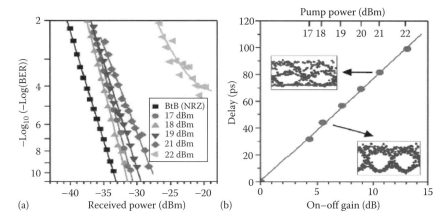

FIGURE 16.21 Experimental results of SBS slow-light-based simultaneous demodulation and delay of 10-Gbps DPSK. (From Yi, L., Jaouën, Y., Hu, W., Zhou, J., Su, Y., and Pincemin, E., *Opt. Lett.*, 32, 3182, 2007. With permission.)

All-optical delay of images using slow light [88]. As shown in Figure 16.22, Camacho et al. have demonstrated that two-dimensional images carried by 2 ns optical pulses can be delayed by up to 10 ns in a cesium vapor cell. The amplitude and phase information of the images are well preserved by interfering the delayed images with a local oscillator, even at very low light levels (e.g., each 2 ns optical pulse contains on average less than one photon). Their work holds the promise of whole-image buffering and processing.

Enhancing the spectral sensitivity of interferometers using slow-light media [89]: As shown in Figure 16.23, Shi et al. have demonstrated that the enhancement factor is equal to the group index of the slow-light medium, which can be as large as 10^7 if appropriate medium is used. Such significant sensitivity enhancement can benefit various applications that require precise location of the output fringes at the output of interferometer configurations.

Slow-light modulator with Bragg reflectors [90]: The authors have shown that the length of the device could be reduced to only 20 μm, 10 times smaller than conventional electroabsorption modulators. This reflects an important step toward functional integration of slow-light photonic circuits for switching and processing.

As the physics of various slow-light schemes are unfolding, the design and development of novel slow-light-based signal processing modules are expected to appear more in the literature. These demonstrations will benefit more and more research areas, ranging from component or device integration to system level functionalities or measurements.

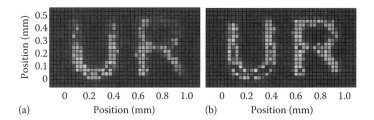

FIGURE 16.22 False color representation of a (a) Delayed and (b) Nondelayed two-dimensional low-light-level image. (From Camacho, R.M., Broadbent, C.J., Khan, I., and Howell, J.C., *Phys. Rev. Lett.*, 98, 043902-1, 2007. With permission.)

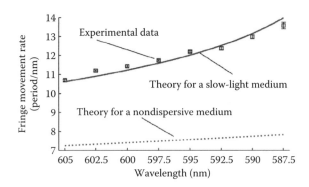

FIGURE 16.23 Theoretical and experimental results of enhancing the spectral sensitivity of an interferometer using slow light. (From Shi, Z., Boyd, R.W., Gauthier, D.J., and Dudley, C.C., *Opt. Lett.*, 32, 915, 2007. With permission.)

16.5 SUMMARY

In conclusion, we have discussed recent advances on slow-light techniques and their applications to optical signal processing. Significant progress has been made during the past few years from designing a tunable delay element based on various slow-light schemes to applying the tunable delay element to the development of novel signal processing modules. In particular, we have focused our attention on analyzing slow-light-induced signal degrading effects and the ways to mitigate them. Phase-preserving slow light is further proposed and demonstrated. We finally show several novel slow-light-based signal processing modules. Unique features such as multichannel operation, variable-bit-rate operation, and simultaneous multiple functions are highlighted. Future research directions are expected to pay further attention to slow-light-induced signal distortions by proposing novel and efficient slow-light schemes. Desirable features mentioned in the beginning of Section 16.4 are worthy of further exploration for multidimensional and multifunctional applications in reconfigurable optical switching and signal processing.

ACKNOWLEDGMENTS

The authors gratefully acknowledge the financial support from the DARPA DSO Slow Light Program, under the contract number N00014-05-0053. The authors also wish to thank the fruitful discussions and collaborations with the following people, including Dr. R. Boyd, L. Christen, I. Fazal, Dr. A. Gaeta, Dr. D. Gauthier, Dr. T. Luo, S. Nuccio, Dr. L.-S. Yan, J.-Y. Yang, Dr. C. Yu, and Dr. Z. Zhu.

REFERENCES

1. R. W. Boyd and D. J. Gauthier, Slow and fast light, in E. Wolf (Ed.), *Progress in Optics*, Vol. 43. Elsevier, Amsterdam, Chap. 6, pp. 497–530, 2002.
2. A. V. Oppenheim, R. W. Shafer, and J. R. Buck, *Discrete-Time Signal Processing*, 2nd edn. Prentice-Hall, Upper Saddle River, NJ, 1999.
3. S. G. Mallet, *A Wavelet Tour of Signal Processing*, 2nd edn. Academic Press, New York, 1999.
4. D. Cotter, R. J. Manning, K. J. Blow, A. D. Ellis, A. E. Kelly, D. Nesset, I. D. Phillips, A. J. Poustie, and D. C. Rogers, Nonlinear optics for high-speed digital information processing, *Science*, 286, 1523, 1999.
5. S. Radic and C. J. McKinstrie, Optical amplification and signal processing in highly nonlinear optical fiber, *IEICE Trans. Electron.* E88C, 859–869, 2005.
6. K. Jackson, S. Newton, B. Moslehi, M. Tur, C. Cutler, J. Goodman, and H. Shaw, Optical fiber delay line signal processing, *IEEE Microwave Theory Tech.*, MTT-33, 193–210, 1985.

7. S. A. Hamilton, B. S. Robinson, T. E. Murphy, S. J. Savage, and E. P. Ippen, 100 Gb/s optical time-division multiplexed networks, *J. Lightwave Technol.*, 20, 2086–2100, 2002.
8. C. R. Doerr, S. Chandrasekhar, P. J. Winzer, A. R. Chraplyvy, A. H. Gnauck, L. W. Stulz, R. Pafchek, and E. Burrows, Simple multichannel optical equalizer mitigating intersymbol interference for 40-Gb/s nonreturn-to-zero signals, *J. Lightwave Technol.*, 22, 249–256, 2004.
9. G. Lenz, B. J. Eggleton, C. K. Madsen, and R. E. Slusher, Optical delay lines based on optical filters, *J. Quantum Electron.*, 37, 525–532, 2001.
10. D. K. Hunter, M. C. Chia, and I. Andonovic, Buffering in optical packet switches, *J. Lightwave Technol.*, 16, 2081–2094, 1998.
11. L.-S. Yan, L. Lin, A. Belisle, S. Wey, and X. S. Yao, Programmable optical delay generator with uniform output and double-delay capability, *J. Opt. Network.*, 6, 13–18, 2007.
12. C. Yu, T. Luo, L. Zhang, and A. E. Willner, Data pulse distortion induced by a slow-light tunable delay line in optical fiber, *Opt. Lett.*, 32, 20–22, 2007.
13. L. Zhang, T. Luo, C. Yu, W. Zhang, and A. E. Willner, Pattern dependence of data distortion in slow-light elements, *J. Lightwave Technol.*, 25, 1754–1760, 2007.
14. B. Zhang, L. Yan, I. Fazal, L. Zhang, A. E. Willner, Z. Zhu, and D. J. Gauthier, Slow light on Gbit/s differential-phase-shift-keying signals, *Opt. Express*, 15, 1878–1883, 2007.
15. F. Liu, Y. Su, and P. L. Voss, Optimal operating conditions and modulation format for 160 Gb/s signals in a fiber parametric amplifier used as a slow-light delay line element, in *Proc. OFC*, Anaheim, CA, paper OWB5, 2007.
16. C. J. Chang-Hasnain and S. L. Chuang, Slow and fast light in semiconductor quantum-well and quantum-dot devices, *J. Lightwave Technol.*, 24, 4642–4654, 2006.
17. C. J. Chang-Hasnain, P. C. Ku, J. Kim, and S. L. Chuang, Variable optical buffer using slow light in semiconductor nanostructures, *Proc. IEEE*, 91, 11, 1884–1897, 2003.
18. C. Liu, Z. Dutton, C. Behroozi, and L. V. Hau, Observation of coherent optical information storage in an atomic medium using halted light pulses, *Nature*, 409, 6819, 490–493, 2001.
19. M. S. Bigelow, N. N. Lepeshkin, and R. W. Boyd, Observation of ultraslow light propagation in a ruby crystal at room temperature, *Phys. Rev. Lett.*, 90, 113903-1–113903-4, 2003.
20. H.-Y. Tseng, J. Huang, and A. Adibi, Expansion of the relative time delay by switching between slow and fast light using coherent population oscillation with semiconductors, *Appl. Phys. B*, B85, 493–501, 2006.
21. K. Y. Song, M. G. Herráez, and L. Thevénaz, Observation of pulse delaying and advancement in optical fibers using stimulated Brillouin scattering, *Opt. Express*, 13, 82–88, 2005.
22. Y. Okawachi, M. S. Bigelow, J. E. Sharping, Z. Zhu, A. Schweinsberg, D. J. Gauthier, R. W. Boyd, and A. L. Gaeta, Tunable all-optical delays via Brillouin slow light in an optical fiber, *Phys. Rev. Lett.*, 94, 153902–153905, 2005.
23. M. G. Herráez, K. Y. Song, and L. Thévenaz, Arbitrary-bandwidth Brillouin slow light in optical fibers, *Opt. Express*, 14, 1395–1400, 2006.
24. Z. Zhu, A. M. C. Dawes, D. J. Gauthier, L. Zhang, and A. E. Willner, 12-GHz-bandwidth SBS slow light in optical fibers, *J. Lightwave Technol.*, 25, 201–206, 2007.
25. K. Y. Song and K. Hotate, 25 GHz bandwidth Brillouin slow light in optical fibers, *Opt. Lett.*, 32, 217–219, 2007.
26. K. Lee and N. M. Lawandy, Optically induced pulse delay in a solid-state Raman amplifier, *Appl. Phys. Lett.*, 78, 703–705, 2001.
27. J. E. Sharping, Y. Okawachi, and A. L. Gaeta, Wide bandwidth slow light using a Raman fiber amplifier, *Opt. Express*, 13, 6092–6098, 2005.
28. Y. Okawachi, M. Foster, J. Sharping, A. Gaeta, Q. Xu, and M. Lipson, All-optical slow-light on a photonic chip, *Opt. Express*, 14, 2317–2322, 2006.
29. D. Dahan and G. Eisenstein, Tunable all optical delay via slow and fast light propagation in a Raman assisted fiber optical parametric amplifier: A route to all optical buffering, *Opt. Express*, 13, 6234–6249, 2005.
30. E. Shumakher, A. Willinger, R. Blit, D. Dahan, and G. Eisenstein, Large tunable delay with low distortion of 10 Gbit/s data in a slow light system based on narrow band fiber parametric amplification, *Opt. Express*, 14, 8540–8545, 2006.
31. L. Yi, W. Hu, Y. Su, M. Gao, and L. Leng, Design and system demonstration of a tunable slow-light delay line based on fiber parametric process, *IEEE Photon. Technol. Lett.*, 18, 2575–2577, 2006.

32. A. Melloni, F. Morichetti, and M. Martinelli, Linear and nonlinear pulse propagation in coupled resonator slow-wave optical structures, *Opt. Quantum Electron.*, 35, 365–379, 2003.
33. M. F. Yanik and S. Fan, Stopping light all optically, *Phys. Rev. Lett.*, 92, 083901-1–083901-4, 2004.
34. J. K. S. Poon, J. Scheuer, Y. Xu, and A. Yariv, Designing coupled-resonator optical waveguide delay lines, *J. Opt. Soc. Am. B*, 21, 1665–1673, 2004.
35. Z. Wang and S. Fan, Compact all-pass filters in photonic crystals as the building block for high-capacity optical delay lines, *Phys. Rev. E*, 68, 066616-1–066616-4, 2003.
36. M. Povinelli, S. Johnson, and J. Joannopoulos, Slow-light, band-edge waveguides for tunable time delays, *Opt. Express*, 13, 7145–7159, 2005.
37. Y. A. Vlasov, M. O'Boyle, H. F. Hamann, and S. J. McNab, Active control of slow light on a chip with photonic crystal waveguides, *Nature*, 438, 65–69, 2005.
38. X. Zhao, P. Palinginis, B. Pesala, C. J. Chang-Hasnain, and P. Hemmer, Tunable ultraslow light in vertical-cavity surface-emitting laser amplifier, *Opt. Express*, 13, 7899–7904, 2005.
39. Q. Xu, S. Sandhu, M. L. Povinelli, J. Shakya, S. Fan, and M. Lipson, Experimental realization of an on-chip all-optical analogue to electromagnetically induced transparency, *Phys. Rev. Lett.*, 96, 123901-1–123901-4, 2006.
40. J. T. Mok, C. M. de Sterke, I. C. M. Littler, and B. J. Eggleton, Dispersionless slow light using gap solitons, *Nat. Phys.*, 2, 775–780, 2006.
41. R. W. Boyd, D. J. Gauthier, A. L. Gaeta, and A. E. Willner, Maximum time delay achievable on propagation through a slow-light medium, *Phys. Rev. A*, 71, 023801-1–023801-4, 2005.
42. J. B. Khurgin, Optical buffers based on slow light in electromagnetically induced transparent media and coupled resonator structures: Comparative analysis, *J. Opt. Soc. Am. B*, 22, 1062–1074, 2005.
43. R. S. Tucker, P. C. Ku, and C. J. Chang-Hasnain, Slow-light optical buffers: Capabilities and fundamental limitations, *J. Lightwave Technol.*, 23, 4046–4066, 2005.
44. J. B. Khurgin, Power dissipation in slow light devices: a comparative analysis, *Opt. Lett.*, 32, 163–165, 2007.
45. R. M. Camacho, M. V. Pack, and J. C. Howell, Slow light with large fractional delays by spectral hole-burning in rubidium vapor, *Phys. Rev. A*, 74, 033801-4, 2006.
46. L. Yi, L. Zhan, W. Hu, and Y. Xia, Delay of broadband signals using slow light in stimulated Brillouin scattering with phase-modulated pump, *IEEE Photon. Technol. Lett.*, 19, 619–621, 2007.
47. B. Zhang, L.-S. Yan, L. Zhang, and A. E. Willner, Broadband SBS slow light using simple spectrally-sliced pumping, in *Proc. ECOC*, Berlin, Germany, paper P025, 2007.
48. E. Shumakher, N. Orbach, A. Nevet, D. Dahan, and G. Eisenstein, On the balance between delay, bandwidth and signal distortion in slow light systems based on stimulated Brillouin scattering in optical fibers, *Opt. Express*, 14, 5877–5884, 2006.
49. A. Zadok, A. Eyal, and M. Tur, Extended delay of broadband signals in stimulated Brillouin scattering slow light using synthesized pump chirp, *Opt. Express*, 14, 8498–8505, 2006.
50. J. Mørk, R. Kjær, M. van der Poel, and K. Yvind, Slow light in a semiconductor waveguide at gigahertz frequencies, *Opt. Express*, 13, 8136–8145, 2005.
51. F. Sedgwick, B. Pesala, J.-Y. Lin, W. S. Ko, X. Zhao, and C. J. Chang-Hasnain, THz-bandwidth tunable slow light in semiconductor optical amplifiers, *Opt. Express*, 15, 747–753, 2007.
52. M. C. Cardakli and A. E. Willner, Synchronization of a network element for optical packet switching using optical correlators and wavelength shifting, *IEEE Photon. Technol. Lett.*, 14, 1375–1377, 2002.
53. D. Petrantonakis, D. Apostolopoulos, O. Zouraraki, D. Tsiokos, P. Bakopoulos, and H. Avramopoulos, Packet-level synchronization scheme for optical packet switched network nodes, *Opt. Express*, 14, 12665–12669, 2006.
54. B. Zhang, L.-S. Yan, J.-Y. Yang, I. Fazal, and A. E. Willner, A single slow-light element for independent delay control and synchronization on multiple Gbit/s data channels, *IEEE Photon. Technol. Lett.*, 19, 1081–1083, 2007.
55. B. Zhang, L. Zhang, L.-S. Yan, I. Fazal, J.-Y. Yang, and A. E. Willner, Continuously-tunable, bit-rate variable OTDM using broadband SBS slow-light delay line, *Opt. Express*, 15, 8317–8322, 2007.
56. I. Fazal, O. Yilmaz, S. Nuccio, B. Zhang, A. E. Willner, C. Langrock, and M. M. Fejer, Optical data packet synchronization and multiplexing using a tunable optical delay based on wavelength conversion and inter-channel chromatic dispersion, *Opt. Express*, 15, 10492–10497, 2007.
57. M. Secondini, Optical equalization: System modeling and performance evaluation, *J. Lightwave Technol.*, 24, 4013–4021, 2006.

58. A. H. Gnauck, C. R. Doerr, P. J. Winzer, and T. Kawanishi, Optical equalization of 42.7-Gbaud bandlimited RZ-DQPSK signals, *IEEE Photon. Technol. Lett.*, 19, 1442–1444, 2007.
59. M. C. Cardakli, S. Lee, A. E. Willner, V. Grubsky, D. Starodubov, and J. Feinberg, Reconfigurable optical packet header recognition and routing using time-to-wavelength mapping and tunable fiber Bragg gratings for correlation decoding, *IEEE Photon. Technol. Lett.*, 12, 552–554, 2000.
60. A. E. Willner, D. Gurkan, A. B. Sahin, J. E. McGeehan, and M. C. Hauer, All-optical address recognition for optically-assisted routing in next-generation optical networks, *IEEE Commun. Mag.*, 41, S38–S44, 2003.
61. B. Meagher, G. K. Chang, G. Ellinas, Y. M. Lin, W. Xin, T. F. Chen, X. Yang, A. Chowdhury, J. Young, S. J. B. Yoo, C. Lee, M. Z. Iqbal, T. Robe, H. Dai, Y. J. Chen, and W. I. Way, Design and implementation of ultra-low latency optical label switching for packet-switched WDM networks, *J. Lightwave Technol.*, 18, 1978–1987, 2000.
62. A. J. Poustie, K. J. Blow, A. E. Kelly, and R. J. Manning, All-optical parity checker with bit-differential delay, *Opt. Comm.*, 162, 37–43, 1999.
63. M. Soljačič, S. G. Johnson, S. Fan, M. Ibanescu, E. Ippen, and J. D. Joannopoulos, Photonic-crystal slow-light enhancement of nonlinear phase sensitivity, *J. Opt. Soc. Am. B*, 19, 2052–2059, 2002.
64. H. Schmidt and R. J. Ram, All-optical wavelength converter and switch based on electromagnetically induced transparency, *Appl. Phys. Lett.*, 76, 3173–3175, 2000.
65. T. Luo, L. Zhang, W. Zhang, C. Yu, and A. E. Willner, Reduction of pattern dependent distortion on data in an SBS-based slow light fiber element by detuning the channel away from the gain peak, in *Proc. CLEO*, Long Beach, CA, paper CThCC4, 2006.
66. M. D. Stenner, M. A. Neifeld, Z. Zhu, A. M. C. Dawes, and D. J. Gauthier, Distortion management in slow-light pulse delay, *Opt. Express*, 13, 9995–10002, 2005.
67. K. Y. Song, M. G. Herráez, and L. Thévenaz, Gain-assisted pulse advancement using single and double Brillouin gain peaks in optical fibers, *Opt. Express*, 13, 9758–9765, 2005.
68. A. Minardo, R. Bernini, and L. Zeni, Low distortion Brillouin slow light in optical fibers using AM modulation, *Opt. Express*, 14, 5866–5876, 2006.
69. Z. Shi, R. Pant, Z. Zhu, M. D. Stenner, M. A. Neifeld, D. J. Gauthier, and R. W. Boyd, Design of a tunable time-delay element using multiple gain lines for large fractional delay with high data fidelity, *Opt. Lett.*, 32, 1986–1988, 2007.
70. A. H. Gnauck and P. J. Winzer, Optical phase-shift-keyed transmission, *J. Lightwave Technol.*, 23, 115–130, 2005.
71. P. Devgan, R. Tang, V. S. Grigoryan, P. Kumar, Highly efficient multichannel wavelength conversion of DPSK signals, *J. Lightwave Technol.*, 24, 3677–3682, 2006.
72. K. Mishina, A. Maruta, S. Mitani, T. Miyahara, K. Ishida, K. Shimizu, T. Hatta, K. Motoshima, and K. Kitayama, NRZ-OOK-to-RZ-BPSK modulation-format conversion using SOA-MZI wavelength converter, *J. Lightwave Technol.*, 24, 3751–3758, 2006.
73. G. P. Agrawal, *Nonlinear Fiber Optics*, 3rd edn. Chap. 3, Academic Press, San Diego, 2001.
74. B. Zhang, L. Yan, L. Zhang, S. Nuccio, L. Christen, T. Wu, and A. E. Willner, Spectrally efficient slow light using multilevel phase-modulated formats, *Opt. Lett.*, 33, 55–57, 2008.
75. M. L. Nielsen and J. Mork, Increasing the modulation bandwidth of semiconductor-optical-amplifier-based switches by using optical filtering, *J. Opt. Soc. Am. B*, 21, 1606–1619, 2004.
76. Y. Liu, E. Tangdiongga, Z. Li, H. de Waardt, A. M. J. Koonen, G. D. Khoe, X. Shu, I. Bennion, and H. J. S. Dorren, Error-free 320-Gb/s all-optical wavelength conversion using a single semiconductor optical amplifier, *J. Lightwave Technol.*, 25, 103–108, 2007.
77. R. W. Tkach, A. R. Chraplyvy, F. Forghieri, A. H. Gnauck, and R. M. Desosier, Four-photon mixing and high-speed WDM systems, *J. Lightwave Technol.*, 13, 841–849, 1995.
78. K. K. Chow, C. Shu, C. Lin, and A. Bjarklev, All-optical wavelength multicasting with extinction ratio enhancement using pump-modulated fourwave mixing in a dispersion-flattened nonlinear photonic crystal fiber, *IEEE J. Sel. Top. Quantum Electron.*, 12, 838–842, 2006.
79. S. L. Jansen, D. Van Den Borne, P. M. Krummrich, S. Spalter, G. D. Khoe, and H. De Waardt, Long-haul DWDM transmission systems employing optical phase conjugation, *IEEE J. Sel. Top. Quantum Electron.*, 12, 505–520, 2006.
80. A. E. Willner, The optical network of the future: Can optical performance monitoring enable automated, intelligent and robust systems? *Opt. Photon. News*, 17, 30–35, 2006.

81. H.-F. Chou, Z. Hu, J. E. Bowers, D. J. Blumenthal, K. Nishimura, R. Inohara, and M. Usami, Simultaneous 160-Gb/s demultiplexing and clock recovery by utilizing microwave harmonic frequencies in a traveling-wave electroabsorption modulator, *IEEE Photon. Technol. Lett.*, 16, 608–610, 2004.
82. V. Kaman, A. J. Keating, S. Z. Zhang, and J. E. Bowers, Simultaneous OTDM demultiplexing and detection using an electroabsorption modulator, *IEEE Photon. Technol. Lett.*, 12, 711–713, 2000.
83. Z. Li, Y. Liu, S. Zhang, H. Ju, H. De Waardt, G. D. Khoe, H. J. S. Dorren, and D. Lenstra, All-optical logic gates using semiconductor optical amplifier assisted by optical filter, *Electron. Lett.*, 41, 51–52, 2005.
84. M. Vasilyev and T. I. Lakoba, All-optical multichannel 2R regeneration in a fiber-based device, *Opt. Lett.*, 30, 1458–1460, 2005.
85. P. K. A. Wai, Lixin Xu, L. F. K. Lui, L. Y. Chan, C. C. Lee, H. Y. Tam, and M. S. Demokan, All-optical add-drop node for optical packet-switched networks, *Opt. Lett.*, 30, 1515–1517, 2005.
86. L. Yi, Y. Jaouën, W. Hu, J. Zhou, Y. Su, and E. Pincemin, Simultaneous demodulation and slow light of differential phase-shift keying signals using stimulated-Brillouin-scattering-based optical filtering in fiber, *Opt. Lett.*, 32, 3182–3184, 2007.
87. Z. Hu and D. Blumenthal, Simultaneous slow-light delay and pulse reshaping of 10Gbps RZ data in highly nonlinear fiber-based optical parametric amplifier with clock-modulated pump, in *Proc. OFC*, Anaheim, CA, paper OWB4, 2007.
88. R. M. Camacho, C. J. Broadbent, I. Khan, and J. C. Howell, All-optical delay of images using slow light, *Phys. Rev. Lett.*, 98, 043902-1–043902-4, 2007.
89. Z. Shi, R. W. Boyd, D. J. Gauthier, and C. C. Dudley, Enhancing the spectral sensitivity of interferometers using slow-light media, *Opt. Lett.*, 32, 915–917, 2007.
90. G. Hirano, F. Koyama, K. Hasebe, T. Sakaguchi, N. Nishiyama, C. Caneau, and C. Zah, Slow light modulator with Bragg reflector waveguide, in *Proc. OFC*, Anaheim, CA, postdeadline paper PDP34, 2007.

17 Slow Light Buffers for Packet Switching

Rodney S. Tucker

CONTENTS

17.1 Introduction 347
17.2 Packet Switch Architectures 348
17.3 Buffers 349
    17.3.1 Buffer Capacity in Packet Switches 349
    17.3.2 Delay Line Buffer Architectures 350
        17.3.2.1 Slow Light Delay Line Buffers 351
        17.3.2.2 FIFO Using Cascaded Delay Lines 352
        17.3.2.3 FIFO Using Adiabatically Slowed Light 354
    17.3.3 Physical Size 355
    17.3.4 Waveguide Losses and Energy Consumption 358
    17.3.5 Resonator Buffers 362
    17.3.6 Electronic Buffers 363
    17.3.7 Comparison of Buffer Technologies 363
17.4 Conclusions 364
References 364

17.1 INTRODUCTION

Packet switches or routers are key items of Internet infrastructure. Routers orchestrate the movement of Internet protocol (IP) packets through the network, and ensure that each packet is forwarded from its source to its correct destination. In today's network, routers are highly sophisticated electronic systems, based on powerful electronic processing, switching, and buffering technologies [1,2]. Incoming packets arrive at the import ports of a router via optical fiber links and are converted to electronic form before entering the router. This conversion is achieved using optical-to-electronic (O/E) converters located at the inputs to the router. Similarly, outgoing packets from the router pass through E/O converters before being launched onto the next optical fiber transmission link in the network.

All of the internal functions of today's routers, including buffering and switching, are carried out using electronics, with the exception of some optical interconnects between boards and racks. Optical packet switching (OPS) (sometimes referred to as photonic packet switching) [3–6] provides a potential alternative to this kind of electronic packet switching or routing. OPS eliminates the need for O/E and E/O converters at the input and output ports. In an optical packet switch, optical components rather than electronic components provide the important functions of buffering and switching. Because an optical packet switch does not require O/E and E/O converters in the data path, the optical data in each packet remains in optical form as it passes through the switch.

There have been many laboratory demonstrations and a number of limited field trial of OPS [7–9]. But while the basic principles have been demonstrated, many obstacles need to be overcome before commercial deployment of OPS can become a reality. One reason for this is the lack of a suitable optical buffer memory that can store high-bit-rate optical packets [10]. Optical fiber delay lines have been used to provide a buffering capability in some experiments, but they are bulky and do not scale well to realistic-size optical packet switches. The demonstrations reported to date have been limited to switches with only a few input and output ports. But practical packet switches require many hundreds or even thousands of input and output ports.

The recent growth in research activity in slow light (see, e.g., Refs. [11,12] and other chapters of this book) has stimulated conjectures that slow light delay lines might be suitable as the storage elements in buffers for optical packet switches. This chapter presents an analysis of the prospects of using slow light delay lines in buffers for OPS. The chapter begins with an introduction to the functional building blocks of high-capacity routers and the requirements of buffers for packet switching. The chapter then focuses on the capabilities and limitations of slow light delay lines for buffering. Two key parameters considered here are the power dissipation in slow light buffers, and the physical size. Both of these parameters are critically important in the design of commercial packet switches.

The energy consumption of telecommunications equipment is growing as the size and capacity of the Internet expands [13]. In fact, there is a concern that the capacity of the network may ultimately be limited by energy considerations rather than by the bandwidth capabilities of its components. The energy consumption of a buffer or memory component can be measured in terms of the energy required to store and retrieve each bit of data. The dissipated power of a buffer is the energy required to write, store, and read out a bit, multiplied by the bit rate of the data passing through the buffer. One of the objectives of this chapter is to compare this energy consumption of optical buffers with the energy consumption of electronic buffers. We calculate the energy per bit in each buffer and show that this is a useful figure of merit for optical buffers. We show that, in general, the energy consumption of optical buffers is larger than the energy consumption of electronic buffers.

17.2 PACKET SWITCH ARCHITECTURES

Figure 17.1a shows a block schematic of the key optical components in an optical packet switch and Figure 17.1b shows the format of an incoming packet, which comprises a header of duration τ_h and a payload of duration τ_p. The total duration of the packet is τ_t. The header contains the address information and the payload contains the data. The header typically occupies 10% or less of the total packet length. In IP networks, the average duration τ_t of a packet at 40 Gbps is around 200 ns. The switch in Figure 17.1a has F input fibers and F output fibers, with each fiber carrying K wavelengths. Each of the $F \times K$ input and output wavelengths represent a port on the switch. Information about the destination and routing of the incoming packets is extracted from headers of the incoming packets using O/E converters on each of the $F \times K$ input wavelengths (details not shown in Figure 17.1a). This information can, in principle, be at a lower bit rate than the payload of the packet. The address information is fed to the forwarding engine, which is generally an electronic processor that is outside the switch fabric. The forwarding engine controls the switching and buffering of all packets as they progress through the router.

It is important to note that the forwarding engine in Figure 17.1 processes address information, but it pays no attention to the data in the payload of the packets. Therefore, the payload data are able to pass directly through the packet switch without requiring any electronic processing. This is the key to OPS—the payload data are not processed by any electronics. Therefore, the data capacity of the packet switch is not limited by the speed of O/E or E/O converters. O/E converters are required to detect the header and to extract routing information. But the header is short compared to the payload, and the bit rate of the header can be lower than the bit rate of the payload.

There are two broad classes of optical packet switch—synchronous and asynchronous. In synchronous switches, all packets are switched through the cross connect in synchronism. To achieve

Slow Light Buffers for Packet Switching

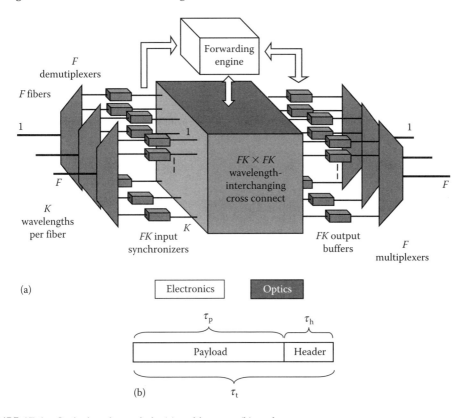

FIGURE 17.1 Optical packet switch; (a) architecture, (b) packet structure.

this, all packets are aligned in time using a bank of FK input synchronizers, as shown in Figure 17.1a. These synchronizers are short adjustable delay lines that provide up to τ_t of delay. Asynchronous packet switches do not require the packets to be aligned in time. Therefore, the synchronizers in Figure 17.1a are not required in asynchronous packet switches.

The switch fabric in the packet switch in Figure 17.1a is an $FK \times FK$ wavelength-interchanging cross connect [14]. This means that any input wavelength at any input port can be switched to any output port and converted to any desired wavelength on that output port. The function of the forwarding engine is to ensure that each packet is routed to the appropriate output port. To achieve this, the forwarding engine controls the cross connect on a packet-by-packet basis.

The packet switch in Figure 17.1a is an optical packet switch. However, regardless of whether the packet switch is optical or electronic, there is always the possibility that packets could collide at the output ports if more than one packet at the input ports needs to be delivered to the same output port. These collisions or contentions can be avoided using buffering. In general, buffering can be placed at the input ports, the output ports, or shared between the inputs and outputs. In Figure 17.1a, the buffers are shown at the output ports. One buffer is assigned to each port of the packet switch and the number of output buffers is $F \times K$.

17.3 BUFFERS

17.3.1 Buffer Capacity in Packet Switches

In conventional electronic packet switches, the buffers are constructed using electronic random-access memory (RAM) chips. RAM is relatively inexpensive and as a result there has traditionally been little incentive for router manufacturers to minimize the size of the buffers. It is common in electronic packet switches to provide up to RTT × B bits of buffer capacity at each port, where RTT is the round trip time between the data source and its destination, and B is the bit rate of the data stream. For example, for transoceanic links with round trip times as high as 250 ms and with data at 40 Gbps, this corresponds to a buffer capacity of 10 Gb per port. Buffers of this size can readily be accommodated using electronic RAM, but an optical buffer of this capacity would require delay lines with a total delay of at least 250 ms. This is many orders of magnitude longer than can be achieved with any optical delay line technology, including slow light.

In practice, it may not be necessary to provide a full 250 ms of buffering at each port. It turns out that [15] the capacity of the buffer on each port could be reduced from RTT × B to RTT × $B/\sqrt{n}$ where n is the number of users sharing the same wavelength. If each end-user on the Internet has an access rate of 4 Mbps and the line rate at each port of the packet switch is $B = 40$ Gbps, the maximum number of end users sharing a wavelength, or port is $n = 10,000$. On this basis, the buffer capacity per port could be reduced from 10 Gb to 100 Mb. If the access rate for each user was to be increased to 1 Gbps, the required buffer capacity would become 1.6 Gb. These buffer capacities are still too large for OPS, but it is conceivable that under certain circumstances the buffer capacity could be made even smaller. Recent work on calculating the minimum buffer size [16,17] suggests that buffer capacities as small as 10–20 packets, or around 200 kb may be acceptable in some circumstances. In the remainder of this chapter we look at slow light technologies with a view toward achieving buffer capacities of around 10–20 packets.

17.3.2 DELAY LINE BUFFER ARCHITECTURES

The basic functional building block of an optical buffer is the simple variable delay line illustrated in Figure 17.2a [18]. In this figure, the broad arrows represent a control signal that is used to adjust the delay time. Some form of external control of the delay time is needed because buffering or storage of data requires that it is possible to read the data out from the buffer under the control of a write

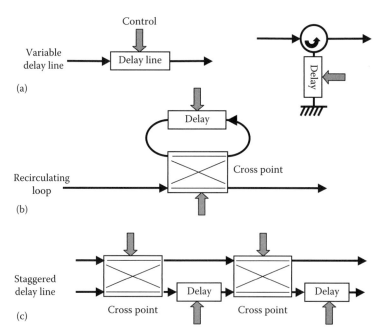

FIGURE 17.2 Optical buffer architectures. (From Tucker, R.S., *J. Lightwave Technol.*, 24, 4655, 2006. With permission.)

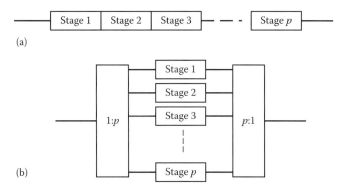

FIGURE 17.3 Serial and parallel buffers. (From Tucker, R.S., *J. Lightwave Technol.*, 24, 4655, 2006. With permission.)

command. In Figure 17.2a, the contents of the delay line can be read from the output port by reducing the delay. Note that a single delay line without control over the delay is not a buffer.

To provide the buffering functionality required in a packet switch, it is necessary to use buffer architectures that are more complicated than the simple variable delay in Figure 17.2a. Figure 17.2b shows a variable (or fixed) optical delay line, combined with a crosspoint, in a feedback (recirculating loop) configuration [18]. Strictly speaking, the delay time in Figure 17.2b need not be controllable because data can be read out by controlling the state of the crosspoint. Figure 17.2c is a feed-forward arrangement with crosspoints and delay lines.

The capacity and performance of the simple buffers in Figure 17.2 can be enhanced by combining the building blocks in Figure 17.2 either in cascade, as shown in Figure 17.3a, or in parallel as shown in Figure 17.3b [18]. If the cascaded stages in Figure 17.3a are simple variable delay lines of the type shown in Figure 17.2a, the cascade is a first-in-first-out (FIFO) buffer, in which the order of packets emerging from the buffer is the same as the order of packets entering it. In general, a FIFO buffer is a read–write device in which the sequence of data at the output is the same as the sequence of the data at the input. A FIFO buffer is similar to a line of people waiting for a coffee at Starbucks. In general, the data rate of the input and output can differ in a FIFO. Serially connected optical delay-line buffers are inherently FIFO devices, but optical FIFO memory generally operates with the same input and output data rates. FIFO buffers are well suited for OPS applications.

17.3.2.1 Slow Light Delay Line Buffers

We now focus on delay line buffers constructed using slow light delay lines. We consider two distinct classes of slow light delay lines—Class A and Class B [19]. In Class A delay lines, the group velocity of the light is slowed across an interface between two waveguides with different group velocities. In Class B delay lines, the velocity of the light is reduced adiabatically and uniformly across a data packet [19]. Class B delay lines are sometimes referred to as "adiabatically tunable" slow light delay lines [20].

Figure 17.4 schematically shows a train of periodic pulses in a Class A slow light delay line [19]. Pulses enter from the right, in Region 1 the group velocity is v_{g1} and at position x_A the pulses enter a slow light delay region with low group velocity v_{g2} (Region 2). The pulses speed are up to v_{g1} when they enter Region 3 at position x_B. The group velocity distribution along the device length is shown in Figure 17.4a and b as a snapshot of the pulses in the device. Note that the physical pulse width and the physical spacing between pulses, both decrease with the group velocity. The length of each pulse in Region 1 is L_{in}, and the length of each pulse in Region 2 is L_b. The reduction in the length of the pulse is in proportion to the decrease in group velocity. Therefore $L_b/L_{in} = v_{g2}/v_{g1} = 1/S$, where S is the slowdown factor. The length of the bit L_b in the slow light medium is a critically important practical

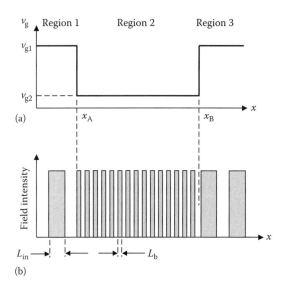

FIGURE 17.4 Class A delay line. (From Tucker, R.S., Ku, P.-C., Chang-Hasnian, C.J., *J. Lightwave Technol.*, 23, 4046, 2005. With permission.)

parameter because it affects the physical size of a buffer. For a given buffer capacity, the size of the buffer is minimized when L_b is minimized. Another important parameter affecting the performance of Class A delay line buffers is the so-called delay–bandwidth product. This is the product of the total delay T (in seconds) through the delay line and the information bandwidth B or throughput (in bits per second) of the delay line. The product of the delay and the throughput is equal to the capacity C, or the number of bits of information that can be stored in the delay line. Thus the capacity is

$$C = TB. \tag{17.1}$$

We will explore this important parameter further in Section 17.3.3.

Figure 17.5 shows a group of three pulses in a Class B slow light delay line [19]. The center curve shows snapshots of the three pulses at various points along the length of the delay line and the upper curve shows the group velocity seen by the three pulses as a function of time, and shows (with dots) the points in time when the snapshots were taken. The lower curve shows the bandwidth in the slow light medium as a function of time. The key difference between a Class A delay line and a Class B delay line is that in a Class A delay line, the group velocity changes as a function of position x, while in a Class B delay line, it changes as a function of time t. As shown in Figure 17.5, the three pulses enter the delay line during Interval 1. During Interval 2, the group velocity of all three pulses is adiabatically decreased from v_{g1} to v_{g2}. At the same time, the bandwidth of the medium is decreased from B_{g1} to B_{g2}. During Interval 3, the three pulses propagate with velocity v_{g2}, and in Interval 4 the pulses are returned to their original velocity and bandwidth. It is important to note that the physical length of each pulse L_b is not decreased in a Class B delay line. This is in contrast with Class A delay lines, in which the length of the pulse is decreased.

17.3.2.2 FIFO Using Cascaded Delay Lines

Figure 17.6 shows how a FIFO buffer can be realized using M cascaded controllable delay lines, each of length L [19]. In Figure 17.4, the group velocity v_g in each of the M delay lines can be set at one of two values: v_{g1} or v_{g2} using the M control signals Control 1–Control M. Light at v_{g1} is at "full speed" and light at v_{g2} is at "low speed". The curve at the bottom of Figure 17.6 is the group

Slow Light Buffers for Packet Switching

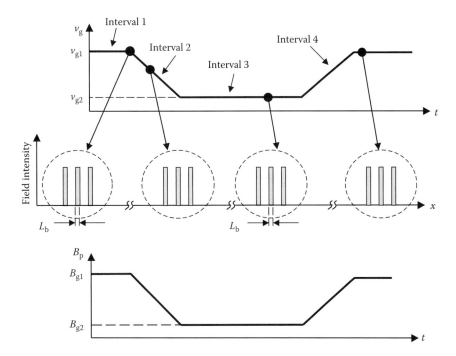

FIGURE 17.5 Class B delay line. (From Tucker, R.S., Ku, P.-C., Chang-Hasnian, C.J., *J. Lightwave Technol.*, 23, 4046, 2005. With permission.)

velocity as a function of the position x along the delay line at one point in time. The FIFO buffer operates as follows. The group velocity in the first delay line (DL1 in Figure 17.6) is always equal to v_{g2}. If a packet arrives at the input of the first delay line while another packet is stored in DL1, then the incoming packet cannot be accepted. To avoid an undesired collision between two packets,

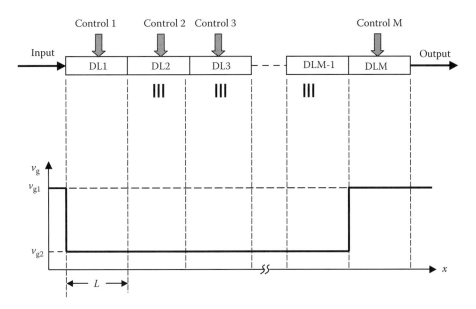

FIGURE 17.6 FIFO buffer using cascaded Class A delay lines. (From Tucker, R.S., Ku, P.-C., Chang-Hasnian, C.J., *J. Lightwave Technol.*, 23, 4046, 2005. With permission.)

it is generally necessary to dump nonaccepted packets using an optical switch or gate (not shown in Figure 17.6). If there is no packet currently stored in DL1, the incoming packet is written into DL1.

To maximize the time that packets can be stored, the group velocity is set to v_{g2} in all delay lines between the most recently accepted packet and the packet closest to the output. Delay lines to the right of the packet closest to the output also have group velocities v_{g2}. When it is necessary to read the packet closest to the output, the group velocity in all delay lines to the right of that packet are set to group velocity v_{g1}. To illustrate the above, Figure 17.6 shows a snapshot of a number of three-bit packets stored in the buffer. The buffer closest to the output is currently in DLM-1, and it has been decided to readout this packet to the output. To do this, the group velocity in DLM has been set to v_{g1}.

The recirculating loop buffer in Figure 17.3b can operate as a FIFO buffer, but if more than one packet is stored in the delay, the order of the packets can be changed. There is a common misconception that less waveguide delay is required for recirculating buffers other than buffer memories. In fact, while a packet is circulating in the loop, no other packets can enter it. Therefore, if it is necessary to buffer a stream of closely spaced incoming packets, multiple recirculating loops are required. This can be achieved using the cascade configuration of Figure 17.3a or the parallel configuration in Figure 17.3b.

17.3.2.3 FIFO Using Adiabatically Slowed Light

Figure 17.7 shows a FIFO buffer using a cascade of M Class B delay lines [19]. The group velocity in each of the M delay lines can be adjusted continuously between an upper limit of v_{g1} and a lower limit of v_{g2} using the M control signals Control 1 to Control M. The FIFO buffer operates as follows. An incoming packet enters the DL1 at velocity v_{g1}. Once the entire packet has entered DL1, the group velocity in DL1 is changed adiabatically from v_{g1} to v_{g2}. The first of the two insets in Figure 17.7 shows this reduction in group velocity with time. The group velocity is v_{g2} in all delay lines between the most recently accepted packet and the packet closest to the output. When it is required to read the packet closest to the output, the group velocity in the delay line where the packet is located, is ramped back to v_{g1} as shown in the second inset.

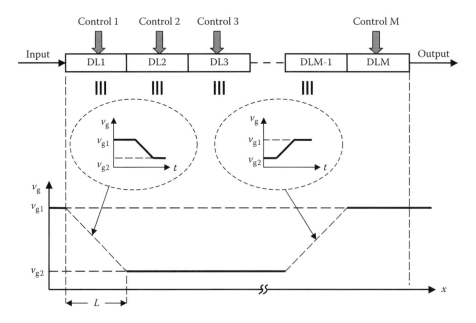

FIGURE 17.7 FIFO buffer using cascaded Class B delay lines. (From Tucker, R.S., Ku, P.-C., Chang-Hasnian, C.J., *J. Lightwave Technol.*, 23, 4046, 2005. With permission.)

A commonly stated misconception in the literature is that Class B delay lines circumvent the delay–throughput product limitation encountered in Class A delay lines. However, in Class B delay lines it is necessary to load an entire packet or segment of a packet into the delay line before the velocity is adiabatically reduced [18,19]. The implication of this fact in the FIFO buffer in Figure 17.7 is that while the packet in DL1 is being slowed, another packet cannot be allowed to enter the FIFO. Thus, a single Class B delay line or FIFO cannot continuously accept input data and, as a result, its effective information bandwidth or throughput is limited [19]. The information bandwidth of Class B delay lines becomes smaller as the group velocity on the delay line is reduced. As the velocity on a Class B delay line tends to zero, the information bandwidth and the capacity C both tend to zero. This could have serious implications in packet switching because packets in telecommunications systems are, by necessity, closely spaced and can arrive at any time. One solution to this problem of limited information bandwidth in Class B delay lines is to use the parallel architecture in Figure 17.3b, with a number of Class B FIFO buffers in parallel. In this arrangement, an incoming packet is directed to a FIFO buffer that can accept it. However, because multiple FIFO buffers are required, the physical size, or footprint, of this solution is large.

17.3.3 Physical Size

An important practical consideration in the design of any buffer is its physical size. As the number of ports on large routers grows into the 100s and the 1000s, it becomes increasingly important to ensure that each buffer is sufficiently small to enable the complete router to be constructed in a manageably small envelope. Therefore, it is important to consider the potential of miniaturization of buffer technologies. In this section we consider the physical size of slow light buffers.

The key parameter that we use to characterize the size of a buffer is the physical size of a stored bit of data. The total length L of a buffer is the size of a stored bit multiplied by the number of bits (i.e., the capacity of the buffer). Slow light appears to be an attractive technology for buffer applications where miniaturization is important. This is because the lower group velocity in a (Class A) slow light waveguide results in the physical size of the individual bits of data being smaller than in regular waveguides. Therefore, the length of the waveguide for a given storage capacity is less that of a regular waveguide.

As shown previously in Figure 17.4, the physical length of a bit as a function of position x in a waveguide decreases as the input pulses experience a sudden transition from a regular waveguide, to a Class A slow light waveguide with a reduced group velocity [19]. The length of an individual bit of data is equal to the bit period times the group velocity. Consequently, the physical length of the bits is reduced from L_{in} on the input line to the smaller length L_b on the slow light waveguide. Therefore the length of the delay line is

$$L = L_b C = L_b TB \qquad (17.2)$$

and the bit length L_b is given by

$$L_b = \left(\frac{C}{L}\right)^{-1} \qquad (17.3)$$

where C/L is the delay–throughput product per unit length. Therefore the physical size of the stored bits in a Class A delay line is directly linked to the delay–throughput product.

For given delay T, and a given information bandwidth B, the length L of the delay line is minimized when the length L_b of each bit is minimized. In designing slow light delay lines, it is useful, therefore, to have an understanding of the minimum achievable L_b. This enables a lower limit on the physical length of the delay line to be determined. The minimum bit length can be estimated with reference to Figure 17.8, which shows the effective refractive index profile against optical frequency ω of an ideal slow light medium. In this ideal characteristic, the effective refractive index n of the waveguide is a linear function of the frequency. It has a minimum value n_{min} at optical

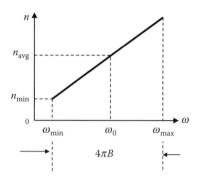

FIGURE 17.8 Effective refractive index versus optical frequency.

frequency ω_{min}, a maximum value at optical frequency ω_{max}, and the average effective refractive index across the signal band is n_{avg}. The center frequency ω_0 of the signal is aligned with the center of the region of linear slope.

The group velocity v_{g2} in the slow light waveguide can be written in terms of the free-space velocity c, the effective refractive index n of the waveguide, and the optical frequency ω:

$$v_{g2} = \frac{c}{n + \omega(dn/d\omega)} \qquad (17.4)$$

For dispersionless slow light propagation, the term $\omega(dn/d\omega)$ needs to be large and independent of the optical frequency across the signal passband. This is achieved using a straight-line effective refractive index as shown in Figure 17.8.

To minimize the length L_b of a stored bit, the optical radian frequency bandwidth of the data signal (shown as $4\pi B$ in Figure 17.8) needs to fully occupy the device bandwidth from ω_{min} to ω_{max} [19]. Therefore, from Figure 17.8,

$$\frac{dn}{d\omega} = \frac{n_{avg} - n_{min}}{2\pi B}. \qquad (17.5)$$

The fundamental lower limit on the bit size L_b is obtained when $n_{min} = 0$ [19]. Substituting $n_{min} = 0$ into Equations 17.4 and 17.5, gives $v_{g2} = (c2\pi B)/(\omega n_{avg}) = B\lambda_{avg}$, where λ_{avg} is the mean wavelength in the slow light medium. Therefore the length of a bit is $L_b = v_{g2}/B = \lambda_{avg}$. In other words, no matter which slow light technology is employed, one can never compress a bit of data to a dimension less than about one wavelength [19]. Miller [21] has derived a similar limitation using a more general approach. Miller has pointed out that the bit length could possibly be made smaller (or equivalently, the delay–throughput product per unit length could be made larger) by using materials with very large dielectric constants, such as metals. However, the loss associated with this kind of material would impose extremely several limitations on the capacity of the delay line (see Section 17.3.4).

Table 17.1 provides an indication of the size required for delay line buffers for OPS. Table 17.1 compares the physical length of fiber-based buffers with ideal slow light buffers for a 40 Tb/s optical router operating with a line rate of 40 Gbps and having 1000 ports (i.e., 1000 separate buffers located at the output ports of the optical cross connect, as shown in Figure 17.1a). Fiber delay lines, have a bit length L_b of 5 mm at 40 Gbps, and ideal slow light delay line with a bit length L_b of one wavelength, or about 1 μm. In Table 17.1, we have assumed a buffer size of 20 packets or about 200 kb per port. This corresponds to a buffer delay of 5 μs/port, or a total delay of 5 ms. The total capacity is 200 Mb on all of the 1000 ports. Table 17.1 shows the total length of the delay lines in all of the 1000 buffers, and the length per port. The total length for fiber delay line buffers is 1 mm, and 200 m for ideal slow light buffers.

TABLE 17.1
Storage Density Per Unit Length and Total Length of Fiber Delay Line and Ideal Slow Light Delay Line, at 40 Gbps

	Fiber	Ideal Slow Light Waveguide
Storage density	1 bit/5 mm	1 bit/μm
Total Length	100 km	200 m
Length per port	1 km	20 cm

In order to provide a compact realization of slow light buffers, it will generally be necessary to devise some kind of folded waveguide arrangement similar to the scheme illustrated in Figure 17.9. In this figure, the parallel waveguides are separated by approximately five wavelengths. If the waveguide is an ideal slow light waveguide, with a bit length L_b of one wavelength, the chip area occupied by each stored bit is $5\lambda^2$ or around $5\ \mu m^2$. Table 17.2 compares the bit area, storage density (per unit area), and the chip area of an ideal slow light waveguide with complementary metal oxide silicon embedded dynamic random access memory (CMOS eDRAM) memory, based on International Technology Semiconductor Roadmap for Semiconductors (ITRS) projected out to the year 2018 [22]. In Table 17.1, the total capacity of the buffer is taken as 200 Mb of data—the total data storage required in the router considered in the previous paragraph. The storage density in an ideal slow light device (i.e., the best that can be achieved using slow light) is $150\ Gb/m^2$, which is three orders of magnitude less than the storage density in CMOS DRAM ($150\ Tb/m^2$). The total chip area required for ideal slow light delay lines is $13\ m^2$, compared to $1.3\ mm^2$ for CMOS. Note that the storage density figures presented here are optimistic because they do not include the physical size of any crosspoints (see Figure 17.2), or any ancillary control circuitry.

Due to a number of nonideal factors such as signal dispersion, practical slow light devices will not perform as well as the ideal devices considered above. Figure 17.10 shows the buffer length L as a function of the bit rate for various slow light delay lines, with buffer capacities C of 100 and 10,000 [18]. The data in Figure 17.10 were calculated under the assumption that at all bit rates, the bandwidth of the slow light medium is adjusted to match the full optical bandwidth of the data (see Figure 17.8). Figure 17.10 includes three slow light delay line technologies: electromagnetically induced transparency (EIT), coupled resonator waveguide (CRW) delay lines (solid and broken lines), and ideal slow light delay lines that are dispersionless and have a minimum bit length of one wavelength.

At low bit rates, all of the buffer lengths are independent of bit rate. This is because the slope of the effective refractive index in Figure 17.8 increases as the bit rate decreases. Above about 500 Mbps,

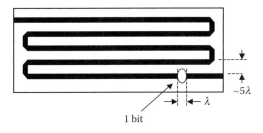

FIGURE 17.9 Compact delay line structure. (From Tucker, R.S., *J. Lightwave Technol.*, 24, 4655, 2006. With permission.)

TABLE 17.2
Bit Area and Storage Density Per Unit Area of Slow Light Buffer and CMOS eDRAM

	Delay Line	CMOS
Bit area	$\sim 5\,\mu m^2$	$\sim 0.005\,\mu m^2$
Storage density	$150\,Gb/m^2$	$150\,Tb/m^2$
Area required for 200 Mb	$13\,m^2$	$1.3\,mm^2$

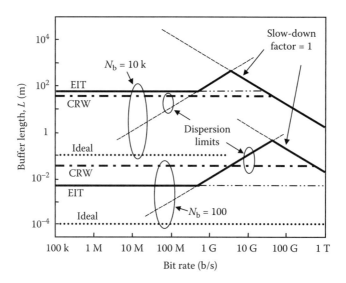

FIGURE 17.10 Buffer length versus bit rate for capacities of 100 b and 10 kb. (From Tucker, R.S., *J. Lightwave Technol.*, 24, 4655, 2006. With permission.)

the buffer length increases in EIT devices due to dispersion [18]. The solid lines with negative slopes in Figure 17.10 represent a slowdown factor of $S = 1$. On this line, the light is not slowed. Thus, when $S = 1$, there is no advantage afforded by slow light technologies over conventional waveguide delay lines such as fiber. From Figure 17.10, it can be seen that the maximum bit rate for $S > 1$ with buffers of capacity 10 kb is around 3 GHz for EIT delay lines and for CRW delay lines 20 GHz. For 100-bit-capacity delay lines, these bit rates increase to 50 and 700 Gbps, respectively.

17.3.4 WAVEGUIDE LOSSES AND ENERGY CONSUMPTION

In this section, we consider the influence of waveguide losses on the capacity of slow light and conventional waveguide delay lines and fiber delay lines. Waveguide loss is a critical and fundamental limitation on the capacity of delay lines, and in this section we show how the limitations caused by waveguide losses can be quantified. As pointed out in Section 17.3.3, limitations on delay line capacity are also caused by dispersion. However, dispersion can be compensated using a number of techniques, including linear dispersion compensation, nonlinear soliton effects [23], or pre- and postcompensation of the data signal [24]. Losses can be compensated, to a degree, by optical gain but there is an upper limit to how much gain can be used. Therefore, waveguide losses lead to a fundamental upper limit on the capacity of a delay line. This upper limit on delay line capacity,

TABLE 17.3
Attenuation Characteristics of Fiber and Slow Light Delay Lines

	Fiber	Low Loss Waveguide	Extremely Low Loss Waveguide
Intrinsic attenuation coefficient	0.2 dB/km	0.5 dB/cm	0.01 dB/cm
Absorption time	100 μs	400 ps	20 ns
Attenuation of 20-packet delay line	0.2 dB	16,500 dB	330 dB

together with the lower limit on the size of a stored bit is centrally important to the design of practical delay line buffers.

Waveguide losses are usually quantified in dB per unit length. An alternative measure of loss is the e^{-1} absorption time, which is the time taken for a pulse on the waveguide to decrease in magnitude by a factor of e^{-1}. The absorption time is a useful measure of loss in slow light devices because it is independent of the slowdown factor. The loss in dB per unit length of a slow light delay line increases in proportion to the slowdown factor [18,25]. Therefore, the attenuation A_{sl} in dB in a slow light waveguide with a slowdown factor S is

$$A_{sl} = SA_i \tag{17.6}$$

where A_i is the intrinsic attenuation for a waveguide with a slowdown factor of unity.

Table 17.3 shows typical values of absorption time for a fiber, a waveguide with an intrinsic attenuation of 0.5 dB/cm, and a waveguide with an intrinsic attenuation of 0.01 dB/cm. Usually, a waveguide loss of 0.5 dB/cm would be considered to be in the "low loss" regime, particularly if the waveguide is to be used in small optical integrated circuits where the maximum waveguide length is on the order of a few cm or less. But, as shown below, this level of loss is unacceptably high for packet buffers in OPS. Data for a waveguide with an intrinsic attenuation of 0.01 dB/cm are included in Table 17.4 for comparison with the 0.5-dB/cm waveguide.

Table 17.3 also shows the total attenuation in each of the three waveguides for delay line with a total capacity of 20 packets, or about 200 kb, that is, a delay of about 5 μs. The total attenuation in a fiber is only 0.2 dB, but for a waveguide delay line (either a slow light delay line or a conventional delay line) with an intrinsic 0.5-dB/cm attenuation, the total attenuation reaches a staggering 16,500 dB!

TABLE 17.4
Comparison of Buffer Technologies

	Intrinsic Attenuation	SNR (dB)	Capacity			
			5 IP Packets (50 kb)		100 IP Packets (1 Mb)	
			E_{bit}	Power	E_{bit}	Power
Slow light	0.05 dB/cm	30	1.2 pJ	50 mW	—	—
Slow light	0.5 dB/cm	30	—	—	—	—
Fiber	0.2 dB/km	30	1.3×10^{-2} fJ	5 mW	6.9 fJ	260 nW
Resonator-based RAM			$Q = 10^9$, ER = 80 dB		—	—
CMOS			0.2 pJ	8 mW	0.2 pJ	8 mW

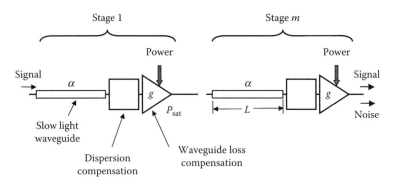

FIGURE 17.11 Optically amplified delay line.

Clearly, exceptionally low-loss waveguides would be required to store these many data packets. For example, if the intrinsic attenuation coefficient is reduced to 0.01 dB/cm, the absorption time increases to 20 ns and the total attenuation decreases to 330 dB. This attenuation is still large, but may be acceptable in delay lines incorporating optical gain.

Optical amplification is a convenient way to overcome losses in delay line buffers. Figure 17.11 shows a schematic layout of a large delay line buffer including optical gain to overcome losses [18,26]. In Figure 17.11, a number of stages of optical gain are placed along the length of the delay line in much the same way that optical amplifiers are placed along the length of a long-distance optical fiber transmission system. Like optical transmission systems, it is feasible to include dispersion compensation stages with each stage of gain to reduce the adverse effects of dispersion in the delay line.

A detailed analysis of the amplified delay line in Figure 17.11 is presented in Ref. [18] for a system using semiconductor optical amplifier (SOA) gain blocks and assuming ideal (lossless) dispersion compensators that remove all residual dispersion introduced by the delay lines. Key considerations in this analysis are the build up of spontaneous noise in the amplifier chain, and the influence of this accumulated spontaneous noise on the signal-to-noise ratio (SNR) at the output of the delay line. Also of importance is the gain saturation in the gain stages of the delay line caused by the signal and accumulated spontaneous noise. The maximum capacity C_{max} (in bits) of lossy delay lines with SOA loss compensation, calculated using this analysis [18], is presented in Figure 17.12. The saturation power on the horizontal axis of Figure 17.12 is the saturation power of the loss-compensating amplifiers in Figure 17.11. The bit rate is 40 Gbps, and we have assumed that the SNR at the output of the buffer is 20 dB. Indicative ranges of capacities required for delay line buffers and synchronizers are shown on the vertical (capacity) axis.

The upper curve in Figure 17.12 is for fiber buffer including fixed-delay fiber delay lines and crosspoint switches using the architecture shown in Figure 17.2c. In this ideal calculation, we ignore the losses in the crosspoints. As a result of the low loss in the optical fibers, the fiber delay line buffer has a reasonably large capacity. For example, if the saturation power of the amplifiers is 10 mW, the fiber delay line buffer has a capacity of 10 Mb, which is sufficient for buffering in OPS. The main disadvantage of this solution, however, is its very large physical size. The three center curves in Figure 17.12 give the maximum capacity for a variety of different waveguide losses.

The maximum achievable capacity for a delay line with an attenuation of 0.01 dB/cm and an SOA saturation power of 10 mW is about 40 Kb, which may be just sufficient for buffering OPS, depending on the buffer size required. For an attenuation of 0.5 dB/cm, the maximum capacity is less than 1 kb. This is much too small for buffering in OPS, and is also too small for use in a synchronizer (see Figure 17.1a). The conclusion from this comparison is that the attenuation coefficient of any waveguide delay line (slow light or nonslow light) needs to be 0.01 dB/cm or less if the delay line

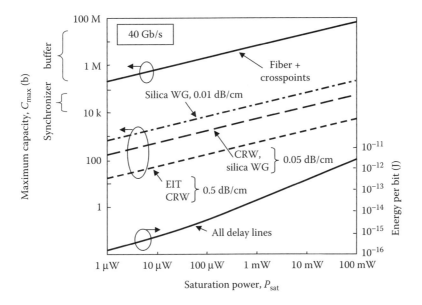

FIGURE 17.12 Maximum buffer capacity and energy per bit versus amplifier saturation power. (From Tucker, R.S., *J. Lightwave Technol.*, 24, 4655, 2006. With permission.)

is to be useful as a buffer in an optical packet switch, and 0.1 dB/cm or less if the delay line is to be used as a synchronizer in an optical packet switch.

The bottom curve in Figure 17.12 is the energy per bit E_{bit}. The scale for E_{bit} is on the right-hand axis. The energy E_{bit} has been calculated by dividing the total pump power into the amplifiers and dividing this by the bit rate [27]. The E_{bit} curve in Figure 17.12 applies to all delay lines, of all losses. For the examples considered in the previous paragraph, where saturation power of the SOAs was 10 mW, the write/store/read energy per bit is 0.5 pW.

Figure 17.13 shows the length L_b of each stored bit against the capacity N_b for the amplified delay line in Figure 17.11, at a bit rate of 40 Gbps. This figure incorporates limitations caused by

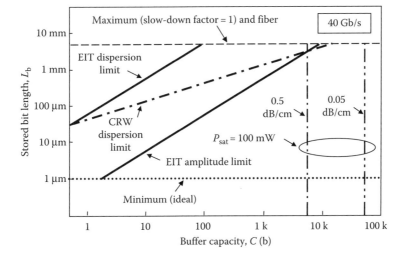

FIGURE 17.13 Stored bit length versus buffer capacity. (From Tucker, R.S., *J. Lightwave Technol.*, 24, 4655, 2006. With permission.)

dispersion (i.e., if no dispersion compensation is included in Figure 17.11) and also incorporates limitations caused by waveguide losses. In Figure 17.13, the calculations have been performed for an amplifier saturation power of 100 mW. The maximum size of a stored bit (broken line) is the length of a 40 Gbps bit in a regular (nonslow light) waveguide (approximately 5 mm). As shown earlier, the minimum L_b that can be achieved with any slow light delay line is one optical wavelength (approximately 1 μm). This lower limit on bit length is indicated in Figure 17.13 with a dotted line. The allowed region of operation of a slow light delay line lies above the ideal minimum length, below the least of the EIT or CRW dispersion limits and the EIT amplitude limit, below the slowdown factor = 1 line, and to the left of the attenuation limits.

17.3.5 RESONATOR BUFFERS

A possible alternative to delay line buffers is an array of high-Q optical resonators that trap the energy of individual data pulses. Optical resonators with Q values in the range 10^5 to 10^9 have been reported in a variety of structures, including photonic crystals (PC) and crystalline whispering gallery resonators. Resonators with Q values in this range can store energy for a tens of nanoseconds or more [28,29]. In order to trap optical energy from an incoming pulse and to release that energy on demand, it is necessary to couple the optical resonator to the input/output waveguide via an adjustable coupling region. The coupling region ensures that the Q of the cavity is low when the input pulse first arrives at the cavity and that the Q increases as energy builds up in the cavity. An expression for the time-dependent coupling is derived in Ref. [30], where it is shown that the maximum storage time in this kind of memory is ultimately limited by the extinction ratio of the adjustable coupling region rather than by the Q of the cavity.

Figure 17.14 shows a possible structure for an optical ring resonator RAM that could store a number of multiple-bit words or packets [30]. This memory is a true RAM because words can be retrieved in random order. A packet to be written in the RAM enters from the top left into a serial-to-parallel demultiplexer (i.e., a row decoder) comprising a cascade of delay lines of length equal to a bit period τ_b, with a crosspoint between each delay line. To store the packet, once all the bits in a data word have entered the cascaded τ_b delay lines, the crosspoints in the serial-to-parallel demultiplexer are simultaneously switched and the parallel bits enter the horizontal waveguides in Figure 17.14 (i.e., the bit lines).

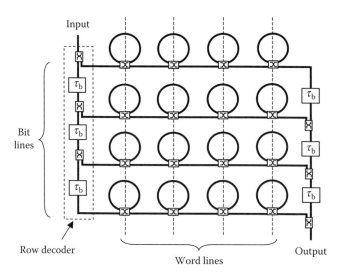

FIGURE 17.14 Optical resonator RAM. (From Tucker, R.S. and Riding, J.L., *J. Lightwave Technol.*, 26, 320, 2008. With permission.)

Similarly, when data are read out from the RAM, bits from the ring resonators are converted from parallel to serial in a parallel-to-serial converter comprising another cascade of delay lines of length equal to one bit period, with a crosspoint between each delay line. The word lines in Figure 17.14 control the variable couplers in the cells. If the variable couplers are electro-optical directional couplers, these word lines are electrical wires.

Like delay line buffers, ring resonator RAM buffers suffer from optical losses. With a resonator Q of 10^7 and a coupling region with high extinction ratio, the retention time of a stored pulse is approximately 50 ns. By way of comparison, the retention time needed to store 20 IP packets at 40 Gbps is around 5 μs, or two orders of magnitude larger than this. In order to achieve a retention time of 5 μs, the resonator Q would need to be around 10^9 and the extinction ratio of the adjustable coupling region would need to be around 80 dB. Cavity Qs of this order may be achievable, but it would be extraordinarily challenging to achieve an extinction ratio of 80 dB in the adjustable coupling region. Therefore, as pointed out earlier, it appears that resonator buffers will not be suitable for packet switching. One solution to the retention time problem might be to incorporate gain in the resonator, but this would be difficult to achieve without experiencing lasing action. The storage density in ring resonator RAM is limited by the size of the resonators and the crosspoint switches. If the diameter of the ring resonator is 100 μm and the resonators are spaced by 50 μm, the storage density is on the order of 40 Mb/m^2. This is more than three orders of magnitude less than ideal slow light delay lines and more than six orders of magnitude less than CMOS RAM.

17.3.6 Electronic Buffers

Projections for the year 2018, based on the 2006 International Technology Roadmap for Semiconductors [22] indicate that embedded DRAM (eDRAM) will achieve storage capacitances of around 1 fF, read/write energies of 1.6×10^{-16} J/bit for each cell, retention times of 64 ms, and read/write cycle times of 200 ps. The effective cell pitch is approximately 80 μm, and with an array area efficiency of 60% [22], the projected storage density of eDRAM is approximately 150 Tb/m^2 (see Table 17.2). This is almost five orders of magnitude larger than in an EIT or CRS delay line with a capacity of 100 b, in which the bit length is around 100 μm and the storage density is around 2 Gb/m^2. It is clear that on the basis of chip area, optical delay lines are not competitive with electronic buffers, except for very small buffer sizes.

17.3.7 Comparison of Buffer Technologies

Table 17.4 compares the energy per bit and power dissipation of optically amplified delay line buffers, resonator-based optical RAM cells, and CMOS eDRAM for buffers with capacities of 5 IP packets (50 kb) and 100 IP packets (1 Mb), at a bit rate of 40 Gbps. The delay line buffers are subdivided into slow light delay lines, with intrinsic attenuation coefficients of 0.05 and 0.5 dB/cm, and fiber delay lines, with an attenuation of 0.2 dB/km [18]. Note that the energy and power figures for slow light delay lines also apply to single-chip nonslow light delay lines such as silicon planar waveguides [18]. Except where noted, the parameters used in Table 17.4 are the same as used for Figure 17.12. Table 17.4 uses output SNRs of 30 dB (compared with 20 dB in Figure 17.12). This is representative of the higher value of SNR that will be required in practical applications where multiple optical devices are cascaded. Each of the entries in Table 17.4 applies to a single buffer of the specified capacity. A router with n ports will require n of these buffers. In all slow light delay lines and the fiber delay lines in Table 17.4, single-wavelength operation is assumed. As pointed out earlier, some efficiency in device area could be achieved through wavelength division multiplexing. However, the energy per bit and total power dissipation cannot be improved by employing wavelength division multiplexing.

The lowest power consumption in Table 17.4 is associated with fiber delay lines. However, because of their physical bulk, fiber delay lines are generally not considered to be practical. The next lowest power consumption is in CMOS buffers and the largest consumption is in slow light delay lines and other planar waveguide delay lines, which consume the same amount of power as slow light delay lines with the same waveguide losses and the same buffer capacity. The blank entries in Table 17.4 indicate where the slow light buffers cannot be realized in practice because the power dissipation is impractically large or because bandwidth limitations are not met.

17.4 CONCLUSIONS

The energy bottleneck in large-scale network routers has become a major issue in the design of future generations of network equipment. At a superficial level, OPS appears to offer an attractive alternative to electronic packet switching. Optical switching is a potentially low power technology, and the concept of OPS is consistent with the dream of an all-optical network. However, the lack of a practical optical buffer technology is a major impediment to the development of OPS.

In this chapter we have shown that even if the capacity of optical buffers is not greater than around 20 packets per port, it will prove very difficult to build practical optical buffers using slow light techniques or conventional waveguide technology. Unfortunately, slow light buffers are bulky, consume a significant amount of energy, and pose significant technical challenges in terms of dispersion compensation. On the basis of power consumption alone, slow light technologies and other optical buffering techniques such as those based on high-Q microresonators are unlikely to be able to compete with electronic buffer technologies in OPS. However, optical buffering techniques may find limited application in synchronizers for OPS, where the maximum delay required is at least an order of magnitude smaller than in buffers.

Future developments in networking protocols may enable the buffering requirements in routers to be further reduced. In fact, there have been some suggestions that with changes to the transmission control protocol (TCP), it might be possible to eliminate the buffers altogether [31].

REFERENCES

1. V. W. S. Chan, Guest editorial: Optical communications and networking series, *IEEE J. Sel. Areas Commun.*, 23, 1441–1443, 2005.
2. D. T. Neilson, Photonics for switching and routing, *IEEE J. Sel. Top. Quantum Electron.*, 12, 669–677, 2006.
3. D. Blumenthal, P. Prucnal, and J. Sauer, Photonic packet switches: Architectures and experimental implementations, *Proc. IEEE*, 82, 1650–1667, 1994.
4. J. Spring and R. S. Tucker, Photonic 2 × 2 packet switch with input buffers, *Electron. Lett.*, 29, 284, 1993.
5. S. J. Yoo, Optical packet and burst switching technologies for the future photonic internet, *J. Lightwave Technol.*, 24, 4468–4492, 2006.
6. R. S. Tucker and W. Zhong, Photonic packet switching: An overview, *IEICE Trans.*, E82-B, 254–264, 1999.
7. D. Chiaroni et al., Physical and logical validation of a network based on all-optical packet switching systems, *J. Lightwave Technol.*, 16, 2255–2264, 1998.
8. A. Okada, T. Sakamoto, Y. Sakai, K. Noguchi, and M. Matsuoka, All-optical packet routing by an out-of-band optical label and wavelength conversion in a full-mesh network based on a cyclic-frequency, Presented at the Optical Fiber Communications Conference (OFC 2001), Anaheim, California, 2001.
9. C. Guillemot et al., Transparent optical packet switching: The European ACTS KEOPS project approach, *J. Lightwave Technol.*, 16, 2117, 1998.
10. D. K. Hunter, M. C. Chia, and I. Andonovic, Buffering in optical packet switches, *J. Lightwave Technol.*, 16, 2081–2094, 1998.
11. R. W. Boyd, M. S. Bigelow, N. Lepeshkin, A. Schweinsberg, and P. Zerom, Fundamentals and applications of slow light in room temperature solids, presented at *IEEE Lasers and Electro-Optics Society, Annual Meeting, LEOS 2004*, 2004.
12. C. J. Chang-Hasnain and S. L. Chuang, Slow and fast light in semiconductor quantum-well and quantum-dot devices, *J. Lightwave Technol.*, 24, 4642–4654, 2006.
13. J. Baliga, Hinton, K., and Tucker R. S., Energy consumption of the Internet, presented at *COIN-ACOFT*, Melbourne Australia, 2007.
14. S. Yao, B. Mukherjee, and S. Dixit, Advances in photonic packet switching: An overview, *IEEE Commun. Mag.*, 38, 84–94, 2000.

15. G. Appenzeller, I. Keslassy, and N. McKeown, Sizing router buffers, presented at *SIGCOMM'04*, Portland, Oregon, 2004.
16. M. Enachescu, Ganjali, Y., Goel, A., and McKeowan, N., Part III: Routers with very small buffers, *ACM/SIGCOMM Comput. Commun. Rev.*, 35, 83–89, 2005.
17. N. Beheshti, Y. Ganjali, R. Rajaduray, D. Blumenthal, and N. McKeown, Buffer sizing in all-optical packet switches, presented at *Optical Fiber Communications Conference OFC/NFOEC 2006*, Anaheim, CA, 2006.
18. R. S. Tucker, The role of optics and electronics in high-capacity routers, *J. Lightwave Technol.*, 24, 4655–4673, 2006.
19. R. S. Tucker, P.-C. Ku, and C. J. Chang-Hasnain, Slow-light optical buffers: Capabilities and fundamental limitations, *J. Lightwave Technol.*, 23, 4046–4066, 2005.
20. J. Khurgin, Adiabatically tunable optical delay lines and their performance limitations, *Opt. Lett.*, 30, 2778–2780, 2005.
21. D. A. B. Miller, Fundamental limit to linear one-dimensional slow light structures, *Phys. Rev. Lett.*, 99, 203903, 2007.
22. *International Technology Roadmap for Semiconductors, 2006 Edition* (2006). [Online]. Available at: http://public.itrs.net/.
23. J. T. Mok, E. Tsoy, I. C. M. Littler, C. M. de Sterke, and B. J. Eggleton, Slow gap soliton propagation excited by microchip Q-switched pulses, presented at *IEEE Lasers and Electro-Optics Society Annual Meeting, 2005, LEOS 2005*, Sydney, Australia, 2005.
24. Q. Yu and A. Shanbhag, Electronic data processing for error and dispersion compensation, *J. Lightwave Technol.*, 24, 4514–4525, 2006.
25. S. Dubovitsky and W. H. Steier, Relationship between the slowing and loss in optical delay lines, *IEEE J. Quantum Electron.*, 42, 372–377, 2006.
26. J. Khurgin, Power dissipation in slow light devices—comparative analysis, *Opt. Lett.*, 32, 163–165, 2006.
27. R. S. Tucker, Petabit-per-second routers: Optical vs. electronic implementations, presented at *Optical Fiber Communications (OFC'2006)*, Anaheim, CA, 2006.
28. J. Guo, M. J. Shaw, G. A. Vawter, P. Esherick, G. R. Handley, and C. Sullivan, High-Q integrated on-chip micro-ring resonator, presented at *IEEE Lasers and Electro-Optics Society Annual Meeting*, Puerto Rico, 2004.
29. K. Vahala, H. Roksari, T. Kippenberg, T. Carmon, and D. Armani, Nonlinear optics in ultra-high-Q micro-resonators on a silicon chip, presented at *Quantum Electronics Conference, 2005. International*, 2005.
30. R. S. Tucker and J. L. Riding, Optical ring resonator random-access memories, *J. Lightwave Technol.*, 26, 320–328, 2008.
31. G. Das, Tucker, R. S., Leckie, C., and K. Hinton, Paced TCP gives higher utilization with no buffers than with small buffers, in *33rd European Conference on Optical Communication 2007*. Berlin, 2007.

18 Application of Slow Light to Phased Array Radar Beam Steering

Zachary Dutton, Mark Bashkansky, and Michael Steiner

CONTENTS

18.1 Introduction 367
18.2 Radar System Background 369
18.3 Squinting in Phase Shifter Beam Forming 371
18.4 TTD Beam-Forming Requirements 373
      18.4.1 Delay Precision 374
      18.4.2 Amplitude Precision 375
      18.4.3 Bandwidth 376
      18.4.4 Other Considerations 377
18.5 Summary 378
Acknowledgments 379
References 379

18.1 INTRODUCTION

All early radars utilized single large antennas such as parabolic reflectors. Such radars operate by transmitting a focused beam in the direction of interest. Generally such antennas mechanically rotate in azimuth and also often in elevation in order to provide full beam coverage. Many of the older radars were relatively low in frequency and required a large antenna. As radars progressed, particularly to higher frequencies, it became viable and advantageous to consider antennas that do not rotate but are electronically steered. In electronically steered systems, the antenna can be steered slowly along the same direction or could have arbitrary search patterns allowing high priority targets to be revisited at different rates. The time on target can be increased relatively easily, and multiple simultaneous functions (such as surveillance, multiple-target tracking, illuminating targets, electronic counter-countermeasures, communication, etc.) are possible. In more advanced systems today, there can be simultaneous transmission as well as reception of multiple beams.

Phased array radars operate by transmitting and receiving microwave radiation through an array of antenna elements as shown in Figure 18.1. By phase shifting each element by a different and precise amount, a narrow beam is coherently steered in a particular direction without the need for mechanical scanning of the antenna. Because clutter suppression is related to the stability of the transmitter, the stability can limit the ability to detect many targets of interest. With targets of interest to the military becoming smaller, and because of size and weight requirements, new radars often use

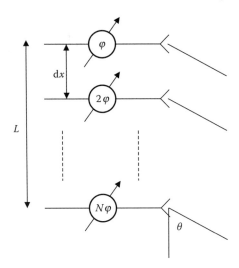

FIGURE 18.1 Conventional phased array radar. Antenna elements are shifted element-to-element by φ radians.

solid-state amplifiers that are part of the phased array transmit/receive (T/R) module itself. Moreover, the technology of solid-state amplifiers has been growing at an enormous pace, mainly due to the investment of the cell phone industry, which also requires miniature microwave amplifiers. The first major phased array system, the AEGIS SPY-1 was built in the 1970s. Over the years, solid-state amplifiers have greatly improved the dynamic range and reduced bulkiness, allowing improved clutter suppression and ease of deployment. Technological progress continues to bring improvements in weight, cost, and performance.

However, phased arrays run into a fundamental limit in regard to one particular performance metric: the range resolution. The limit comes about from an effect known as squinting, which arises anytime high bandwidth signals are used. The range resolution of a sensor utilizing pulses of a particular bandwidth B is $R = c/2B$, where c is the speed of light propagation in vacuum. Target classification and identification has become increasingly important for military applications in modern radar systems. Although many older surveillance and tracking radars were able to detect a target and track the location, detailed spatial information is required for classification. For a 1 GHz bandwidth, 15 cm resolution is achievable.

To be more specific, considering a one-dimensional array of antennae at positions x_j, the appropriate phase shifts to steer a beam into direction θ_0 are $\varphi_0^{(j)} = \omega_0 x_j \sin\theta_0/c$, where ω_0 is the (central) RF carrier frequency. (The most commonly used ones are S-band, $\omega_0 \sim (2\pi)$ 3 GHz, and X-band $\omega_0 \sim (2\pi)$ 10 GHz, so we will concentrate, in this chapter, on parameters associated with these bands.) For sufficiently large signal bandwidth B, applying phase shifts $\varphi_0^{(j)}$ leads to squinting. The proper, frequency-dependent shift required for beam steering is $\varphi^{(j)} = \omega x_j \sin\theta_0/c$, where $\omega = \omega_0 + \delta$ and $\delta \sim B$ represents a particular frequency component of the signal. By neglecting the frequency-dependent part of $\varphi^{(j)}$, $\delta x_j \sin\theta_0/c$, different frequencies get steered into different angles (a chromatic aberration) leading to beam loss and distortion.

Examination of $\varphi^{(j)}$ quickly reveals that one needs to apply phase shifts which vary linearly with frequency δ to steer all frequency components coherently into a single direction. However, this linear frequency phase shift is equivalent to a slow group velocity, as noted in the context of the various slow light methods described in this book. Thus, if one were to apply varying pulse delays $\tau^{(j)}$ at each element, for example via slow light, one automatically obtains the frequency-dependent phase shifts necessary for wideband beam steering. This is referred to as true-time delay (TTD) (Frigyes, 1995) in the context of phased array radar. Another, perhaps more physical, viewpoint is that steering is fundamentally achieved by delaying signals in different elements according to their transverse position, such that the plane wave is redirected into a particular direction. In this scenario, applying phase shifts $\varphi_0^{(j)} = \omega_0 \tau^{(j)}$ is simply an approximate method of doing this, valid when the variation in required phase shift across the spectrum of the signal $\sim Bx_j \sin \theta_0/c \ll 2\pi$. In order to controllably steer the beam using TTD, one must be able to quickly and precisely switch the signal delays in each element. The requirements for implementing TTD depend on a number of factors related to the radar design. Several basic concepts of radar design relevant to the specifications of TTD are discussed in this chapter.

TTD has been implemented with limited success via the use of time delay units (TDUs) directly in the RF domain. RF TDUs implemented electronically are often undesirable because of the long lengths of coaxial cable or strip line. These become very bulky and costly with sufficiently long delays. Because of these limitations, TDUs are implemented only at the subarray level in practice, whereby traditional phase shifting is applied to individual elements, and time delays are applied to groups of adjacent elements. One strategy to circumvent the practical limitations of RF TDUs is to use optical TTD, where the RF signal is converted to a modulation of an optical carrier, which can then be delayed. Optics intrinsically has a much smaller spatial scale (alleviating some weight and heat dissipation issues) and a much larger bandwidth. However, many efforts when implemented have encountered limitations. The most straightforward technique, using a series of binary switchable paths (Madamopoulos and Riza, 2000), suffers from loss (typically 1.0–1.5 dB per switch). This creates an inherent trade-off between loss and delay precision (for example, seven levels of switching would generate 128 possible delays, but 7–10 dB of loss). A more elegant technique, known as a fiber-optic prism (Esman et al., 1993), uses highly dispersive fiber to tune the delay through the fiber path by changing the wavelength. This technique, while promising, suffers from long latency, as the fibers are typically several hundred meters long and one cannot adjust the delays faster than the time for light propagation through these fibers ($\sim \mu s$). Additionally, the need for fast and accurate adjustment of the wavelength over a broad range (tens of nanometers) can add latency and possible systematic errors to the system. Finally, in an optical TTD system, it is, in principle, possible to use wavelength division multiplexing to simultaneously and independently steer each beam's signal at a different wavelength. But wavelength-dependent delays preclude this simple multiple beam architecture. Also worthy of note is a hybrid system (Riza et al., 2004) that combines switching and fiber-Bragg gratings to achieve delay tunability.

The slow light techniques discussed in this book share the important characteristic that the optical signal delays are quickly and easily controllable by adjustment of an auxiliary pump or control field. This makes many of them applicable to fast and agile beam steering in phased array radars. In order to assess the performance of various techniques (atomic systems, optical fiber techniques, solids, semiconductors, photonic bandgap materials, etc.) in this context, one must consider it with regard to providing appropriate delay ranges and precisions, bandwidths, amplitude and phase stability, dynamic range, multiple beam capabilities, and ease of hardware integration into a radar system.

In this chapter, we first consider the performance of beam steering by phase shifters for a variety of parameters to identify regimes in which TTD are necessary. We then analyze the beam forming with TTD in the presence of delay errors, amplitude errors, and finite bandwidths and identify the necessary characteristics to obtain a certain baseline level of performance. Most of our results are presented for X-band parameters, however, we give corresponding numbers for an S-band system and the results are typically quite easy to generalize to any frequency. Using our analysis, one can assess how various slow light techniques would be for this very promising application.

18.2 RADAR SYSTEM BACKGROUND

There are fundamentally two types of radars, those that transmit a set of pulses and those that transmit and receive continuously. Most modern advanced radars are pulsed Doppler and operate by the first method, that is, by transmitting a pulse sequence as seen in Figure 18.2. In between the pulses

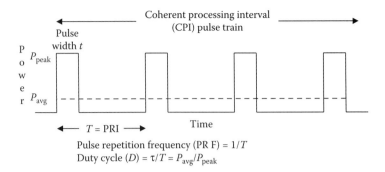

FIGURE 18.2 Single CPI RF pulse train.

the radar receiver is turned on and the data is digitally recorded. The data are passed to the radar signal processor in order to detect and track targets of interest. There are several parameters used in pulsed Doppler radars that follow from the basic pulse sequence: the pulse repetition frequency (PRF) is the rate at which the pulses are transmitted; the pulse repetition interval (PRI) T is the reciprocal of the PRF; the pulsewidth τ is the transmit time; and the duty cycle $D = \tau/T$. The peak power is P_{peak} and the average power is $P_{avg} = P_{peak} D$. There are several ranges of PRFs and duty cycles that are common, depending on the application. Pulsed Doppler radars are typically classified into one of the three ranges: high PRF >50 kHz, medium PRF $= 5$–50 kHz, and low PRF <5 kHz, with modern phased arrays usually in the medium-to-high PRF range. Duty cycles for pulsed Doppler are typically 0.05–0.5. The maximum range that can be detected in any single PRI is limited by the PRF. That is, when the true target range R is greater than the unambiguous range $R_u = 2T/c$, the target return will not occur during the listening period corresponding to the pulse that is transmitted. When the PRF is sufficiently high that the returns of targets of interest often fold into the next PRI, the radar is termed range ambiguous. There is a variety of techniques in signal processing to determine the range, even if the target folds into the next PRI. One common method is to transmit a second coherent processing interval (CPI) of pulses, but at a slightly different PRF. Because of the change in PRF, the target return will now be at a different range relative to the second transmit pulse. This new range relative to the second transmit pulse is called the ambiguous range. By correlating the detections given by the two PRFs, one can unwrap the ambiguous range and determine the true range. A similar process occurs in the Doppler domain. Targets can be ambiguous in Doppler and also require unwrapping for targets greater than the blind speed, $V_b = \lambda/(2T)$. By using multiple CPIs at different frequencies, the unambiguous Doppler can be determined in a manner similar to unwrapping the range for a range ambiguous radar. Sometimes both, range and Doppler are ambiguous, but this can also be handled by utilizing multiple CPIs that have both, differing frequencies and PRFs, and correlating both, range and Doppler using multiple CPIs. It should be noted that Doppler resolution improves with time-on-target (Skolnik, 1990).

The waveform transmitted is of two types. It may be a sinusoid or a more complex waveform called a pulse compression waveform. A sinusoid has a time–bandwidth product of near unity whereas pulse compression waveforms have time bandwidths greater than unity. The purpose of pulse compression is to increase the range resolution for a given peak power. Pulse compression is often implemented either via a linear frequency ramped signal or via a phase coded pulse, similar to that found in communication spread spectrum techniques such as phase-shift-keying (PSK). After a signal is transmitted and received, it is sampled and sent to the signal processor. The signal processor digitally correlates the phase-coded pulse with the received signal. This process, known as matched filtering, optimizes the signal-to-noise ratio. Additionally, since a signal is pulse compressed after it

Application of Slow Light to Phased Array Radar Beam Steering

is received, the dynamic range requirements before the analog-to-digital converter are only a function of the peak power and not the integrated power out of the matched filter. Hence, one effectively spreads energy in time and frequency to have the same range resolution as a shorter pulse, but with peak energy and dynamic range requirement significantly reduced (Skolnik, 1990).

Another important parameter is the latency of the system (i.e., the speed and agility with which the delays can be tuned) which will determine how quickly the beam can be steered. Depending on the different system configuration, it would be desirable to have latency times less than anywhere from 10 μs to 100 ns. This largely depends on the reciprocity of the time delays or phase shifters. Nonreciprocal phase shifters (where the value depends on whether transmit or receive is occurring) have to be reset between transmit and receive on each pulse, hence the requirements for latency for nonreciprocal phase shifters are very stringent. With reciprocal phase shifters, the values need to be changed only once every CPI rather than every pulse. Assuming a 20 kHz PRF and 100 pulses per CPI, then each pulse period T in the pulse train is 50 μs and 100 pulses have 5 ms. Hence a latency of 10 μs is a fractional loss of 10 μs/5 ms = 0.2%, which is not significant. On the other hand, if it is necessary to change the settings on each pulse, then the fractional loss would be 10/50 μs = 20%, which is very significant. Because slow light pulse delays are generally controlled via pump powers, the latency tends to be quite small and slow light performs very well in this regard. However, this must be considered in particular systems, such as stimulated Brillouin scattering in long (~km) fibers.

18.3 SQUINTING IN PHASE SHIFTER BEAM FORMING

As alluded to above, phase shifters introduce errors for high bandwidth signals due to chromatic aberration (squinting). To quantify this, we consider here a one-dimensional ($N = 128$) element, X-band (angular frequency $\omega_0 = (2\pi)$ 10 GHz carrier) system, with half-wavelength element spacing $x_j = j\lambda_0/2$. While radars are usually two-dimensional planar arrays, the squinting effects in each dimension are separable. Thus, if squinting results in 1 dB loss in one-dimension, then there would be 2 dB loss if one scanned the beam over the same angle range in each dimension (note that often the steering requirements may be much more severe in one dimension though, for example, for a ship-based radar scanning the horizon). We consider scanning angles up to ±60°. The array factor generated for a beam steered toward angle θ_0, as a function of angle θ and signal frequency ω, is the coherent sum of all the elements:

$$A(\theta, \omega) = \sum_j a_j \exp[i(\omega u - \omega_0 u_0) x_j / c]$$

where $u = \sin(\theta)$, $u_0 = \sin(\theta_0)$ and a_j is the element weighting. We will consider here the typical case of Taylor weighting (Trees, 2002) with sidelobe level -40 dB (a uniform weighting gives -15 dB sidelobes). This low sidelobe level is important for clutter suppression algorithms in the signal processing. Note that $\varphi_0^{(j)} = \omega_0 u_0 x_j / c$ represents the phase shift applied to a particular element j. To account for a wideband signal of bandwidth B we consider a uniform integration of the power over the relevant frequencies. This assumes that the signal replica used in the radar's matched filter (Skolnik, 1990, p. 10.3) is optimized to include the effects of the array, that is, is the expected received signal from a point target. We then get a beam profile:

$$|A_B(\theta)|^2 = \frac{1}{B} \int_{\omega_0 - (2\pi)B/2}^{\omega_0 + (2\pi)B/2} |A(\theta, \omega)|^2 d\omega.$$

The solid curve in Figure 18.3a shows the beam pattern for steering to 45° for a narrowband signal ($B = 0$). The 3 dB beamwidth is seen to be on the order of 1° (generally $\sim 100/N$ degrees) and the sidelobe levels are indeed -40 dB as designed. The dashed and dotted curves then show the results for $B = 300$ MHz and 1 GHz, respectively. As B increases, one sees a severe loss of the mainlobe level (MLL) and a spreading of the beam. In Figure 18.3b we plot the MLL versus the relative bandwidth of the signal B/ω_0 for several different steering angles. Because larger phase shifts are needed for larger steering angles, the errors introduced are correspondingly larger. Figure 18.3c shows the loss versus steering angle for $B = 300$ MHz. Taking an upper limit of 1 dB MLL loss over the entire $\pm 60°$ scan as a benchmark performance requirement, the squinting problem limits an X-band system to $B < 220$ MHz for this size array. Since the relative bandwidth determines the loss, an S-band system would be limited to $B < 70$ MHz.

To obtain better cross-range resolution, arrays use more elements to decrease beamwidth $\sim 100/N$. However, doing this makes the mainlobe more sensitive to the chromatic dispersion of a particular bandwidth signal B. In Figure 18.3d, we plot the critical point where the MLL loss becomes 1 dB for 60° beam pointing versus the number of elements N. There is clearly a direct $1/N$ scaling in the bandwidth. This plot shows the direct trade-off between cross- and range-resolution present in phased arrays, as the cross-range resolution is determined by the beam width $1/N$ while the range-resolution scales as $1/B$.

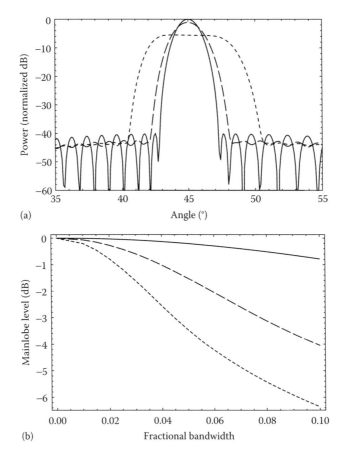

FIGURE 18.3 Squinting in phase shifter beam forming. $\omega_0 = 10$ GHz center frequency (X-band). (a) Array factors $|A_B(\theta)|^2$ for $B = 0$ (solid curve), 300 MHz (dashed), and 1 GHz (dotted). One clearly sees decrease in the MLL and spreading of the beam. (b) MLL (in dB) versus relative bandwidth B/ω_0 for steering angles $\theta_0 = 10°$ (solid), 30° (dashed), and 60° (dotted).

Application of Slow Light to Phased Array Radar Beam Steering

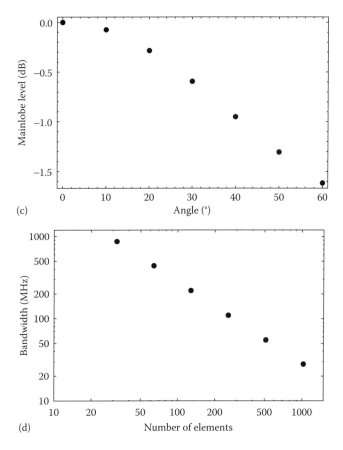

FIGURE 18.3 (continued) (c) MLL versus angle θ_0 for $B = 300$ MHz. (d) Bandwidth B at which MLL loss becomes 1 dB at $\theta_0 = 60°$ versus number of elements N.

18.4 TTD BEAM-FORMING REQUIREMENTS

TTD eliminates the squinting by providing phase shifts that vary linearly with frequency. In particular, the array factor becomes

$$A^{(\text{TTD})}(\theta, \omega) = \sum_j a_j \exp[i\omega(u - u_0)x_j/c]$$

Now the actual signal frequency ω (rather than the center frequency ω_0) determines the steering phase shift applied $\varphi^{(j)}(\omega) = \omega u_0 x_j/c$. The physical delay applied to achieve this is $\tau_d^{(j)} = \varphi^{(j)}(\omega)/\omega = u_0 x_j/c$. Calculation of the array factor $A_B^{(\text{TTD})}(\theta)$ reveals that the perfect beam forming is recovered even for large bandwidths B. The solid curve in Figure 18.4a is an example for $N = 128$ elements, X-band, with $B = 1$ GHz. In a TTD system, a large B can reduce the sidelobe variations, but leaves the mainlobe intact.

In order to scan the beam, one must be able to adjust the delays at each element over the required range. Note that only the relative delay affects the beam steering, so one would design the system such that the element at one end ($j = 1$) would not need any delay adjustment, while the far end element ($j = N$) would have the maximum required adjustable delay range $\tau_d^{(\max)} = u_0 L/c$, where $L = N\lambda_0/2$ is the total size of the array. In the extreme limit of 90° steering, $u_0 = 1$ and $\tau_d^{(\max)}$ is simply the time for light to propagate a distance L. For an X-band system with $N = 128$ elements,

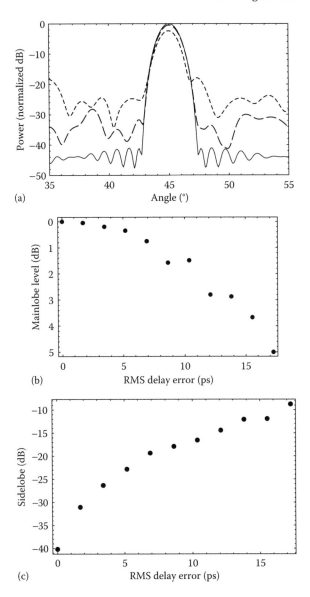

FIGURE 18.4 Effect of delay precision in beam forming. $\omega = (2\pi)$ 10 GHz center frequency with $B = (2\pi)$ 1 GHz bandwidth. (a) Array factor $|A_B^{(TTD)}(\theta)|^2$ of an $N = 128$ element array with perfect TTD (solid curve) and array factors with RMS fluctuations in the delays of $\tau^{(e)} = 3$ ps (dashed) and $\tau^{(e)} = 11$ ps (dotted). (b) MLL loss at $\theta_0 = 30°$ versus delay fluctuation $\tau^{(e)}$. (c) Maximum SLL at $\theta_0 = 30°$ versus delay fluctuation $\tau^{(e)}$.

and 60° steering, $\tau_d^{(max)} = 5.5$ ns, while an S-band system with the same number of elements, due to the larger λ_0, would require $\tau_d^{(max)} = 18.3$ ns.

18.4.1 Delay Precision

Unintended fluctuations of the individual element delays will degrade the coherent summation needed for beam forming and so one must be able to precisely control them. The dashed and dotted curves in Figure 18.4a plot array factors where each element is given a random error in delay, with a Gaussian distribution of width $\tau^{(e)}$. One sees a degradation of the MLL. It also elevates the sidelobe level and introduces some random fluctuations in the sidelobes, which is an important consideration when performing clutter suppression. Figure 18.4b plots the MLL loss as a function of $\tau^{(e)}$ for $N = 128$. We confirmed that this loss was independent of N and that 1 dB of loss occurs at $\tau^{(e-\text{crit})} = 8\,\text{ps}$. This time-scale is related to the time for light propagation across one element which is $\lambda_0/2c = 1/2f = 50\,\text{ps}$ (though the exact value $\tau^{(e-\text{crit})}$ tolerance depends on our choice of a critical loss, chosen to be 1 dB in this analysis). Note, however, that the relative delay control required does depend on N. For example, for $N = 128$, $\tau^{(e-\text{crit})}/\tau_d^{(\text{max})} = 0.0015$ (or 28 dB dynamic range) and every factor of two in N will add 3 dB of dynamic range requirement. An S-band radar would have $\tau^{(e-\text{crit})} = 27\,\text{ps}$ (though a correspondingly larger delay range, the dynamic range requirement would be the same).

Note from the beam patterns in Figure 18.4a that the sidelobe level rises dramatically with delay fluctuations. Figure 18.4c shows the maximum sidelobe level as a function of the time fluctuation level $\tau^{(e)}$. In particular, at $\tau^{(e-\text{crit})} = 8\,\text{ps}$, the sidelobe level has gone from the designed (Taylor weighted) level of $-40\,\text{dB}$ to approximately $-21\,\text{dB}$. This has a large effect on clutter suppression and so, for many applications the sidelobe requirement will demand more precise delay precision than the mainlobe requirement.

In addition to such random errors, one must also consider the angular precision with which one wishes to steer the beam, as this will drive the design of the mechanism to tune the delays. For example, to steer in steps of $1°$ for a usual $120°$ span would require 120 distinct delay values at each element. For a slow light system where delay is driven by a pump field, the pump must then be able to be adjusted to any of 120 distinct powers.

In traditional phased array radars, a large amount of calibration and correction is done once the system is constructed, to correct errors in element spacing, element amplitudes, etc. One can use the same strategy in a TTD system so that the repeatable, systematic errors in particular delays of particular elements can be corrected in a fielded system. It is only random shot-to-shot fluctuations which must be kept below this critical precision $\tau^{(e-\text{crit})}$.

18.4.2 Amplitude Precision

The other important characteristic is the precision with which the amplitude is preserved by the TTD process. In slow light based on electromagnetically induced transparency (EIT), decoherence will lead to signal attenuation which is related to the delay (Field et al., 1996). Similarly, in fiber-based gain processes, such as stimulated Brillouin scattering, there will be a delay-dependent gain, though there are methods to mitigate this problem (Zhu and Gauthier, 2006). In a TTD radar system, one will need to compensate for these delay-dependent amplitude effects with variable attenuators and/or gain processes (again as is done in calibrating a traditional phased array). However, there will be some residual amplitude errors that will degrade the beam forming. Figure 18.5a compares the array factor obtained with a perfect TTD and one with relative amplitude errors of $a^{(e)} = 0.2$. For this calculation, each amplitude a_j is multiplied by a factor $(1 + \alpha)$, where α is chosen from a Gaussian distribution of width $a^{(e)}$. We see that the amplitude errors have no effect on the mainlobe, as the elements are all still perfectly phased and have symmetric positive and negative amplitude errors. However, the amplitude errors clearly inhibit the sidelobe suppression being performed by the Taylor weighting. Figure 18.5b plots the largest sidelobe level versus $a^{(e)}$. We lose about 3 dB of suppression for $a^{(e)} = 0.05$ (or 13 dB dynamic range). This is a remarkably weak sensitivity compared to the much more stringent delay precision requirements. In addition to amplitude fluctuations, one can also have random phase fluctuations. However, these are mathematically equivalent to the delay fluctuations using the corresponding phase error $\varphi^{(e)} = \omega_0 \tau^{(e)}$. Using this, we find that the critical phase error level for 1 dB loss is $\varphi^{(e)} = 0.08$.

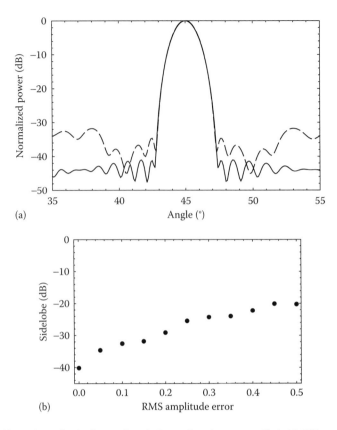

FIGURE 18.5 Effect of amplitude fluctuations in beam forming. $\omega_0 = (2\pi)$ 10 GHz center frequency with $B = (2\pi)$ 1 GHz bandwidth. (a) Array factor of an $N = 128$ element array, steered to $\theta_0 = 30°$, with perfect TTD (solid curve) and with amplitude fluctuations $a^{(e)} = 0.2$ (dashed). (b) Maximum SLL versus amplitude fluctuation $a^{(e)}$.

18.4.3 Bandwidth

One important consideration in applying slow light methods to beam steering is that, because they rely on introduction of dispersive features in the index of refraction, they inevitably (by Kramers–Kronig relations) have a finite bandwidth. Generally speaking, the leading order bandwidth effect is due to parabolic frequency dependence of the amplitude gain or loss that leads to pulse broadening (Boyd et al., 2005). For certain systems, such as EIT in atomic systems, this bandwidth can be quite small ($\sim$1 MHz) for reasonable pump powers. It is partly for this reason that slow light in semiconducting materials and optical fibers has been recently pursued. The bandwidth requirements for radar application are quite clear, since it is signal bandwidth restriction from squinting (and resulting limitation on range resolution) that first motivated TTD.

A typical 127-bit pulse compression sequence with a 1 ns bit length is shown in Figure 18.6a. As mentioned in the background discussion earlier, these sequences have a time–bandwidth product much greater than unity. After being subject to a finite bandwidth of 1 GHz due to slow light delay mechanism, higher frequency components will be attenuated, resulting in a distorted sequence as shown in Figure 18.6b.

Simulations of the beam forming taking into account a finite bandwidth, done by weighting different frequencies according to some attenuation or gain law, show that this has no effect on the array factor other than to attenuate (or apply gain) to the entire pattern. Thus, one can estimate the MLL loss by simply calculating the transmission of the slow light system for a bandwidth of interest.

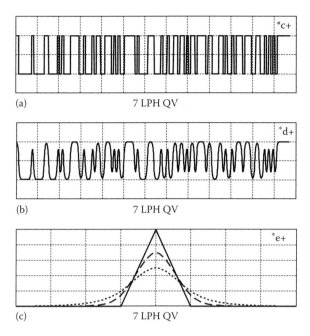

FIGURE 18.6 Bandwidth limitations on radar pulse sequences. (a) Typical compression sequence: 7-bit maximal length sequence with 1 ns bit length, (b) effect of slow light finite bandwidth of 1 GHz on the sequence, and (c) effect of finite bandwidth on the range resolution given by cross-correlation of compression sequences. Solid curve: autocorrelation of perfectly transmitted sequence (a). Dashed curve: cross-correlation of perfect (a) and 1 GHz bandwidth-limited (b) sequences. Dotted curve: cross correlation of perfect (a) and filtered 0.5 GHz bandwidth-limited sequences.

The bandwidth distortion will have a much more direct effect when one considers the signal processing used to obtain range information. Ultimately, the range resolution will be restricted to $R = c/2B$, where B is the bandwidth of the slow light mechanism. We note that, just as known amplitude and phase errors can be calibrated and corrected, known pulse distortion or dispersion effects can sometimes be compensated. However, one will always be limited if a particular frequency component is attenuated significantly with respect to the system noise. This is illustrated in Figure 18.6c where the solid curve shows autocorrelation between two perfectly transmitted sequences, the dashed and dotted curves show cross-correlation between perfect and filtered sequences with bandwidth of 1 and 0.5 GHz, respectively. The dashed curve is the one we wish to use in practice. For these calculations, the bit length is 1 ns and the bandwidth filter applied (B) is a Lorentzian with linewidth of either 1 or 0.5 GHz. This corresponds to the range resolution of approximately $c/2B = 15$ cm (1 GHz) or 30 cm (0.5 GHz).

18.4.4 OTHER CONSIDERATIONS

In addition to the aforementioned parameter requirements for the beam forming, there are several other features that would be desirable in a TTD radar system. First, the overall attenuation loss of the slow light mechanism itself will be an important factor, just as it is for any optical TTD technique, such as the simple switching scheme (Madamopoulos and Riza, 2000) which suffers 1.0–1.5 dB loss per switch. Similarly, the conversion efficiency from optical to RF carriers and vice versa will be an important part of the system, and will be different for different slow light techniques (depending on the wavelength of operation). Both these effects will enter in the total loss budget, noise figure, and dynamic range of the system. The dynamic range required depends highly on the application and the requirement is determined by both, the level of the minimum detectable signal and the strongest interference or clutter expected. If, for example, one desires to detect a small cross-section target −10 dB at a maximum range where the clutter return is expected to be 60 dB on a per pulse basis, then the dynamic range would need to be greater than 70 dB. There are several means by which dynamic range can be increased beyond the dynamic range limitations of the TTD or other analog hardware: (1) Dynamic range limitations at the T/R element level can be increased when combining the signal coherently over many elements; (2) Pulse compression codes are often used that spread the transmit energy over many range cells. After digitizing the signal, the received signal is digitally correlated or "match" filtered against a time-reversed replica of the pulse compression signal; (3) The same signal is transmitted numerous times forming a CPI. Generally speaking, an overall system dynamic range of 110 dB is typical for many applications. If one has a 30 dB integration gain via 1000 elements, a 100 length pulse compression code providing 20 dB pulse compression gain, and 100 pulses per CPI providing 20 dB integration gain, then one requires only 110 − 70 = 40 dB dynamic range at the element any level, although losses will increase this.

Secondly, large radar systems have very large usage demands and it is highly desirable that they can switch many beams (∼100) simultaneously and independently. In phase-shifting systems, this can be accomplished by using several carrier frequency offsets, causing corresponding offsets in the beam positions. This is often implemented by transmitting a succession of pulses at different frequencies, etc., so that the overall pulse is made up of several subpulses. Upon receive, channelized receivers are utilized to process the different frequencies. This has substantial advantages in that the radar search time can be reduced by a factor of the number of subpulses, as long as the energy available is more than sufficient to spread across the subpulses. If energy is not available, then the advantages are less clear. In optical systems, this could be accomplished, for example, by wavelength division multiplexing in fibers.

Finally, the practical considerations of weight, power consumption, cost, and hardware integration will be extremely important. Large radar systems today have ∼10, 000 elements and so the scalability of the slow light methods to a large number of elements will be a stringent requirement. Optical TTD systems, due to the smaller wavelength, are attractive in terms of weight and space. However, because slow light systems generally rely on a pump laser field, the power consumption requirements must be carefully considered.

18.5 SUMMARY

In conclusion, we have considered how tunable optical time delays via slow light could be used to improve the range resolution of phase array radar systems. By considering a basic model for phased array beam steering, we have calculated the bandwidths at which squinting is introduced into traditional phase shifter systems (using <1 dB loss of the mainlobe as a base requirement). We found that for an X-band system with $N = 128$ elements and ±60° steering capability, squinting attenuates the mainlobe for bandwidths $B > 220$ MHz, and that this cutoff scales as $1/N$. TTD can eliminate this problem, provided one can obtain delay ranges of 5.5 ns for $N = 128$, and delay precision with dynamic range 28 dB. The delay range requirement scales directly with N and inversely with carrier frequency. The sidelobe levels also rise dramatically with delay fluctuations and so must be taken into account in evaluating the slow light mechanisms performance for applications where low sidelobes are required. By contrast, the beam forming is much less sensitive to amplitude fluctuations of the elements. The mainlobe level is not affected by amplitude fluctuations and relative amplitude fluctuations of 13 dB end up raising the sidelobe levels to 3 dB. Finally, the bandwidth of the slow light mechanism will limit in a direct way the bandwidth of the pulses (especially high time–bandwidth pulse compression sequences) that can survive and thus provide the desired range resolution. In addition, the dynamic range, latency, and multiple beam capabilities must also be considered in evaluating the ability of a given slow light mechanism to be used in radar beam steering.

As various slow light techniques continue to advance in bandwidth, delay range, and amplitude control, application in phased array radar systems could become possible in the near future.

ACKNOWLEDGMENTS

This work was funded by the Office of Naval Research and the Defense Advanced Research Projects Agency.

REFERENCES

Boyd, R.W., Gauthier, D.J., Gaeta, A.L., and Willner, A.E., 2005, Maximum time delay achievable on propagation through a slow-light medium, *Phys. Rev. A*, 71: 023801.

Esman, R.D., Frankel, M.Y., Dexter, J.L., et al., 1993, Fiber-optic prism true time-delay antenna feed, *IEEE Photon. Technol. Lett*. 5(11): 1347–1349.

Field, J.E., Hahn, K.H., and Harris, S.E., 1991, Observation of electromagnetically induced transparency in collisionally broadened lead vapor, *Phys. Rev. Lett*. 67: 3733–3736.

Frigyes, I., 1995, Optically generated true-time delay in phased-array radars, *IEEE Trans. Microwave Theory Tech*. 43: 2378–2386.

Madamopoulos, N. and Riza, N.A., 2000, Demonstration of an all-digital 7-bit 33-channel photonic delay line for phased-array radars, *Appl. Opt*. 39: 4168–4181.

Riza, N.A., Arain, M.A., and Khan, S.A., 2004, Hybrid analog-digital variable fiber-optic delay line, *IEEE J. Lightwave Technol*. 22(2): 619–624.

Skolnik, M., 1990, *Radar Handbook*, 2nd edn., McGraw-Hill, New York.

Trees, H.V., 2002, *Optimum Array Processing, Part IV*, John Wiley & Sons, New York.

Zhu, Z. and Gauthier, D.J., 2006, Nearly transparent SBS slow light in an optical fiber, *Opt. Express* 14: 7238–7245.

Index

A

Antiresonant reflective optical waveguides (ARROW), 50
Average t-matrix approximation (ATA), 136
Averaged PMD model, fiber NB-OPA
    effective nonlinear parameter, 159
    gain spectra, 159–161
    white Gaussian noise source, 159

B

Bloch's theorem, 266

C

Cavities and slow light comparison
    bandwidth, 68–69
    coupled cavity, 69
    FOM implications
        linear refractive index, 70
        nonlinear refractive index, 69–70
    intensity enhancement
        gain and loss effects, 66
        intensity enhancement, 68
        Q-factor, 67
Coherent population oscillation (CPO), 13
Coherent potential approximation (CPA), 136
Coherent processing interval (CPI), 370
Coupled Fabry–Perot resonators, 310–311
Coupled resonator optical waveguides (CROWs)
    absorption
        couplingûsplit modes, 105
        lossy resonators, 106–107
        Q-factor resonators, 105–106
        transmission coefficient, 106
    Bloch function, 104
    complimentary structures, 83–84
    discrete spectrum
        advantage, 114–115
        angle-cleaved fibers, 113–114
        bichromatic light, 113
        conceptual transition, 110
        microsphere resonator, 112
        model transfer function, 111
        WGM resonator, 115
    dispersion relation, 80–81
    fundamental restriction and fabrication, 107
    light propagation, 103–104
    linearly polarized light, 103
    properties, 102
    resonator chains interference
        eigenfrequencies, 108–109
        subnatural structure, 108
        WGM resonators, 107
    resonator-stabilized oscillators, 109–110
    whispering gallery mode (WGM), 105
Coupled resonator structures (CRS)
    characteristic parameters, 313
    disadvantages of cascaded Bragg gratings, 309–310
    dispersion relation, 311–312
    group velocity, 312–313
    Moiré grating, 310–311
    periodic chain parameters, 311
    photonic SL structures, 310–311
    SL optical buffer length, 313–314
    slow down factor, 312
    strength of, 314
Coupling coefficient
    directional waveguide couplers, 142–143
    guided-wave structure, 141–142

D

Differential phase-shift-keying (DPSK)
    data-pattern dependence and mitigation
        bit-error-rate (BER) measurement, DB port, 333
        NRZ-DPSK intensity patterns, 332–333
        SBS gain peak detuning, 333–334
    signal delaying, 331–332
Dirac delta functions, 259
Discrete spectrum systems
    advantage, 114–115
    angle-cleaved fibers, 113–114
    bichromatic light, 113
    conceptual transition, 110
    microsphere resonator, 112
    model transfer function, 111
    transfer function, 110–111
    WGM resonator, 115
Disordered optical slow-wave structures
    calculation models
        coupling coefficients, 128–130
        diagonal terms, 130–131
    coupling matrix
        spectral characteristics, 126
        states $\rho(\omega)$ calculation, 127–128
    disordered structures
        ballistic transport regime, 133
        band-edge factors, 134
        fabrication technology, 135
        inter-resonator coupling coefficients, 134–135
        random distribution, 131–132
        tight-binding lattice, 133–134
    fields localization
        ATA and CPA, 136
        band-edge, 137–138
        Monte-Carlo iterations, 136–137
        quasi ballistic propagation, 139
        single-resonator fields, 136
        tight-binding approximation, 135
    formalism, 124–125
    spectrum solutions

general principles, 125
M and coupling coefficients, 125–126
tight-binding optical waveguide
bandsolver tools, 123
disordered light velocity, 122–123
engineered dispersion, 120
optical signal processing, 121–122
slow-wave dispersion relationship, 120–121
states density, 123–124
Dispersion management
delay-bandwidth product, 9–10
double Lorentzian susceptibility, 10
group velocity, 9–10
multiple pulse delays, 11
refractive index, 9–10
Dispersive fiber (DCF), 52
Dynamic tuning of coupled resonators, light-stopping system
dispersion suppression, 283–284
general conditions, 279–280
light spectrum tuning, 278–279
loss tuning
experimental progress, 286
theory, 284
microresonator requirements, 285
photonic crystal
dielectric rods square lattice, 281–282
finite difference-time-domain (FDTD) method, 282–283
microcavities, 286
silicon microring resonators, 285
tunable bandwidth filter
cavity–waveguide coupling, 281
coupled-cavity structure, 281–282
transmission matrix method, 280
tunable fano resonance, 280–281

E

Electromagnetically induced transparency (EIT), 101
Bose-Einstein condensation, 49
cryogenic temperature, 49–50
eigenvector, 4
Feynman diagrams, 5
group velocity, 6–7
Hamiltonian, 4
hollow-core fibers, 50–52
index of refraction, 5–6
optical absorption coefficient, 6
optical pulse delay, 7
Rabi frequency, 4
three-level system, 4–8
Electronic buffers, 363
Electronical-to-optic (E/O) converters, 347
Erbium-doped fiber amplifier (EDFA), 47
Excitation-induced dephasing (EID), 19–20

F

Finite difference-time-domain (FDTD) method, 282–283

Four-wave mixing (FWM), SWS
classical theory, 210
coupled resonator optical waveguides (CROW), 210
enhancement factors, 215
frequency conversion, 214
normalized bandwidth and conversion gain, 212
optical parameters, 213
optical waveguide, 211
power transfer, 210, 214
quasi-phase-matched (QPM) schemes, 214
wave mixing, 215

G

Gap solitons, slow light
dispersion effect, 223
dispersive broadening, 223
experiment, 229–233
Kerr nonlinearity, 223
Kramers–Kronig relations, 223
optical fiber Bragg gratings (FBGs), 224–229
power meter and oscilloscope, 229
spectral full-width at half maximum, 229
transmission spectrum, 230, 232
Group delay calculations
birefringent fibers, 169–171
isotropic fibers, 167–168
Group velocity dispersion (GVD)
properties, 84
TE polarization, 94–95

K

Kramers–Kronig relation
atomic resonances, 295
nonlinear photonic SL devices, 315

L

Light-stopping system, coupled resonators
dispersion suppression, 283–284
general conditions, 279–280
light spectrum tuning, 278–279
loss tuning
experimental progress, 286
theory, 284
microresonator requirements, 285
photonic crystals
dielectric rods square lattice, 281–282
finite difference-time-domain (FDTD) method, 282–283
microcavities, 286
silicon microring resonators, 285
tunable bandwidth filter
cavity–waveguide coupling, 281
coupled-cavity structure, 281–282
transmission matrix method, 280
tunable fano resonance, 280–281

M

Mach–Zehnder interferometer (MZI), 315–316
Moiré grating, 310–311

Index

Multichannel synchronizer
   bit-patterns and BER measurement, 339–340
   brute-force approach, 338
   independent delay control and synchronization, 338–339

N

Narrowband optical parametric amplifier (NB-OPA), 150
Nonlinear wave mixing
   coherent backward scattering
      atomic or molecular density, 250
      density matrix equation, 248–249
      double-Lambda scheme, 251
      equation of motion, 249
      Hamiltonian interaction, 248
      nonlinear light steering, 251
      phase-matching condition, 251
      spectroscopy, 250
   forward Brillouin scattering
      energy conservation, 245
      optical field copropagating geometry, 244
      phase-matching condition, 244
      phonon–photon Hamiltonian interaction, 244
      ponderomotive nonlinearity, 244
      Rabi frequency, 246

O

OPA, *see* Optical parametric amplification
Optical fiber Bragg gratings
   linear properties
      dispersion relation, 225
      frequency propagation characteristics, 226
      group velocity, 226
      linear differential equations, 224
      Maxwell equations, 224
      photonic bandgaps, 224
      quadratic dispersion, 226
      refractive index, 225
   nonlinear properties
      Jacobi elliptic functions, 227
      Kerr nonlinearity, 227
      nonlinear coupled mode equations (NLCMEs), 227
      nonlinear optical effects, 229
      nonlinear Schrödinger description, 228
      wave propagation, 228
Optical Kerr effect, 315
Optical packet switching, 347–348
Optical parametric amplification (OPA), 149
Optical spectrum analyzer (OSA), 45
Optical waveguides
   coherent population oscillations
      erbium-doped fiber, 48
      group velocity dispersion (GVD), 47
      two-level system, 46
   electromagnetically induced transparency
      Bose–Einstein condensates, 49
      cryogenic temperature, 49–50
      hollow-core fibers, 50–52
   stimulated scattering
      SBS, 38–43
      SRS, 43–46
   wavelength conversion and dispersion
      advantages, 54
      slow light methods, 52
      temporal pulse shapes, 52–53
Optical-to-electronic (O/E) converters, 347

P

Packet switch architectures, 348–349
Packet switching, slow light buffers
   buffer capacity, 349–350
   buffer technologies, 363–364
   delay line buffer architectures
      FIFO using adiabatically slowed light, 354–355
      FIFO using cascaded delay lines, 352–354
      optical buffer architectures, 350–351
      serial and parallel buffers, 351
      slow light delay line buffers, 351–352
   electronic buffers, 363
   electronical-to-optic (E/O) converters, 347
   optical packet switching, 347–348
   optical-to-electronic (O/E) converters, 347
   packet switch architectures, 348–349
   physical size
      buffer length *vs.* bit rate, 357–358
      compact delay line structure, 357
      refractive index *vs.* optical frequency, 355–356
      slow light waveguide, 355
      storage density and total length, 356–357
   resonator buffers, 362–363
   waveguide losses and energy consumption
      delay line attenuation characteristics, 359–360
      delay line capacity limitations, 358–359
      maximum buffer capacity *vs.* amplifier saturation power, 360–361
      semiconductor optical amplifier (SOA), 360
      slowdown factor, 359
      stored bit length *vs.* buffer capacity, 361–362
Pendulum bobs coupled spring, 140–141
Periodic coupled resonator structures
   coupled resonator devices, 92
   CROWs/optical gain, 96
   delay, bandwidth and loss
      input/output waveguides, 92
      periodic structure types, 90–91
      trade-offs dispersion relation, 91
   finite-size effects
      antireflection (AR) structure, 89
      lossless coupling, 89–90
      transmission spectrum, 88–89
   general description
      complimentary structures, 83–84
      CROW dispersion relation, 80–81
      GVD properties, 84
      SCISSOR dispersion relation, 81–83
   passive microring CROWs
      disadvantages, 95–96
      full-width half-maximum (FWHM), 94

GVD, 94–95
optical microscope, 93
polymer materials, 92–93
transverse electric (TE) polarized light, 93–94
standing-wave resonators
FP CROWS, 84–87
two-channel FP SCISSORs, 87–88
Periodically poled lithium-niobate (PPLN) waveguide, 54
Photonic crystal waveguides
cavities and slow light comparison
bandwidth, 68–69
coupled cavity, 69
FOM implications, 69–70
intensity enhancement, 66–68
coupling
key issue, 72
propagating components, 73
velocity-dependent coefficient, 73
EIT-based schemes, 59
linear interaction enhancement
dispersion curve, 65–66
Mach–Zehnder modulator, 66
two-beam interference, 65
propagation losses
band-edge scales, 71–72
benefits and enhancements, 71
roughness and distortions, 72
slow light effects
2D, 63
band-edge, 63–65
bandstructure terminology, 61
Brillouin zone, 61–62
coupled resonator structures, 60–61
k and k–G components, 62
Photonic crystals
defect modes, 310–311
dielectric rods square lattice, 281–282
finite difference-time-domain (FDTD) method, 282–283
microcavities, 286
Polarization mode dispersion (PMD), 154
Pulse repetition frequency (PRF), 370
Pulse repetition interval (PRI), 370

R

Reconfigurable signal processing
false color representation, 341
multichannel synchronizer
bit-patterns and BER measurement, 339–340
brute-force approach, 338
independent delay control and synchronization, 338–339
phase-preserving slow light
DPSK data-pattern dependence and mitigation, 332–334
DPSK signal delaying, 331–332
spectrally efficient slow light, 334–335
simultaneous multiple functions, 340
slow-light modulator with Bragg reflectors, 341
slow-light-based signal processing modules, 336
slow-light-based tunable optical delay lines
applications, 324–325
data-pattern-dependent distortion, 326–328
distortion mitigation, 329–331
figures of merit (FOMs), 328–329
optical pulse manipulation, 321
potential target identification properties, 322
slow-light techniques, 323–324
spectral sensitivity enhancement, interferometer, 341–342
variable-bit-rate OTDM multiplexer
eye diagrams of data stream multiplexing, 338
power penalty reduction vs. fractional delay, 337–338
slow-light vs. fiber based OTDM MUX, 336–337
Resonator buffers, 362–363

S

Self-induced transparency (SIT) effect, 4
Semiconductor optical amplifier (SOA), 360
Semiconductor-based slow light and fast light
gain regime
advantages, 28
continuous wave (CW), 30
COP/FWM, 29–30
pumpûprobe scheme, 29
small-signal absorption, 28–29
quantum dots
coherent absorption, 23
inhomogeneous-broadened spectrum, 21–22
pumpûprobe scheme, 22
three-dimensional (3D) confinement, 21
quantum wells
density-matrix equation, 17
excitation-induced dephasing (EID), 19–20
FWM components, 17–18
pumpûprobe scheme, 15–16
radiofrequency (RF) phase, 15
spin subsystems, 20
theoretical calculation, 19
transverse electric (TE) polarization, 16
waveguide geometry, 16–17
room-temperature operation
Fermi factors, 23
foregoing description, 27–28
HWHM linewidth, 26–27
p-doped sample, 27
quasi Fermi levels, 25
reverse and forward bias regime, 23–24, 26
RF phase shifts, 26
spin coherence
double-V EIT, 33
DyakonovûPerel (DP) mechanism, 31
key requirements, 30
pumpûprobe scheme, 31–32
TE-polarized pump, 32–33
Side-coupled integrated sequence of spaced optical resonators (SCISSORs)
complimentary structures, 83–84
dispersion relation

band gaps, 82–83
Hamiltonian approach, 81
propagating waveguide modes, 82
Silica-stimulated Brillouin scattering
    modulated pump
        complex amplitude, 182
        gain spectral distribution, 182, 184
        group index change, 183
        Lorentzian distribution, 181
        power spectral density, 185
        pulse amplitude, 184
        refractive index change, 183
        signal amplitude linear transformation, 181
        Stokes and anti-Stokes bands, 184
    monochromatic pump
        As_2Se_3 chalcogenide fibers, 180
        Cornell experiment, 177–178
        dispersion-shifted fiber (DSF), 176
        Ecole Polytechnique Fédérale in Lausanne (EPFL), 175
        electro-optic modulator (EOM), 175
        group index change, 179
        group velocity, 180
        pulse amplitude, 179–180
        pulse delay time, 177
    multiple pumps
        gain and loss spectral distributions, 186–187
        signal frequency, 186
        spectral transmission, 189
        superposed resonance, 187
        time delay, 189
        zero-gain spectral resonances, 188
Slow and fast light propagation
    averaged PMD model, 159–161
    experimental results
        delay vs. parametric gain, 166
        nonreturn to zero (NRZ) modulation format, 166
        polarization dependent gain, 165
    propagation parameter longitudinal variations
        estimation procedure, 162
        results, 162–164
        uniqueness and spatial resolution, 161–162
    silica-stimulated Brillouin scattering
        full width at half maximum (FWHM), 174
        Lorentzian resonance, 174
        modulated pump, 181–186
        monochromatic pump, 175–181
        multiple pumps, 186–190
    SRS-assisted OPA
        birefringent fiber, 153–159
        isotropic fibers, 151–153
Slow light
    atomic vapors
        dispersion management, 9–11
        electromagnetically induced transparency, 4–8
        first experiments, 4
        two-level systems, 8–9
    coherent media propagation
        delay time, 243
        eikonal equation, 241
        interference and frequency stabilization, 243–244
        Maxwell equation, 240
        polarization, 240
        residual absorption, 242
    coupled resonator optical waveguides (CROWs)
        absorption, 105–107
        Bloch function, 104
        discrete spectrum, 110–115
        fundamental restrictions and fabrication, 107
        light propagation, 103–104
        linearly polarized light, 103
        properties, 102
        resonator chains interference, 107–109
        resonator-stabilized oscillators, 109–110
        whispering gallery mode (WGM), 105
    disordered optical slow-wave structures
        calculation models ($\rho(\omega)$), 128–131
        coupling matrix, 126–128
        disordered structures, 131–135
        fields localization, 135–139
        formalism, 124–125
        spectrum solutions, 125–126
        tight-binding optical waveguide, 120–124
    electromagnetically induced transparency, 236
    gap solitons
        dispersion effect, 223
        dispersive broadening, 223
        experiment, 229–233
        Kerr nonlinearity, 223
        Kramers–Kronig relations, 223
        optical fiber Bragg gratings (FBGs), 224–229
    group velocity, kinematics
        dispersion equation, 237
        Fourier transformation, 237
        Maxwell equation, 237–238
        modulation, 236
        phase velocity, 236
        polarization, 237
        refractive index, 237
    nonlinear wave mixing
        coherent backward scattering, 247–253
        forward BRILLOUIN scattering, 244–246
    optical waveguides
        coherent population oscillations, 46–49
        electromagnetically induced transparency (EIT), 49–52
        wavelength conversion and dispersion, 52–54
    photonic crystal waveguides
        band-edge, 63–65
        bandstructure terminology, 61
        Brillouin zone, 61–62
        cavities and slow light comparison, 66–70
        coupled resonator structures, 60–61
        coupling, 72–73
        EIT-based schemes, 59
        k and k–G components, 62
        linear interaction enhancement, 65–66
        propagation losses, 71–72
    stimulated Brillouin scattering (SBS)
        acoustic wave, 38
        anti-Stokes absorption, 42–43
        FWHM linewidth, 39

linewidth tunable laser, 40
optical data buffering application, 39–40
single-mode optical fibers, 42
temporal evolution, 41
stimulated Raman scattering (SRS)
chip-scale photonics devices, 46
highly nonlinear fiber (HNLF), 44
optical phonons, 43
pump peak power functions, 45
pump–probe geometries, 43–44
three-level Λ atom gas
coherent population trapping (CPT), 238
dark state, 239
density matrix equations, 239–240
energy levels, 239
Hamiltonian, 238–239
probe laser frequency, 240
susceptibility, 240
Slow light application, phased array radar beam steering
antenna elements, 367–368
electronically steered system, 367
frequency-dependent phase shift, 368
optical true-time delay (TTD) system, 369
phase shifter beam forming, squinting
array factor, 371
beam pattern, 371–372
beam profile, 371
loss vs. steering angle, 372–373
mainlobe level (MLL) vs. relative bandwidth, 372
radar system background
pulse compression waveforms, 370–371
pulse sequence, 370
pulsed Doppler radar, 369–370
system latency, 371
range resolution, 368
solid-state amplifier, 368
true-time delay (TTD) beam-forming requirements
amplitude precision, 375–376
array factor, 373–374
attenuation loss, 377
bandwidth, 376–377
delay precision, 374–375
dynamic range, 377–378
Slow light buffers, packet switching
buffer capacity, 349–350
buffer technologies, 363–364
delay line buffer architectures
FIFO using adiabatically slowed light, 354–355
FIFO using cascaded delay lines, 352–354
optical buffer architectures, 350–351
serial and parallel buffers, 351
slow light delay line buffers, 351–352
electronic buffers, 363
electronic-to-optic (E/O) converters, 347
optical packet switching, 347–348
optical-to-electronic (O/E) converters, 347
packet switch architectures, 348–349
physical size
buffer length vs. bit rate, 357–358
compact delay line structure, 357
refractive index vs. optical frequency, 355–356

slow light waveguide, 355
storage density and total length, 356–357
resonator buffers, 362–363
waveguide losses and energy consumption
delay line attenuation characteristics, 359–360
delay line capacity limitations, 358–359
maximum buffer capacity vs. amplifier saturation power, 360–361
semiconductor optical amplifier (SOA), 360
slowdown factor, 359
stored bit length vs. buffer capacity, 361–362
Slow light schemes, bandwidth limitation
atomic resonances
absorption coefficient, 295
dielectric constant, 294
forbidden gap, 298
group velocity dispersion (GVD), 296–297
Kramers–Kronig relation, 295
Lorentzian absorption, 297
loss dispersion, 295–296
cascaded gratings, double-resonant photonic SL structures
atomic vs. photonic schemes, 304
bit rate dependent slow down factor, 303
cascaded Bragg gratings, 301–302
cutoff bandwidth, 304
delay line length, storage capacity, 304–305
group velocity dispersion (GVD), 302
maximum bit rate, 302–303
optical buffer length, 303–304
slow down factor dispersion, 301
coupled photonic resonator structures
characteristic parameters, 313
disadvantages of cascaded Bragg gratings, 309–310
dispersion relation, 311–312
group velocity, 312–313
Moiré grating, 310–311
periodic chain parameters, 311
photonic SL structures, 310–311
SL optical buffer length, 313–314
slow down factor, 312
double-resonant atomic SL structures
double atomic resonance, 298–299
group velocity dispersion (GVD), 300
maximum bit rate, 300
residual absorption, 299
storage capacity, 301
Taylor expansion, 299–300
transparency bandwidth, 300–301
nonlinear photonic SL devices, dispersion limitation
group velocity dispersion (GVD), 317
MZI ultrafast all-optical switch, 315–316
nonlinear index change, 316–317
refractive index, 315
switching condition, 316
switching power density, GaAs photonic crystal, 317–318
photonic resonances, 297–298
slow down factor, 293–294
tunable double-resonant atomic SL structures

Index

absorption spectrum, 306–307
background absorption, 305
cascaded grating, 308–309
characteristic parameters, 309
cutoff bit rate, 308
delay buffer length, 309–310
EIT scheme dispersion, 307–308
energy storage and slow light propagation, 307
harmonic wave, 305–306
residual absorption, 307
spectral hole burning, 305
third-order group velocity dispersion (GVD), 308
Slow-light-based tunable optical delay lines
applications, 324–325
data-pattern-dependent distortion
OPA-based slow-light scheme, 328
origin of, 326
pattern-dependent gain and delay, 326–327
quantification, 327
signal Q-factor, 327–328
distortion mitigation
data-pattern dependence reduction, 329–330
fiber Bragg gratings, 331
NRZ-OOK signal quality, 330
figures of merit (FOMs), 328–329
optical pulse manipulation, 321
potential target identification properties, 322
slow-light techniques, 323–324
Slow-wave structure (SWS)
cross-phase modulation
effective phase modulation, 207
enhancement factor, 206
Kerr-induced birefringence, 207
phase shift, 206, 208
pulsed pump sequence amplitude, 207
slowing factors, 206
four-wave mixing (FWM)
classical theory, 210
coupled resonator optical waveguides (CROW), 210
enhancement factors, 215
frequency conversion, 214
normalized bandwidth and conversion gain, 212
optical parameters, 213
optical waveguide, 211
power transfer, 210, 214
quasi-phase-matched (QPM) schemes, 214
wave mixing, 215
fundamentals
complex reflection coefficient, 199
group velocity, 198
infinitely long periodic structures, 196
Kerr nonlinearity, 197
physical single cells, 196
slowing factor, 198
synthetic summary, 196
modulation instability
coupled equation system, 216
dispersion, 217
gain spectra, 218
instability phenomena, 216
lossless nonlinear optical system, 215

phase shift, 217
slowing factor, 218
time-domain, 219
nonlinear phase modulation
frequency shift, 200
Gaussian pulse, 201
nonlinear Kerr effect, 200
power enhancement factor, 200
power spectral density, 202
sensitivity, 200
spectral broadening, 201
nonlinear spectral response
group delay, 208–209
self-pulsing effect, 210
time-domain, 209
transfer function, 208
transmission, 208–209
optical resonator, 196
self-phase modulation (SPM) and cromatic dispersion
dispersion regimes, 202–203
power limiting, 203–204
soliton propagation, 205–206
SRS-assisted OPA
birefringent fibers
delay spectra, 158
gain limit, 156, 158
linear propagation, 154
polarization mode dispersion (PMD), 154, 156–157
probability distribution functions (PDFs), 156–157
pump averaged Jones vector, 155
isotropic fibers
Fourier transform, 152
gain and delay spectra, 153
nonlinear coefficient, 152
nonlinear susceptibility, 151
Raman susceptibility, 152
self-phase modulation (SPM), 152
Stimulated Brillouin scattering (SBS)
acoustic wave, 38
anti-Stokes absorption, 42–43
FWHM linewidth, 39
linewidth tunable laser, 40
optical data buffering application, 39–40
single-mode optical fibers, 42
temporal evolution, 41
Stimulated Raman scattering (SRS), 150
chip-scale photonics devices, 46
highly nonlinear fiber (HNLF), 44
optical phonons, 43
pump peak power functions, 45
pump–probe geometries, 43–44
Stopping and storing light
band structures
Bragg and resonator gaps, 267
BSQW and SCISSOR structures similarity, 265
BSQW potonic bandstructure, 269–270
Fresnel coefficients, 266
quadratic equation, 268–269
resonance frequency, 267
SCISSOR potonic bandstructure, 266–267
transfer matrix, 265–266

Bragg-spaced multiple quantum well
(BSQW) structures
  exciton resonance, 270
  FabryûPerot fringes, 271
  intermediate band (IB) dispersion, 272–273
  limitations, 272
  operational principle, 271
  photonic band structure, 269–270
  pulse propagation, 270–271
  resonant photonic bandgap structure (RPBG), 269
  time dependent output pulse intensity, 273
  trapping scheme, 272
physical slow-light systems, 257
quantum well structure
  electric field, 261
  free electron mass, 259
  Fresnel reflection and transmission coefficients, 261–262
  nonradiative and radiative decay, 260
  polarizability, 259–260
  quality factor, 260
  schematic plot, 258
  second-order differential equation, 259–260
  stationary fields, 258
  transfer matrix, 261–262
SCISSOR structure
  coupling equations, 262
  fundamental resonance frequency, 264
  microresonator and channel waveguide coupling, 262
  propagation equations, 263
  quality factor, 265
  radiative and nonradiative coupling, 264–265
  reflection and transmission coefficients, 263–264
  schematic plot, 258
  transfer matrix, 263
  two-channel SCISSOR unit, 262–263
  vs. QW structure, 264
SWS, see Slow-wave structure

T

Taylor expansion, 299–300
Tight-binding optical waveguide
  bandsolver tools, 123
  disordered light velocity, 122–123
  engineered dispersion, 120
  optical signal processing, 121–122
  slow-wave dispersion relationship, 120–121
  states density, 123–124
Transverse magnetic (TM) modes, 128
True-time delay (TTD) beam-forming requirements
  amplitude precision, 375–376
  array factor, 373–374
  attenuation loss, 377
  bandwidth, 376–377
  delay precision, 374–375
  dynamic range, 377–378
Tunable bandwidth filter
  cavityûwaveguide coupling, 281
  coupled-cavity structure, 281–282
  transmission matrix method, 280

V

Variable-bit-rate OTDM multiplexer
  eye diagrams of data stream multiplexing, 338
  power penalty reduction vs. fractional delay, 337–338
  slow-light vs. fiber based OTDM MUX, 336–337
Vertically stacked multi ring resonator (VMR), 102